ASTRONOMY
FUNDAMENTALS & FRONTIERS

FOURTH EDITION

ROBERT JASTROW
MALCOLM H. THOMPSON

JOHN WILEY & SONS, New York Chichester Brisbane Toronto

The designer was Rafael Hernandez
The editor was Donald H. Deneck
The drawings were designed and executed by
John Balbalis with the assistance of the Wiley
Illustration Department and Eric Heiber.
Cobb-Dunlop Publisher Services supervised production.

COVER: A false-color image showing the pattern of emission of radio waves from a giant elliptical galaxy (IC-4296) at a distance of 120 million light-years. The image was created from radio observations made at the Very Large Array (VLA), 27 connected radio telescopes near Socorro, New Mexico (Chapter 5). The range of colors from blue through green to yellow, orange and red represents increasing intensities of radio emission. The image reveals two energetic streams of gas (in red and orange) emerging from the galaxy, which is not visible here. More detailed radio maps show that the energetic streams come from an extremely compact, pointlike source which coincides with the center of the galaxy. See Chapter 12 and Color Plate 14.

COVER: The National Radio Astronomy Observatory, operated by Associated Universities, Inc. under contract with the National Science Foundation. Observers: N. Killeen, R. D. Ekers.

Cover design: Jerry Wilke

Library of Congress Cataloging in Publication Data

Jastrow, Robert, 1925-
Astronomy: fundamentals and frontiers.
Includes index.
1. Astronomy. I. Thompson, Malcolm H. II. Title.
QB43.2.J37 1984 520 83-21768
ISBN 0-471-89700-0

Printed in the United States of America

10 9 8 7 6 5 4 3

FOREWORD

In the introduction to the present textbook the authors briefly discuss the Copernican revolution. We are reminded that the ideas of Copernicus, Kepler, Galileo and Newton profoundly influence human culture, and we recall how long it took for these ideas to be generally understood and accepted. The addition of knowledge that is taking place in our day in astronomy occurs at a spectacularly rapid pace. There can be no doubt that the task of making the results of astronomical research generally available is a most important one. The challenge is strongly felt by the astronomical community.

Professors Jastrow and Thompson have written an astronomy textbook with the liberal arts student in mind at every point of the development. There is a very good balance between the discussion of basic methods and description of the results of astrophysical research, including the most recent advances.

I believe that the authors have chosen wisely in emphasizing the central problems of modern astrophysics, dealing only briefly with some of the chapters of classical astronomy. In placing the discussions of stellar structure and evolution of galaxies, and large scale cosmology before the discussion of the solar system and problems of structure and history of planets, the authors are guided by considerations of evolution in time rather than location in space. This way of arranging the subject material has obvious advantages.

If this textbook is as widely used and read as it deserves the authors will have made a very significant contribution in the area of communication between scientist and the community. For the book tells compellingly about basic research, important to us all, and of its results, its spirit and its excitement.

Bengt Stromgren

PREFACE TO THE FOURTH EDITION

The pace of new discoveries in astronomy has quickened in the interval between the publication of the third and fourth editions of ASTRONOMY: FUNDAMENTALS AND FRONTIERS. Much of the new progress can be explained by the acquisition of new observational tools, such as the orbiting telescopes that open up the infrared, ultraviolet, x-ray and gamma ray regions, and the Very Large Array and Very Long Baseline Interferometry facilities in radio astronomy. Charge coupled devices and other new types of detectors have enhanced the sensitivity and usefulness of existing ground-based optical telescopes. Finally, new computerized methods of processing data have enhanced the value of observations at all wavelengths. The picture of stellar evolution presented in the third edition has not changed a great deal as a result of these observational developments. However, dramatic changes have occurred in the study of galaxies and the large-scale structure and evolution of the Universe. Evidence for massive galactic halos, superclusters and voids, the nonluminous "hidden mass," and new observations of active galactic nuclei and cosmic jets have required the addition of large amounts of new material to the corresponding chapters on galaxies and cosmology.

The new developments in instrumentation—including innovative designs for large ground-based optical telescopes, such as the multiple mirror telescope and the segmented mirror telescope, and the impressive VLA facilities and VLBI links, in radio astronomy, with their extraordinary resolutions approaching 10^{-4} arc seconds—are so interesting as to merit fairly complete explanations of their working principles. The chapter on telescopes and detectors has been expanded into two full chapters as a result of the additions in these areas.

The Voyager spacecraft missions have led to great advances in our knowledge of the properties of the giant planets and their moons. These remarkable flights have yielded exciting images of alien worlds. The images and other data collected by Voyager reveal planetary bodies with strange geologies and surface conditions. We have incorporated the new information, all of it unavailable at the time of the third edition, into the sections on the giant planets and their moons and into the color insert.

A new chapter has been added on the evolution of astronomical thought from the Greeks to Newton. The theme of this opening chapter is the Copernican revolution, and its impact on man's understanding of his place in the cosmic order. The entire development of astronomy in the centuries after Newton has been, in a sense, a continuation of that revolution in human thought whose roots can be traced back to the work of a handful of early astronomers living more than 2000 years ago.

The organization of the fourth edition continues the unusual arrangement in which material on stars and galaxies is presented before the discussion of the solar system. This arrangement has the advantage of providing students with early information on star formation and element synthesis in stars, that permits deeper insights into the evolution and structure of the major and minor bodies of the solar system than the more conventional treatments beginning with a discussion of the earth and planets. However, the section on the solar system has been written in such a way that with only minor omissions of material, this section can be taught before the sections on stars and galaxies.

Several friends and colleagues have been kind enough to prepare critical reviews of the manuscript of the fourth edition. We are particularly indebted to Professor Neville J. Woolf of the University of Arizona and Professor John Thorstensen of Dartmouth College for their careful reading of the entire manuscript. The book has benefitted enormously from their thoughtful suggestions and criticisms. We are also indebted to Dr. Richard Stothers of the Goddard Institute for Space Studies, for valuable suggestions relating to the chapters on the history of astronomy and on stellar evolution.

We are deeply grateful to Doris Cook who worked with us throughout the preparation of the fourth edition and provided invaluable counsel on all matters of clarity and organization. She also brought the manuscript through its many successive stages of draft and revision. The fourth edition could not have come into being without her collaboration. We also wish to thank Margaret Newfield and Dorothy Davis, who typed and copy-edited several chapters.

The high competence and cooperative spirit of the Wiley staff continue to make this publishing relationship professionally pleasant and rewarding for both authors. We wish particularly to express our appreciation to our editor, Donald H. Deneck, whose gentle pressure and unfailing good humor brought our project to fruition far earlier than would otherwise have been the case. Words will not express the debt we owe to Don for his efforts. We also wish to thank Maureen Conway and her colleagues for their extremely competent guidance in carrying the manuscript through the complexities of the production process.

Robert Jastrow
Malcolm Thompson

PREFACE TO THE FIRST EDITION

Astronomy, more than any other physical or behavioral science, offers the nonscience student a mind-expanding educational experience. The steadily increasing enrollments in introductory astronomy courses reflect a growing awareness of this fact on the part of liberal arts students. The historical and philosophical elements in astronomy, always a large factor in the appeal of this subject for the nonscience major, have been strengthened by new discoveries in stellar evolution and cosmology. Progress in these fields during the last twenty years has filled in many details of the sequence of events that led from the explosive beginnings of the Universe through the birth of innumerable stars and planets to the formation of the sun and the earth. When the latest advances in astronomy are combined with developments in the life sciences, the result is a chain of cause and effect that stretches back over 10 billion years and links the earth and its life forms to events that occurred early in the history of the Universe. At that point—on the threshold of the appearance of life on our planet—the direct contribution of astronomy ends, and the story is taken up by other branches of science.

The advances in astronomical knowledge provide many points of contact between this subject and other scientific disciplines. Modern astronomy, fascinating in itself, seems increasingly to be a fragment of a mosaic that, when viewed from a distance, forms an image of the human observer. The appeal of astronomy to the nonscientist is further strengthened by the fact that its subject matter forces the imagination to contemplate larger expanses of space and time than fall within the province of any other scientific discipline. These qualities make the study of astronomy a uniquely attractive means of introducing the liberal arts student to the physical sciences.

We have focused on the needs and interests of the liberal arts student in our choice of topics, as well as in the style of writing and in the level of required mathematical skills. No mathematics is used beyond the level of elementary algebra, and technical terms are avoided. Each chapter opens with the statement of a central theme to which the previous chapters of the book are clearly related. The remainder of the chapter is an explicit development of this central theme.

The central problems of twentieth century astronomy are emphasized. Full chapters are included on the Hertzsprung-Russell diagram, nuclear reactions in stars, stellar evolution, galactic structure and evolution, radio galaxies, Seyfert galaxies and quasars, cosmology, the history of the moon, and the evolution of planetary atmospheres. Much discussion is devoted to recently opened areas of research such as infrared astronomy, x-ray and gamma-ray astronomy, gravitational waves, pulsars, and black holes in space. The choice of topics and the allotment of space to each topic reflect much of contemporary research publication. An unusual feature is the inclusion of a final chapter on the evolution of life in the Cosmos.

A complete discussion of these topics in a one-semester textbook necessitated the omission of some areas, such as astrometrics, which are centers of active research on contemporary astronomy but are not as directly related to the book's central line of development. Celestial mechanics is treated very briefly in the chapter on the solar system. The motions of the earth in space, tides, eclipses, and celestial coordinates are described in an introductory, separately paginated section.

A basic innovation in the book is its presentation of material on stars and galaxies before the discussion of the solar system. This is the reverse of the traditional presentation, in which astronomical knowledge is given in the order in which it was acquired in human history, starting with the earth and then radiating outward to the moon, planets, stars, and galaxies. Our book embeds the study of the solar system in the context of a general study of stars and planets, and more accurately reflects the impact of the Copernican Revolution on the history of astronomy.

The new organization of material has the advantage that it permits the instructor to use astrophysical knowledge when discussing the structure, chemical composition, and origin of the earth, moon, and planets. Astrophysics reveals how elements are made in the stars; why some are more abundant than others; and how these elements condensed to form the clouds out of which the sun, moon, and planets were born. An astrophysical background is required to discuss conditions at the beginning of the solar system, when the earth and other planets were newly formed. Instructors who prefer the traditional organization, but like other features of the book, can start their course as usual with "The Solar System," whose opening chapters summarize the astrophysical background needed for studying the solar system.

The rapid pace of change in modern astronomy makes the task of the textbook author very difficult if he wishes to present a balanced view of recent developments. We are deeply indebted to a number of friends and colleagues, closely associated with these developments, who have been willing to spare time from their research for careful reviews and detailed criticisms of portions of the manuscripts dealing with subjects of which they have a profound knowledge. We are particularly grateful to Dr. Richard Stothers for many informative discussions on stellar evolution and numerous detailed criticisms of Chapters 1 to 8; to Professor Lodewijk Woltjer for a careful commentary on the Chapters dealing with galaxies and cosmology; to Professors Paul Gast and Robert Phinney and to Dr. Vivien Gornitz for their comments on the Chapters relating to the solar system, the earth, the moon and the planets; and to Dr. S. I. Rasool for illuminating discussions of the planets in general and critical review of Chapter 17 in particular. We also profited greatly from conversations with Dr. Patrick Thaddeus on topics in radio astronomy and interstellar chemistry. Professor Neville Woolf gave us the benefit of his reading and criticism of the entire manuscript, and offered valuable comments on the balance of the contents between traditional and contemporary areas of astronomy. Drs. J. W. Hogan, R. Stewart, and Dennis Hegyi reviewed the manuscript for pedagogical effectiveness and offered many helpful suggestions based on classroom testing of the materials.

The completion of the manuscript would not have been possible without the devoted editorial and secretarial assistance of Misses Judith Silverman and Ruth McCarthy. Finally, with particular pleasure we express our thanks to Donald Deneck and Dennis Hudson in particular, and to the extremely capable Wiley editorial, production, picture research, illustration and design departments for their enthusiastic support and cooperation in bringing the raw material of our text into finished form.

Robert Jastrow
Malcolm Thompson

CONTENTS

ASTRONOMY
FUNDAMENTALS & FRONTIERS

FOURTH EDITION

EARTH AND SKY

PART ONE

EVOLUTION OF ASTRONOMICAL THOUGHT

1

The earth seems vast and immobile; throughout two million years in the prehistory of man it has provided the stage for all human experience, with the heavens seemingly no more than a backdrop of moving lights. Modern astronomy was born in the contrary realization that space is vast, and the world of man is small.

THE ANCIENT IDEAS

To anyone who follows the motion of the sun day after day, and the motions of the moon and the stars night after night, it is obvious that the earth is the center of the Universe, and that the heavenly bodies revolve about it daily, paying homage to the abode of man. Every day the sun moves across the vault of the heavens; every night the moon and stars travel in their stately procession across the sky.

In ancient times, observers of the heavens marveled at this nightly movement of the stars and pondered what its cause could be. As they followed the constellations from night to night, they saw further that their patterns were unchanging; the stars of the Big Dipper and the Little Dipper moved across the sky as a unit then, and still do today. Reflecting on these facts, the early astronomers decided that the stars must be attached firmly to an enormous sphere surrounding the earth. The sphere revolved around the earth at a constant speed every twenty-four hours; as it turned, the stars turned with it. In the center of the sphere of the stars rested

The stars of the constellation Orion setting behind buildings of Lick Observatory.

the earth, solid and unmoving, fittingly placed at the hub of the Universe (Figure 1.1).

The early Greek philosophers read into the even, circular motion of the heavens an indication of a larger order in the Universe, and of a permanence totally unlike the chaos of their daily lives. Some speculated that the circular motion and constant speed were signs of a harmony in the Universe that would become apparent some day, when man resolved nature's complexity into simple elements. This notion exists today, in what physicists call Laws of Nature.

A few astronomers in ancient Greece thought differently about the movement of the heavenly bodies; they reasoned that if the earth, and not the heavens, turned on its axis every twenty-four hours, that would create an apparent motion of the sky. The stars could be fixed in space, but a person on the rotating earth would see them moving past him in the opposite direction, just as the landscape seems to move past a person on a merry-go-round. Heraclides, a Greek astronomer who lived in the fourth century B.C., was the best known advocate of that point of view.

FIGURE 1.1
The celestial sphere containing the stars; it revolved around the earth every twenty-four hours.

Anaxagoras, who lived a century later than Heraclides, had another radical idea. He heard that a meteorite had fallen from the sky, and saw that this meant there were earthlike rocky materials in space. He then conjectured that the moon was made of rock and seemed a very bright object because it was illuminated by the sun. In this way Anaxagoras explained the phases of the moon.

Aristarchus, who lived in the third century B.C., was even more in advance of his time; he proposed that the earth might be moving in orbit around the sun, in addition to turning on its axis. This second motion of the earth would explain why different constellations appear in the night sky at different times of the year.

Although the ideas of Heraclides and Aristarchus are familiar to us today, in their time they were ridiculed or ignored, and after a while they disappeared from view. It seemed nonsensical to most people in early times to postulate that the massive earth could spin on its axis like a child's top, or sail through space like a ship. Surely, everything not anchored to the earth would be left behind. An arrow shot straight up into the air would come down miles away; a stone dropped from a tower would never reach the ground beneath; and rocks and trees would be hurled from the spinning earth like mud flung from the rim of a turning wagon wheel. Since no one ever saw such effects, the earth must be stationary, and the sun, moon, and stars must revolve around it daily. All human experience proved that this must be so.

Planets: The Wandering Stars

One aspect of the heavens did not agree with the picture of a stationary earth surrounded by a rotating sphere of stars. Five bright starlike points of light did not behave like normal stars; instead of maintaining fixed positions relative to their neighbors, they wandered about in the heavens, at times near one star and at times near another. The Greek astronomers, puzzling over the fact that the five mysterious objects were unlike any other stars, called them Wanderers, or *planetes* in Greek.[1] In English, they became known as the planets.

Today, a planet means a massive ball of rock and iron like the earth or Mars, or a huge sphere of hydrogen and helium like Jupiter; but the Greeks, lacking telescopes, had no idea that the objects they called planets could be massive bodies like the earth. To them, a planet was just a starlike point of light that moved about strangely in the sky.

[1] The planets or wandering stars were also known earlier to the Mesopotamians.

Bemused by the erratic behavior of the planets, the astronomers observed their positions year after year, and after a time they perceived a pattern in the seemingly random movements. Each planet followed a looping path in the night sky, first curving around from east to west, and then back from west to east. The planets looped their way across the background of the stars very slowly, and years of patient study were required to reveal their gyrations, but in a modern planetarium, which speeds up the motions of celestial bodies so that years go by in minutes, the loops appear clearly as in Figure 1.2.

FIGURE 1.2
This photograph, taken in the Munich planetarium, shows the motions of the planets simulated over a period of seventeen years.

If the planets were attached to the great sphere that revolved in the heavens, they would move across the sky in an orderly procession with the rest of the stars. Clearly, they were not attached to the sphere of the stars, and must be located somewhere else. But where? And why did they move in loops? Reflecting on these questions, an astronomer named Eudoxus reasoned that planets must be attached to transparent crystal spheres that rolled across the heavens. Later, two astronomers named Apollonius and Hipparchus had an ingenious idea. They reasoned that each planet must be attached to a wheel that rolled across the sky on the rim of still another wheel. As the wheel rolled, the planet would describe a looping path in the sky, just as some planets are observed to do

(Figure 1.3). Those additional circular motions, superimposed on the main circular motion of the planet around the earth, were called epicycles.

Wheels on Wheels: The System of Ptolemy

The idea proposed by Apollonius and Hipparchus worked well in principle, but when the astronomers tried to make their picture of epicycles and rolling wheels fit the observed motions of the planets accurately, they found they had to introduce complicated combinations of wheels rolling on wheels. The astronomer Ptolemy, who lived in the second century A.D., and did the best job of fitting, found that no less than forty wheels were needed to produce epicycles that would describe the movements of the sun, the moon, and the five planets known at that time.

Ptolemy's ideas were widely adopted because his forty wheels gave the best results, but people found his model of the heavens very complicated. When Alfonso X, King of Castile and Aragon, became acquainted with the Ptolemaic system, he remarked, "If the Lord had consulted with me, I would have recommended something simpler." And Milton, who had to teach the Ptolemaic system as a schoolmaster in the seventeenth century, wrote in disgust about Ptolemy and the astronomers who followed him,

> *"How they contrive*
> *To save appearances, how gird the Sphere*
> *With Centric and Eccentric scribbled o'er*
> *Cycle and Epicycle, Orb in Orb . . ."*

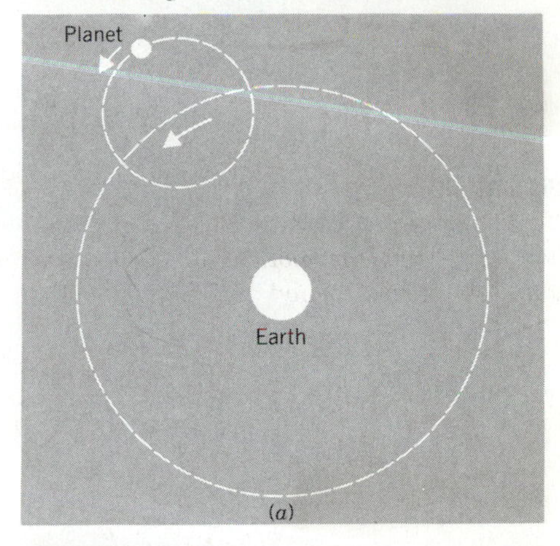

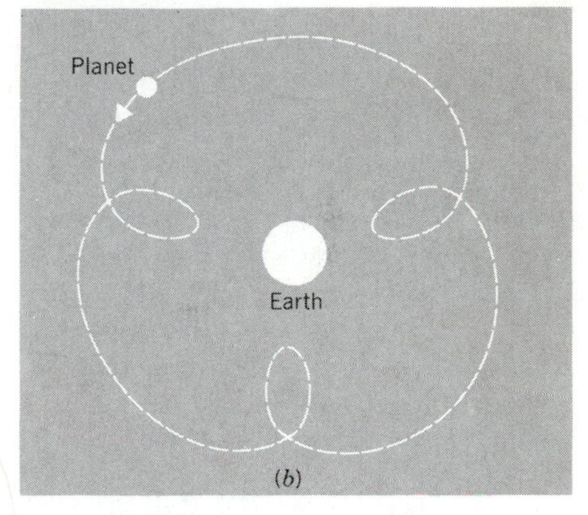

FIGURE 1.3
(a) An explanation of the looping paths of a planet: the planet is attached to the rim of a wheel, whose axis rolls along the rim of another wheel with the earth at its center. (b) The looping path described by the planet on the rolling wheel.

THE REVOLUTIONARY PROPOSAL OF COPERNICUS

Ptolemy's picture of the Universe was the best the human brain could devise for a very long time. In spite of its cumbersome machinery it went unchallenged for nearly fourteen centuries, until, finally, around 1500, a Polish churchman named Copernicus (Figure 1.4) made a strange proposal. *The earth moves,* he said, reverting to the discredited theories of Heraclides and Aristarchus. *The earth moves around the sun every year, and turns on its axis every day.* No longer is the earth at the center of the Universe; now, Copernicus wrote, it is the sun: "rightly called the Lamp, the Mind, the Ruler of the Universe, gathering his children the Planets which circle around him."

When Copernicus began to think about the motion of the earth, he was aware of the ideas of Heraclides and Aristarchus even though

FIGURE 1.4
Nicholas Copernicus (1473–1543)

these astronomers had been dead for nearly two thousand years. Why did Copernicus resurrect ideas that had been dead for twenty centuries? He described his reasons later, in the introduction to his treatise on the new theory of the Universe. For one thing, he wrote, it was more reasonable to assume that the earth turned on its axis every twenty-four hours than to believe that the entire Universe traveled with incredible velocity in the opposite direction every day. And he was also disturbed because Ptolemy's model of the heavens required the planets to change their speeds, slowing down and speeding up again as they moved around their circles. That seemed to violate the traditional view of an orderly, unchanging circular motion in the Universe.

Copernicus was troubled by other aspects of the traditional picture as well. For example, if the planet Mars moved around the earth, as Ptolemy said it did, Mars should have about the same brightness throughout the year; but, in fact, the brightness of Mars varied greatly at different times. We know today why that is so. Both Mars and the earth move around the sun, and when Mars is on the opposite side of its orbit from the earth, it is very far away and therefore relatively faint; but when Mars is on the same side of the sun as the earth and closer to us, it appears relatively bright (Figure 1.5).

Others had noticed this difficulty, but were not disturbed. Copernicus felt differently, and set to work to build a model of the Universe that would do better.

FIGURE 1.5
Mars at two points in its orbit. When Mars is closer to the earth, it is brighter. The variation in the brightness of Mars, which is confirmed by observation, was cited by Copernicus as one of the proofs of his theory that the planets revolve around the sun.

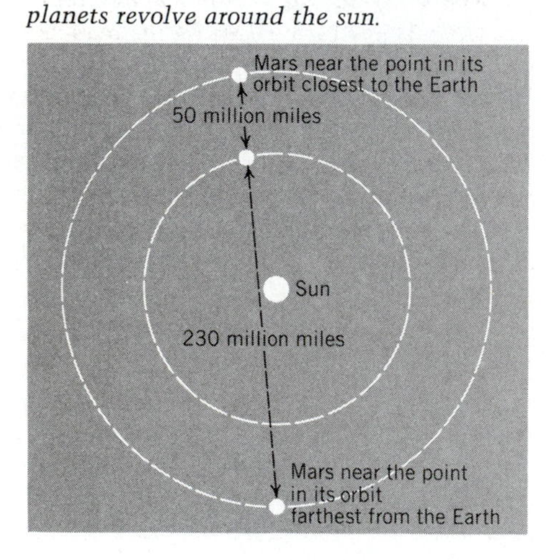

The Copernican Theory of a Moving Earth

Copernicus had reason to be pleased with his idea of a moving earth. For one thing it enabled him to surmount the great obstacle of ancient astronomy and explain the looping paths of the planets as well as Ptolemy, and in a much more natural manner. His explanation was simple. Imagine, said Copernicus, that the earth is moving around the sun, and the other planets are also moving around the sun, but at different rates, some faster than the earth and some slower. From the perspective of a person standing on the earth, the positions of the other planets will seem to loop back and forth, but the looping motions are apparent and not real. In this way, the strange motions of the planets are accounted for without the need for rolling wheels in the sky (Figure 1.6).

This explanation of the planetary movements is far simpler than Ptolemy's, and esthetically more pleasing. However, when Copernicus came around to matching the detailed observations on the posi-

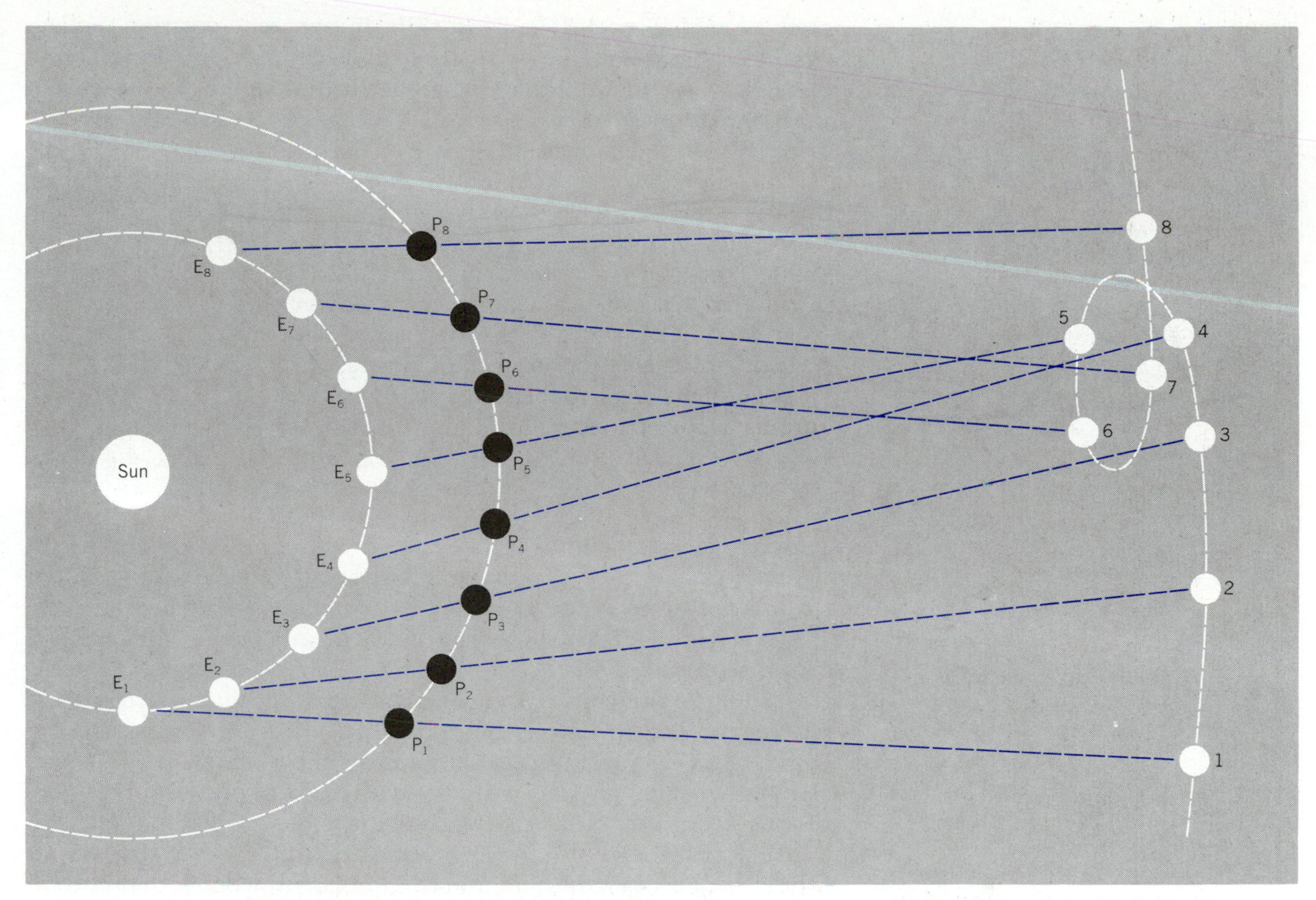

FIGURE 1.6
In Copernicus' picture of the solar system, the reversal of the motions of the planets came about very naturally, without the need for rolling wheels in the sky. The diagram shows how this happens. As the earth and Mars move around the sun, at first (positions 1, 2 and 3) the line of sight from the earth to Mars sweeps in the same direction as the planets themselves. As the motions of the two planets become parallel (positions 4, 5 and 6), the line of sight from the earth to Mars moves backward because the earth is passing by Mars on the inside track. Finally, when the planets reach positions 7 and 8, the earth is no longer moving parallel to Mars, and the line of sight sweeps forward once more, in the direction of Mars' motion.

tions of the planets, he found that his new theory lost its simplicity because he had to introduce rolling wheels in the sky once more to secure the same accuracy as Ptolemy. The reason is that Copernicus was trying to describe the paths of the planets by circles, whereas actually they move in the path of the somewhat flattened circle known as an ellipse. In the end, Copernicus needed about as many rolling wheels as Ptolemy—forty-eight by one count—against Ptolemy's forty. When all was done, and his celestial machinery was complete, it turned out to be as complicated as Ptolemy's, and no more accurate.

Furthermore, his theory of the moving earth failed its most important experimental test. According to the theory, the pattern of the fixed stars should shift during the course of the year as the earth travels around the sun, because people on the earth will see the stars from different perspectives on opposite sides of the earth's orbit (Figure 1.7). This shift in apparent position, called parallax, is seen when a person holds a pencil at arm's length, and closes first one eye and then the other. If the pencil is drawn closer, the apparent shift in its position increases; if it is held at a greater distance, the shift decreases.

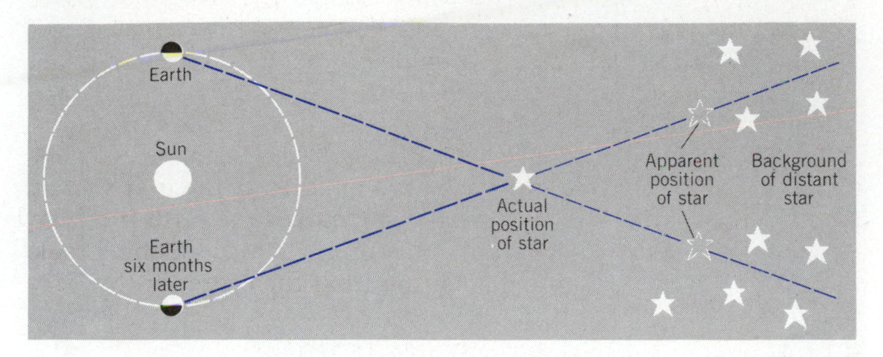

FIGURE 1.7
The change in the apparent position of a nearby star against the background of distant stars, as the earth goes around the sun in its orbit. This shift in position is called the star's parallax.

When the astronomers looked for the shift in the positions of the stars during the year, they did not see it. Today we know the reason; the shift is there, but it is extremely small because the stars are very far away. The naked eye cannot detect the change; only a powerful telescope will reveal it. This explanation had been grasped by Aristarchus, two millenia earlier, but most astronomers rejected it. Few believed the stars could be so distant, and the Universe so vast, as to make the parallax effect undetectable. Most astronomers thought the fault lay with Copernicus, and regarded his theory as a failure. And of course the old problems remained that objects would fly off the spinning earth, or be left behind in the earth's headlong rush around the sun. Small wonder that Luther called Copernicus "fool" for trying to turn astronomy on its ear when the facts were against him, and for contradicting the Bible to boot, since, as Luther angrily pointed out, "The Holy Scriptures tell us that Joshua commanded the sun to stand still, and not the earth."

Implications for Man's Position in the Cosmos

Yet, in spite of its defects, the Copernican theory took root in men's minds. There was an air of freshness about it; it opened the door to new ideas that went far beyond the science of astronomy. Men saw the implications in the sun-centered Universe. Consider the following facts, they said: the five planets revolve around the sun; the earth also revolves around the sun; since the earth and planets behave in the same way, they must be the same kinds of objects.

The earth a planet? This was a surprising conclusion. Prior to that time, many astronomers had believed that planets were hard, polished spheres made of a jewel-like substance, perfect and unchanging, while the earth was made of ordinary substances like mud, rock, and water. If the earth and planets are similar objects, the followers of Copernicus reasoned, the planets may also be made of mud, rock, and water. That unsettling thought led to another; if the planets are made of the same material as the earth, perhaps they bear life; perhaps there are people on them. People had speculated on such notions before, but now there seemed to be evidence for them.

These were radical ideas. They presaged the latest developments in twentieth-century science, which unite life on the earth to life in the Cosmos. The astronomy of Copernicus was the first step in the Copernican revolution; Copernicus removed the earth from the center of the Universe, and put the sun in its place. Others took the second step; they removed the sun from the center of the Universe and put nothing in its place. *There is no center,* they said; the Universe is infinite, and contains an infinite number of stars. Each star is a sun like ours; each may have a family of planets.

These ideas led finally to the modern picture of a Universe populated by innumerable suns, innumerable earths, and, perhaps, innumerable forms of life. That thought expresses the essence of the Copernican revolution. While man stands at the summit of creation on the earth, in the cosmic order his position is humble. (Figure 1.8).

FIGURE 1.8
Stars in the Milky Way. Every point of light in the photograph is a star; every star is a sun.

FIGURE 1.9
Galileo Galilei (1564–1642)

CONFIRMATION OF THE COPERNICAN THEORY BY GALILEO

Copernicus was not as bold and revolutionary as his theories. Timid, retiring, fearful of ridicule, he circulated a brief early draft of his ideas, but he kept his main work from publication for more than thirty years. Time passed, and word of the new theory spread through Europe. The Church took notice, and a Cardinal wrote him, "Learned man, I beg you to communicate your discovery to the learned world." Yet Copernicus hesitated. Finally, an enthusiastic young disciple named Rheticus succeeded in wresting his manuscript away, and it went off to the printer. Still Copernicus was apprehensive, and wrote in his dedication to Pope Paul II, "I may well presume, most Holy Father, that certain people, on learning that in my book *On the Revolutions of the Heavenly Spheres* I ascribe certain movements to the Earth, will cry out that, holding such views I should at once be hissed off the stage." He never found out whether his fears were justified, for he died just as his great work was published.

The theory that the earth moves was rescued from relative obscurity by an Italian professor of mathematics named Galileo Galilei (Figure 1.9), who was born in 1564, not long after Copernicus' death. Galileo also came to believe that the earth could not be at the center of the Universe, and he communicated his ideas with wit and in a sparkling literary style that made the movement of the earth a major topic of conversation in Italy. The impact of Galileo's arguments was strengthened by his use of the newly invented telescope, which, turned on the celestial bodies, revealed secrets never known to man before (Figure 1.10). The telescope converted the ideas of Copernicus from theory to fact.

Galileo's Discoveries with the Telescope

The telescope was probably invented sometime between 1600 and 1608. So far as is known, Galileo was the first to raise the newly invented instrument towards the heavens. Looking at the planet Jupiter, he made a remarkable discovery: Four small bodies of light, invisible to the naked eye, were moving in orbit around Jupiter, just as the earth moves in its orbit around the sun. These were the four largest of Jupiter's moons; today they are known as the Galilean satellites. Galileo saw the significance in his discovery; Jupiter and its moons formed a miniature solar system, like the sun and its family of planets. This picture was in conflict with the ancient idea, supported by the Church, that the earth is the center of the world and everything turns around it; for here was another body, far from the earth, with its own family of satellites circling around it. In

FIGURE 1.10
Galileo's telescope.

Galileo's mind, the moons of Jupiter provided strong support for the theory of Copernicus.

Galileo looked at the planet Venus, and discovered that it showed phases like those of the moon, ranging from the "crescent" Venus to the "full" Venus. Galileo realized that the phases of Venus must be caused by the fact that the planet moves around the sun. Figure

1.11 shows how he came to this conclusion. When Venus is directly between the sun and the earth, the planet's dark side faces us and it is invisible. As Venus moves around its orbit, more and more of the illuminated face can be seen from the earth. First, the "crescent" Venus appears; the crescent grows to the "quarter" Venus and the "half" Venus; finally, on the other side of the planet's orbit, the nearly "full" Venus becomes visible.

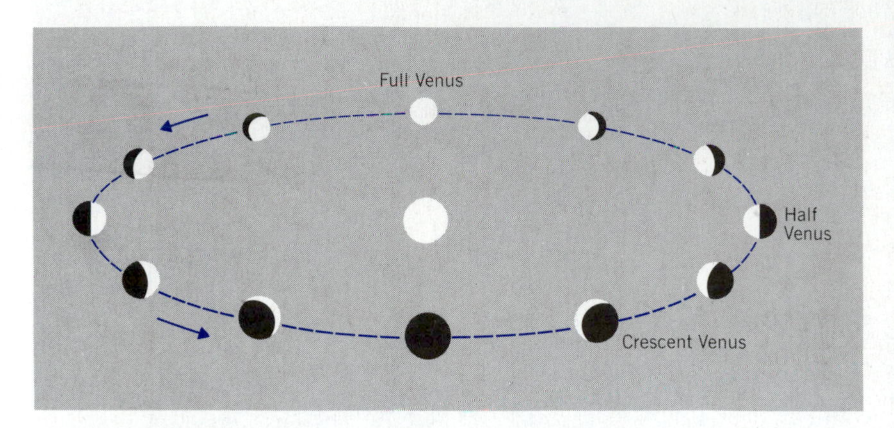

FIGURE 1.11
Venus passes through phases like the phases of the moon as it revolves around the sun. Galileo's observations of the phases of Venus through his telescope provided additional proof that the planets revolve around the sun.

Galileo's observations of Venus provided more evidence against the idea that the earth is at the center of the Universe. They lent further support to the theory of Copernicus that the sun is at the center, and the planets, including the earth, revolve around it. Galileo had not proven that Copernicus was right; he had only shown that *one* planet moved around the sun. Still, his discovery tended to upset the ancient view that the earth was at the center of all motions in the Universe.

Galileo also studied the moon (Figure 1.12), and, although his instrument had a magnification of only 30 powers, he could see that the surface of the moon, far from being of a polished smoothness, as had been thought, was "uneven, full of irregularities, hollows and protuberances, just like the surface of the earth itself."

That discovery was startling to Galileo's contemporaries. If the moon, a celestial body floating serenely through the heavens, contained "lofty mountains and deep valleys," it must be made of the same earthly stuff as our planet. Copernicus had shown that the earth resembled the planets; now Galileo showed that the earth resembled the moon. Clearly, the difference between the earth and the heavenly bodies could not be as complete as the ancients had imagined it. This conclusion was reinforced by another discovery. Galileo looked at the sun, and found its bright surface marred by dark, ugly patches—the sunspots. Now it appeared that the two most important heavenly bodies, the sun and the moon, were flawed by crude imperfections.

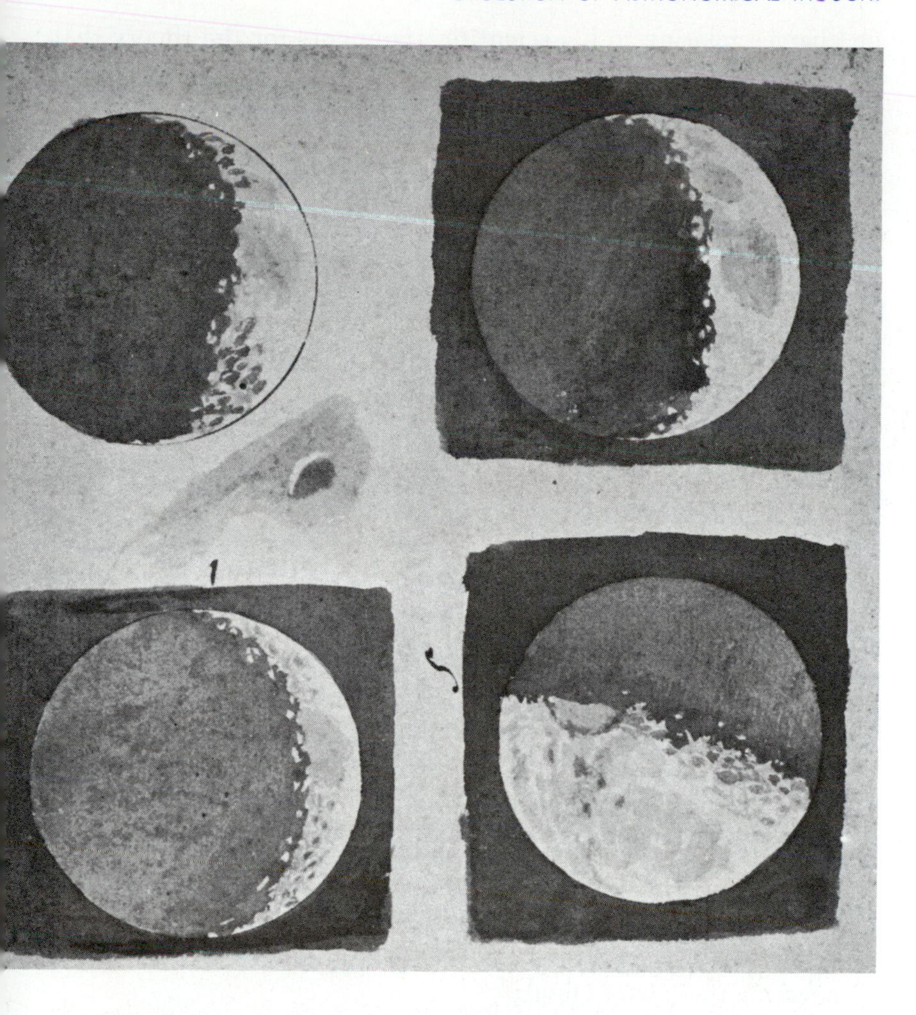

FIGURE 1.12
Galileo's drawings of the moon, showing the mountainous lunar surface.

Opposition to Galileo. When Galileo first described his discoveries they were greeted with admiration by prominent members of the Church. The eminent Jesuit astronomer, Father Clavius, looked through the telescope and announced he had confirmed the existence of the moons of Jupiter; and Cardinal Barberini, the future Pope Urban VIII, praised Galileo's great work. But now there was a dangerous air of novelty in his findings, and the Church began to look at Galileo's researches with a less friendly eye.

In 1616, the Holy Office convened a committee of theologians in Rome to obtain their opinion on the Copernican theory that "the sun is at the center of the world . . . [and] the earth is not at the center of the world, nor immovable." They declared unanimously that these views were "foolish, absurd . . . and heretical." A few weeks later, Copernicus' book was placed on the Index and Galileo was admonished to abandon his opinions on the movement of the earth. In 1632, he was summoned to Rome to face the Inquisition on charges relating to his scientific arguments for the theory that the earth revolved around the sun. He was tried, convicted, and forced on his knees to swear that he believed the sun revolved around the earth.

In 1979, Pope John Paul II looked back to the events of three centuries ago and said about the trial, "Galileo had to suffer a great deal . . . at the hands of men of the church. I hope that theologians, scholars and historians, animated by a spirit of sincere collaboration, will study the Galileo case more deeply and, in loyal recognition of wrongs from whatever side they come, will dispel the mistrust that still opposes, in many minds, a fruitful concord between science and faith."

ISAAC NEWTON

Although Galileo believed in Copernicus' theory that the planets moved around the sun, he did not know *why* they moved in this way. What kept the planets in their orbits? Why didn't they fly off into space?

Galileo thought he knew the answers to those questions. Like most astronomers in his time, he was wedded to ancient notions of the perfection of the circle, and thought that the circle was a natural self-perpetuating form of motion that required no force. The planets moved in circles around the sun because that was in the nature of things.

Today we know this is not so. Gravity is needed to keep the planets tied to the sun, and the moons tied to the planets. Galileo's erroneous views on circular motion probably prevented him from discovering the law of gravity.

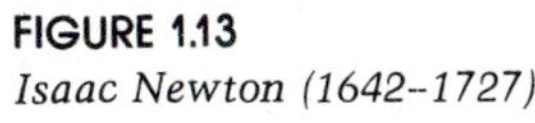

FIGURE 1.13
Isaac Newton (1642–1727)

Isaac Newton (Figure 1.13) was one of the first persons to strip away the old misconceptions about circular motion. Newton realized that if no forces act on an object, it does not move in a circle, but a straight line. If the path of a body is a circle, some force must be acting on the body to bend its path into a curved line.[2]

Newton's Reasoning on the Motion of the Moon

In his youth Newton began to think about the motion of the moon. He knew that the moon must fly off from the earth at a tangent, unless a force was applied to bend it around into its orbit.

[2] Several scientists including Kepler, Hooke and Borelli anticipated Newton by suggesting, in 1666, that the attractive force of gravity is exerted by the sun on the planets, and bends their paths into ellipses. Only Newton was able to prove the result mathematically.

In the third book of his great work, *The Mathematical Principles of Natural Philosophy,* Newton included a figure intended to apply this idea to the moon's motion (Figure 1.14). The figure shows a mountain on whose top a cannon is mounted with its barrel directed horizontally. When a ball is fired from the cannon, it moves forward under the impetus of the pressure exerted by the hot gases in the barrel. At the same time, it is subjected to the downward attraction of gravity, which pulls it toward the center of the earth. The combination of the forward motion and the motion downward under gravity is a curved path, which terminates when the projectile hits the ground.

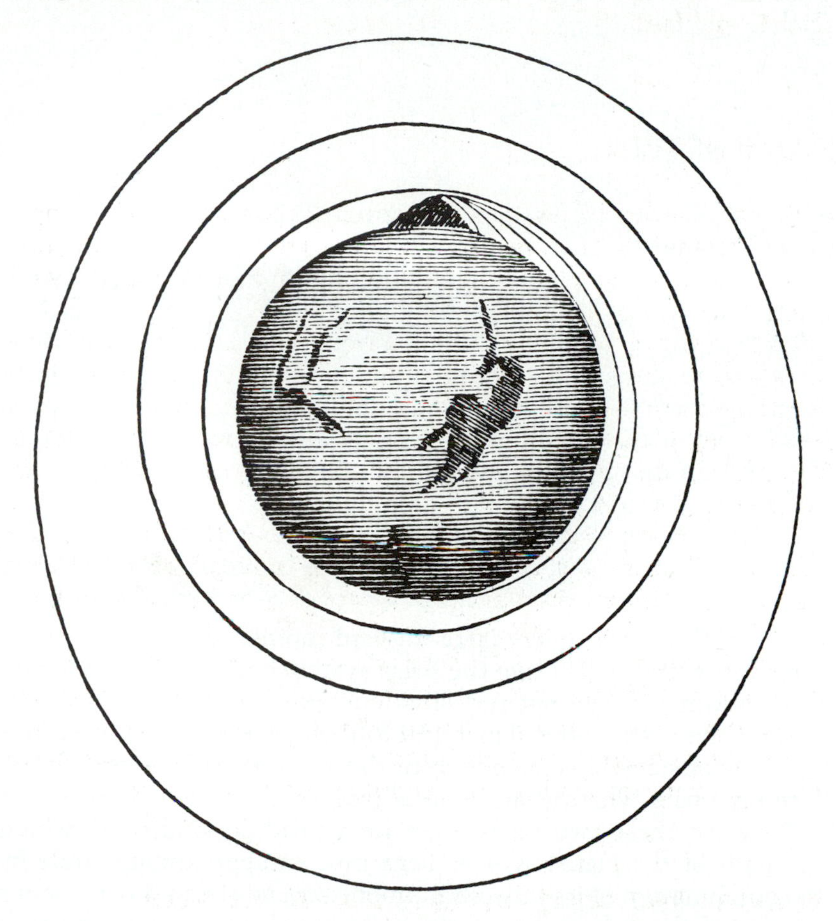

FIGURE 1.14
Newton's diagram illustrating the motion of the moon, from the third volume of Principles of Natural Philosophy. *The cannon placed on a mountaintop fires a shot that travels a curved path, compounded of its horizontal forward velocity and a downward motion produced by the gravitational attraction of the earth.*

If the charge of explosives is increased, the forward velocity increases and the ball traverses a greater distance before it is pulled to the ground by gravity. It is conceivable that the charge of explosives can be so great that the ball will travel entirely around the earth without striking the ground, although falling freely toward the ground at every moment. The combination of the forward motion produced by the discharge of the cannon, and the downward deflection produced by gravity, curves the path of the projectile into a circular orbit around the earth. The cannonball has become a satellite.

Newton constructed this imaginary experiment to explain the motion of the moon around the earth, but it is also the explanation for the motion of the planets around the sun (Figure 1.15). If a planet has no forward momentum carrying it around in its orbit, then it will fall into the sun, drawn by the sun's gravity. Suppose now that the planet is given a small forward momentum. Then it will fall in toward the sun on a curved path at a close distance, and move outward again in an orbit which has the shape of an ellipse.

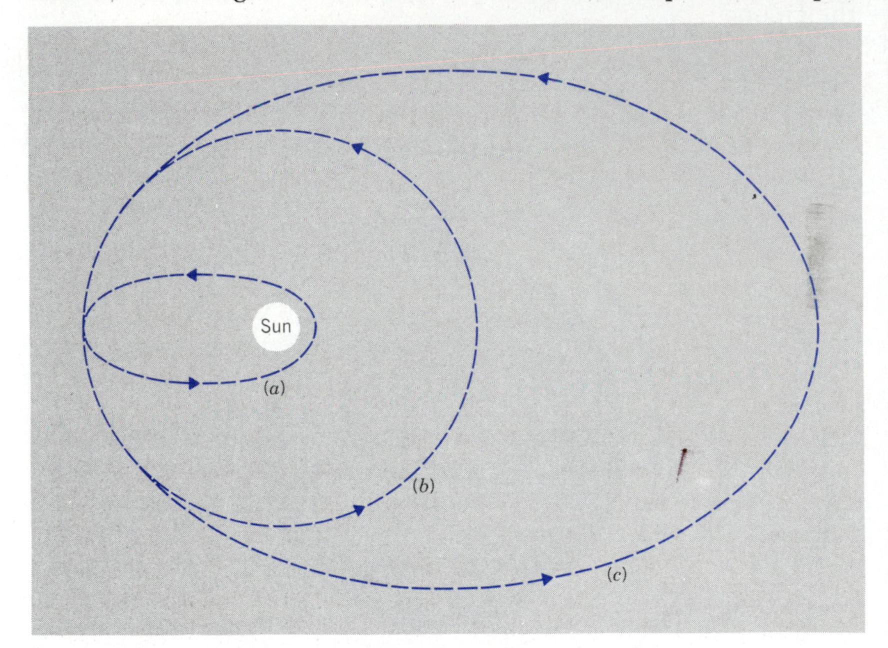

FIGURE 1.15
Types of planetary orbits.

If the planet has a very large forward momentum, it will escape the sun's gravity and leave the solar system entirely. If the forward momentum is large, but not large enough to carry the object out of the solar system, the planet will follow a curved trajectory carrying it a considerable distance from the sun, but will eventually return on a path which is again an ellipse.

Between these two cases must lie a third possibility, in which the path of the planet will be bent into an approximate circle by the combination of its forward momentum and the pull of the sun's gravity. If the two motions are in the proper proportion, the path will be a perfect circle.

The Contribution of Kepler

In his work on the nature of gravity Newton used the results of a man who lived in the same period as Galileo, a remarkable astronomer and mathematician named Johannes Kepler (1571–1630). Kepler (Figure 1.16) had analyzed the motions of the planets, and from them he had derived three general laws of planetary motion.

FIGURE 1.16
Johannes Kepler (1571–1630)

1. The orbit of each planet is an ellipse about the sun.
2. The line from the sun to the moving planet sweeps over equal areas in equal intervals of time. This law says, in effect, that a planet moves fastest when it is closest to the sun.
3. The square of the time (T) in which a planet completes one circuit around the sun is proportional ($\propto$) to the cube of its distance to the sun (R): $T^2 \propto R^3$.

The first two of these were published in 1609. The third law appeared in 1618. Kepler arrived at these laws because he had available to him the superb measurements of Tycho Brahe on the positions of the planets. Kepler joined Brahe as his assistant in 1600, and was assigned the orbit of Mars as his first task. Kepler tried to explain the observed positions of Mars in the sky by assuming that the planet traversed a circular orbit around the sun, following the traditional point of view that the circle was the natural path for celestial bodies. He was forced to place the sun in an off-center position in order to improve the agreement between his theory and the observations. With the aid of this assumption, and after four years of calculations, he reduced the discrepancy between theory and observation to eight minutes of arc. Eight minutes of arc is the angle between the eyes of a person at a distance of one city block. Yet this minute discrepancy signaled the fall of the old astronomy and the rise of the new science.

Copernicus had simplified the older, Ptolemaic system by placing the sun at the center of the solar system, but his description of planetary motions still relied on the idea of combining circular motions. Kepler was the first astronomer to overthrow the circle and the epicycle, and establish the ellipse as the correct description of the motion of the planets. In so doing, he opened the way to Newton's subsequent linking of the elliptical orbit and the inverse square law of gravity.

The accuracy of Tycho Brahe's observations played a critical role in this development. Ptolemy and Copernicus had based their epicycle theories on naked-eye observations of the positions of the planets which were accurate to ten minutes of arc. They were only able to fit the epicycle to the observed motions of the planets if they allowed this amount of error. The discrepancy of eight minutes of arc which Kepler found in his calculations for Mars was small enough to have satisfied Ptolemy and Copernicus. But Brahe claimed that his observations had an error of no more than two minutes of arc, and Kepler believed him. He continued his labors for two more years; finally, after many false starts, and 900 pages of 16th-century algebra and geometry, he discovered that if Mars moved, not in a circle, but in an *ellipse* around the sun, then the positions of the planet could be predicted with a discrepancy of less than two minutes of arc at each point.

It was a fortunate circumstance that Brahe assigned Mars to Kepler on his arrival in Prague, because the orbit of Mars is more elliptical than the orbit of any other nearby planet. The earth's orbit is close to a perfect circle; it departs from circularity by two percent. The orbit of the planet Venus departs from a circle by only 0.7 percent. The orbit of Mars, however, is different; it departs from circularity by 9 percent. Because the departure was as great as 9 percent Kepler was unable to fit his calculations of the Mars orbit into the conventional ideas about circular orbits.

Kepler's second law proves that the force holding the planets in their orbits emanates from the sun. However, it does not play as pivotal a role in the discovery of the law of gravity as the other two. The third law, $T^2 \propto R^3$, led Newton directly to the law of gravity.

As an illustration of the third law, consider the orbits of the earth and Mars. The earth is 93 million miles from the sun, and revolves about the sun in 365 days or approximately 30 million seconds. Mars is 142 million miles from the sun, or roughly 1.5 times farther than the earth. Since $T^2 \propto R^3$ we have

$$\frac{T^2\ (\text{Mars})}{T^2\ (\text{Earth})} = \frac{R^3\ (\text{Mars})}{R^3\ (\text{Earth})} = 1.5^3 = 3.5$$

or

$$\frac{T\ (\text{Mars})}{T\ (\text{Earth})} = \sqrt{3.5} = 1.9$$

hence the Martian year is 1.9 times the length of a year on the earth, or approximately 690 days.

The Law of Gravity

Newton claimed that he was a very young man when he first had the remarkable thoughts on the universality of gravity. He had returned home from Cambridge University in 1665, at the age of 23, to avoid the Great Plague which struck the city in 1665 and 1666. He remained at home for two years, and, according to his recollections in later years, it was during this period of isolation that he first reflected on the motion of the moon, which revolves around the earth in an approximately circular orbit as the planets revolve around the sun.

The moon would fly off at a tangent, were there not a force attracting it to the earth and constraining it to move in a circular path. What is this force? Could it be the same as the force of gravity which the earth exerts on objects at its surface? Newton conjectured that the two forces were identical. If this proposal were correct it would unite the earth and the heavens in one mechanical system. The boldness of the step cannot be exaggerated.

Newton turned to the motions of the moon and planets for a test of his hypothesis. Kepler had already written in 1609, ". . . the attractive force of the earth . . . extends to the moon and even farther." But Newton went beyond this conjecture. From Kepler's third law of planetary motion he derived the basic property of gravity. *The gravitational force between two objects varies as one over the square of the distance between them.*[3] That is, if the distance between two objects is doubled, the force of gravity diminishes to one-quarter of its former value; if the distance is tripled, the force of gravity decreases to one-ninth of its previous value; and so on. This relationship is known as the inverse-square law of gravity.

Suppose that the masses of the two objects are m_1 and m_2, respectively, and the distance between them is R. Then the full mathematical statement of Newton's law of gravity is

$$\text{Force} = G\frac{m_1 m_2}{R^2}$$

In this formula, G is the constant of proportionality, called the *universal constant of gravitation.* In the cgs (centimeter-gram-second) system, G has the value of 6.7×10^{-8}.

Armed with the $1/R^2$ law for the force acting on the planets, Newton could now test the universality of gravity. He reasoned: Suppose the force of gravity acts between all objects in the Universe; then this force, which emanates from the sun to the planets, will also emanate from the earth, and will attract objects to the earth as well. Perhaps this one force pulls objects to the surface of the earth and also holds the moon in its orbit.

Newton followed this line of reasoning further: Since the force of the sun's gravity falls off as $1/R^2$, the force of gravity around the earth must also fall off as $1/R^2$. The moon is 240,000 miles from the earth, or 60 times farther removed from the center of the earth than

[3] According to Newton's law of motion, Force = mass × acceleration, $F = ma$. If m is the mass of the moon, the force holding the moon in its orbit must be mv^2/R, where v is the speed of the moon in its orbit, R is the radius of the orbit, and v^2/R is the centripetal acceleration of a body moving in a circle. The velocity of the moon is the circumference of its orbit divided by the time T required to complete one circuit: $v = 2\pi R/T$. Therefore, $F = m\,4\pi^2 R/T^2$. But according to Kepler's third law, $T^2 \propto R^3$. Inserting this relationship into the formula for the force, we obtain $F \propto 4\pi^2 m/R^2$. That is, the force of gravity is proportional to one over the square of the distance.

a body on the surface of our planet. Therefore, if the earth's gravity is the force that is acting on the moon, that force should be weaker at the moon's distance by a factor of $(60)^2$, or 3600.

Newton calculated the force required to keep the moon in its circular orbit around the earth, and compared it to the force of gravity acting on objects at the surface. He found that the ratio was close to 1/3600. Since this result agreed with the law of gravity, Newton concluded that it was indeed the earth's gravity that was pulling on the moon. In this way, Newton extended the domain of gravity from the earth to the moon.

Years later, Newton took the final step. Kepler's first law of planetary motion stated that the planets move in elliptical orbits. Newton proved mathematically that the ellipse is just the path that will be followed by a body moving under an inverse-square law of attractive force. This proof united all the known laws of planetary motion in one system. All were a consequence of a single force—a force that held the planets in their orbits around the sun, held the moon in its orbit around the earth, and made objects fall to the earth's surface.

Thus Newton extended the law of gravity from the earth to the entire solar system. Encouraged by his success, he conjectured that the law of gravity was valid throughout the Universe. In this one flight of thought, Newton united the heavens and the earth. His achievement laid the cornerstone of modern science.

Main Ideas

1. The ancient ideas and the system of Ptolemy.
2. The Copernican theory.
3. Implications of the new astronomy for man's position in the Cosmos.
4. Confirmation of the Copernican theory by Galileo.
5. Isaac Newton: The motion of the moon.
6. The contribution of Kepler.
7. Newton's law of gravity.

Important Terms

Copernican theory
elliptical orbit
epicycles
Galilean satellites
Galileo
inverse square law of gravity
Kepler
Newton
Ptolemaic theory
parallax
spiral galaxy
sunspots

Questions

1. With the aid of a drawing, demonstrate how epicycles can explain the retrograde motion of planets about an earth-centered solar system.
2. What difficulties did Copernicus see in the Ptolemaic theory of the heavens? How did his theory correct for these difficulties? How did the Copernican theory fail the test of comparison with observation?
3. If the motions of celestial objects actually carried them around a stationary earth, how far from the earth would an object have to be in order to circumnavigate the earth at a speed greater than that of light? Use the formula that states: distance equals velocity multiplied by time.
4. With the aid of a diagram, show why the phases of Venus, as discovered by Galileo, indicate that Venus is an "inferior planet", i.e., a planet whose orbit is inside the orbit of the earth.
5. How did Galileo's other discoveries with the telescope support the Copernican theory? List his major discoveries and explain their bearing on that theory.
6. Write a brief argument showing that an object moving in a circle would have to "feel" a force pulling it toward the center of the circle. In answering this question, begin with the idea that if no force were acting on the object it would move in a straight line.
7. If Newton's Law of Universal Gravitation is applied to calculate the force on a ball falling near the surface of the earth, what should be used for the distance between the ball and the earth?
8. Explain Newton's reasoning which led him to the idea that gravity is universal, i.e., acts in the heavens as well as on the earth.
9. The successive proposals of Copernicus, Galileo, Kepler and Newton were "bitter pills" for the philosophers of the time. But each proposal constituted a step in the evolution of thought regarding the Universe. What change in view did each proposal require? Specifically:

   A. What was at the heart of contemporary philosophical objection to each proposal?
   B. What specific change in view did the acceptance of each require?
   C. Together, the proposals form the basis of a fundamental principle that underlies our modern view of the Universe. What is the principle?

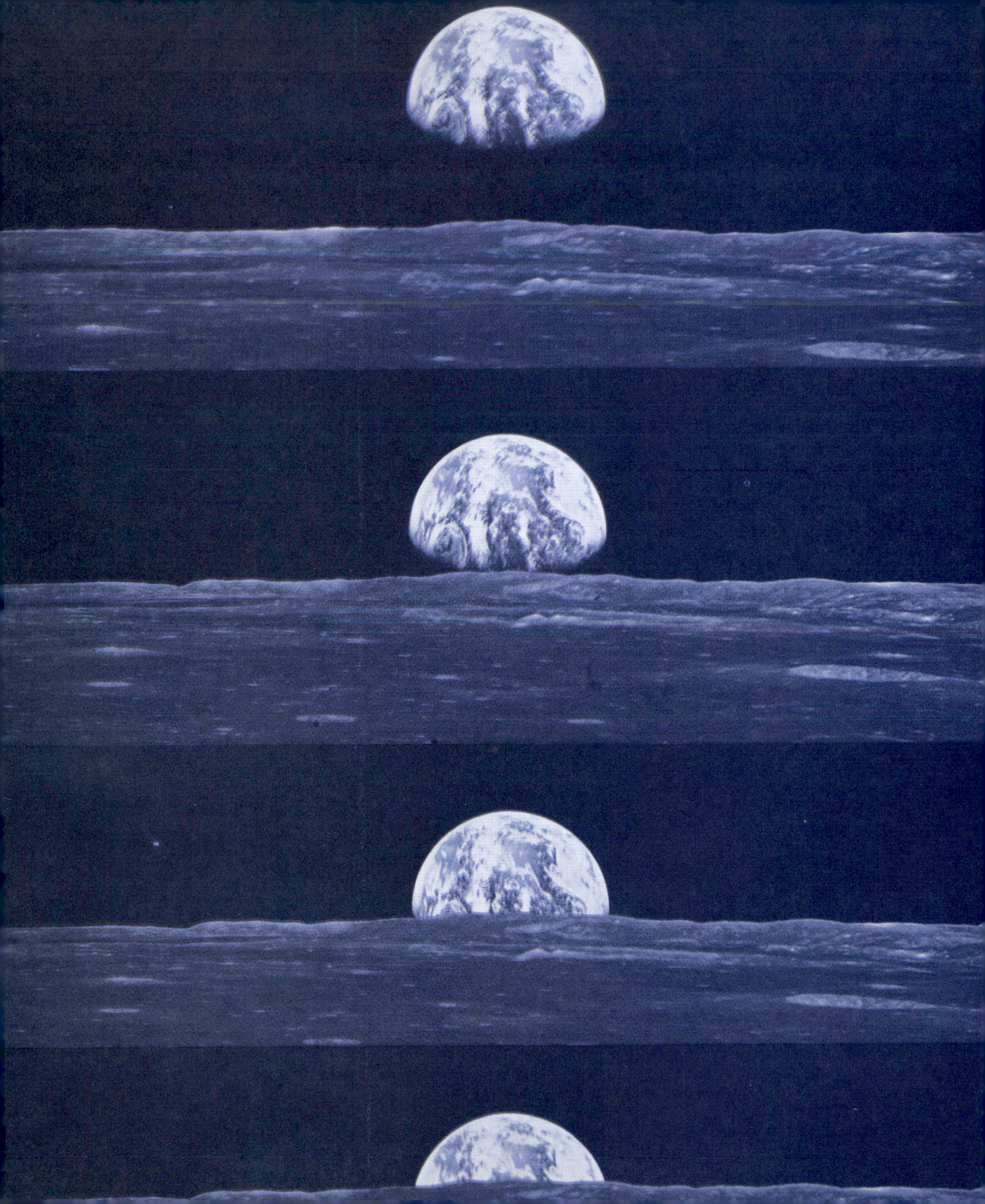

THE EARTH IN SPACE

2

From the time *Homo sapiens* first stood on the earth and looked upward, human curiosity has been aroused by the motions of the celestial bodies—some regular, others seemingly erratic. As early observers followed the movements of the sun, the moon, the stars, and the planets, patterns emerged that had practical value as well as philosophical interest. These patterns led to primitive calendars for predicting when to plant without being caught by a false spring; and the positions of the sun and the stars became tools for navigating over the surface of the earth. And then there were the mysterious, terrifying eclipses, perhaps signs of divine displeasure. What was their meaning? Could they be predicted? It seemed that forces beyond human comprehension must be at work in these phenomena. These were the roots of astronomy—the philosophical, the practical, and the mystical. All depended on visual phenomena created by the motions of the earth and its celestial neighbors.

MOTIONS OF THE EARTH

Although the motions of the earth on its axis and in its orbit around the sun produce the largest changes in the appearance of the heavens, they are not the earth's only motions. The sun and its family of planets belong to a mammoth cluster of billions of stars called the Milky Way Galaxy. The stars in the Galaxy move randomly around the center of the Galaxy, and they have random, irregular motions as well, each star buzzing around in space like an atom in a

The setting earth, photographed from the moon.

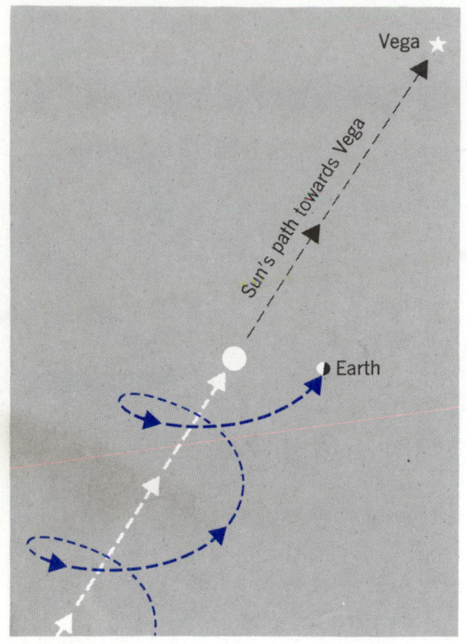

FIGURE 2.1
The earth's spiraling motion through space.

gas of particles. The sun's velocity at this moment is carrying it toward Vega—a bright star directly overhead in the autumn sky around 9 P.M.—at a speed of 12 miles a second. The earth accompanies the sun in its flight toward Vega, tracing a spiral through space (Figure 2.1).

The sun and the other stars in the neighborhood of our solar system revolve slowly around the center of the Galaxy. In the course of this motion, our solar system completes one full turn around the Galaxy in 250 million years, covering a distance of nearly one million trillion miles in that time at an average speed of 200 miles per second.

The entire Galaxy, carrying the earth with it, is moving through space relative to other nearby galaxies. As a consequence of this motion, our galaxy and the Andromeda Galaxy—our nearest large galactic neighbor—are approaching one another at a speed of 180 miles per second. The earth is carried along with our galaxy in this motion.

The entire Galaxy belongs to a group of a dozen or so galaxies that form a small cluster. This local group of galaxies, carrying the earth with it, is moving toward a larger group of galaxies called the Virgo cluster at a speed of approximately 250 miles a second, apparently drawn to the Virgo cluster by its enormous gravitational pull.

Thus the motion of the earth in space is very complicated. It rotates about its axis, revolves around the sun, moves with the sun and other nearby stars around the center of the Galaxy, and moves with the Galaxy on its journey toward Andromeda. Finally, our galaxy and Andromeda, as well as our other galactic near neighbors, are moving in the direction of the Virgo cluster. Perhaps the earth participates in still other movements as well. We cannot say we know all the motions of the earth.

Rotation of the Earth

The rotation of the earth is also more complicated than early observers thought it was. At the present time the earth's axis of rotation is pointed in a direction that makes an angle of 23.5 degrees with a perpendicular to the plane of the earth's orbit—*the ecliptic* (Figure 2.2). However, the tilt of the axis changes with time, varying between 22 degrees and 24 degrees. The variation goes through a complete cycle every 41,000 years. The cause of the variation in the tilt is the gravitational pull of the moon and Jupiter.

Precession of the Axis. In addition, the earth's axis shifts its direction in the course of time. Its movements are similar to the movement of a spinning top, rotating very rapidly about its axis

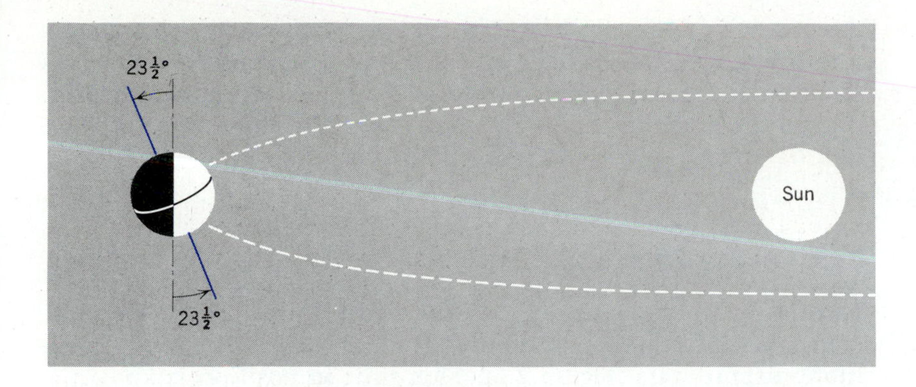

FIGURE 2.2
The tilt of the earth's axis.

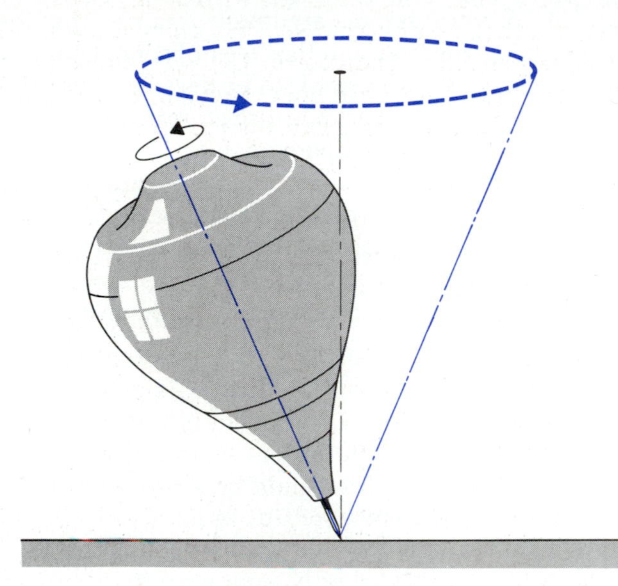

FIGURE 2.3
The precession of the axis of a spinning top.

while the axis itself revolves slowly about the vertical. This slow change in the direction of the axis of a spinning object, whether the earth or a gyroscope, is called the *precession* of the axis (Figure 2.3). The precession of the earth's axis is produced by the gravity of the sun and moon, which tug at the equatorial bulge of the earth. Under the influence of these forces, the axis of rotation revolves in a cone, completing a circle once every 26,000 years.

At present, the axis is pointed toward the North Star; 5000 years ago it pointed in the direction of the star Thuban in the constellation Draco, and 12,000 years from now it will point in the direction of Vega, which will be the replacement for the North Star at that point in the distant future (Figure 2.4).

Nutation. In addition to the slow movement of the axis of rotation, completed once in every 26,000 years, there is also a smaller movement of the axis, something like a wobble superimposed on

FIGURE 2.4
Precession of the earth's axis.

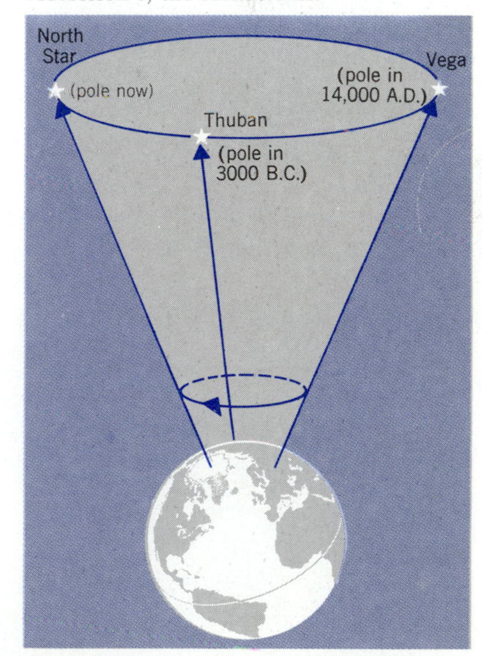

the steady 26,000-year drift. The wobble of the axis of rotation of a spinning object is called *nutation* (Figure 2.5). As a result of the nutation of the earth's axis of rotation, the axis varies in its inclination to the plane of the earth's orbit by somewhat more than one four-hundredth of a degree every 19 years.

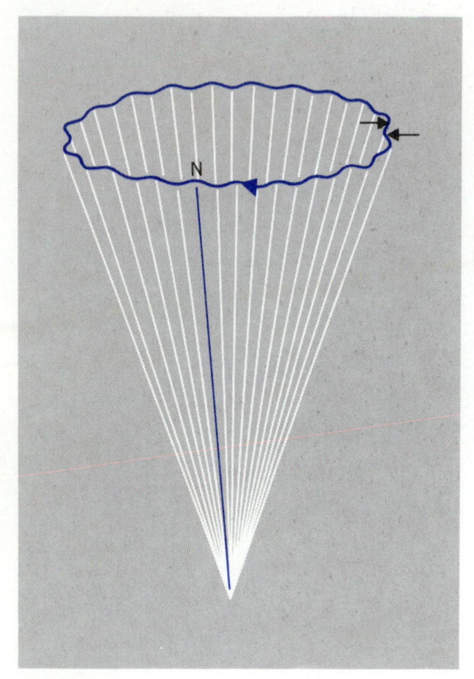

FIGURE 2.5
Nutation.

Proof of the Earth's Motion

During the course of a year, as the earth moves around its orbit, stars nearer to the sun should shift their position relative to the more distant stars (Figure 2.6). This shift in position is known as *parallax* (p. 12). A very strong argument against the theory of Copernicus had been that no parallax was observed. Actually the parallax was there but was too small to be detected with the naked eye or even with early telescopes, because all stars, even the closest, are exceedingly far away in comparison with the diameter of the earth's orbit. It was not until 1838 that the telescope reached a state of perfection such that in the hands of a very skillful observer —the German astronomer, Bessel—the shift in the position of a nearby star could finally be measured. Bessel's proof of the earth's motion was obtained in 1838, 295 years after the death of Copernicus.

FIGURE 2.6
The change in the apparent position of a nearby star against the background of distant stars.

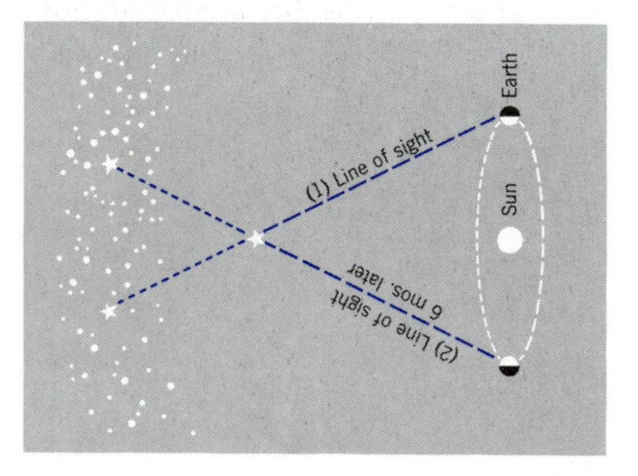

Proof of the Earth's Rotation. The clearest proof of the earth's rotation is provided by artificial satellites. Suppose a satellite is launched from Cape Canaveral into an orbit 100 miles above the earth carrying it toward the southeast, at an angle of 30 degrees to the equator (Figure 2.7*a*). This would be a typical launch trajectory from the Cape. Once the satellite is launched, the plane of its orbit stays fixed in space, since no forces are exerted on the satellite to change it. (This is only approximately true, but the change is small enough to be ignored in the present discussion.) Therefore, if the earth is not rotating, the satellite should pass over Cape Canaveral once in every orbit. But it does not; it passes over Alabama on the completion of its first orbit, over Louisiana at the end of the third, and so on (Figure 2.7*b*).

Because the earth is rotating beneath the satellite, it passes over places in the United States that are in this case 1000 miles farther west on each orbit. This fact, which has been observed in all of the hundreds of satellites that have been launched, directly proves the rotation of the earth.

Aberration of Starlight. If the earth is really moving around the sun, its speed is about 1/1000 of the speed of light. This motion shows up with respect to the stars as a slight shift in position of all stars regardless of their distance. The effect is similar to the one observed by a person walking or driving through the rain, who sees the raindrops apparently coming from a slightly different direction according to the direction in which he or she is travelling. The effect on the position of the stars, called the aberration of starlight, was discovered by James Bradley in 1729, about a century before parallax was measured. The size of the apparent motion of the stars from observation is, in fact, some 25 times larger than the parallax effect for the nearest stars.

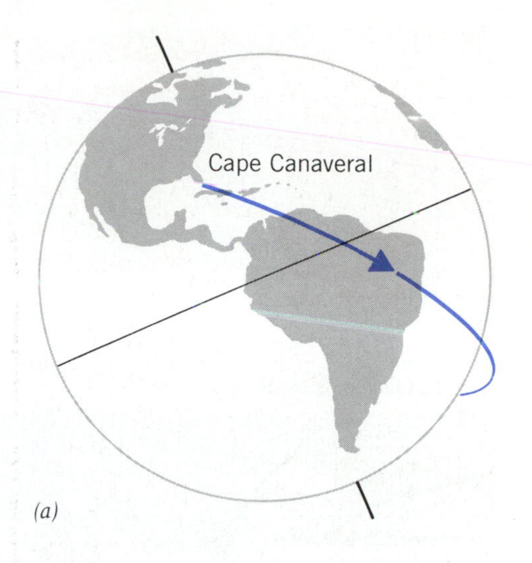

(a)

THE MOON AND THE EARTH

The moon accompanies the earth on its journey through space, revolving around our planet every 27.3 days in a nearly circular orbit. The relative positions of the sun, earth, and moon change continually as a result of the motion of the moon around the earth and the earth around the sun, creating phenomena of great beauty as well as practical importance.

The serene passage of the moon's silver orb across the heavens has excited the admiration of poets since the dawn of language:

> *"The starry host, rode brightest, till The moon,*
> *Rising in clouded majesty, at length*
> *Apparent queen, unveil'd her peerless light,*
> *And o'er the dark her silver mantle threw."*
>
> Milton

But the moon also exerts a strong influence over the practical affairs of men because of its exceptional size and mass in comparison to its parent planet. No other satellite in the solar system approaches the moon in this respect. The large size of the moon enables it to cast a giant shadow over parts of the earth on occasion, when the movements of the moon and the earth bring them into a straight line with the sun, in the phenomenon known as a *solar eclipse.* The massiveness of the moon creates a powerful gravitational pull that causes large areas of the land and the oceans to bulge outward along the earth-moon line by amounts ranging from a few inches to many feet, in the phenomenon known as the *tides.*

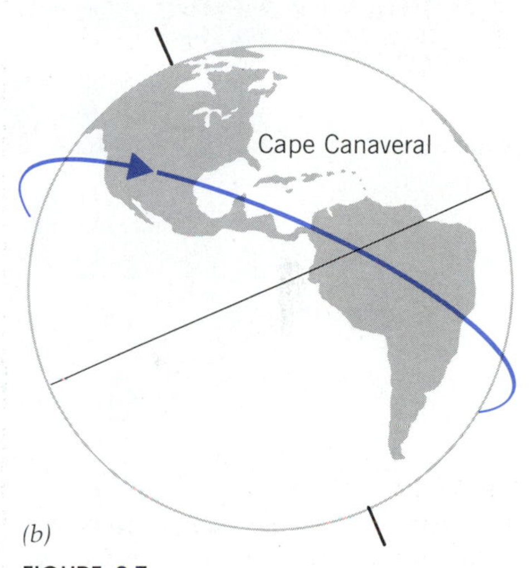

(b)

FIGURE 2.7
Successive orbits of a satellite provide proof of the earth's rotation. a) The satellite is launched from Florida. b) It passes over Alabama after one orbit.

The Phases of the Moon

The light we see coming from the moon is reflected sunlight. Half the lunar surface is always in sunlight, but to us the moon's face appears to change shape because we see varying portions of the sunlit side as the moon revolves around the earth. The moon is invisible when it is between the earth and the sun, partly because the side facing the earth is in shadow, but also because in this position the moon, being near the sun in the sky, is lost in the solar brilliance. When the moon is a little off the earth-sun line, we can see the thin crescent of the illuminated edge. When the moon is on the opposite side of the earth from the sun, its fully lighted face dominates the nighttime sky.

The shapes successively assumed by the moon are referred to as its *phases.* The phases of the moon proceed from new moon to crescent, to first quarter, to gibbous, to full moon, and then, in reverse order for the remaining half of the cycle, through gibbous, third quarter, and crescent to new moon again (Figure 2.8). In the

FIGURE 2.8
The phases of the moon.

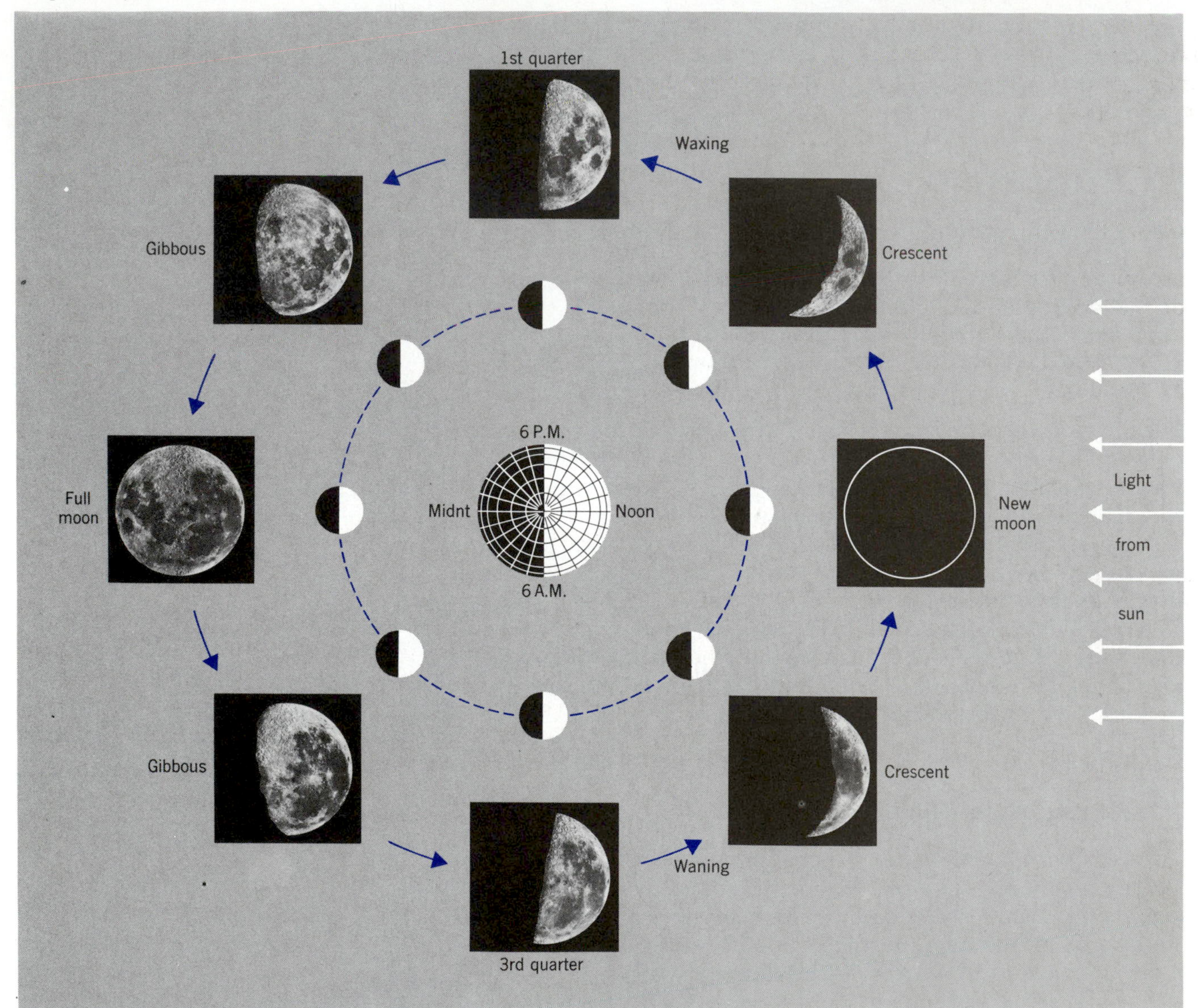

quarter phases, exactly one-half of the moon's visible face is illuminated. For this reason the moon in its quarter phases is often described as being half full, or as a half moon.

As the moon progresses through the first half of its cycle from new to full, it is called a *waxing* moon. In the second half of the cycle it is called a *waning* moon. The cycle takes 29.5 days.

To astronauts looking back at earth from the moon, the earth goes through the same sequence of phases that we observe the moon going through. Any particular phase of the earth, however, will be opposite the phase of the moon. If we see the moon in the third quarter, the astronauts see a first quarter earth.

Moonrise and Moonset. The times at which the moon rises and sets can be worked out from studying Figure 2.8. These times depend on the phase of the moon. For example, suppose the moon is in its first quarter. Suppose an observer is standing on the earth at a point in which the sun is directly overhead, that is, the local time is noon, as shown in Figure 2.9. In this figure the earth is viewed from above the North Pole, and the earth is therefore rotating in a counterclockwise direction. As the earth rotates, the observer sees the moon rising above his eastern horizon. The moon appears to be half full with the illuminated half up. This same observer sees the moon directly overhead at 6 P.M. Around midnight the moon sets on his western horizon with the illuminated half facing down. The actual time of setting is delayed somewhat by the fact that the moon's own motion, which is counterclockwise around the earth, carries it backward toward the east with respect to the stars during the night.

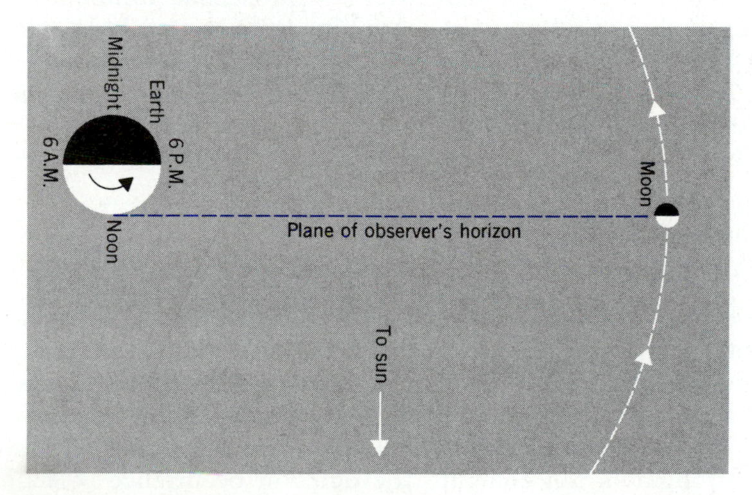

FIGURE 2.9
Rise of the first-quarter moon.

The times of moonrise and moonset for all phases of the moon are given in the table at the right.

These times are not precise because they neglect the moon's motion as noted above and also because they assume that the moon's

Phase	Rise	Set
New	6:00 A.M.	6:00 P.M.
New crescent	9:00 A.M.	9:00 P.M.
First quarter	Noon	Midnight
First gibbous	3:00 P.M.	3:00 A.M.
Full	6:00 P.M.	6:00 A.M.
Last gibbous	9:00 P.M.	9:00 A.M.
Last quarter	Midnight	Noon
Old crescent	3:00 A.M.	3:00 P.M.

orbit is a perfect circle in the same plane as the earth's orbit and that the observer is standing on the equator. The actual times of the moon's rising and setting found in newspapers may differ from the times in the table by more than one hour at the latitudes included within the continental United States.

Solar Eclipse

When the moon passes between the sun and the earth, it blocks all or a part of the sun's rays from some regions of the earth. This event, called a solar eclipse, is the most awesome of all astronomical phenomena if the rays of the sun are entirely blocked from the observer. As the moon slides over the disk of the sun, an unnatural darkness descends on the earth, and the birds cease their chatter. For several minutes the moon blots out the sun entirely, the temperature of the air drops, and the solar corona—the sun's outer atmosphere—appears as a pearly, shimmering halo around the moon's black shadowing disk. The brighter stars and planets become visible in the sky at this time. During the next hour the face of the sun reveals itself again, growing from a thin crescent to its full brilliance as the moon continues on its way.

The ancients regarded eclipses as harbingers of important events. Many ancient societies kept careful eclipse records, but the Babylonian astronomers were exceptional among ancient peoples in the attention that they devoted to the compilation of these events. References to eclipses can be found in cuneiform inscriptions on Babylonian tablets that date back to the second millenium B.C. An example of a reference to what appears to be an eclipse is found in a tablet describing unusual events, such as the appearance of a bearded woman, a four-horned sheep, and a talking corpse. Figure 2.10*a* shows a tablet in this series.

FIGURE 2.10
(a) A Babylonian tablet recording an unusual event interpreted as a total solar eclipse. (b) Lines from the tablet including the statement, "the day was turned into night."

(b) [. . . . .]

The fourteenth line of the tablet (Figure 2.10*b*) says that,

> *On the twenty-sixth day of the month Sivan in*
> *the seventh year the day was turned to night,*
> *and fire in the midst of heaven (. . . .).*

The event it describes could equally well be interpreted as a thunderstorm or an eclipse; but because the rest of the tablet is concerned with truly exceptional events, it is very likely that this passage describes a total eclipse, in which the "fire in the midst of heaven" was the flickering solar corona during the few minutes of totality.

Astronomers can compute the dates of occurrence of eclipses with great precision—for thousands of years in the past as well as in the future. They know the orbits of the earth and the moon with sufficient accuracy to be able to calculate the times at which the sun and the moon will be in line with a particular point on the earth. Starting from the observed positions of the earth and the moon at the present moment, and using their knowledge of the orbits, they trace the motions of the earth and the moon forward and backward in time, predicting the exact moments and the locations on the earth at which an eclipse is to be visible. The astronomical calculations confirm the occurrence of an eclipse in eleventh-century B.C. Babylon in the early afternoon of July 31, 1062 B.C.

Total Eclipses. The moon's path around the earth is an ellipse that carries it out to a maximum distance of 253,000 miles and in to a minimum distance of 221,000 miles once each month. If an eclipse occurs when the moon is close to the earth so that its apparent diameter is great enough to block all the rays of the sun, the resultant eclipse is said to be *total* (Figure 2.11). This combination of circumstances occurs once every two or three years, on the average, at some point on the earth. A total eclipse can be seen at a given spot on the earth only once every 360 years.

The period of totality of a solar eclipse is usually two to three minutes, and cannot last for more than seven minutes. The total solar eclipse that occurred in the United States on March 7, 1970, was visible along an arc running across the Gulf of Mexico and up the east coast (Figures 2.12*a* and *b*). The duration of totality was three minutes along the central line of the eclipse swath. The next solar eclipse visible from the United States will occur August 17, 2017.

FIGURE 2.11
Positions of the earth and moon during a total eclipse.

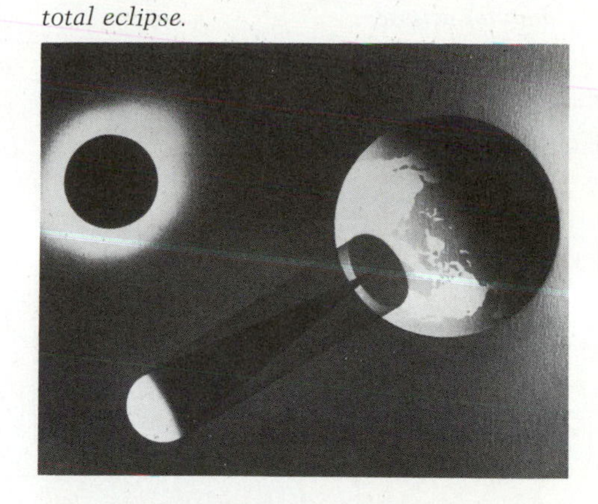

FIGURE 2.12(a)
The total eclipse of March 7, 1970.

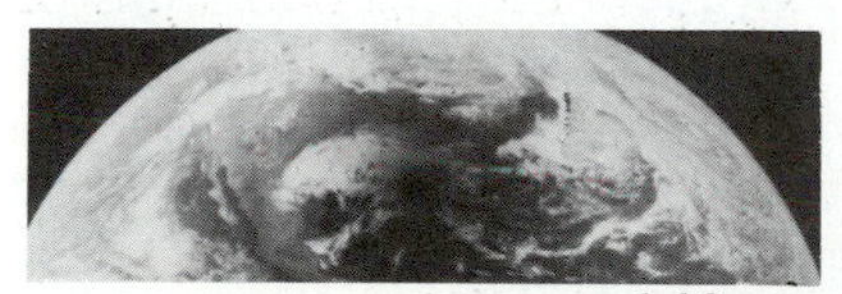

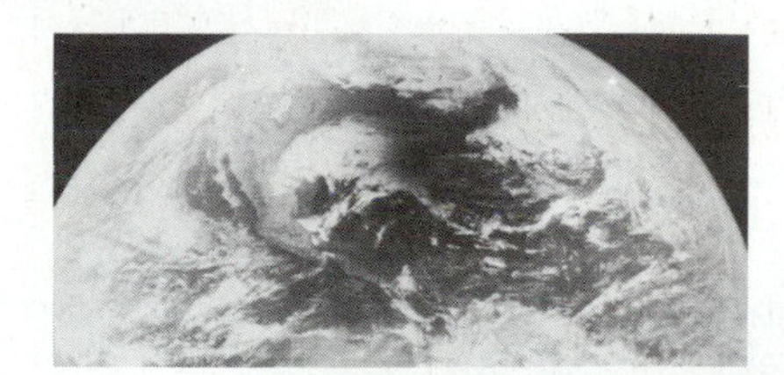

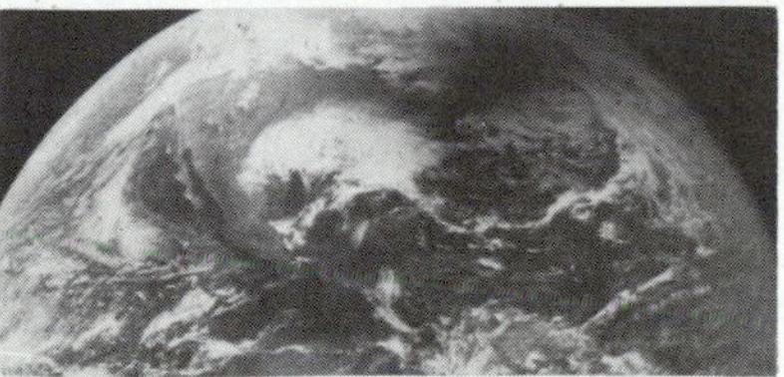

FIGURE 2.12(b)
The shadow of the moon moving over the earth's surface from Florida to the North Atlantic, photographed from a satellite during the eclipse of March 7, 1970.

FIGURE 2.13
Positions of the earth and moon during an annular eclipse.

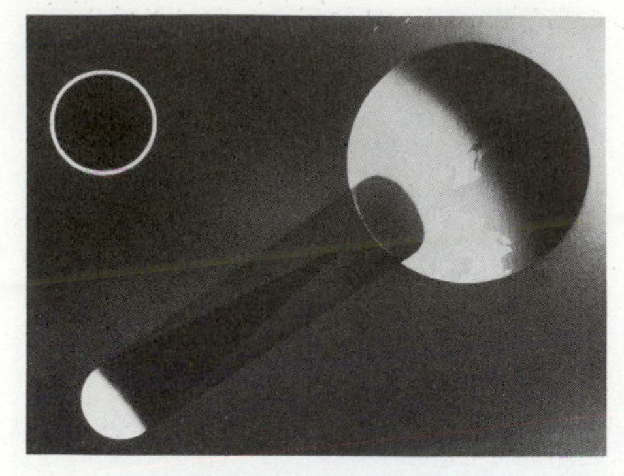

The Annular Eclipse. If a solar eclipse occurs when the moon is at its greatest distance from the earth, the moon's apparent diameter will be smaller than the diameter of the sun, and the outer rim of the sun's disk will remain visible throughout the eclipse, even for a location on the earth that lies exactly on the line between the centers of the moon and the sun (Figure 2.13). From such a location, at the midpoint of the moon's passage between the sun and the earth, the moon will appear to be a black disk surrounded by the bright ring of the sun's outer layers. This type of eclipse is called *annular.* Annular eclipses occur with a slightly higher average frequency than total eclipses. Figure 2.14 shows the passage of the moon across the face of the sun during an annular eclipse.

FIGURE 2.14
Appearance of the sun during an annular eclipse.

The Partial Eclipse. A partial eclipse occurs when the moon is close to the sun-earth line, but not close enough to completely block the sun's rays from any region on the earth. On the regions of the earth that lie within the path of the partial eclipse, the moon's disk will bite into the sun but will never cover it entirely. Partial eclipses occur more frequently than total or annular eclipses. Figure 2.15 shows the sequence of events in a partial eclipse. The moon begins to block out the sun at (*a*), reaches maximum coverage at (*b*), and recedes to a small bite again at (*c*). The total time span from (*a*) to (*c*) is about two hours.

Lunar Eclipse

Lunar eclipses occur at full moon when the earth, moon, and sun are in a straight line and the moon lies behind the earth, that is, in the full-moon phase (Figure 2.16). They are visible at any place on the

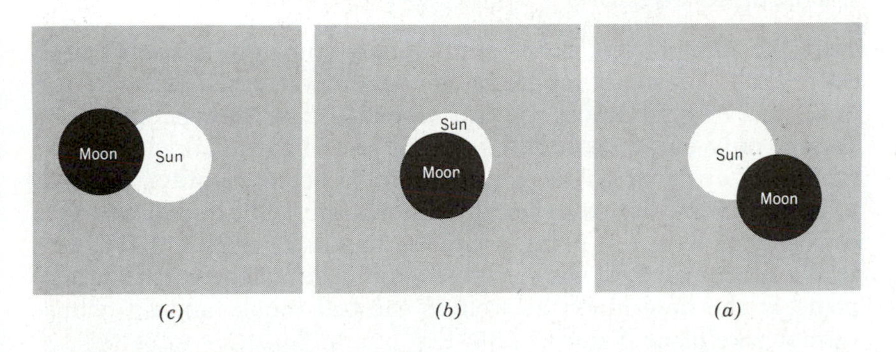

FIGURE 2.15
Appearance of the sun as the moon crosses its face during a partial eclipse.

FIGURE 2.16
A lunar eclipse.

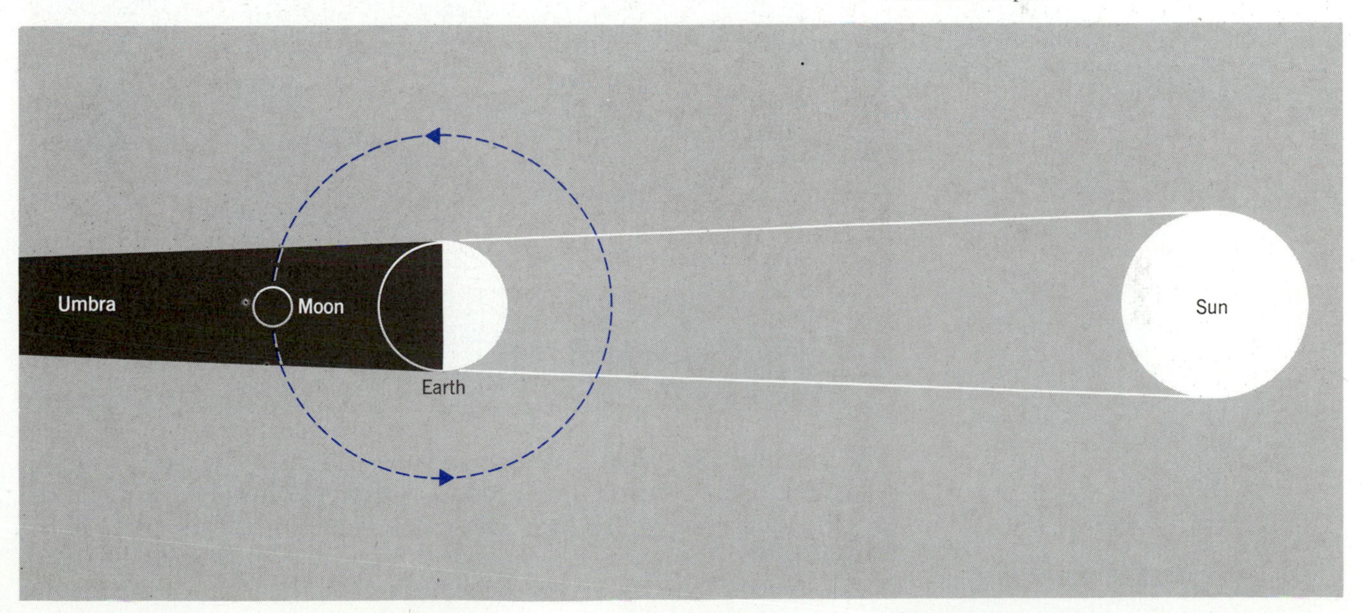

earth at which the moon is also visible, in contrast to the solar eclipse, which usually is visible only in a narrow band on the earth's surface, less than 100 miles across. During a total lunar eclipse the moon receives no light except that which is bent around the edges of the earth by refraction in the earth's atmosphere, which acts like a lens, focusing sunlight to the moon. The blue component in this refracted light is reduced in intensity by scattering in the atmosphere of the earth, and the moon is illuminated by sunlight depleted in the blue wavelength, making it appear a dull coppery-red color. As the lunar eclipse develops, the earth's shadow sweeps across the face of the moon at a speed of about 2000 miles an hour.

The earth's shadow as it appears on the moon is clearly circular. The Greeks noticed this fact 2400 years ago and concluded then that the earth is spherical and not flat.

The Eclipse Seasons

From the study of the phases of the moon, you would expect a solar eclipse to occur every month at the new moon, and a lunar eclipse to occur every month at the full moon. In fact, they occur on the average only every six months, during the semiannual *eclipse seasons.* The explanation for their failure to occur at other times is connected with the fact that the moon's orbit is tilted at an angle of 5.2 degrees to the plane of the earth's orbit (Figure 2.17). As a result, during most of the year the moon is out of the earth's orbital plane at the new moon as well as the full moon, and an eclipse cannot take place (Figure 2.18).

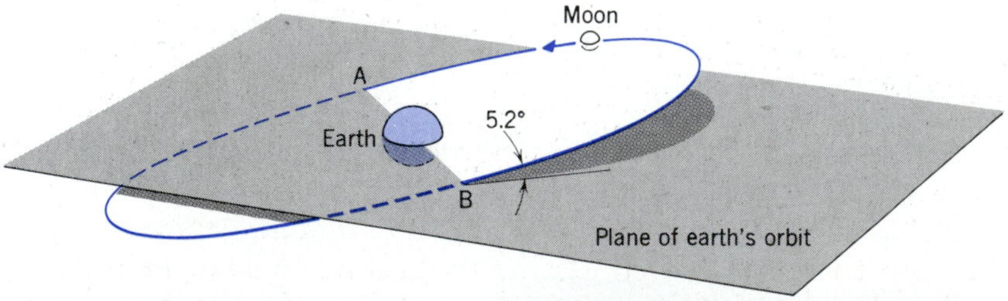

FIGURE 2.17
The tilt of the plane of the moon's orbit to the plane of the earth's orbit.

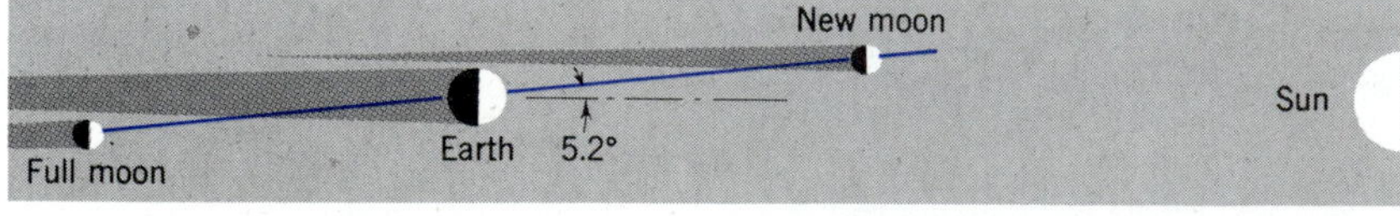

FIGURE 2.18
Unfavorable conditions for an eclipse.

Twice each month, the moon passes through the earth's orbital plane. The points at which it passes through this plane are marked by *A* and *B* in Figure 2.17. These points are called the nodes of the orbit, and the line *AB* is known as the line of nodes. As the earth and moon together revolve around the sun, the plane of the moon's orbit stays fixed in space,[1] and the line of nodes also keeps a fixed direction in space (Figure 2.19). Twice each year, when the line of nodes points to the sun, an eclipse can occur. At other times of the year, when the line of nodes does not point toward the sun, solar and lunar eclipses cannot occur.

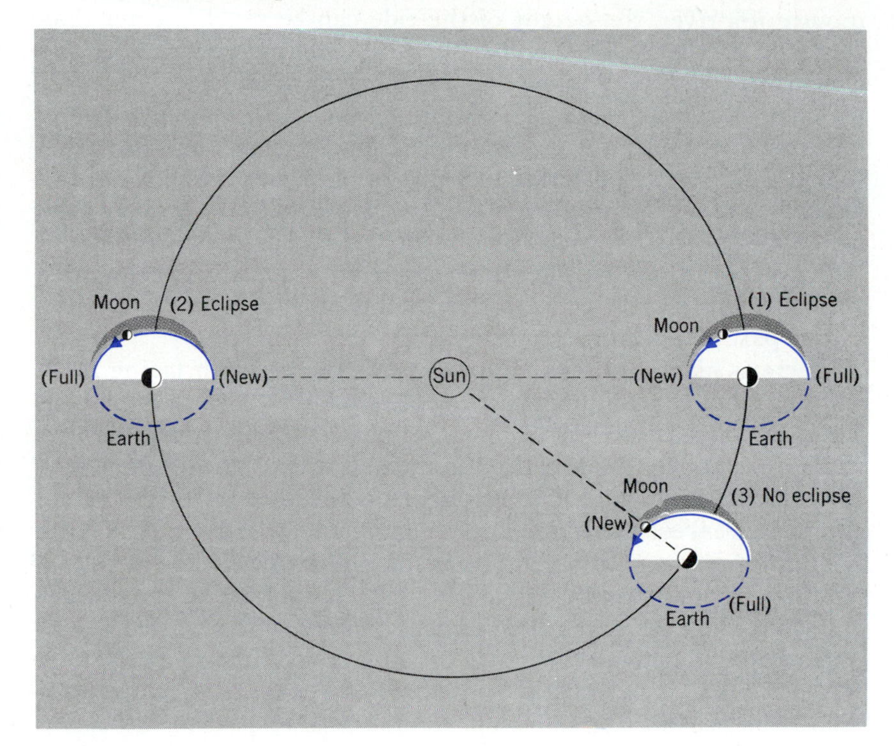

FIGURE 2.19
Eclipse seasons occur semiannually at (1) and (2) when the moon's line of nodes points to the sun. No eclipses can occur at other times of the year. The figure neglects the precession of the planes of the moon's orbit (footnote 1).

Tides

The Lunar Tide. The moon's gravity pulls on all parts of the earth as our satellite circles in its orbit. Different regions of the earth feel the pull of lunar gravity to a slightly different degree, however, because they are located at slightly different distances from the moon. The side of the earth facing the moon, for example, is 8000 miles closer than the side facing the other way. Since the force of gravity becomes weaker with increasing distance, the moon's pull on the near side of the earth is greater than its pull on the rest of the earth. The side of the earth facing the moon bulges out as a result of this effect. The bulge toward the moon is called the lunar tide.

[1] This is not precisely true; actually, the plane of the moon's orbit precesses slowly about the normal to the ecliptic, completing one revolution in about 19 years. In one year the shift in orientation is small enough so that the description in the text is accurate.

The height of the lunar tide is greatest when a large body of water, such as one of the major oceans, faces in the direction of the moon, since water can flow freely in response to the extra force. Tides in the open ocean or sea are approximately two feet high. If the ocean or sea tide is channeled into a narrow opening, for example the mouth of a river, the height of the tide can be much greater than two feet. In the Bay of Fundy, on the border between Maine and Canada, the tide sometimes reaches a height of 50 feet.

When a continent faces the moon, the solid rock resists the moon's gravity. The tide is not zero, however, because even the solid rocks of the earth's interior yield to some degree when a force is applied to them. The response of the continents to the moon's pull, called a *land tide,* is usually no more than a few inches in height. Nonetheless, it can be easily measured with modern instruments, which are able to detect changes in the force of gravity as small as one part in a billion.

There is also a tidal bulge on the side of the earth facing away from the moon, equal in height to the tide facing toward the moon. This fact seems puzzling at first, but has a simple explanation. As mentioned above, the moon pulls most strongly on the side of the earth facing it; it pulls less strongly on the interior of the earth; and it is weakest of all on the side facing away from it. In Figure 2.20, the moon's pull is represented by the arrows *A* and *B,* whose lengths are proportional to the strength of the force of the moon's gravity at the two places. The force *A* tends to pull the ocean away from the solid body of the earth, causing the ocean tide on the near side. On the far side, because the force *B* is weaker than at any other point, the ocean tends to be *left behind* with respect to the underlying earth, causing the tidal bulge on that side. In other words, the tide raised by the moon is the result of a *differential pull* of the moon on various parts of the earth.

The height of the lunar tidal bulge depends on the rate at which the moon's pull falls off with increasing distance from the moon. This rate is about the same on the near and far sides of the earth, hence the tidal bulge, and, therefore, the height of the tide, is the same on the two sides (Figure 2.20).

FIGURE 2.20

Explanation of the tides: the lengths of the arrows indicate the relative strength of the moon's gravitational pull on various parts of the earth.

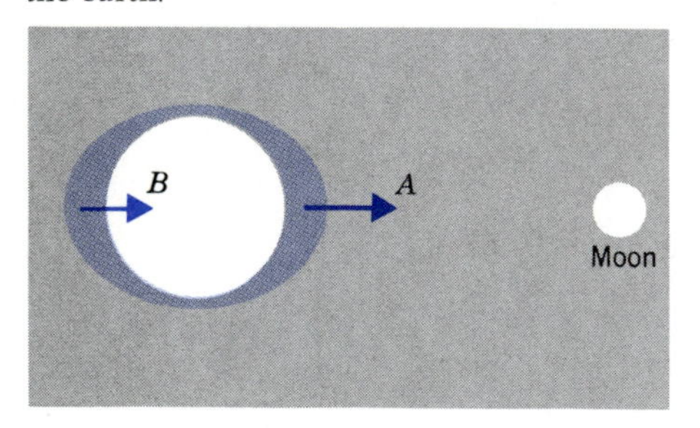

The Solar Tide. The sun's gravity also pulls on the part of the earth facing the sun more strongly than it pulls on the other parts of the planet. Thus, like the moon, the sun tends to raise a tide on the earth. Although the sun is much more massive than the moon, it is also much further away. As a result, the tide-raising force of the sun is less than half as strong as that of the moon. However, twice each month the sun and moon work together to produce exceptionally large tides. These are the times when the moon is new and the moon and sun are on the same side of the earth:

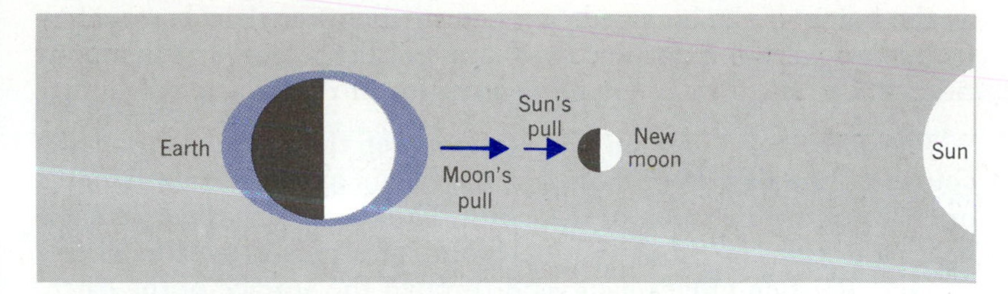

and when the moon is full and sun and moon are on opposite sides.

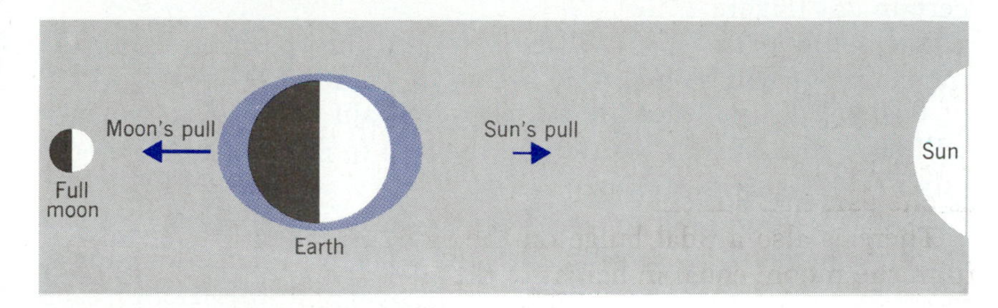

The tides produced on these occasions, called *spring tides,* are 20 percent higher than the average tides.

On two other occasions in each month, the sun and moon are at right angles with respect to the earth. These situations occur when the moon is in the first or third quarter.

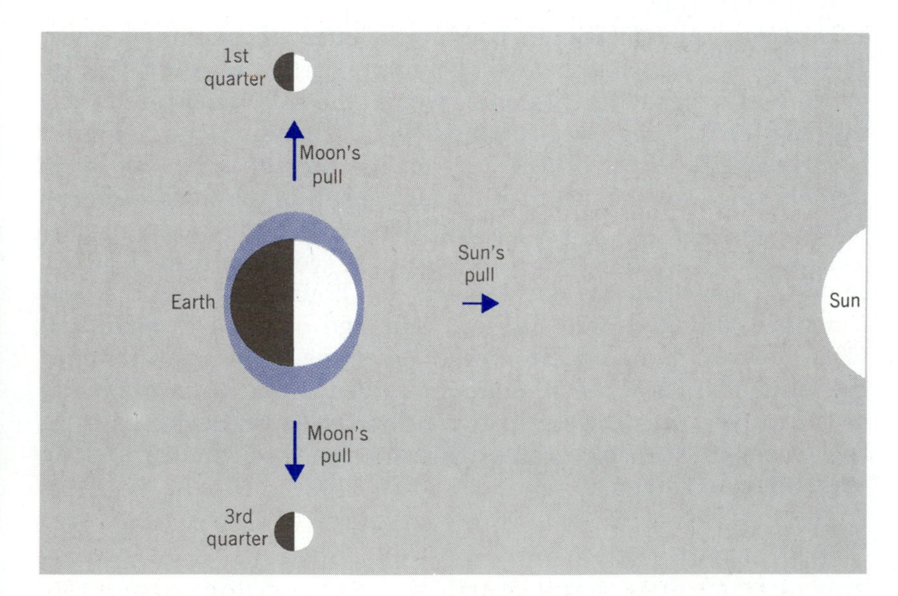

At these times the sun's gravity works to draw away the water from the points on the earth's surface to which it would be pulled by the moon's tidal force. Since the sun's tidal force is not as strong

FIGURE 2.21
Rotation of the earth beneath the tide.

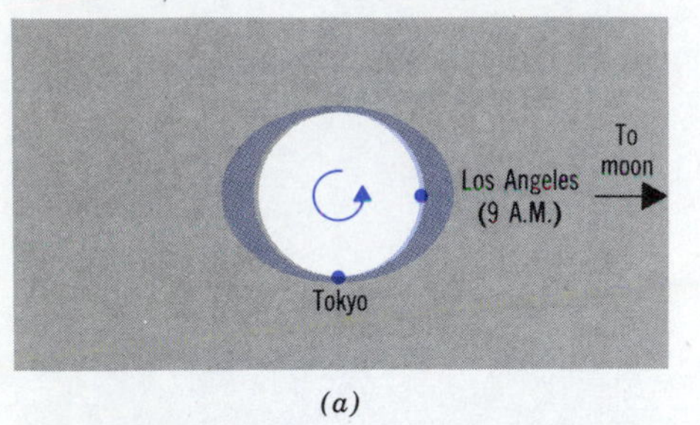

(a)

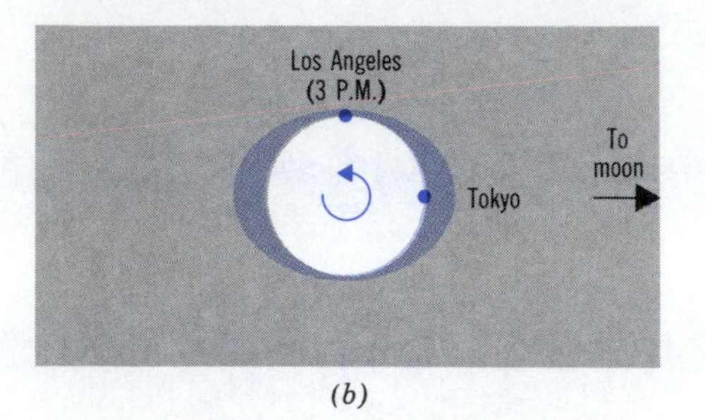

(b)

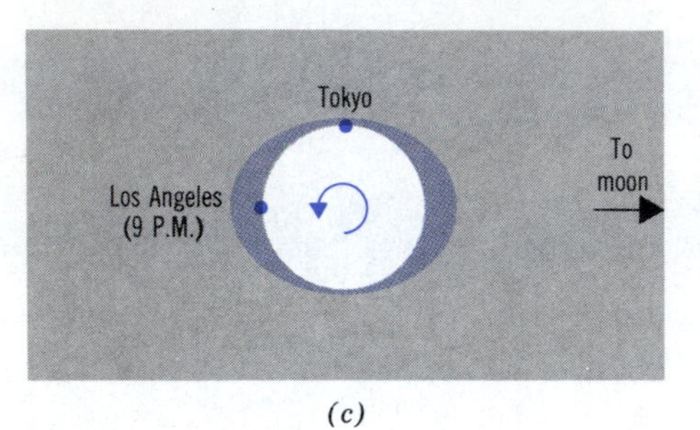

(c)

as the moon's, a lunar tide occurs, but it is lower than it would be if the sun were not pulling away some of the water. The resultant tides, called *neap tides,* are 20 percent lower than the average tide.

The Effect of Tides on the Rotation of the Earth. During the course of one day, as the earth turns on its axis, the tide remains oriented in the direction of the moon. Thus, the earth rotates underneath the tide, or, from the viewpoint of a person on the earth's surface, the tide moves backward through the surface of the earth. For example, suppose it is high tide in Los Angeles at 9 A.M. on a certain day (Figure 2.21*a*).

Six hours later, Japan will have moved into position under the moon, and it will be high tide in Tokyo (Figure 2.21*b*).

Los Angeles will not see a high tide again until 9 P.M., when it will have rotated into position on the line to the moon, but on the other side (Figure 2.21*c*).

The daily motion of the tides across the oceans and seas, as well as through the solid body of the lands, creates a large amount of friction. A huge quantity of energy is dissipated in this friction each day. If the energy could be recovered for useful purposes, it would be sufficient to supply the electrical power requirements of the entire world several times over. The energy is actually dissipated in the turbulence of coastal waters plus a small degree of heating of the rocks in the crust of the earth, and cannot be diverted to constructive work.

However, the tidal friction has another effect of great importance. The tides act as a brake on the spinning earth, slowing down its rotation at a steady rate. In other words, the tides tend to increase the length of the day. Because the earth is so massive, and its inertia is so great, the effect of the tides on the length of the day is extremely small. At the present time, as a result of the tides, each day is one-hundred millionth of a second longer than the preceding day. This small effect projected backward over geological time leads to the conclusion that the day was only 22 hours long, and the year contained 400 days, 300 million years ago in the Devonian era, when our vertebrate ancestors were emerging from the water onto the land. The fossil record contains evidence that there were 400 days in a year in the Devonian era, indicating that tides have, in fact, been slowing down the earth's rotation throughout this long period of earth history.

THE MOON, SUN, AND STARS AS CALENDARS

The determination of an accurate calendar was probably the earliest practical motivation for the refinement of observational techniques in astronomy. The position and appearance of the moon, sun, and stars change regularly in a one-year cycle, and any one of these objects can be used as the basis for a calendar. The cyclic changes in the shape of the moon are more obvious than the annual changes in the positions of the sun and stars, and, for this reason, the moon was the favorite calendrical timepiece of most people in early times.

The Moon as a Calendar

Keeping track of the phases of the moon from new moon to full moon to new moon again, observers of the sky must have noted very early in human prehistory that by counting 12 full moons from the start of spring in one year, they would arrive at the start of spring in the following year; that is, 12 cycles of the moon added up to one cycle of the seasons. But the 12-month lunar calendar has a defect. A cycle of the moon lasts 29.5 days as viewed from the earth.[2] Therefore, 12 lunar cycles contain $12 \times 29.5 = 354$ days. The earth, however, completes one full turn around the sun in 365¼ days; that is, the year is 365¼ days long. Thus, when 12 months have gone by in the lunar calendar, the earth is still 11¼ days short of completing a full circuit around the sun. If the beginning of spring in one year is marked by the appearance of the full moon on a certain date, and 12 lunar cycles are counted off, the result will be 11¼ days short of the true start of spring in the following year.

For the first agriculturists, the discrepancy between the 12-month lunar year and the solar year could create a serious problem. If the lunar calendar were followed as much as three years in a row without correcting for its errors, it would predict the start of spring 34 days too early, in the winter cold of February. To solve this problem, societies that relied on the lunar calendar fell into the habit of throwing in an extra month now and then, making 13 months in the year, in order to bring the lunar calendar and the cycle of the seasons back into agreement. For example, an official document issued to the inhabitants of ancient Israel around 100 A.D. stated: "We make known to you that the lambs are small and the young of the birds are tender and the time of the corn-harvest has not yet come, so that it seems right to me and my brothers to add to this year thirty days."

[2] The actual period of revolution of the moon in space is 27.3 days. This is its sidereal period, that is, its period measured by the fixed stars. The value of 29.5 days is the synodic period. The 2.3-day difference is due to the fact that the earth's orientation in space changes by roughly $(29.5 \div 365.25) \times 360$ degrees or 29 degrees during one cycle of lunar phases, and the moon must revolve through this additional angle in order to present the same appearance—for example, new moon to new moon—to an observer fixed on the earth.

FIGURE 2.22
Borneo tribesmen measure the length of the sun's shadow.

The Sun as a Calendar

The observation of the sun from day to day provides a more reliable calendar than the cycles of the moon. The observation consists in determining the highest position of the sun each day—that is, the elevation of the sun above the horizon at local noon—and recording the changes in this highest position during the course of a year. But how does the observer fix the time of local noon accurately? Ancient peoples and primitive people today could drive a stick into the ground and observe the length of its shadow; when the shadow is at its shortest during any particular day, the time is noon (Figure 2.22). The length of the noontime shadow will itself vary from day to day during the course of the year. When the noontime shadow is longest, the sun is at its lowest elevation in the sky, marking the beginning of winter. The day on which the noontime shadow is shortest marks the beginning of summer. On the modern calendar these days usually correspond to December 21 and June 21, respectively.

The Stars as a Calendar

In certain parts of the world, knowing the time of the year was important for another reason. In these regions the planting season was determined by the overflowing of a major river, rather than the arrival of the spring season. Mesopotamia and Egypt were parts of the world in which agriculture was ruled by the river. Mesopotamia occupied the flat terrain between the Tigris and Euphrates Rivers, where the country of Iraq stands today, while ancient Egypt included the territory occupied today by modern Egypt and the Sudan—the lands through which the Nile flows on its way to the Mediterranean. In both areas the terrain in the neighborhood of the river was relatively dry and flat. Each year, when the spring came, the rivers rose, overflowed their banks, flooded the surrounding territory, and deposited a rich layer of silt over large areas, renewing the fertility of the land. This annual flood was the most significant event of the year; it was essential to be able to predict when it would recur, because prompt preparations had to be made for the planting that followed.

The Egyptians probably used the moon and sun as calendars in their first efforts to predict the onset of the annual flood, but sometime during the third millenium B.C., an astute Egyptian observer of the heavens noticed that, by a fortuitous coincidence, the bright star Sirius happened to be a marker star for that event, for in most years Sirius became visible above the eastern horizon at Memphis, the capital of ancient Egypt, just before sunrise on the day when the Nile began to rise.

Observations of other stars showed that they, too, appeared above the horizon on fixed days of the year. Thus, observations of the stars could serve as well as the observation of the sun in revealing the completion of a year. Probably through their observations of the stars, the Egyptians found that one year contained 365¼ days.

The Modern Calendar

Many ancient civilizations, including Rome, used the lunar calendar with an extra month added now and then as needed. Sometimes officials charged with adjusting the calendar were negligent; for example, when Julius Caesar came into power in Rome in 46 B.C. the official calendar year had fallen out of line with the season of the year by about three months. Caesar knew of the Egyptian discovery that the length of the year was 365¼ days, and he devised a new calendar to fit this discovery. The new calendar contained 12 months, as did the traditional lunar calendar, but the months were of unequal length, with 30 days in some, 31 days in others, and 28 days in February. The entire calendar was constructed so that the number of days in the year added up to 365.

If he had stopped at this point, the calendar would have lagged behind the march of the seasons nearly one month in every century. In order to correct for this error, Caesar introduced the leap year; that is, he allotted an extra day to the month of February every four years, giving it 29 days in place of its normal complement of 28. In this way, the calendar achieved an average year of 365¼ days.

The calendar devised by Caesar came to be known as the *Julian calendar.* It worked well and was used by the Western world for many centuries, but it still was not perfect, because the accurate length of the year is 365 days, 5 hours, 48 minutes, and 48.7 seconds, or 11 minutes and 11.3 seconds less than 365¼ days. This small discrepancy, accumulating year after year, adds up to approximately three quarters of a day per century. Since Roman times, the year would have slipped behind the seasons by 15 days or more than two weeks.

By the sixteenth century the problem had become serious. In 1582 Pope Gregory XIII introduced a new calendar in all Roman Catholic countries, in which leap years would not occur at the turn of the century unless the turn-of-the-century year were divisible by 400. Thus, 1900 was not a leap year, although 1896 and 1904 were; but the year 2000 will be a leap year, and the next turn-of-the-century leap year, after 2000, will be 2400.

In the system proposed by Pope Gregory, known as the *Gregorian calendar,* the remaining error was only 26 seconds per year. This small discrepancy, which makes the Gregorian calendar fall behind the seasons by one day in 3300 years, is not important in most activities.

THE CELESTIAL SPHERE

The motion of the earth makes it difficult to fix the positions of celestial bodies. In observing the stars, the astronomers as well as the amateur stargazer is confronted by the problem of the man trying to keep his eye on a distant object as he rides a merry-go-round. The solution adopted by modern astronomers is based on a concept borrowed from the ancients who thought of the sky as a spherical dome covering the heavens, across which the sun, moon, planets, and stars moved in stately procession. This rigid dome, spanning the vault of the heavens, is called the celestial sphere (page 51).

Latitude and Longitude

The description of position on the surface of the earth in terms of latitude and longitude serves as a useful introduction to the celestial sphere and the methods used for specifying the positions of stars in the sky.

Latitude and longitude are defined in terms of two types of planes that cut the earth's surface. The starting point for the definitions is the rotation of the earth. The axis of rotation intersects the earth's surface at the *North* and *South Poles.*

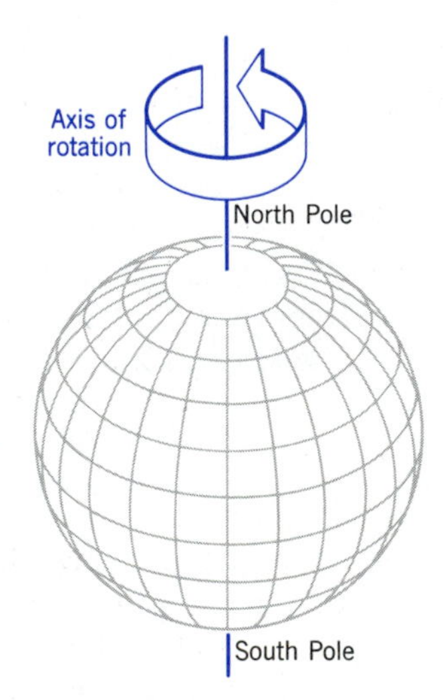

The *equatorial plane* is perpendicular to the axis of rotation and cuts the earth's surface midway between the poles. The line in which it cuts the earth's surface is the *equator.*

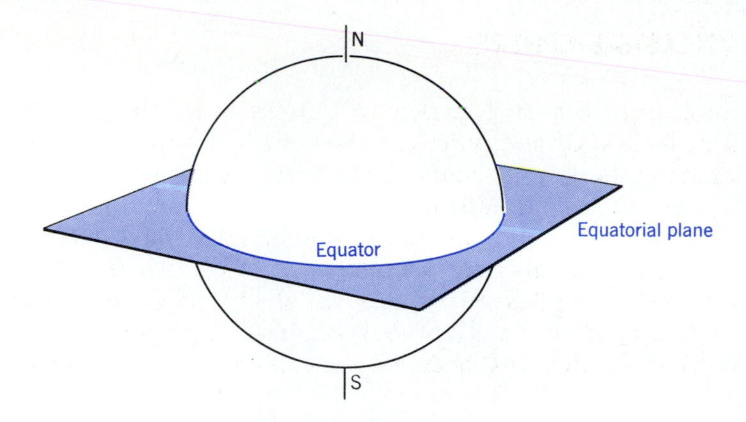

A plane perpendicular to the equatorial plane, and passing through the *North* and *South Poles,* is called a meridian plane. The circle in which it cuts the earth's surface is called a *meridian* or meridian circle.

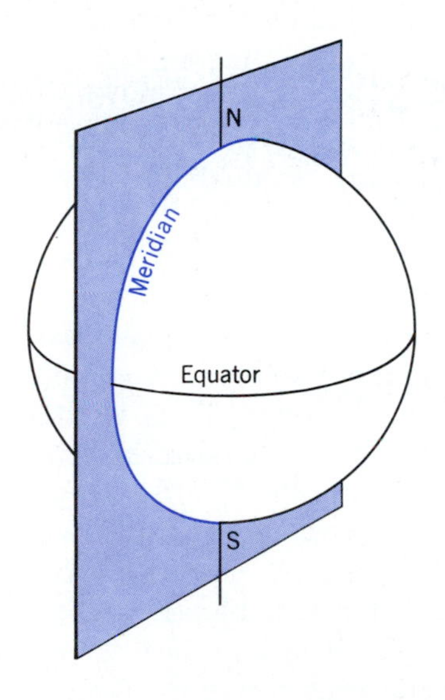

Longitude. The meridian circle passing through a particular point is called the local meridian of that point. The particular meridian that passes through the earth's surface at the original site of the Greenwich Observatory in England is called the *prime meridian,* or Greenwich meridian. The longitude of a point on the earth's surface is the angle between the meridian plane passing through that point and the plane of the prime meridian passing through Greenwich.

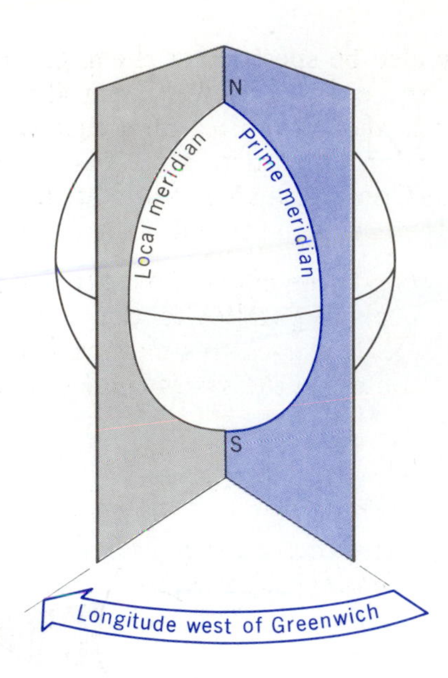

The definition is ambiguous as written, because the angle could be either the one shown in the diagram, which is measured between the "front" surfaces of the plane, facing the viewer, or it could be the angle between the "rear" surfaces, facing away from the viewer. The sum of these two angles equals 360 degrees. According to the convention, the angle between the "front" surfaces, which is the smaller of the two angles, is defined as the longitude. With this convention, the longitude is always less than or equal to 180 degrees.

All longitudes are expressed in degrees east or west of the prime meridian. If the meridian of a point, for example, is 30 degrees west of the prime meridian, its longitude is written as 30°W or 30W. If the meridian of the point is 30° east of the prime meridian, its longitude is written as 30°E or 30E. If, for example, a ship starts at the Greenwich meridian and moves steadily westward, its longitude increases from 0°W to 180°W. At the moment immediately after reaching 180°W, the longitude changes abruptly to 180°E.

The two longitudes, 180°E and 180°W, represent the same meridian, which is the continuation of the Greenwich meridian on the other side of the earth.

If the ship continues its westward motion beyond 180°W, its longitude decreases, and becomes zero when the ship returns to the Greenwich meridian.

Sometimes the longitude is expressed in other ways. It may be written as the angle from the prime meridian to the local meridian, running counterclockwise (eastward) when the earth is viewed from above the north pole. In this system, longitudes may range to 360°, and the point in the example above, with a longitude of 30°W in the conventional system, would have a longitude of 330°.

Longitude may also be specified as the angle running clockwise (westward) when the earth is viewed from above the north pole, in which case in the above example the longitude would be simply 30°. However, the definition first given, in terms of the smaller angle between the Greenwich meridian and the local meridian, is the one most frequently encountered.

Latitude. The second type of plane used to define position on the earth's surface is a plane parallel to the equatorial plane. Such planes, passing through the earth's surface between the equator and the poles, cut circles on the surface called *parallels of latitude.*

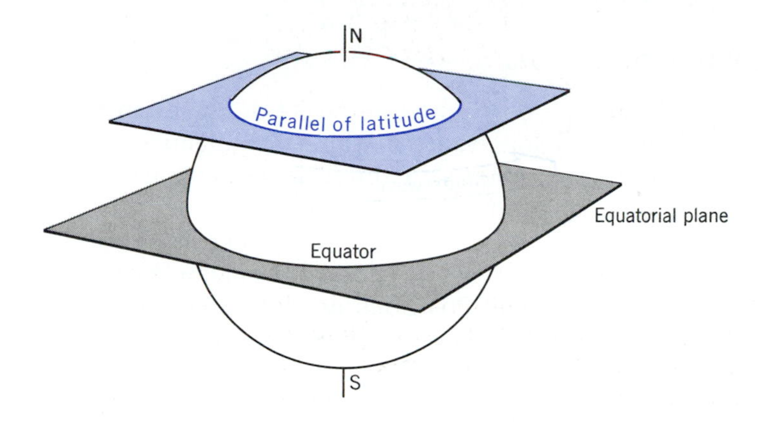

To define the latitude of a point, draw two lines in the plane of the local meridian, from the center of the earth to the equator and to the point in question. The latitude is the angle between these lines.

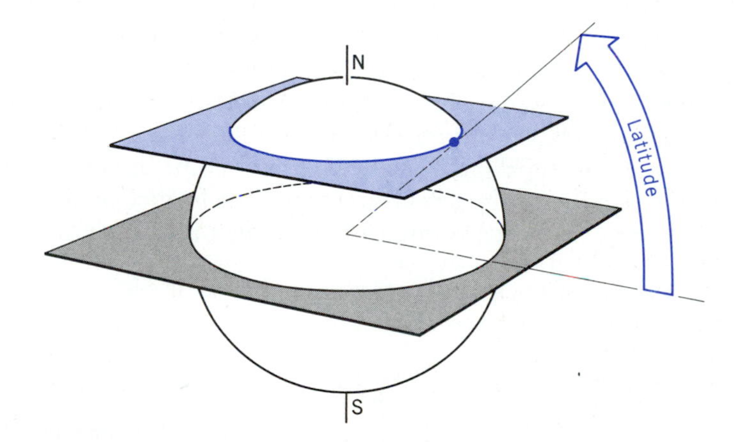

Latitudes are measured in degrees north and south of the equator, and range from 0° to 90°N or 90°S (also written 90N or 90S) at the poles.

Relation Between Distance on the Earth's Surface and Latitude/ Longitude Coordinates. Ninety degrees of latitude extend over a quarter of the earth's circumference, or roughly 6200 miles. Thus, one degree of latitude is equivalent to a distance of 6200 ÷ 90 or 69 miles on the earth's surface.

One degree of longitude corresponds to a variable distance on the earth's surface, according to the latitude. Since a parallel of latitude at latitude λ has a circumference equal to $2\pi R \cdot \cos \lambda$ (R = earth's radius), one degree of longitude at that latitude corresponds to a distance equal to 69 cos λ miles.

The Positions of Stars

Just as we locate the position of an object on the earth by specifying its latitude and longitude, in the same manner astronomers locate the positions[3] of stars in the heavens by marking off lines of latitude and longitude on the celestial sphere. The position analogous to latitude is measured in degrees north or south of the celestial equator, defined as the projection of the earth's equator onto the celestial sphere. Positions north or south of the equator are given plus or minus signs, respectively. The latitude of a star is called its *Declination,* abbreviated Dec.

The celestial analogue to longitude, however, cannot be defined in the same way as longitude on the earth, because the earth's lines of longitude rotate with it and sweep across the celestial sphere. Coordinates on the celestial sphere must be fixed in space and not moving with the earth, since they are intended to describe the locations of fixed stars. Astronomers have agreed to select a fixed point on the celestial equator to be used as the starting point for marking off degrees of celestial longitude, just as the Greenwich meridian is used as the zero-point for measuring degrees of longitude on the surface of the earth. According to their agreement, the chosen place on the celestial equator is marked by the direction of a certain point located in the constellation Pisces. The zero-point direction is defined more precisely as the direction of the line formed by the intersection between the earth's equatorial plane and the plane of the

[3] *Position* and *location* are used interchangeably with *direction* in this context. Normally the position of an object means its location in space, including, in the case of stars, not only the direction of the star along the line of sight from the earth, but also its distance from the earth. However, the celestial sphere is assumed to be infinitely far away. Two stars that are at different distances but lie on the same line of sight from the earth will have identical coordinates on the celestial sphere.

earth's orbit. This point is called the vernal equinox.[4] Because of the precession of the earth's axis of rotation, the chosen direction is not fixed in space but moves slowly through a circle, completing one circuit every 26,000 years (see page 28). An accurate statement of the coordinates of a celestial body includes a reference to a specific year, indicating that these coordinates are based on the position of the longitude zero-point in that year.

Longitude on the celestial sphere is marked off in degrees running eastward around the celestial equator, ranging from zero degrees at the position of the vernal equinox to 360 degrees on the return to this point. With the selection of a zero-point on the celestial equator, we have completed the definition of the coordinate system required to locate stars and other objects on the celestial sphere (Figure 2.23).

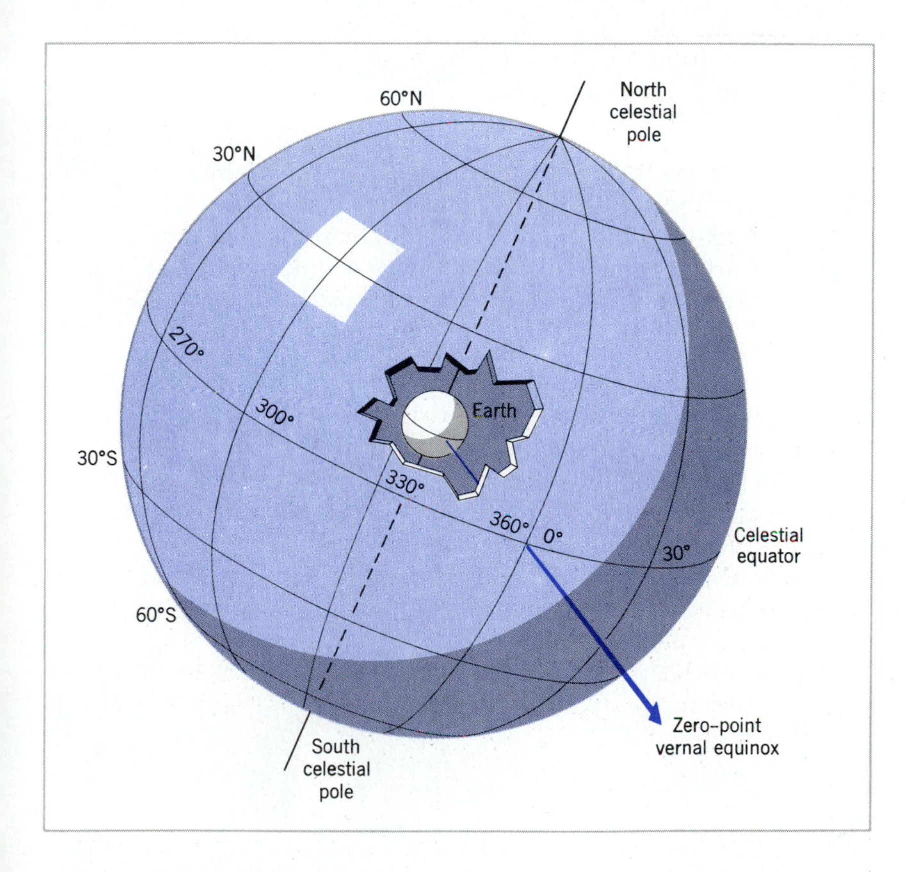

FIGURE 2.23
The celestial sphere.

[4] The *vernal* (of the spring) *equinox* (night equal to day) is the point on the celestial sphere at which the sun is to be found on March 21. Its name derives from the fact that when the sun is at the vernal equinox, spring begins in the Northern Hemisphere and the day and night each have 12 hours.

FIGURE 2.24
The apparent movement of the stars in the course of one night, as a result of the earth's rotation.

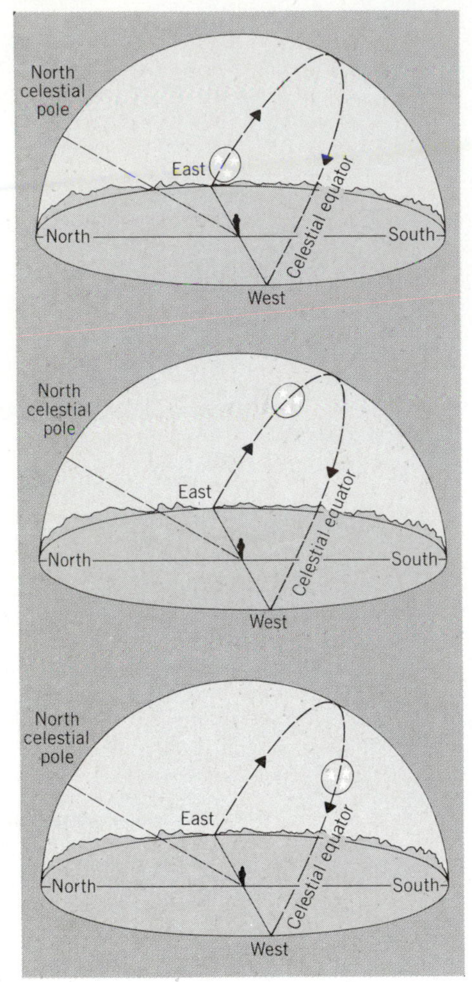

Sidereal Time: Right Ascension

The stars revolve overhead as the earth turns on its axis, rising in the east and setting in the west (Figure 2.24). Once in every rotation of the earth, an observer on the ground looks out into the same direction of the sky and sees the same stars in the same positions. For this reason, the period of rotation of the earth is called the *sidereal day* (Latin: *sider*—star). The length of the sidereal day is 23 hours, 56 minutes, and 3 seconds.

A day is usually defined as having precisely 24 hours. Why is the sidereal day—or period of rotation of the earth—shorter by approximately four minutes? The answer is that the 24-hour day is a *solar* day, defined as the interval from noon to noon, that is, the interval from the time the sun is at its highest point in the sky on one day to the time at which it reaches its highest point on the following day. One solar day is longer than the period of rotation of the earth because during the time in which the earth turns on its axis, it also moves along its orbit. As a result, the direction from the earth to the sun changes by a small amount from one day to the next (Figure 2.25). It is clear from Figure 2.25 that in the period from noon to noon the earth completes slightly more than one turn; consequently, the solar day is slightly longer than the sidereal day.

The difference turns out to be approximately four minutes because the earth completes a full orbit of 360 degrees in 365 days, and thus advances along its orbit through an angle of (360/365) de-

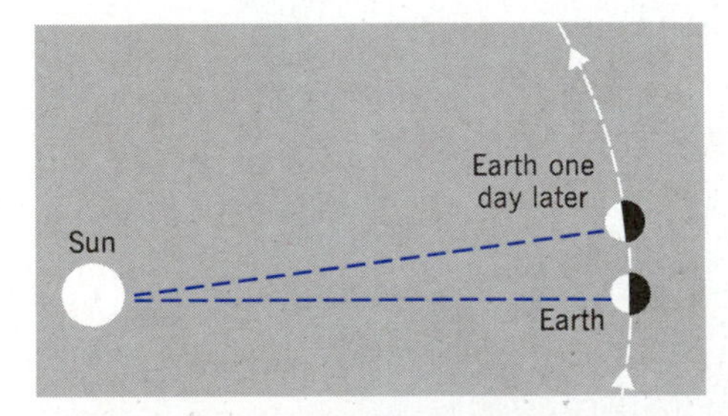

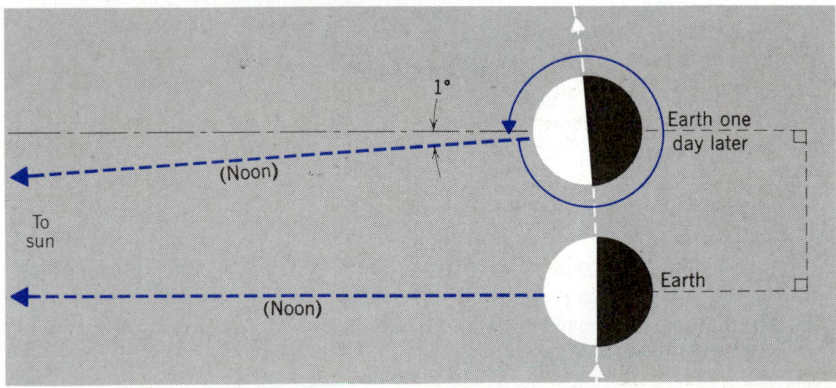

FIGURE 2.25
The daily shift in the apparent position of the sun at noon, as the earth moves in its orbit.

grees, or nearly one degree, in each day of the year. Consequently, the direction of the sun at noontime shifts by about one degree each day. The earth rotates through 360 degrees on its axis in approximately 24 hours or 1440 minutes, corresponding to a rate of rotation of one degree in four minutes. Thus, four minutes elapse while the earth rotates the extra degree required to complete a solar day.

Sidereal Time Versus Solar Time. If you observe the position of a particular star at, say, 9 P.M. on a certain evening, and look for that star at 9 P.M. on the following evening, you will find that it is slightly to the west of the position it had on the previous night, because the earth has rotated four minutes or one degree past a complete turn. However, at 8:56 P.M. on the second night you would find this star at its original position, and again at the same position at 8:52 P.M. on the night after that. If you adjusted your watch to run four minutes fast each day, each star would appear in the same position after an apparent lapse of 24 hours. The watch would then be keeping sidereal time. A sidereal watch—running four minutes fast each day—would be a great convenience to a person who observes the stars a great deal, because all the stars would return to their identical positions in the sky every 24 hours. In sidereal time a star appears on the horizon, rises to its highest point in the sky, and sets at the same time throughout the year. Astronomical observatories find it useful to have clocks that keep sidereal time, but an astronomer will soon get into trouble if he decides when to go home on the basis of a sidereal clock,[5] because he will be home four minutes earlier every day, and after six months he will expect his dinner at breakfast time. Clocks that keep solar time are preferable for the regulation of practical affairs.

Right Ascension. The position of a star on the celestial sphere was defined above in terms of celestial latitude and longitude, each expressed in degrees. Astronomers frequently use a mixed system of coordinates in which celestial latitude is given in degrees north or south of the celestial equator, but celestial longitude is given in hours, minutes, and seconds of sidereal time, calculated from the celestial longitude in degrees by applying a ratio based on the equivalence of 360 degrees to 24 sidereal hours. The celestial longitude of a star expressed in these units is called its *Right Ascension* (R.A.). The Right Ascension runs eastward around the celestial equator from zero hours at the zero-point on the equator

[5] An observatory sets its sidereal clock to read zero hours when the zero-point of longitude on the celestial equator is "overhead," that is, at its highest point in the sky. Different observatories have sidereal clocks set to different times, just as people live in different time zones. Of course, all sidereal clocks run at the same sidereal rate.

to 24 hours at the same point after completion of a full circle. If a star lies 90 degrees east of the zero-point, for example, its Right Ascension is 6 hours. If the star lies 210 degrees east, its Right Ascension is 14 hours. Table 2.1 lists the twenty brightest stars in the sky with their coordinates in Declination and Right Ascension, as they would appear in a standard star catalog.

TABLE 2.1
The Twenty Brightest Stars in Order of Decreasing Brightness

	RA		DECLINATION		APPROXIMATE DISTANCE
Name	**Hours**	**Minutes**	**Degrees**	**Minutes**	**(Light-years)**[a]
Sirius	6	42.9	−16	39	8.8
Canopus	6	22.8	−52	40	98
Arcturus	14	13.4	+19	27	36
Alpha Centauri	14	36.2	−60	38	4.3
Vega	18	35.2	+38	44	26
Capella	5	13.0	+45	57	46
Rigel	5	12.1	− 8	15	900
Procyon	7	36.7	+ 5	21	11
Achernar	1	35.9	−57	29	150
Hadar	14	0.3	−60	8	
Altair	19	48.3	+ 8	44	16
Betelgeuse	5	52.5	+ 7	24	700
Aldebaran	4	33.0	+16	25	68
Alpha Crucis	12	23.8	−62	49	350
Spica	13	22.6	−10	54	230
Antares	16	26.3	−26	19	400
Pollux	7	42.3	+28	9	35
Fomalhaut	22	54.9	+29	53	23
Deneb	20	39.7	+45	6	1400
Beta Crucis	12	44.8	−59	25	500

[a] The light-year = 5.8 trillion miles, the distance traveled by light in one year. These distances are not known to any great precision, and the larger distances may well be in error by as much as 30%.

The convenience of Right Ascension is connected with the concept of sidereal time. An astronomer knows that on any night of the year a star he is observing will be "overhead," that is, at its highest point in the sky, when the time on his sidereal clock is the same as the star's Right Ascension.

THE CHANGING NIGHT SKY

On a clear autumn evening in most locations in the United States, the brightest stars in the sky have the appearance shown in Figure 2.26. Deneb, Vega, and Altair form a conspicuous triangle overhead, and two ensembles of stars nearer the horizon are grouped in the characteristic shapes of the constellations Capricornus and Sagittarius.[6]

But on a spring night in the same location, most of these stars are gone, and their places are taken by an entirely different group dominated by the Dog Star, Sirius—the brightest star in the sky—and by the constellations Canis Minor, Gemini, Leo, Orion, and Canis

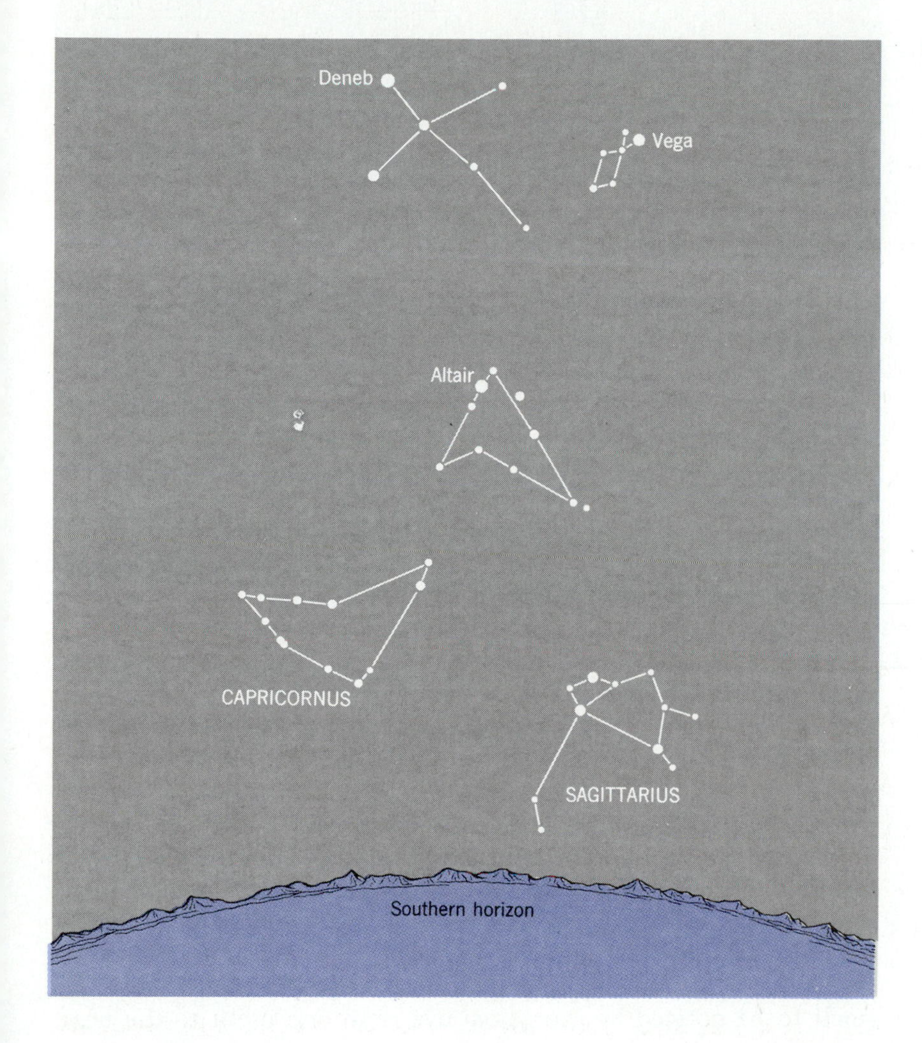

FIGURE 2.26
The autumn sky.

[6] The Goat and the Archer.

Major[7] (Figure 2.27). Year after year the same stars and constellations return, each in its proper season. Every winter evening the constellation Canis Major appears; every summer Canis Major disappears, and other constellations, such as Leo, appear in its place.

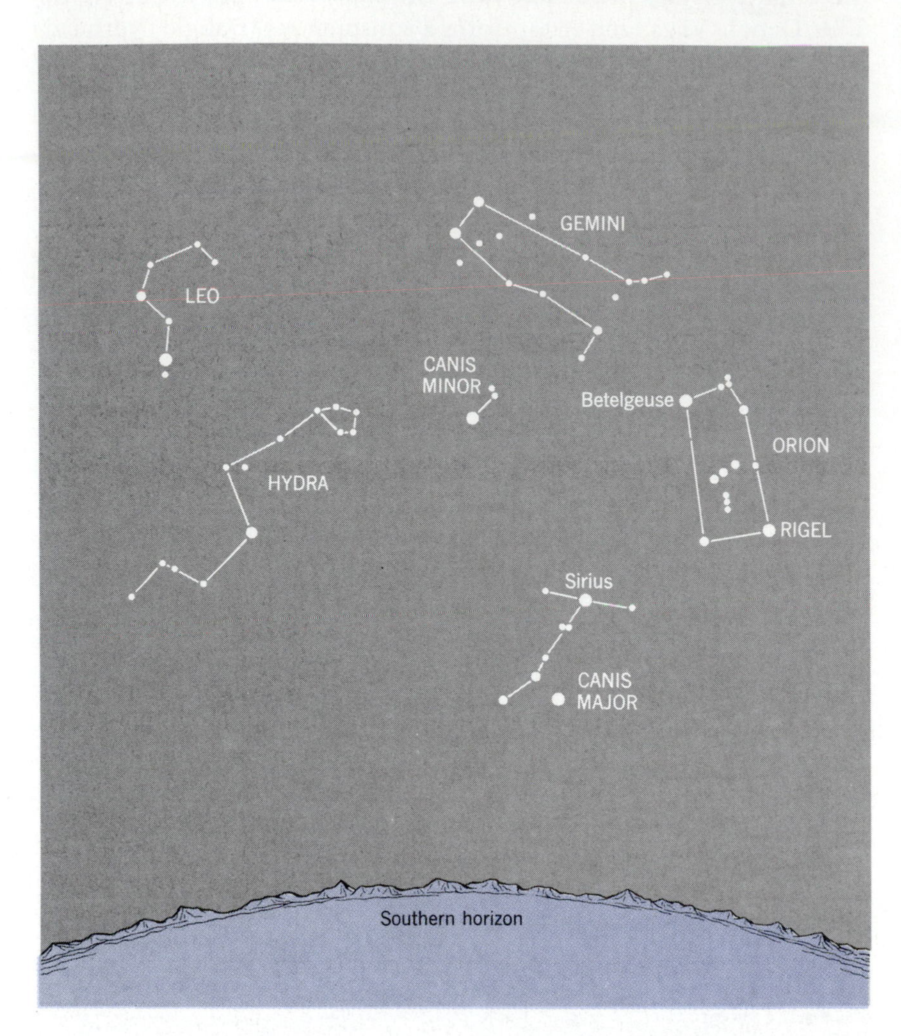

FIGURE 2.27
The spring sky.

The explanation for these changes is connected with the revolution of the earth around the sun. Each night a person looks out into space along a slightly different direction than on the previous night because the earth has moved a small distance along its orbit during the past 24 hours (Figure 2.28). The change in direction is only one degree every 24 hours, and its effect on the position of a star is too small to be noticed by the naked eye from one night to the next,

[7] Little Dog, Twins, Lion, Hunter, and Great Dog.

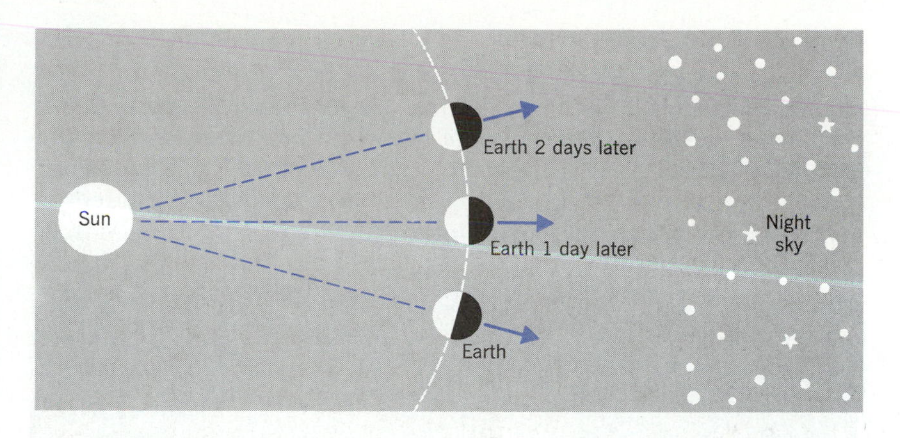

FIGURE 2.28
The changing view of the night sky as the earth moves in its orbit.

but as the earth continues along its orbit during the course of weeks, the change in direction becomes substantial. After six months the earth has moved to the other side of its orbit, and the observer of the night sky, looking out at a different part of the Universe, sees a different family of stars.

On page 53 we noted that a star rises four minutes earlier every night as a consequence of the difference between the sidereal and solar days. The complete alteration in the appearance of the night sky in six months is the result of a large number of four-minute daily changes. Four minutes a day summed over six months adds up to 12 hours, signifying that a star visible in the night sky at one time in the year will be in the day sky, and invisible, six months later.

Figure 2.29*a* shows the earth at the point of its orbit that corresponds to the month of September. At this time, the night side of

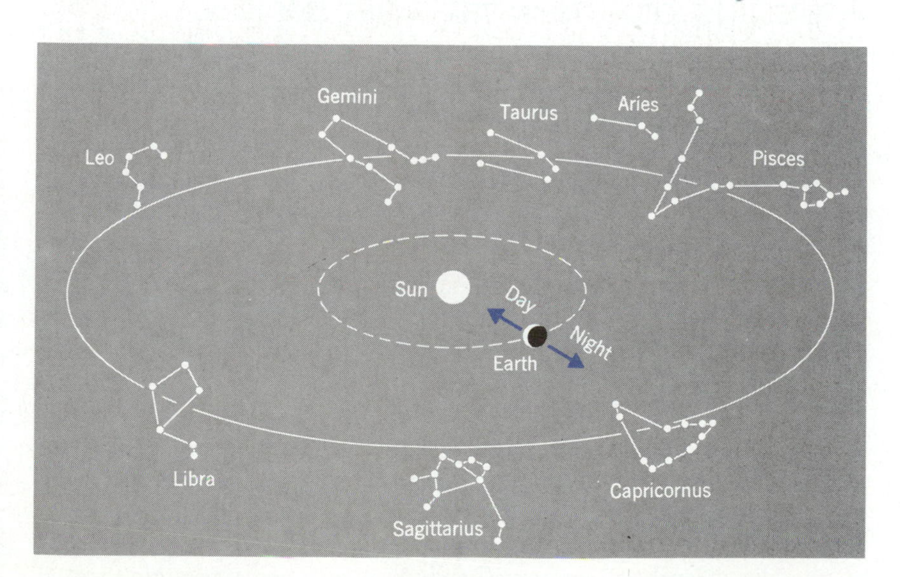

FIGURE 2.29
Changes in the appearance of the sky during the course of the year in (a) September and (b) March (p. 58). The arrows on the night side point toward the constellations that are on the meridian at 9 P.M. in the indicated months.

the earth faces a part of the sky in which the constellations Capricornus and Sagittarius are located, and a person observing the night sky will see these two constellations in that month. Figure 2.29*b* shows the earth in the position of its orbit that corresponds to the month of March. At this time the night side of the earth looks out into the part of the sky containing the constellation Leo. Therefore, Leo will be visible to a person observing the night sky during March.

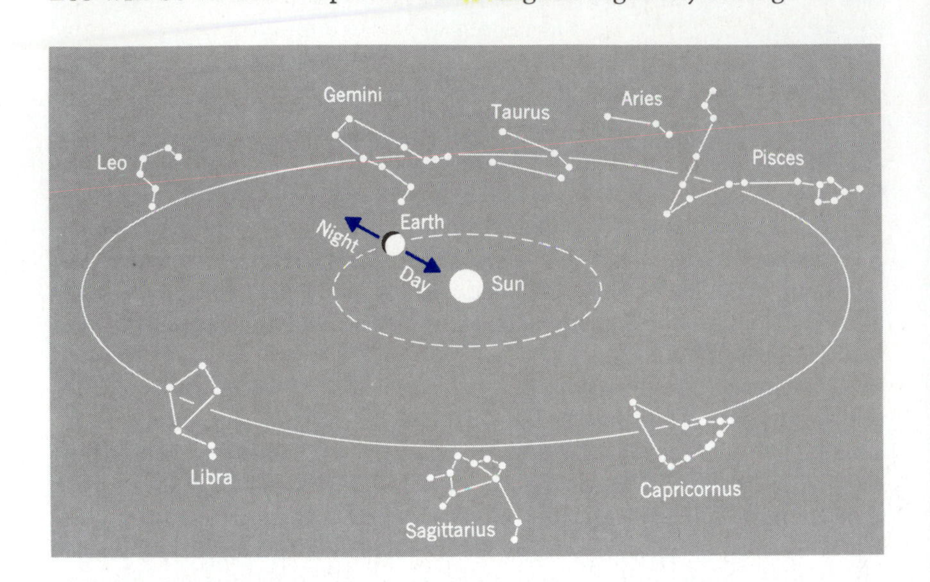

FIGURE 2.29b
The night sky in March.

INTERESTING OBJECTS IN THE NIGHT SKY

The changes described above can be verified readily by the untrained observer if he watches the prominent constellations during the course of several months. We assume that the reader is embarking on the study of astronomy at the beginning of the autumn or the spring term, and choose September and February skies as an example. Among the objects visible in these months are some that play an important role in the story of stellar evolution to be unfolded in Part Two. These objects, described below, are easy to find provided that the night is clear. The descriptions of their positions assume that the observer is out in the field around 9 P.M. on a September or a February evening.

September. As a start, orient yourself with respect to north and south, using the North Star, a compass, or your memory as to where the sun rises and sets. Face south and tilt your head back as far as you can, so that you are looking at the *zenith,* which is the point in the sky directly above your head. You will see a triangle formed by three bright stars (Figure 2.30). It is a huge triangle, extending from the zenith nearly halfway to the southern horizon, and it is impossible to miss. The triangle is "upside down," that is, its base is at the top, near the zenith, and the vertex is pointed downward toward the southern horizon. The two stars at the top, forming the base, are very conspicuous because they are the only bright stars near the zenith. These two stars are called Deneb and Vega. The star at the bottom, forming the vertex of the triangle, is called Altair. Altair also is easy to spot because it is a part of a short, straight line of three stars. (Altair is in the middle of the line of three and is brighter than the two stars on either side.)

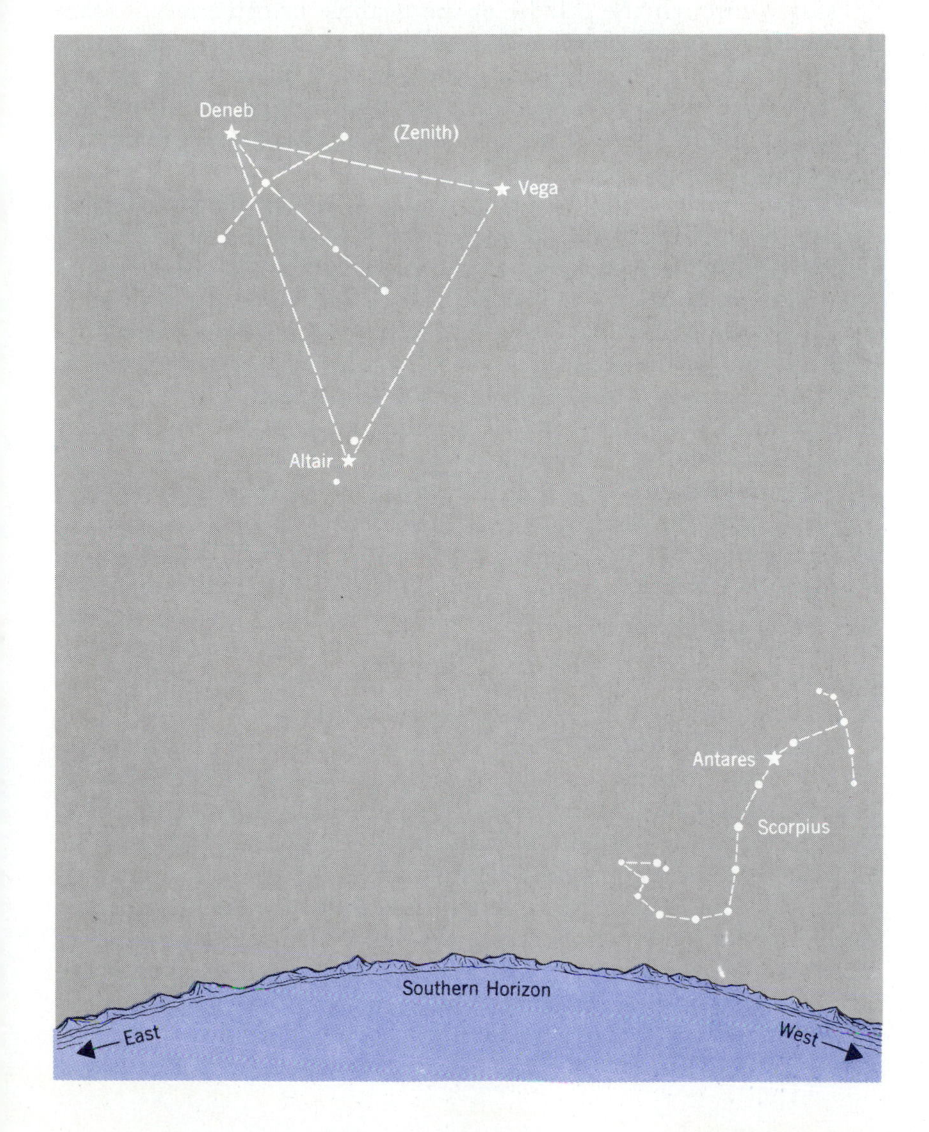

FIGURE 2.30
The summer triangle formed by Deneb, Vega, and Altair.

Of the three stars in the triangle, Vega appears to shine most brightly. Altair is next, and Deneb is the dimmest. The actual brightnesses of the three stars differ radically from their apparent brightnesses, because they lie at widely different distances from us. Deneb is intrinsically the brightest of the three stars, 1600 times brighter than Vega and 6400 times brighter than Altair, but it appears to be the dimmest star in the triangle because it is farthest away—1400 light-years from the solar system versus 26 light-years for Vega and 16 light-years for Altair. Because the intensity of light from a star decreases in proportion to the square of its distance from us, the fact that Deneb is $1400/16 = 88$ times farther away than Altair means that its brightness is cut down by a factor of 88^2 or approximately 7700.

Passing through the middle of the triangle of three bright stars is a luminous band, which is the Milky Way. The Milky Way will be visible only if the night is clear and you are away from the lights of large cities. It is formed by the overlapping light from 100 billion or so stars that are packed densely together in the midplane of our galaxy.

Turning partly to the west from the direction of the triangle, and looking down toward the horizon, you will see the constellation Scorpius, which will be easily recognizable as a coiled chain of stars resembling a scorpion's tail. It is near the horizon a little bit west of the downward extension of a line joining Deneb and Altair. The brightest star in this constellation is Antares, a distinctly red star. The name Antares is derived from Greek and means "rival of Mars." Because of its red color, ancient astronomers thought Antares was another red planet rivaling Mars.

You can find Antares in another way if you have trouble locating it according to the directions above. Look at Deneb again, and you will see that it marks the top of a cross whose long arm points downward to the southwest through the middle of the huge triangle. If you follow the direction of this arm of the cross with your eye, moving in a straight line, it will bring you to the reddish star, Antares, near the horizon. Figure 2.30 shows the cross that runs from Deneb through the triangle toward Antares.

Antares, which lies 400 light-years from the sun, is intrinsically 30,000 times brighter than the sun, about 10 times more massive, and several hundred times larger. It is one of a group of stars called red giants because of their color and size. Originally, Antares was a brilliant blue star but, as with all stars when they grow old and consume their fuel, it increased in brightness and reddened in color as it swelled to its present size. When the sun reaches this stage in its life, in which the hydrogen at its center has been burned and converted to helium, it will also become a red giant, thousands of times brighter than it is today.

Turning to the north, you will see the Big Dipper in the northwest. The bowl of the dipper is less than one-third of the way from the horizon to the zenith, and the handle extends toward the west. The various stars that make up the Big Dipper lie at quite different distances from us. The stars of the Big Dipper provide an example of the fact that the stars making up a constellation may be unrelated to one another and situated at widely different distances from the sun. Usually, the stars in a constellation happen to fall into a closely knit or clearly recognizable pattern only because of the accidental circumstances of the particular line of sight from which we on earth view these stars. Of the seven stars in the Big Dipper, five stars—the middle five of the seven—are all about 75 light-years away from the sun. They form a single group moving through space with approximately the same velocity and direction. The first star of the seven, marking the end of the handle, and the last star, marking the end of the bowl, are completely unrelated to the other five and to one another. The star marking the end of the handle is 109 light-years from us and moving in a direction opposite to that of the middle five, while the star marking the end of the bowl is 250 light-years away and also moving in a direction opposite to that of the middle five stars.

Because of the differences in their directions of movement, the stars of the Big Dipper will not retain the famous form they have today for very long. In 50,000 years the handle will become badly bent, and the bowl will be as flattened as if a child had been banging it on the table. Figure 2.31 shows the Big Dipper as it is today, with arrows marking the direction of motion of each of its seven stars, the Big Dipper as it was 50,000 years ago, and the Big Dipper as it will look 50,000 years hence.

If your eyesight is good, you will see that the next to last star in the handle of the Dipper consists of two stars close together forming a so-called *optical double.* One of the two stars shines about twice as brightly as the other. The brighter of the two is called Mizar and the less bright one is called Alcor. They are sometimes called the Horse (Mizar) and Rider (Alcor). Actually, the two stars are quite far apart in space, but happen to lie along approximately the same line of sight from the solar system, which makes them seem close together.

To make the situation even more complicated and interesting, each of the two stars, Mizar and Alcor, is itself a pair of stars, too close together to be resolved except with the aid of a good-sized telescope. They are so close to one another that the force of gravity ties them permanently together. Stars close enough together to be tied to one another by their own forces of gravity are called *binary stars.* The Mizar binary in the handle of the Big Dipper is the first that was ever observed. Both stars of this binary are brighter and

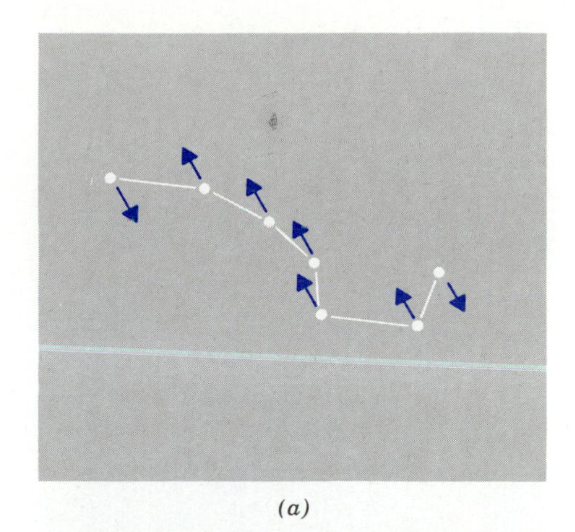

(*a*)

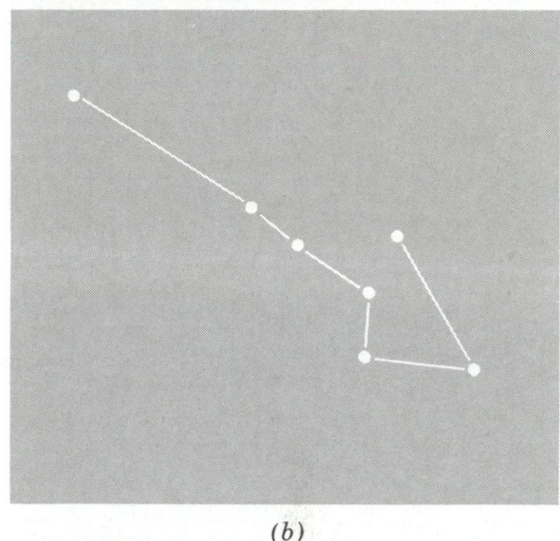

(*b*)

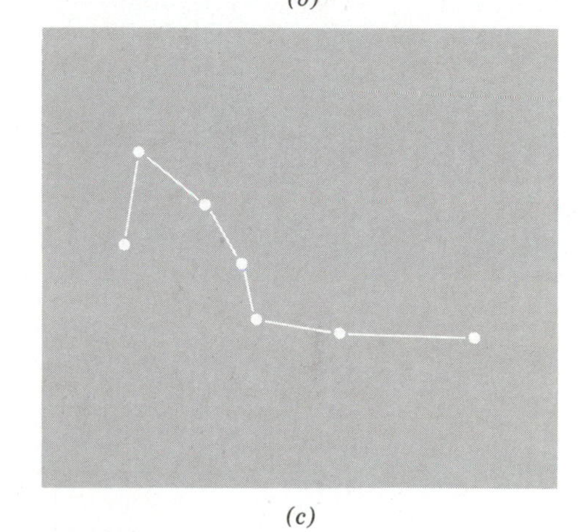

(*c*)

FIGURE 2.31
The Big Dipper (a) today; (b) 50,000 years ago; (c) 50,000 years in the future.

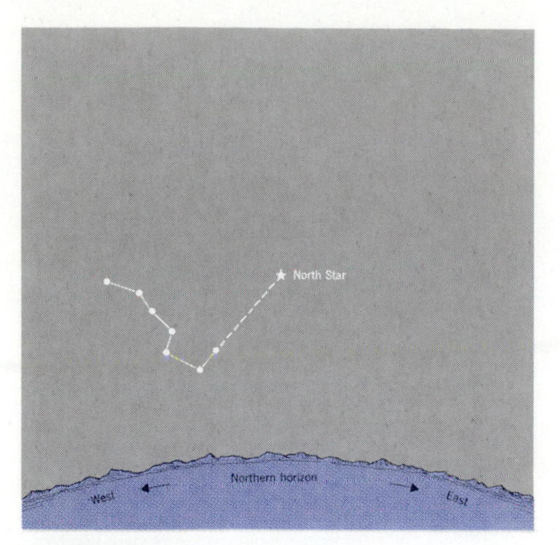

FIGURE 2.32
Finding the North Star.

FIGURE 2.33
Changes in the position of the Big Dipper during 24 hours.

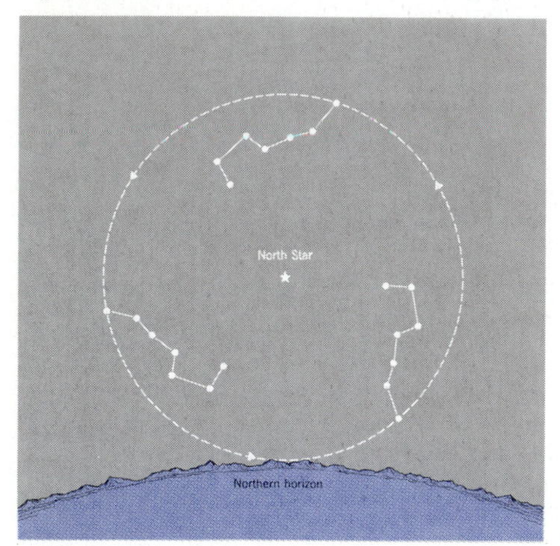

more massive than the sun. Because of their higher surface temperature, their color is white in comparison to the sun's yellow-white.

One of the most important stars in the sky is the North Star, or Polaris, which happens to lie almost exactly on the earth's axis of rotation, making it a useful star for navigation. To find the North Star, draw a line through the two stars in the Dipper that form the end of the bowl and extend it away from the bowl about five times the distance between these two stars (Figure 2.32).

Because the North Star lies on the earth's axis of rotation, it appears stationary as the earth rotates, but all the stars near it seem to move in circles around the North Star during the course of the night. A photograph of the night sky, taken with a time exposure lasting several hours, shows the paths of the stars circling the North Star. A constellation near the North Star, such as the Big Dipper, seems to turn over during the course of 24 hours as a result of this rotation (Figure 2.33). On the other hand, if you turned away from the North Pole and looked to the south, you would find that stars quite far to the south would move along relatively flat paths that never carried them very high in the sky. These stars, like the stars in the north, are also describing circles around the earth's axis of rotation. However, most of their circular path is blocked from your view by the earth (Figure 2.34).

The North Star has an additional interest because it is an example of a group of stars whose light output, or brightness, varies regularly instead of remaining constant as it does for most stars. In the case of Polaris, the brightness increases and decreases by a factor of 9 percent over four days. The variation is produced by a rhythmic breathing, or pulsation, in the size of the star; when it is contracting it shines more brightly and when it is expanding its brightness is at a minimum.

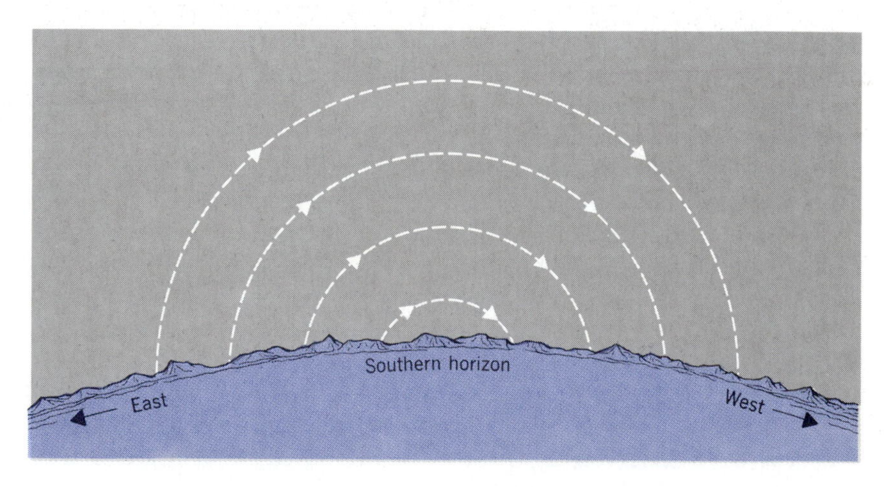

FIGURE 2.34
Star tracks over the southern horizon.

February. As before, it is assumed that you are out in the field around 9 P.M. Orient yourself again with respect to north and south, and, facing south, look for a point in the region of the sky approximately midway between the horizon and your zenith. You will see a famous constellation known as Orion, or the Hunter, shown in Figures 2.35, 2.36, and 2.37; Orion is the most prominent feature of the winter and spring skies, replacing the summer triangle of the September sky in this respect.

Three close stars in the center of Orion form a straight line running from southeast to northwest. The stars are not exceptionally bright, but the tight line of three is readily recognizable. The three stars are the belt of the Hunter (Figures 2.36 and 2.37). Immediately below the belt lies the sword of the Hunter and in the middle of the sword is the spectacular Orion Nebula (Color Plate 5). The Orion

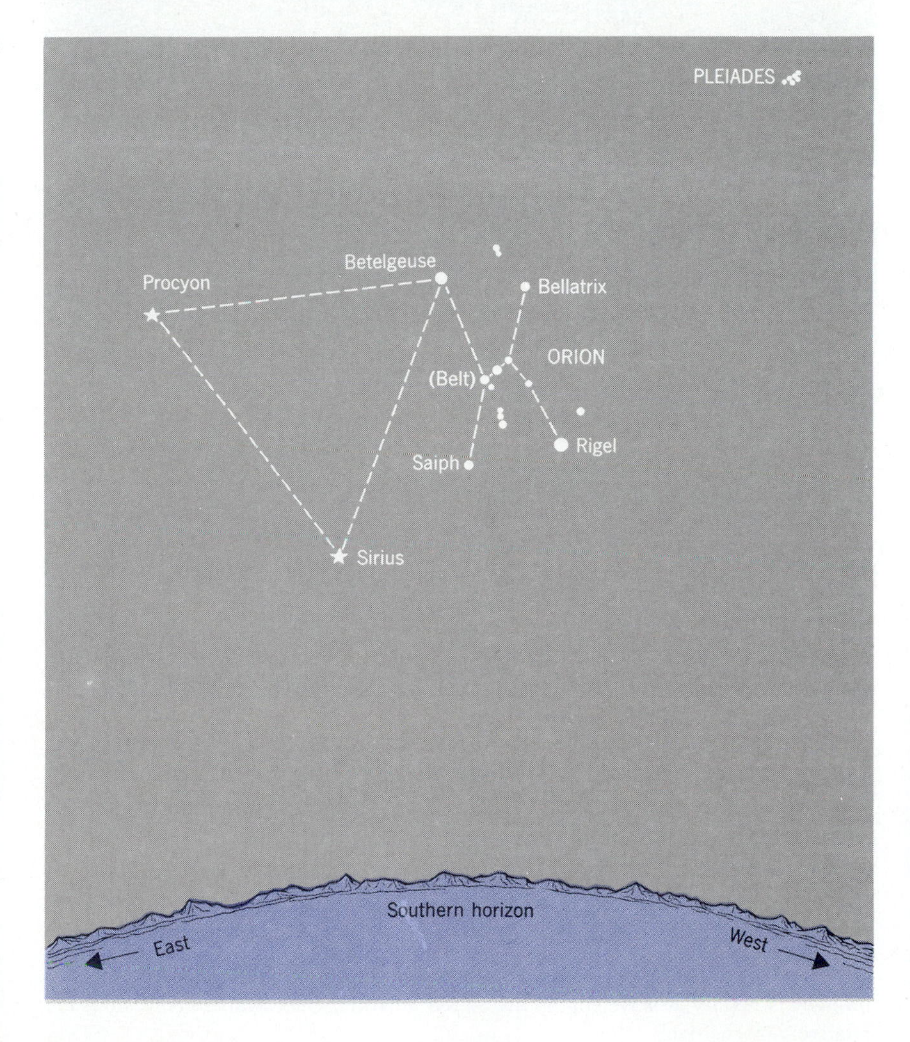

FIGURE 2.35
The constellation Orion, or the Hunter, and other prominent stars and constellations in the winter sky.

FIGURE 2.36
The Hunter.

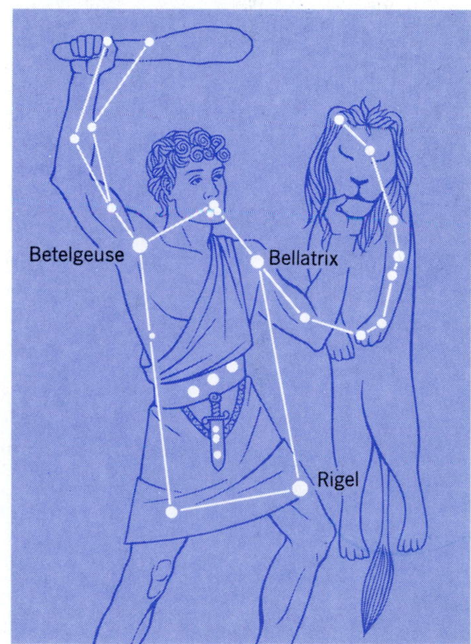

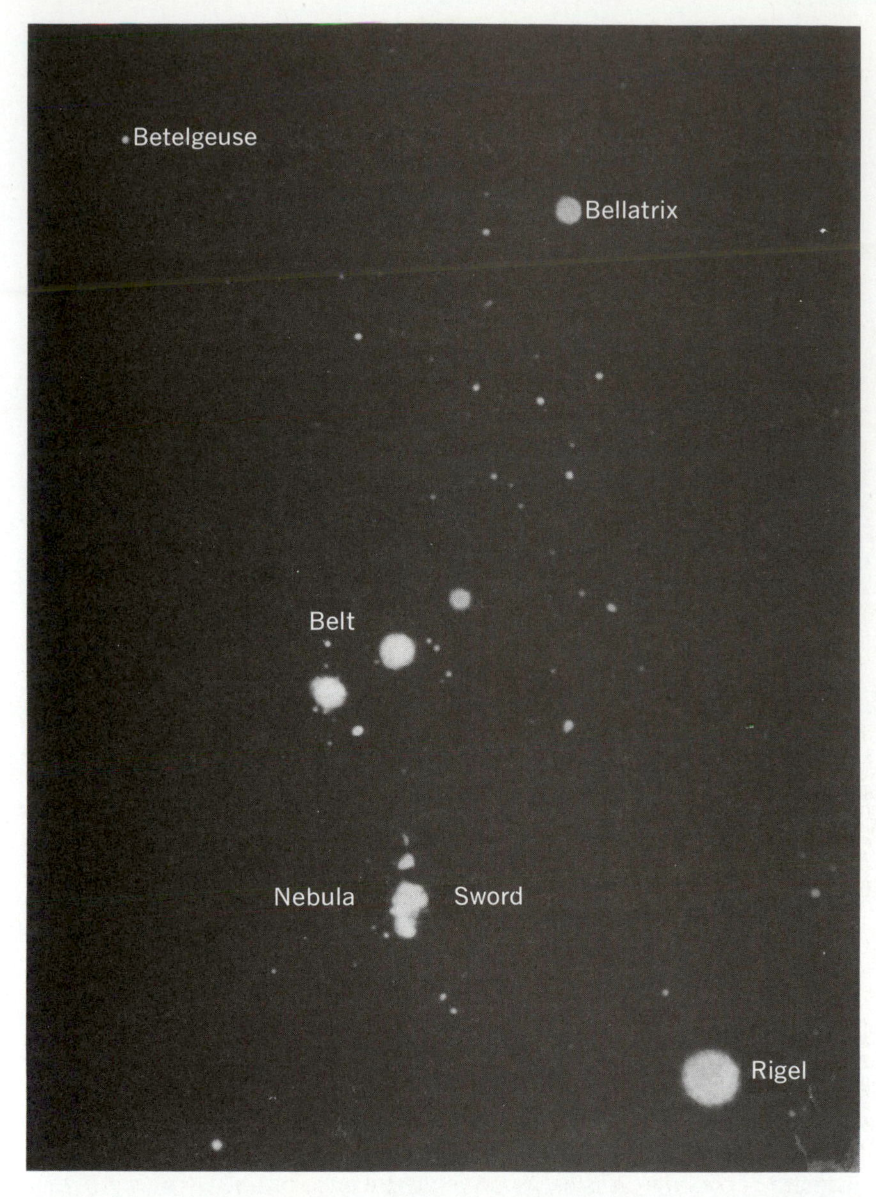

FIGURE 2.37
A photograph showing the location of the Orion Nebula in the sword of the Hunter.

Nebula is a region filled with a dense concentration of interstellar gases, rich in newborn and young stars (Chapter 8). It is faintly visible to the naked eye. Figure 2.37 shows the position of the nebula in the central region of Orion.

Some distance above the belt stars and slightly to the left, on the right shoulder of the Hunter, is a bright star with a reddish hue, called Betelgeuse.[8] This star is an exceptionally large and luminous

[8] Pronounced Beet-il-jooz. The photograph shows it as a rather faint star because its light is mainly at wavelengths in the red, to which photographic emulsions are relatively insensitive.

red giant, similar to Antares but nearly 10 times brighter. Betelgeuse is located at a distance of 700 light-years from us and is intrinsically 100,000 times brighter than the sun, 20 times more massive, and 500 times larger. If placed at the center of the solar system, this giant star would engulf the earth and extend beyond the orbit of Mars. Betelgeuse, like Antares, was a brilliant blue star in the prime of its life, but as its fuel became exhausted it reddened and expanded into its present form.

To the west and a slight distance downward lies a fairly bright star named Bellatrix, with approximately the same brightness as the stars in the belt. Bellatrix is located in the left shoulder of the Hunter. Unlike Betelgeuse, it is considerably hotter than the sun, with a surface temperature in the neighborhood of 20,000°K, and a blue color in contrast to the red color of Betelgeuse. Bellatrix and similar stars are called blue giants. They are usually massive stars, still in the prime of their lives, with a plentiful supply of fuel remaining.

A line drawn diagonally downward from Betelgeuse, across the line of the belt stars toward the southwest, will intersect a very bright blue-white star called Rigel, in the left thigh of the Hunter. Rigel, like Bellatrix, is a blue giant in its prime, far more massive, larger and brighter than the sun. A study of Rigel with a telescope reveals it to be a binary—a system of two stars circling about one another under the pull of mutual gravity. The small component, although invisible to the naked eye, is intrinsically far brighter than the sun.

Returning to the belt stars, a line drawn through these three stars and extended downward to the southeast intersects an extremely brilliant star, the brightest star in the heavens, known as Sirius (Figure 2.35). Sirius is also called the Dog Star because it is located in the constellation Canis Major—the Great Dog. Sirius is only 20 times brighter intrinsically than the sun, but it dominates the stellar multitudes in brilliance because it is so close to us, being our fifth closest neighbor. Sirius is also a binary, and of a particularly interesting sort; the brilliant, visible component is circled by a small, invisible companion star, detectable only with large telescopes. The invisible companion is as massive as the sun, but 300 times fainter, and its diameter is only 20,000 miles, closer to the size of an earthlike planet than a star. This invisible companion to Sirius belongs to a class of stars known as the white dwarfs. These are dying stars that have exhausted their fuel and have collapsed to a small radius and extraordinary density under the inward force of their great weight. Slowly, with the passage of time, they cool to black dwarfs and disappear from view. It is believed that the sun will follow this course some six billion years from now.

Directly north of Sirius, and at approximately the same elevation above the horizon as Betelgeuse, is a fairly bright star called Procyon

FIGURE 2.38
A photograph of the Pleiades.

(Figure 2.35). Procyon, Sirius, and Betelgeuse form the nearly equilateral winter triangle in the sky, similar to the summer triangle, but less conspicuous because the February sky contains so many bright stars. Procyon lies at a distance of about 11.5 light-years from the sun and is intrinsically 7.5 times brighter than the sun. It, too, has a companion white dwarf orbiting it.

A line drawn diagonally upward to the northwest from the belt of The Hunter will intersect the small cluster of stars known as the Pleiades (Figure 2.38 and Color Plate 10). Six stars can be seen in the Pleiades with the naked eye, but a telescopic study reveals approximately 100 stars within the entire cluster. The stars in the Pleiades cluster are very young, the entire cluster having been formed out of a single concentration of interstellar gas approximately 60 million years ago. They are among the youngest stars visible in the sky.

This brief tour of the autumn and winter skies demonstrates that the heavens are populated by stars of many colors, sizes, and ages. The meaning behind the existence of this puzzling variety of stars eluded astronomers until recent years, when rapid progress occurred through the discoveries made in astronomy and other branches of the physical sciences. Some important parts of the story of the stars are now known, and form the subject matter of Parts Two and Three of this book, but major mysteries remain, and new discoveries are reported by astronomers almost daily. The new developments in astronomy cannot be grasped without an understanding of basic concepts such as the nature of light, the structure of the atom, and the release of nuclear energy. Before embarking on a voyage through these interesting waters, you should first acquire a general perspective on the subject of astronomy by familiarizing yourself with the contents of the Universe.

Main Ideas

1. Motions of the earth: rotation, revolution, other motions.
2. The earth-moon system; phases of the moon.
3. Cause of eclipses.
4. Calendars based on the sun, moon, and stars.
5. Celestial coordinates.
6. Identifying constellations.

Important Terms

axis of rotation
binary stars
celestial sphere
constellation
declination
eclipse seasons
equator
latitude
longitude

lunar eclipse
lunar tide
meridian
nutation
parallax
precession of the axis
right ascension
sidereal time
solar corona
solar eclipse
solar tide
tilt of the axis
zenith

Questions

1. A 35mm camera is pointed at the North Celestial Pole, the shutter is opened and the film is exposed for 8 hours. Describe the appearance of the developed picture.
2. Assume that the earth's axis of rotation is pointed in a direction perpendicular to the plane of the earth's orbit. Assuming the earth's orbit to be a perfect circle, describe the seasonal changes of climate on the earth.
3. Suppose the precession of the earth's axis had a period of 10 years, rather than about 26,000 years. If the longest day of the year (the first day of summer) is June 22, 1983, what will the climate be like on June 22, 1988? June 22, 1993?
4. The equatorial bulge of the earth is produced by the earth's rotation on its axis. If the earth's axis of rotation were as described in question 2, and the moon's orbit were in the same plane as the earth's orbit, would the earth's axis precess? Why?
5. Define longitude and latitude. If you were in Chicago and wished to decrease your latitude, in which direction would you travel? If you wished to increase your longitude, in which direction would you travel?
6. Explain the difference between solar time and sidereal time. Which runs faster, a solar clock or a sidereal clock?
7. Why do we see different stars at different seasons of the year? Sketch the major constellations of the sky during the spring and fall.
8. Suppose you were an observer on one of Saturn's satellites. Saturn is about 10 times further from the sun than the earth. In the course of Saturn's orbit around the sun, how much larger would the parallax of a star appear to be, compared to its parallax as observed from the earth? Explain your answer.
9. What would be the frequency of solar and lunar eclipses if the moon's orbit were in the plane of the earth's orbit (the ecliptic)? Explain your answer.
10. The moon and sun each subtend angles of approximately $\frac{1}{2}$ degree on the sky. How large an angle on the sky would the earth subtend for an observer on the moon? For simplicity, assume that the earth is four times the diameter of the moon.
11. Write a computer program that uses as its input the standard time for any day of the year, calculates the sidereal time, and displays the answer.

CONTENTS OF THE UNIVERSE

3

In recent years the history of man's origin has been extended backward in time from the beginning of life on the earth to the beginning of time in the Universe. The result is a new and more complete story of Genesis, which provides fresh perspectives on some of the most profound questions to occupy the mind of man: What am I? How did I get here? What is my relationship to the rest of the Universe?

The ideas involved in this bold inquiry into the origin of the Universe are not a part of everyday thinking. Before the mind can understand them, it must first stretch its concepts of space far beyond their normal limits, and exercise the imagination to vault over the galaxies and span the dimensions of the Universe. It must comprehend the scale and contents of the Universe.

THE SUN AND ITS NEIGHBORS

The sun, its family of planets with their moons, and a large number of smaller bodies, form the *solar system.* The earth travels around the sun at an average distance of 93 million miles. Between the earth and the sun lie the planets Mercury and Venus. Outside our orbit lie the earthlike planet Mars, the giant planets Jupiter, Saturn, Uranus, and Neptune, and the frozen world of Pluto (see Figure 3.1). Pluto, the outermost planet, travels in an elliptical orbit at distances from the sun ranging between 3 and 4 billion miles. The diameter of the solar system—defined by the farthest sweep of Pluto's orbit—is approximately 8 billion miles.

Beyond the orbit of Pluto, space contains nothing but a few atoms of hydrogen and occasional comets, until we reach the stars that

Hercules cluster of galaxies. Nearly every spot of light in this photograph is a galaxy containing billions of stars.

FIGURE 3.1
The solar system.

are the sun's neighbors. These stars are an average of 30 trillion miles away.

An analogy will help to clarify the meaning of such enormous distances. Let the sun be the size of an orange; on that scale the earth is a grain of sand circling in orbit around the sun at a distance of 30 feet; Jupiter, 11 times larger than the earth, is a cherry pit revolving at a distance of 200 feet or one city block; Saturn is another cherry pit two blocks from the sun; and Pluto is still another grain of sand at a distance of 10 city blocks from the sun. The nearest stars are orange-sized objects more than a thousand miles away.

An orange, a few grains of sand some feet away, and then some cherry pits circling slowly around the orange at a distance of a city block; more than a thousand miles away is another orange, perhaps with a few specks of planetary matter circling around it. That is the void of space.

The Nearest Stars

The sun's closest neighbor is, according to information available at the present time, the star Alpha Centauri. Alpha Centauri is 24 trillion miles from our solar system, or slightly closer than the average distance between stars in our neighborhood. It is actually a triple star—a family of three stars formed simultaneously out of a single cloud of gas and dust. Ever since their birth, the three stars have circled one another under the attraction of gravity. The largest of the three stars in Alpha Centauri resembles the sun and possesses a similar surface temperature and color. The other two are smaller, redder stars. The middle-sized stars of the triplet, somewhat smaller than the sun and orange in color, circles around the largest star in a close waltz at a distance of two billion miles. One turn around takes 80 years for this pair. The third member is a very small, faint, red star, a tenth as massive as the sun, which circle the two other members of the triplet at a distance of a trillion miles, completing one turn in a million years.

Alpha Centauri is the closest star to us that is bright enough to be visible to the naked eye. However, the sun may have still closer neighbors—very small, dimly luminous stars—too faint to have been detected thus far. There may also be burned-out stars that have exhausted their fuel in the space between the sun and Alpha Centauri. Finally, there may be many bodies the size of planets, too small to glow by their own nuclear energy, in the space around us. All these possibilities await the future exploration of the regions outside the solar system.

The next nearest neighbor of the sun beyond Alpha Centauri is Barnard's Star, 30 trillion miles away. Barnard's Star is smaller than the sun, and almost 300 times fainter. The temperature at the surface of Barnard's Star is 5500 degrees Fahrenheit versus 11,000 degrees Fahrenheit at the surface of the sun, and its color is orange-red rather than yellow. Barnard's Star, unlike Alpha Centauri, is a single star.

Fifty other stars exist within roughly 100 trillion miles of the sun. Some are yellow stars, resembling the sun in size and temperature; a few are larger and brighter than the sun and blue-white in color; most are faint, reddish stars. These stars are our neighbors in space. All have been named, although only a few of the names are familiar. Twenty-six of the 50 belong to multiple stars—doubles or triples. On the average, about one-half the stars in the Universe are multiple stars.

Our Galaxy

The sun and its neighbors are only a few among more than 100 billion stars that are banded together by gravity in an enormous cluster,

FIGURE 3.2
The galaxy NGC 4565, resembling our galaxy viewed edge-on.

FIGURE 3.3
The galaxy NGC 7331, resembling our galaxy in a three-quarter view.

FIGURE 3.4
The galaxy M74, resembling our galaxy viewed face-on.

called the Galaxy. Most, if not all, the stars in the Universe are held within such clusters. These other clusters also are called galaxies. Our own galaxy, singled out because it contains the sun, is written with a capital "G."

THE CONTENTS OF THE GALAXY

The Appearance of the Galaxy

The stars in the Galaxy revolve about its center as the planets revolve about the sun. The sun itself participates in this motion, completing one circuit around the Galaxy in 250 million years. The Galaxy is flattened by its rotating motion into the shape of a disk, whose thickness is roughly one-twentieth of its diameter. The sun is located in the disk about three-fifths of the way out from the center to the edge. A small, spherical clump of stars, called the nucleus of the Galaxy, bulges out of the disk at the center.

The general appearance of the Galaxy is shown very clearly in photographs of three other galaxies that are similar to ours and that happen to be oriented in space so that we see them at several different angles. If you could stand outside the Galaxy and view it edge-on, it would look very much like NGC 4565 (Figure 3.2).[1] Viewed face-on, the Galaxy would look like M74[1] (Figure 3.4). Viewed obliquely, the Galaxy would look like NGC 7331 (Figure 3.3). The arrows in Figures 3.2 and 3.4 illustrate the position that the sun would have if these were, in fact, photographs of our galaxy. Our galaxy, like M74, has spiral arms in which most of the bright stars of the Galaxy are located, including the sun.

When we look into the sky in the plane of the galactic disk, we see so many stars that they are not visible as separate points of light but blend together into a luminous band stretching across the sky (Figure 3.5). The irregular lanes of black running through the center of the Milky Way in Figure 3.5 are caused by the extensive clouds of dust, concentrated in the central plane of the Galaxy, which block out the light of many stars.

Additional details can be seen in a montage of many separate but overlapping photographs of the Milky Way (Figure 3.6). Notice the extended black area in the center of the montage. This region, called the Great Rift, is a particularly striking example of the dense clouds of dust that prevent the light of the stars in the Milky Way from reaching us.

[1] NGC—New General Catalog, a catalog of extended objects—galaxies, star clusters, and nebulas—compiled by Cambridge University astronomers in 1890. M—Messier, an earlier compilation published by Messier, a French astronomer, in 1784.

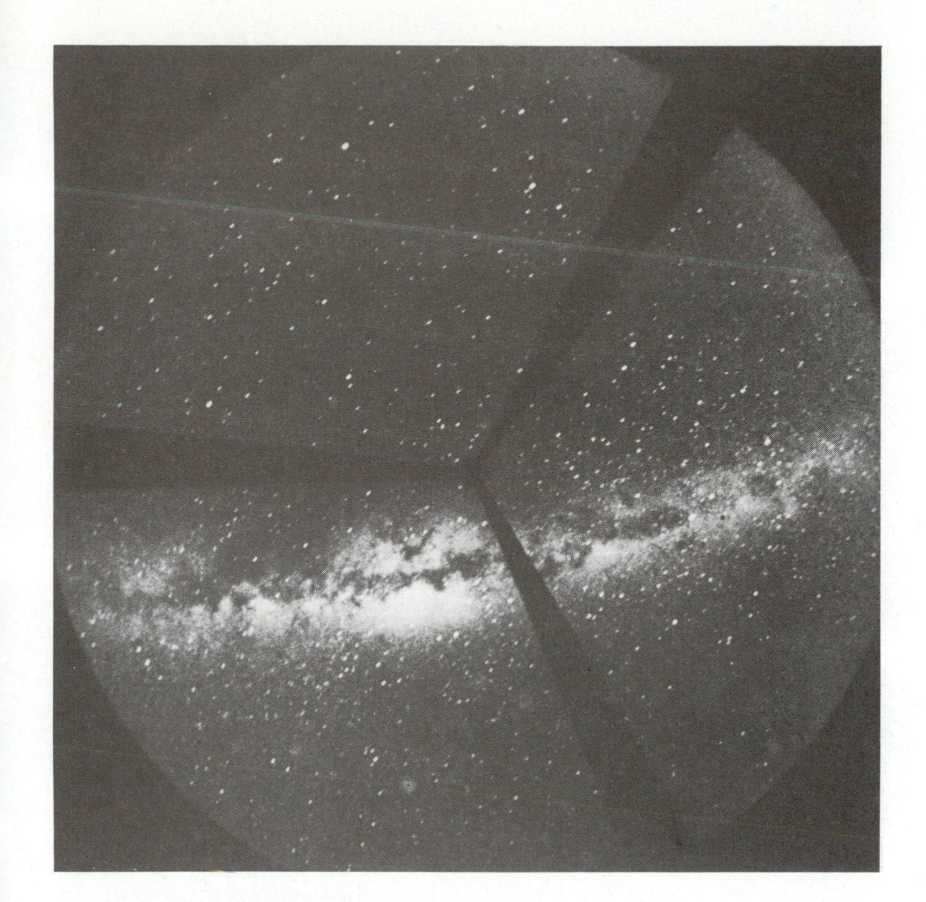

FIGURE 3.5
Edge-on view of our galaxy.

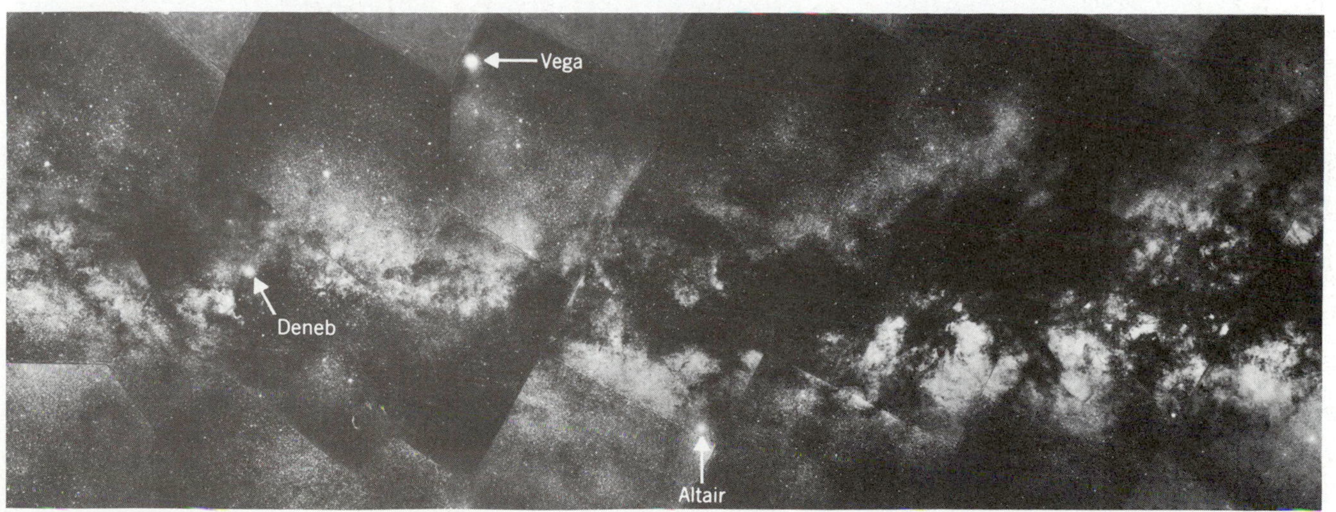

FIGURE 3.6
A montage of Milky Way photographs.

If the montage in Figure 3.6 is extended to include the entire Milky Way, it gives a nearly complete view of the Galaxy as seen by an observer located in our solar system. This edge-on view of our galaxy is shown in Figure 3.7. It presents in an extended detailed view the image of the Galaxy that was captured by a single camera in Figure 3.5. Figure 3.7 also shows the positions of the 7000 brightest stars in the sky, drawn in by an artist. These are the stars that are visible to the naked eye under the best conditions. In addition, we have lettered in the names of several objects of special interest. Three of them are stars—Deneb, Vega, and Altair. The other objects of special interest are external galaxies. The Andromeda Galaxy is the most distant object visible to the naked eye. The Magellanic Clouds, visible only in the Southern Hemisphere, are two dwarf galaxies anchored permanently to our galaxy by its gravitational force, each containing only 10 billion stars.

FIGURE 3.7
The Milky Way: an edge-on view of our galaxy.

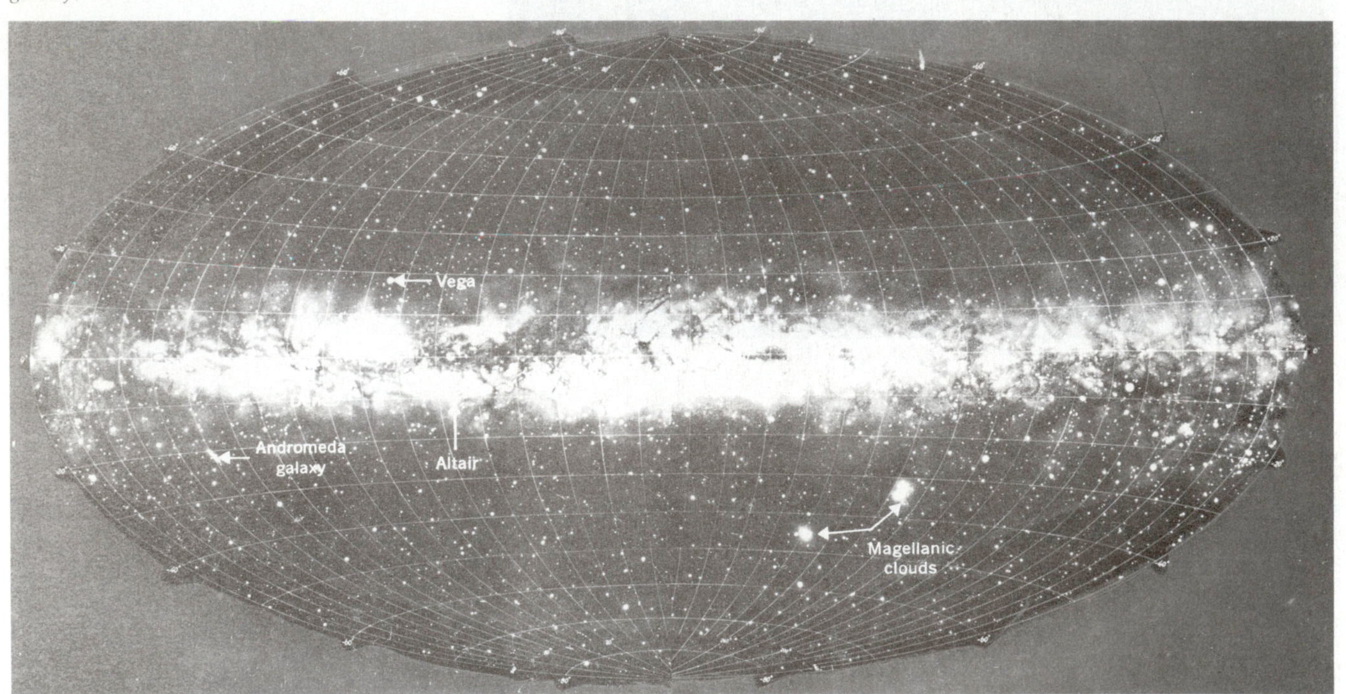

An inspection of the Milky Way with even a modest-size telescope or pair of binoculars reveals the immensity of the number of stars concentrated in this region. In the photograph shown in Figure 3.8, obtained with the enormous light-gathering power of one of the world's large telescopes—the 120-inch instrument at Lick Observatory in California—more than 10,000 stars can be seen, although this photograph shows only a minute portion of the sky.

FIGURE 3.8
The Milky Way: 10,000 stars in a small segment.

The glowing region in the middle of this photograph is called the North American Nebula because of its resemblance to the outline of the North American continent. Notice that very few stars are visible in the dark regions—the "Gulf of Mexico" and "Atlantic Ocean"—that outline the right edge of the North American Nebula. In these regions of the sky, an exceptionally thick concentration of obscuring dust conceals many of the stars in the Milky Way from our sight. The only stars that are visible in this region are the ones that lie on our side of the dust clouds, relatively near to the sun.

Deneb is located just beyond the right edge of the photograph. This fact should enable you to locate this small portion of the Milky Way in the montage of the entire Galaxy in Figure 3.7.

The Emptiness of the Galaxy

The vastness of the space between the stars is difficult to comprehend. The same analogy that we used to clarify the meaning of the size of the solar system is helpful in attempting to comprehend the emptiness of the Galaxy. Suppose again that the sun is reduced from its million-mile diameter to the size of an orange. The Galaxy, on this scale, is a cluster of 100 billion oranges, each orange separated from its neighbors by an average distance of more than 1000 miles. In the space between, there is nothing but a tenuous distribution of atoms and a few molecules and dust grains. That is the emptiness of space in the Galaxy.

Distances Between Stars; the Light-Year

The stars within the Galaxy are separated from one another by an average distance of 30 trillion miles. To avoid the frequent repetition of such awkwardly large numbers, astronomical distances are usually expressed in units of the light-year, defined as the distance covered in one year by a ray of light traveling 186,000 miles per second. This distance turns out to be approximately 5.8 trillion miles; hence, in these units, the distance from the sun to Alpha Centauri is 4.3 light-years, the average distance between the stars in the Galaxy is 5 light-years, and the diameter of the Galaxy is 100,000 light-years.

Neighboring Galaxies

Although the stars within our galaxy are very thinly scattered, they are, nonetheless, relatively close together in comparison to the space that separates our galaxy from neighboring galaxies. The distance to the next nearest galaxy comparable in size to ours is 2 million light-years, or 20 times the diameter of our galaxy. It is difficult to imagine the emptiness of intergalactic space. Once outside the Galaxy, we encounter a region empty of stars and nearly empty of dust.

No vacuum ever achieved on earth can match the vacuum of the space outside our galaxy. But if we go far enough away from the Galaxy, we come to other galaxies, clusters of billions of stars held together, like ours, by the force of gravity. These galaxies are island universes—isolated clusters containing vast numbers of stars and, perhaps, planets—each separated from the others by the void of intergalactic space.

The closest galaxy comparable to the Milky Way Galaxy in size is the Andromeda galaxy, which is 2 million light-years from us.

This galaxy happens to resemble our own closely in size and shape; it is a disk-shaped spiral of stars, gas, and dust, containing approximately 100 billion stars in all, the entire collection of matter slowly spinning around a central axis like a gigantic pinwheel.

Andromeda is the only major galaxy visible to the naked eye, and it is the most distant object that can be seen without the aid of a telescope. However, it is not conspicuous, despite the fact that its intrinsic brilliance is 100 billion times that of the sun. Because of its enormous distance, Andromeda is barely visible to the naked eye, under the best conditions, as a very faint patch of light.

But if it is photographed with even a modest-sized telescope, the faint patch is seen to have a structure that reminds one of our galaxy, with a brightly glowing center, a distinct impression of spiral arms, and dark lanes presumably formed by obscuring clouds of dust. The photograph of the Andromeda galaxy shown in Figure 3.9, taken with a 48-inch telescope, indicates these features. The arrows point to two dwarf galaxies very close to Andromeda, which are similar to the Magellanic Clouds in being captives of the attraction of the mass concentrated in their larger neighbor. The

FIGURE 3.9
The Andromeda galaxy, our nearest galactic neighbor, comparable in size to the Milky Way.

rectangle encloses the area shown in greater detail in Figure 3.10.

If Andromeda is photographed through a larger telescope such as the 200-inch one on Mount Palomar in California, whose light-gathering power and resolution of detail are thousands of times greater than those of the eye, the luminous cloud resolves into billions of individual stars. The photograph in Figure 3.10, taken through the 200-inch telescope, constitutes the area enclosed by a rectangle in the previous photograph. This picture shows many of the separate stars in the upper edge of the Andromeda galaxy. The many stars that appear in the photograph outside the boundaries of the Andromeda galaxy appear to be stars in intergalactic space, but this impression is misleading. Actually, these stars lie in our own galaxy, but happen to be in the line of sight from the sun to Andromeda.

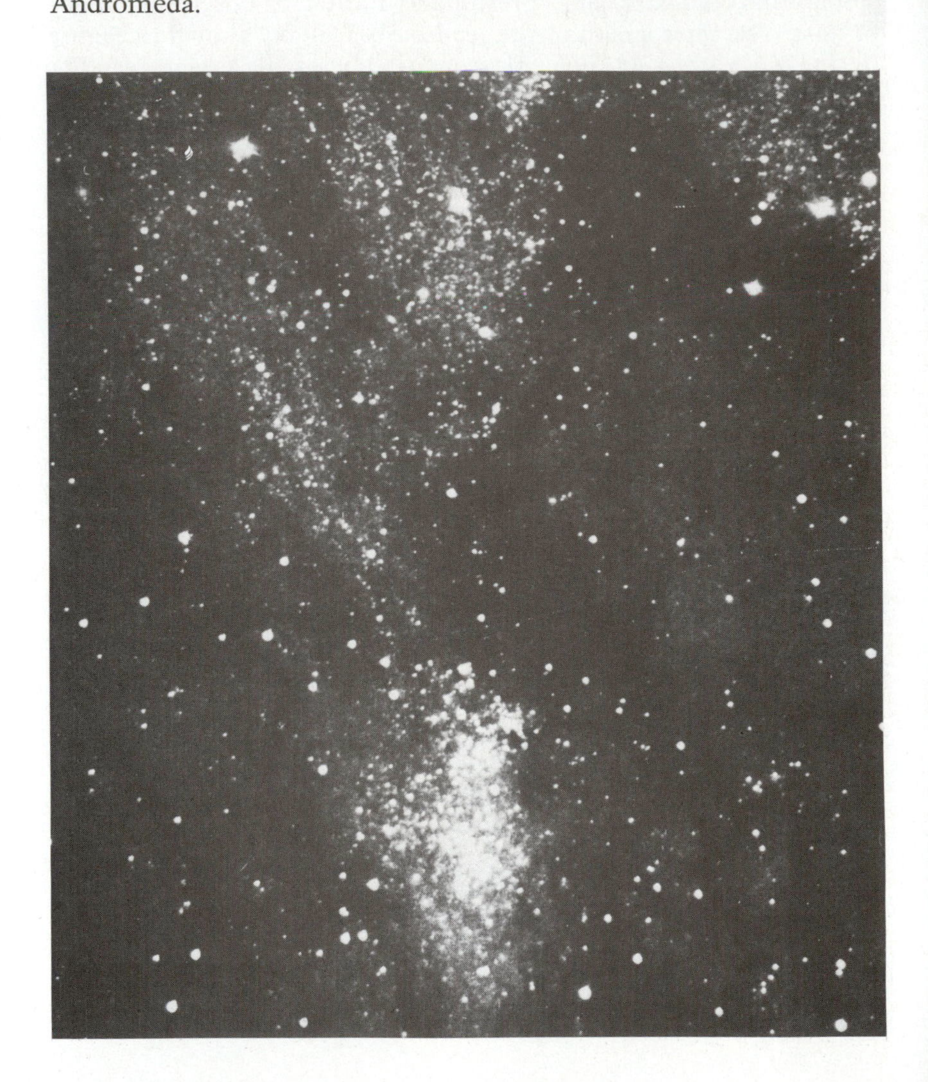

FIGURE 3.10
Detail of the Andromeda galaxy, showing individual stars.

The Local Group

Approximately twenty other galaxies, including the two dwarf galaxies known as the Magellanic Cloud, exist within 3 million light-years of ours. Astronomers call these galaxies the Local Group (Figure 3.11). Of the galaxies in the Local Group, only three—ours, Andromeda, and M33—have the spiral form shown in Figure 3.4.

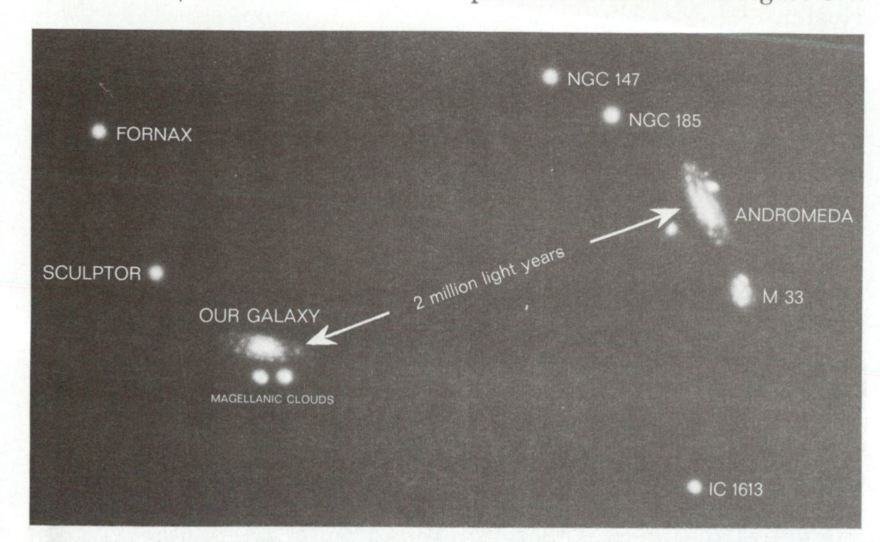

FIGURE 3.11
The Local Group. Only major galaxies are shown.

Our remaining galactic neighbors are five to ten times smaller in diameter, 100 to 1000 times less populous, and either elliptical or irregular in shape. A photograph of one of the smaller, more irregular galaxies in the Local Group—the large Magellanic Cloud—is shown in Figure 3.12.

FIGURE 3.12
One of the Magellanic Clouds, a small captive galaxy of the Milky Way.

Clusters of Galaxies

Enormous though a single galaxy is, it does not constitute the largest collection of matter known in the Universe. Galaxies themselves occur in clusters, held together, once again, by the force of gravitational attraction that each galaxy in the cluster exerts on the others. The Local Group, described above, is an example of such a cluster.

Some clusters contain only a few galaxies. An example is the cluster of galaxies in the direction of the constellation Pegasus, shown in the photograph in Figure 3.13. (Astronomers have long used constellations, or groups of stars that seem to form figures in the sky, to impose some order on the huge number of stars in the night sky. Much like signposts, they tell us what part of the heavens

FIGURE 3.13
Stephan's Quintet. The four galaxies marked by arrows form a small cluster. The fifth member of the quintet, at lower left, is closer to us and does not belong to the cluster.

we are gazing at.) In Figure 3.13, the spiked objects[2] and the small circular spots are individual stars situated in our own Galaxy along the line of sight to the Pegasus Cluster. The large, luminous objects marked by arrows are four galaxies in a small cluster.

The Local Group, of which our galaxy is a member, is an aggregate of galaxies including the Milky Way, Andromeda, and numerous fainter objects. It is not certain that the Local Group is a clearly defined cluster, i.e., a group of galaxies bound together by gravity.

The first group of galaxies outside the Local Group that clearly is a cluster is located in the direction of the constellation Virgo. It contains 2500 galaxies, is located approximately 60 million light-years from the Local Group, and is one of the larger clusters known.

The Local Group appears to be a part of an even larger aggregate of galaxies which has the Virgo Cluster at its center. This giant aggregate is called the Local Supercluster.

About 300 million light-years from our galaxy, in the constellation Hercules, is another giant group of galaxies called the Hercules Cluster, which contains about 10,000 galaxies, each with 10 billion to 100 billion stars. The photograph facing page 69 shows one small region in the Hercules Cluster, containing approximately 50 galaxies. In this photograph the spiked objects and the small spots of light are stars situated in our galaxy along the line of sight to the Hercules Cluster; every other object is a galaxy of stars. Superclusters like the one we belong to and the one in Hercules constitute the largest known aggregates of matter in the Universe.

The "Edge" of the Universe

The largest telescopes can photograph galaxies that are billions of light-years from the earth. If we built still larger telescopes, could we see farther out to space? Is there any limit to the range of our observations? A surprising answer has come out of recent astronomical discoveries.

According to the latest evidence, described in Chapter 13, the Universe is about 15 billion years old. Therefore, we can only see galaxies that are less than 15 billion light-years away from us; if a galaxy is more than 15 billion light-years away, the light from it has not yet had time to reach the earth.

Fifteen billion light years is the radius of the *observable Universe.* The true Universe may be infinitely large and contain an infinite number of galaxies and stars, but we can never hope to see them all from the earth.

[2] The "spike" is caused by the struts in the tube of the telescope.

THE FORCES THAT HOLD THE UNIVERSE TOGETHER

All objects in the Universe, from the smallest atomic nucleus to the largest galaxy, are held together by only three fundamental forces: a nuclear force, the force of electromagnetism, and the force of gravity.

Most powerful is the *nuclear* force[3] which binds neutrons and protons together into atomic nuclei. This strong force of attraction pulls the particles of the nucleus together into a very compact body with a density of one billion tons per cubic inch. Although it is an exceedingly strong force, it has a very short range. The nuclear force will not attract two particles if they are more than one ten-trillionth of an inch apart.

Next strongest is the *electrical* (electromagnetic) force, which is approximately 100 times weaker than the nuclear force. This force binds electrons to nuclei to form atoms, and it binds atoms together into solid matter. It grows weaker with increasing distance between two particles, although unlike the nuclear force, it does not disappear entirely at any point.

Least powerful is the force of *gravity.* The gravitational force is exceedingly weak, about 10^{38} times[4] weaker than the force of electricity. Gravity, like electricity, falls off in strength with increasing distance, but never disappears entirely. In spite of its intrinsic weakness, gravity holds the moon in orbit around the earth, the earth and other planets revolving around the sun, and the sun and other stars clustered together in our galaxy.[5]

In some ways gravity is the simplest force of the three. Its simplicity lies in the fact that the force of gravity acting between two objects always pulls them together and never pushes them apart.

The electrical force is more complicated, because its action on some pairs of particles is to pull them together, while on others its action is to push them apart. The explanation of these two kinds

[3] Until recently it was believed that a fourth independent force existed, called the *weak force,* which played a role in nuclear reactions such as beta decay, involving electrons, positrons, and neutrinos. However, recent evidence indicates that the weak force and the electromagnetic force are two manifestations of a single basic force.

[4] $10^{38} = 1$ followed by 38 zeros. If this number were to have a name it would be 100 trillion trillion trillion.

[5] The disparity in strength between gravity and electricity is demonstrated by the following example. A fist-sized piece of iron weighs about one pound. That is, all the mass of the entire earth pulls the iron to the ground with a force of one pound. But a few turns of copper wire, through which an electric current is flowing (an electromagnet) will suffice to lift this piece of iron up from the ground against the gravitational force of the whole earth.

of action has come out of laboratory studies of electricity during the last 200 years. These studies show that two types of electric charge exist, positive and negative. Protons, for example, carry a positive charge, and electrons carry a negative charge. The studies also show that two particles bearing the same kind of charge—that is, both positive or both negative—repel each other. If two particles carry opposite charges, they attract one another. The nucleus of the atom, which carries positively charged protons, can hold electrons in orbit around it for this reason.

The Dominant Force in Outer Space

In spite of the strength of the nuclear force, it has a negligible effect on the motions of stars and planets in space, which are controlled by the relatively weak force of gravity. The nuclear force, powerful though it is, plays no role in space because its effect is confined to extremely small distance, whereas the force of gravity falls off relatively slowly with increasing distance,[6] and can still be felt across the diameter of a solar system, or even across the distances that separate stars from one another or galaxies from neighboring galaxies.

Why is the force of electromagnetism not dominant over gravity in controlling the motions of the stars and planets? This force can also extend to great distances, just as the force of gravity does, and it is many times more powerful. The answer comes out of the fact that when both kinds of electricity, positive and negative, are present, their effects tend to cancel. If an object has precisely the same amounts of positive and negative electricity, it is electrically neutral and exerts no electrical force at all on other objects located at a great distance from it.

Many electrons and atomic nuclei move freely in the spaces around every star or planet. If a star or a planet picks up an excess of protons from the space around it and becomes positively charged, it immediately exerts a force of attraction on all the negatively charged electrons moving nearby. These electrons migrate to the star or planet and are captured by it; the star or planet continues to attract electrons in this way until it has picked up enough charge to cancel the excess positive charge it possessed at the start. It is now electrically neutral, and no longer exerts an electrical force on other objects.

If the star or planet picks up an excess of electrons initially, it becomes a center of attraction for all positively charged nuclei moving through the space around it, and will pick up nuclei until its negative charge is canceled and it is electrically neutral once more.

[6] See Chapter 1.

The force of gravity cannot be canceled in this way because there appears to be only one sign of "gravitational charge." Mass always pulls other mass, and two masses never repel one another. There may be a second kind of gravity, or antigravity, in the Universe, but thus far physicists have been unable to find evidence for it.

Undiscovered Forces

Are there other forces in nature? Is there a fourth force and a fifth? Why are there only three forces? No one knows the answers to these questions. It is even more interesting to ask whether there might be connections among the fundamental forces of gravity, electromagnetism, and the nuclear force. Is it possible that there are fewer than three basic forces? Perhaps there is an underlying unity in the Universe that runs even deeper than the unity the physicists have uncovered thus far. Einstein spent the last 25 years of his life in the effort to discover a connection between gravity and electromagnetism which, his instinct told him, must be there. Yet, his instinct, superb though it was in revealing the truths of the special theory of relativity and the general theory of relativity, seems to have failed him in his last endeavor; for he never succeeded in finding a connection between these two forces that was satisfactory to him and to other physicists. Perhaps there is no connection; or perhaps it will require another genius, as great as Einstein, but born in a later age when more is known, to uncover this connection. These are among the unanswered questions of physics today.

Main Ideas

1. Scale of distances in the solar system.
2. Scale of distances in the Milky Way Galaxy; organization of the Milky Way Galaxy.
3. Distribution of galaxies in our neighborhood; clusters of galaxies.
4. Radius of the observable Universe.
5. Basic forces acting on matter in the Universe; a brief account of the role of basic forces in the organization of the Universe.

Important Terms

Andromeda galaxy
cluster of galaxies
dwarf galaxy
electromagnetic force
galaxy
gravity
light-year
Local Group
molecule
Milky Way
multiple star
nuclear force
observable Universe
planet
solar system
star

Questions

1. If the sun were the size of an orange, on that scale how big would the solar system be? How far away would the nearest star be?
2. What is the speed of light? How big is a light-minute (the distance light travels in a minute)? How big is a light-year? Show your calculations.
3. The average radius of Pluto's orbit is about 3.5 billion miles. How long would it take light to cross Pluto's orbit?
4. What is the average distance between stars in light years? In miles? The average diameter of a star is about one million miles. Compute the ratio of star diameter to star separation. Express your answer as a power of ten. Based on your calculation, estimate the chance of two stars colliding.
5. What is the Milky Way? What is its diameter in light-years? How does it appear in a telescope when viewed from the earth? How would it appear in a telescope if viewed from the Andromeda Galaxy?
6. Imagine each disk-shaped galaxy like ours to be the size of a dinner plate. On this scale, what would the distance be from our galaxy to Andromeda? What would the average distance be between stars in the Galaxy? What would the extent of the observable Universe be, assuming its true radius to be 15 billion light-years?
7. What are the basic forces that hold the Universe together. Arrange them in the order of strongest to weakest.
8. List the hierarchy of structures from the smallest subatomic particles to the largest structures in the Universe. After each structure, indicate the basic force that holds it together.
9. If the Universe is defined as everything that exists, comment on the meaning of the "edge" of the Universe.
10. Suppose the Universe were one year old. How far out could an observer in such a universe see?
11. Assuming that the Universe is infinite and contains an infinite number of stars, what intensity of starlight would you expect to observe at the surface of the earth, coming from all the stars in the Universe? Would the sky still be dark at night? Explain.
12. It has been suggested that our Universe may be an atom in a larger cosmos, and each atom in our world a universe of its own. What philosophical and scientific arguments can you offer pro and con regarding this suggestion?
13. Write a computer program that asks the user a speed of travel, and then calculates and prints the travel time to each of the objects in the list: the moon, Mars, Jupiter, Pluto, the nearest star, the center of the Galaxy, and the Andromeda Galaxy. Include a check for speeds greater than the speed of light.

STARS

PART TWO

TOOLS OF THE ASTRONOMER

4

Most of the information we possess about the Universe comes to us in the form of light from distant stars and galaxies. Light is the principal link between the earth and the stars. This light passes through "windows" in the earth's atmosphere to the surface of our planet, where it is collected by the telescope and analyzed by the spectroscope. The telescope and spectroscope are the essential tools of the astronomer.

THE TELESCOPE

The precise circumstances of the invention of the telescope are obscure, but it is known that opticians in Germany and Holland were building telescopes in 1608, and by 1609 some of the Dutch instruments had found their way to Paris. There, a friend of Galileo saw one and wrote to Galileo about it. According to his own statements, Galileo figured out its principle himself and built several telescopes, grinding his own lenses for that purpose. The military value of the invention struck him first, and he made a gift of his instrument to the powerful government of the city of Venice, explaining that it could protect the city against seaborne invasions by bringing into view "sails and sailing . . . two hours before they were seen with the naked eye." The senate of Venice immediately doubled Galileo's salary.

Artist's concept of a space telescope in orbit.

How the Telescope Works

The working of the telescope depends on the fact that a lens can create an image of a distant object. A lens is glass, or other transparent material, shaped so that it brings rays of light to a focus.

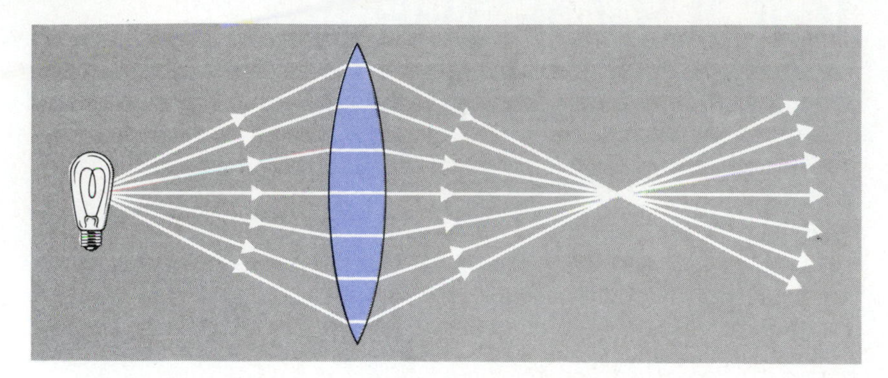

Beyond the point at which the rays of light converge, they diverge again in the same manner as they diverged from the original object. If you place your eye beyond this point, and look toward the lens, the diverging rays of light enter the eye and fall on the retina, conveying a message to the brain that they are emanating from a copy of the original object located at the new point of divergence.

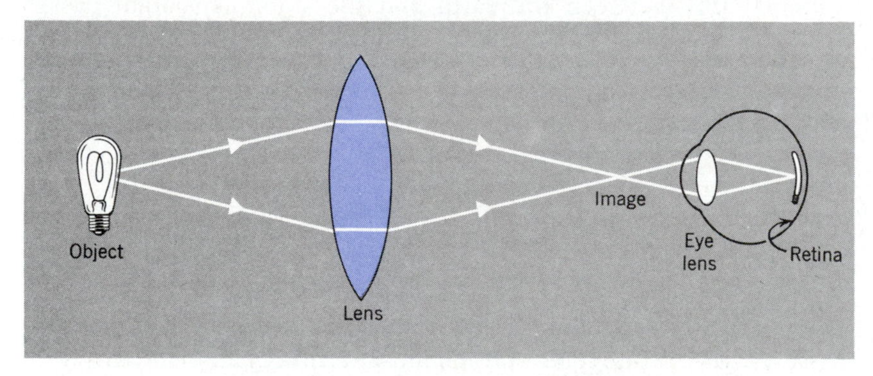

No object is actually there, but we say that an *image* of the object has been formed by the lens at this point.

Now we are ready to apply our understanding of lenses and magnification to the telescope. The telescope consists of two lenses, the objective and the ocular. The light from a distant object enters the objective lens first. This is the larger of the two lenses. The objective lens forms an image of the distant object, just as in the preceding diagrams. The image is formed inside the tube, close to the other lens, or ocular.

The ocular lens serves as a magnifying glass which enlarges the image formed by the objective, just as it would enlarge a real object placed there, such as a postage stamp or an insect. The person using the telescope looks through the ocular, and sees a magnified view of the image. Because the image is magnified, the distant object appears to be much closer than it actually is (Figure 4.1).

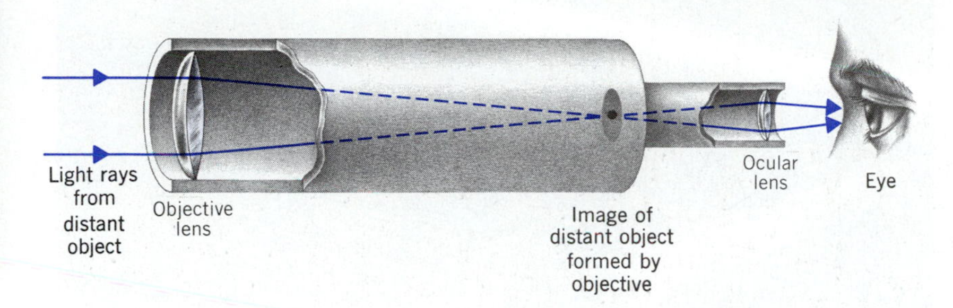

FIGURE 4.1
The principle of the telescope. Light rays from a distant object enter the objective lens at left. The objective lens forms an image of the distant object at the other end of the telescope, near the ocular. The ocular acts as a magnifying lens. The eye looks through the ocular and sees the image magnified, making the distant object seem closer than it really is.

Reasons for Building Large Telescopes

The instrument with which Galileo made the great discoveries described in Chapter 1 was about 18 inches long and had a one-inch lens. Since Galileo's time, astronomers have built larger and larger telescopes. Some telescopes in use today are 50 feet long, with lenses three feet in diameter. A great deal of trouble and expense is involved in grinding these mammoth lenses to nearly perfect shapes. Why did astronomers bother to construct such instruments?

Blurring of a Telescope Image by Diffraction. One reason has to do with the sharpness of the image. In general, the bigger a telescope is, the clearer the image produced. The reason for this is that when a beam of light enters a small telescope lens, the rays passing through the lens near its edge are bent by a different angle than the rays passing through the rest of the lens. This effect is called diffraction (Figure 4.2). As a result of diffraction, all the rays do not come to a focus at the same point, and the image is blurred. In telescopes with large lenses, the amount of light passing through the body of the lens is greater relative to the amount passing near the edge. Thus the ratio of "good" light to "bad" light is greater, and the image is clearer. In other words, the diffraction effect is smaller for large telescopes.

FIGURE 4.2
Bending of light at the edge of a lens.

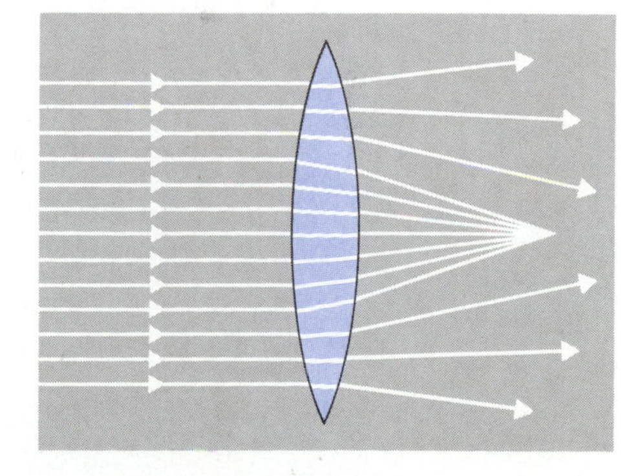

The second reason for building large telescopes is that they collect more light than small telescopes, and therefore reveal stars and galaxies that are too faint or too far away to be seen with a small instrument. The largest telescopes used by amateurs have lenses about one foot in diameter. These enable us to see stars as far away as the edge of our galaxy, about one million trillion miles from the earth. The telescope at the Yerkes Observatory in Wisconsin has the largest glass lens in the world—40 inches in diameter. This telescope can see hundreds of galaxies beyond our own (Figure 4.3).

FIGURE 4.3
The 40-inch refracting telescope at Yerkes Observatory.

Blurring of an Image by the Earth's Atmosphere. The larger a telescope lens is, the sharper the image should be, because the diffraction effect is smaller for a large lens. In theory, a sufficiently large telescope should provide images so sharp that we can tell from the earth whether there is life on Mars. In that case, why did NASA go to the expense of sending a spaceship to Mars to search for life on its surface?

The answer has to do with the blurring effect of the earth's atmosphere, which limits the amount of detail that can be seen from the earth, no matter how large a telescope is used. Rays of light from Mars or other objects in the heavens are bent slightly by the earth's atmosphere as they travel through it on their way to the telescope. This bending of the rays in the atmosphere is caused by the fact that light travels at a slightly slower speed in air than in a vacuum. In a vacuum it travels at 186,235 miles per second. When a ray of light from a star enters the earth's atmosphere, it slows down. As a result it changes its direction, just as a band marching down a football field will change its direction if the marchers on one side of the band slow down by taking shorter steps than the marchers on the other side. The band will wheel around the slow-moving marchers, and turn. The line of marchers will "bend."

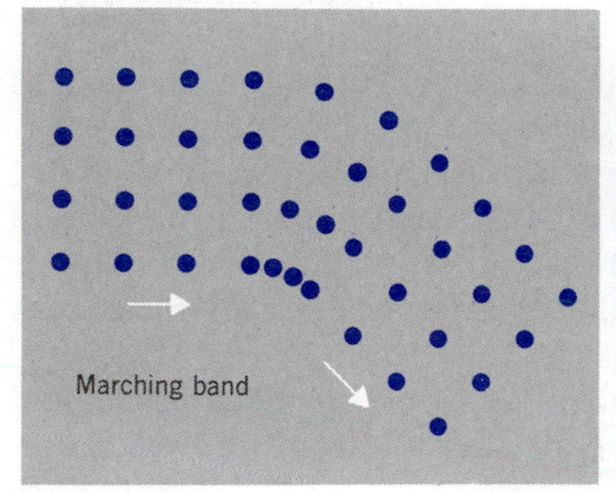

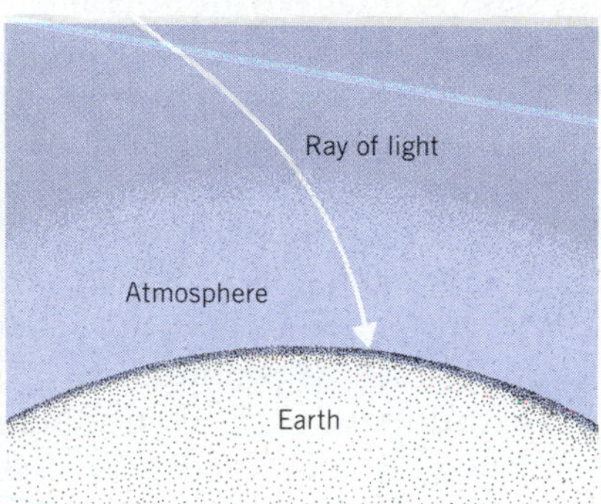

The bending of rays of light in the atmosphere would have no effect on the clarity of the star image formed by the telescope if the air were perfectly quiet; it would only shift the apparent position of a star by a small amount. The atmosphere, however, is continuously in motion, and as a result the position of the image dances about continually, many times a second. It moves rapidly back and forth on the retina of the eye if the astronomer is looking through the telescope, blurring the image. It also moves across the photographic film if he uses a camera, blurring the photograph.

Astronomers take most of their photographs with long exposures so that they can capture as much light as possible from faint, distant objects. As a result, photographs taken through large telescopes are badly blurred by the atmospheric jiggling of the image. The effect of the atmosphere is evident in the contrast between the photograph of the rim of the Alphonsus crater on the moon (Figure 4.4*a*) taken with the 100-inch telescope at the Mount Wilson Observatory in California, and a photograph of the same region taken from a small camera on a spacecraft, as the spacecraft approached the moon's surface (Figure 4.4*b*).

FIGURE 4.4(a)
A section of the floor and surrounding rampart of Alphonsus crater, photographed by the 100-inch telescope at Mount Wilson. No feature smaller than one mile in diameter can be seen.

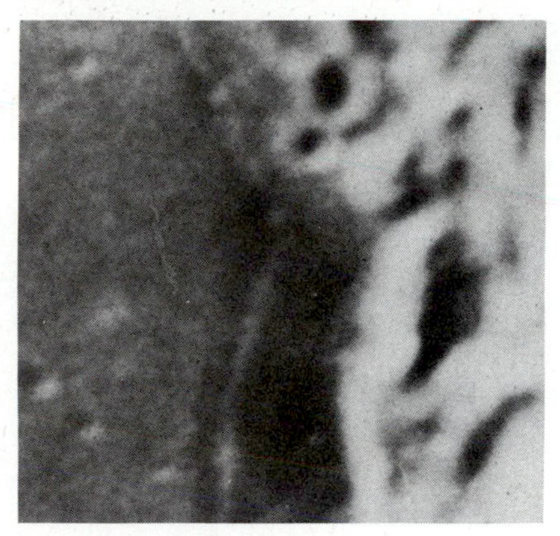

FIGURE 4.4(b)
The same area, photographed from a height of 115 miles above the moon by a camera on the Ranger 9 spacecraft. Surface features as small as 300 feet in diameter may be seen.

Good and Bad "Seeing." A clear, cold night, when the stars are twinkling brightly, is a bad night for the astronomer, who knows that when the stars twinkle most, the image in the telescope is moving rapidly and all photographs will be very blurred. Astronomers refer to the conditions of the atmosphere as the "seeing." If the atmosphere is quiet on a particular night and the stars shine steadily, the "seeing" is said to be good that night. When the stars twinkle, the "seeing" is never very good.

For small telescopes, the blurring produced by the atmosphere is

FIGURE 4.5
Clavius crater; (a) photographed through the 36-inch Lick telescope, (b) photographed through the 200-inch Palomar Mountain telescope.

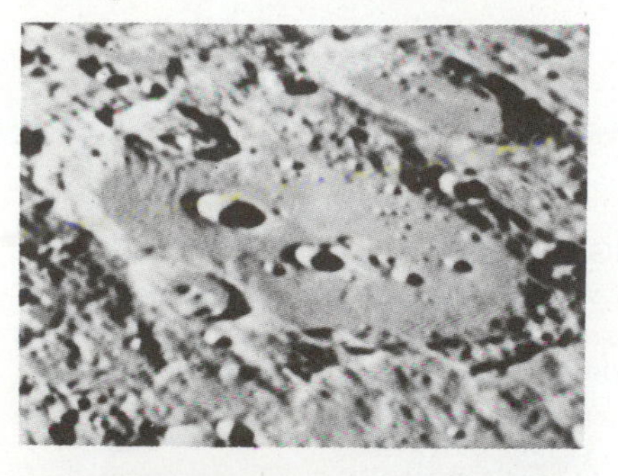

greater than the blurring that results from diffraction. An inexpensive telescope usually has an objective lens with a diameter of about 2 inches. The effect of diffraction on a lens of this size prevents the observer from seeing details on the moon that are less than 10 miles in size. A hole on the moon big enough to swallow up the city of Paris would be invisible through a 2-inch telescope.

For larger lenses or mirrors, the relative effect of the edge is less than for smaller lenses, and the diffraction is correspondingly less important. With a telescope containing a lens 12 inches in diameter, for example, features on the moon as small as 1.5 miles across can be seen. A good 12-inch telescope costs a thousand dollars or more, and is the largest telescope ordinarily used by anyone except a professional astronomer.

Instruments used in astronomical research are both larger and considerably more expensive. The 40-inch telescope at Yerkes Observatory, for example, cost approximately $300,000 (Figure 4.3).

Figure 4.5 shows the region of the crater Clavius on the moon taken with telescopes of two different sizes. The upper photograph was taken through a 36-inch telescope. The somewhat sharper image of the crater Clavius below was taken with the 200-inch telescope.

Atmospheric Blurring vs. Diffraction. The blurring effect of the atmosphere is the same regardless of the size of the telescope. However, the blurring effect due to diffraction decreases as the telescope becomes larger. Figure 4.6 shows the separate effects of the atmosphere and diffraction on the clarity of an image for telescopes of various sizes. The graph shows that the two effects are approximately equal for a 12-inch telescope. Since the two effects add, the sharpness of the image can always be improved by increasing the telescope diameter; however, the improvement becomes less important for instruments considerably larger than 12 inches.

FIGURE 4.6
Atmospheric and diffraction blurring for telescopes of various sizes.

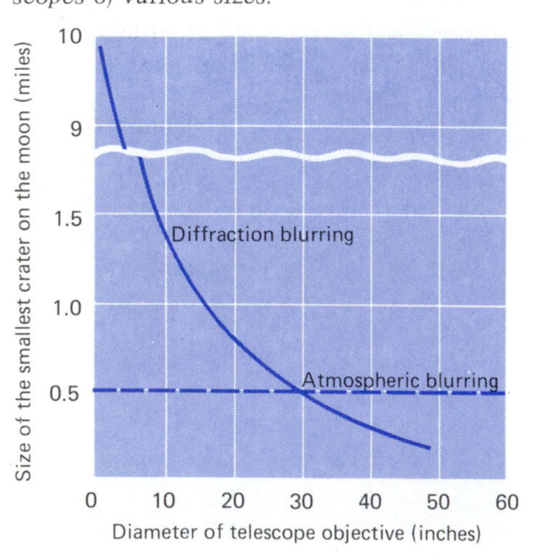

Reflecting Telescopes

Telescopes with larger lenses than about four feet cannot be made because the glass sags under its own weight and spoils the image. Larger and more powerful telescopes have been built, but they are based on a different principle, involving the use of mirrors rather than lenses. This tradition of telescope design dates back to the time of Isaac Newton. When Newton was 23 years old, he had an ingenious idea for making a telescope without going to the trouble of grinding a glass lens. Newton was not the first to have this idea; others had thought of it years before him, but Newton was the first to make it work.

Newton reasoned that the important aspect of a telescope lens was simply its ability to converge rays of light, that is, to bring them to a focus. A concave spherical mirror has the same property, thought Newton; it will reflect rays of light from a distant object in such a way as to bring them together near a single point (Figure 4.7).

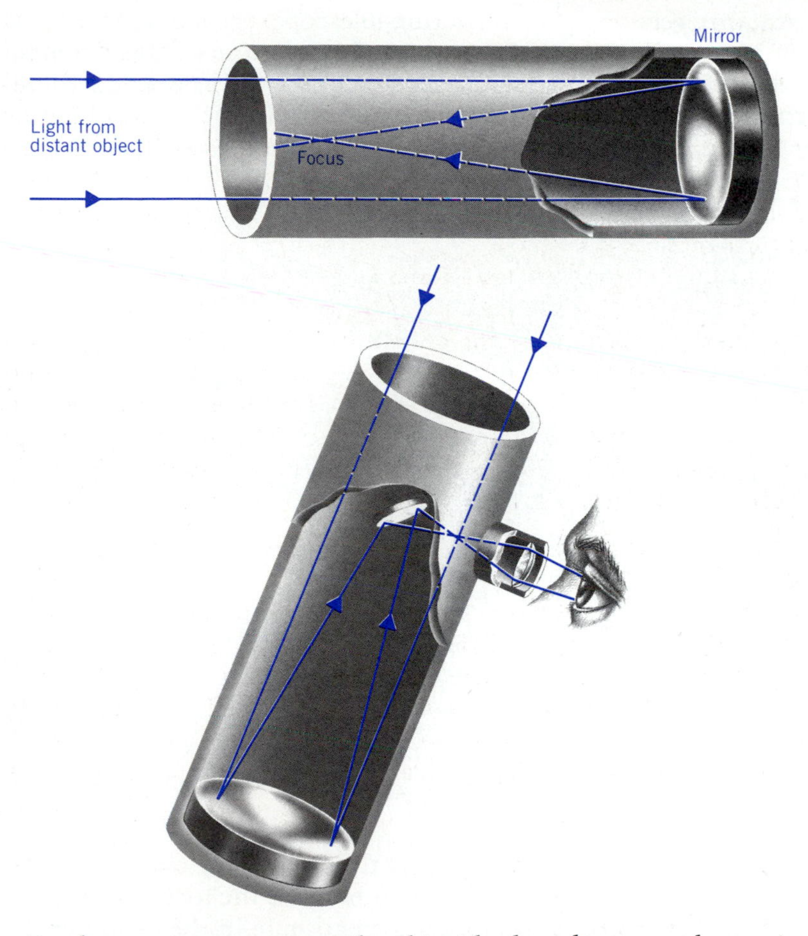

FIGURE 4.7
Focusing of light by a concave mirror.

FIGURE 4.8
Newton's design for a reflecting telescope.

But how can you get your head inside the telescope tube to view the image? Here Newton's ingenuity showed itself. He placed a little mirror at an angle of 45 degrees, in a position such that it intercepted the converging rays just before they came to a point. Figure 4.8 shows where he placed the little mirror.

The mirror intercepted the converging rays of light and reflected them to one side, where they emerged through an opening in the telescope tube. A magnifying lens placed in the opening served as an ocular, permitting the observer to look at the magnified image, as in an ordinary telescope. The large concave mirror at the base of the instrument, that brings the rays to a focus is often called the *primary* mirror. The smaller mirror that intercepts the rays and deflects them to one side is called the *secondary* mirror.

FIGURE 4.9
One of the original reflecting telescopes built by Newton in 1671. The objective was two inches in diameter.

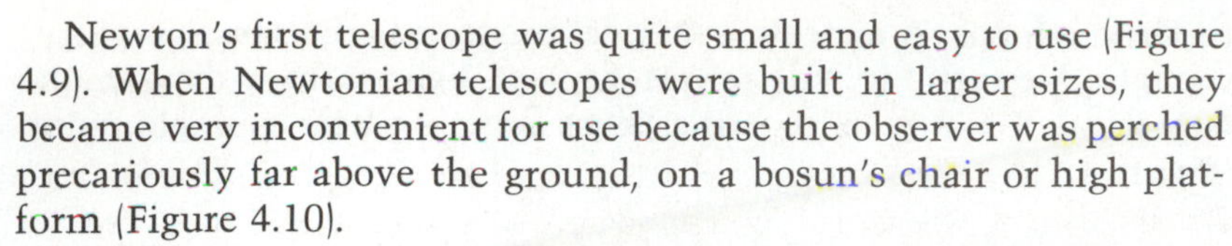

Newton's first telescope was quite small and easy to use (Figure 4.9). When Newtonian telescopes were built in larger sizes, they became very inconvenient for use because the observer was perched precariously far above the ground, on a bosun's chair or high platform (Figure 4.10).

Another version of the reflecting telescope, even more ingenious than Newton's, and much more comfortable to use, was invented by the French astronomer Cassegrain in 1672. He placed a small mirror opposite the main one, so as to reflect the rays of light as they converged back down the telescope tube toward the observer.

FIGURE 4.10
An eighteenth-century telescope built according to Newton's design.

He made the secondary mirror slightly convex; by opening up the angle of the rays to a slight degree, it brought the focus outside the telescope. Finally, Cassegrain cut a small hole in the center of the primary mirror, permitting the converging rays of light to pass through to the ocular and then to the eye (Figure 4.11).

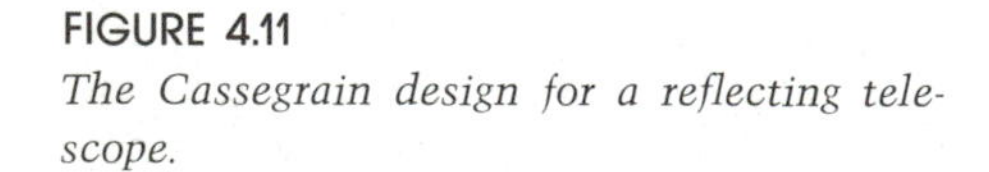

FIGURE 4.11
The Cassegrain design for a reflecting telescope.

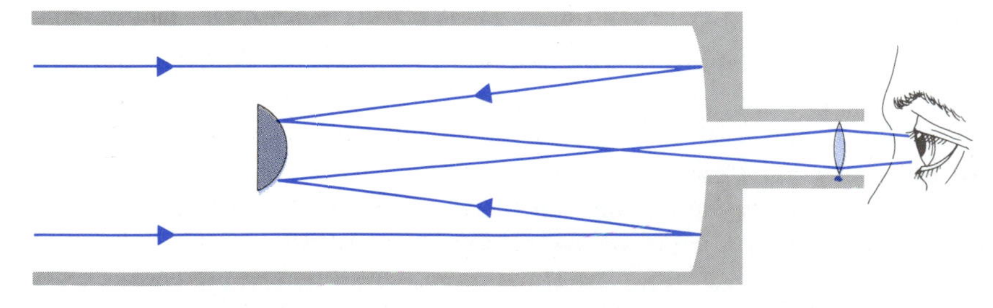

Although Newton criticized Cassegrain's idea in scathing terms, the Cassegrain reflector offered a great advantage over Newton's design because the astronomer could sit at the bottom end of the telescope tube, remaining on the ground in safety while studying the sky. In addition, the combination of a concave and a convex mirror tended to cancel out distortions near the edge of the image, and made a sharper image.

A later version of Cassegrain's telescope combined it with Newton's design by adding a third mirror which deflected the light to the side. This design was even more convenient, because the astronomer could turn his telescope to look at any part of the sky without moving (Figures 4.12 and 4.13).

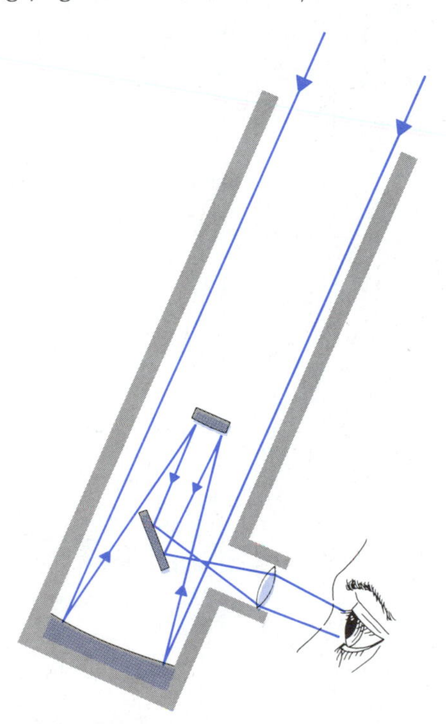

FIGURE 4.12
The combined Cassegrain-Newton design.

FIGURE 4.13
This telescope, built according to the Cassegrain-Newton design, was in use around 1845.

The 200-Inch Telescope

The most famous telescope in the world is the 200-inch telescope on Palomar Mountain. The mirror of this telescope is made of Pyrex glass covered with aluminum. Grinding and polishing of the mirror required eleven years. Five tons of glass were removed during the polishing. The final mirror surface was accurate within one-millionth of an inch.

The 200-inch telescope is one of the marvels of the modern world. It weighs 500 tons but is so delicately mounted that it can be moved by an electric motor a few inches in size.

FIGURE 4.14(a) (opposite)
The 200-inch Hale Telescope.

In the photograph (Figure 4.14*a*) the telescope tube is the open framework of girders angled upward toward the left. The solid structure angled toward the right is the telescope mount, which permits the telescope to be turned in any direction. The path of light rays is shown by arrows (Figure 4.14*b*). The rays are shown converging on the "prime focus." A mirror can be placed at the prime focus to reflect rays down to the bottom of the telescope, as in the Cassegrain design, or a camera can be mounted at the prime focus to record the light directly there.

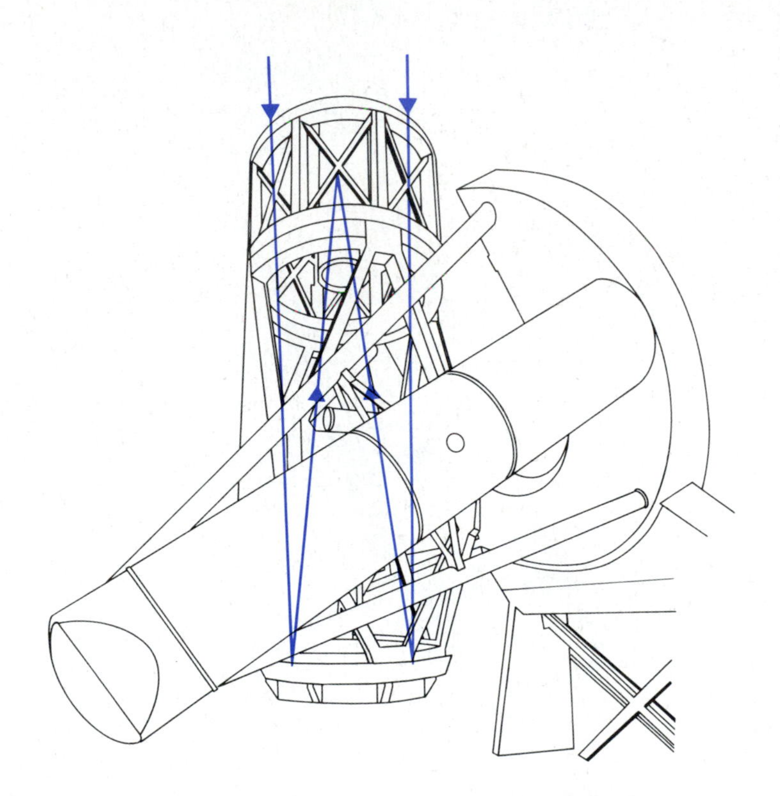

FIGURE 4.14(b)
A diagram showing the path of light rays through the 200-inch telescope. Light from the night sky enters the dome through the wide slit (A) at the top, travels down the open "tube" of (B) to the primary mirror (C), and is reflected back up the tube to the prime focus (D); or it may be deflected down or to the side by a secondary mirror placed in the tube.

The Multiple Mirror Telescope

How can a telescope collect large amounts of light while retaining a sharp image? The Multiple Mirror telescope (Figures 4.15*a* and *b*), on Mount Hopkins in Arizona solves this problem by using six smaller mirrors in place of one large one. An ingenious method permits the mirrors to be coordinated with the extraordinary precision needed to maintain a sharply defined image. The concept of the Multiple Mirror telescope is one of the most innovative ad-

FIGURE 4.15(a)
The six mirrors of the Multiple Mirror telescope on Mount Hopkins.

FIGURE 4.15(b)
The Multiple Mirror telescope located in its 4½-story building on the summit of Mount Hopkins. The entire building turns on a rotating platform.

vances in the science of telescope production since the time of Newton.

Figure 4.16 shows how the Multiple Mirror telescope works. Light rays from a star or galaxy enter each of the six telescope mirrors separately, travel down to the primary, are reflected up to a secondary mirror, and then reflected down again to a third mirror at the center of each primary. Finally, the light rays enter a pyramidal arrangement of mirrors at the center of the entire mount and are brought to a single focus.

The success of the Multiple Mirror telescope depends on the precision with which the six separate images can be combined in a single point. For this purpose, a computer analyzes the images of a guide star and sends instructions to the various secondary mirrors to effect small changes in their tilt angles, until the image of the guide star has achieved maximum sharpness. Further adjustments in the positions of the secondary mirrors are made repeatedly throughout the observing session, to compensate for changes due to temperature variations and shifts in the position of the telescope mount.

The six 72-inch mirrors of the Multiple Mirror telescope are equivalent to a single instrument with a diameter of 176 inches. This effective diameter makes the Multiple Mirror telescope the third largest operating telescope in the world.

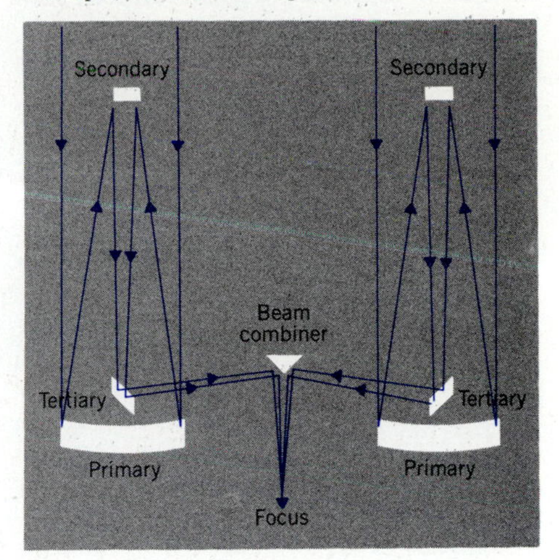

FIGURE 4.16
The paths of light rays passing through a Multiple Mirror telescope.

Other New Telescope Designs

The University of Texas has prepared plans for a telescope with a single 300-inch mirror of very unusual design. In contrast to the mirror of the 200-inch telescope, which is approximately two feet thick, the new mirror will be only four inches thick and therefore will be relatively flexible, like a soft contact lens. Normally this flexibility would lead to distortions that would blur the image and destroy the value of the mirror. However, in the new design the rear surface of the large but very thin mirror is attached firmly to many small pads that can be moved in and out in response to commands from a computer. Movement of the pads changes the shape of the mirror and restores it to its proper form continuously, in spite of distortions produced by changes in temperature as well as the shifting of weight as the telescope moves in its mount.

Astronomers at the University of California have proposed another innovative design for a telescope with light-gathering power equivalent to a 400-inch instrument. In this design, which is a blend of the single mirror and multiple mirror construction, 36 separate 72-inch mirrors of hexagonal shape are joined in a mosaic, like tiles on a bathroom floor. Adjustments to the position of each hexagonal mirror in the mosaic are made continuously by a com-

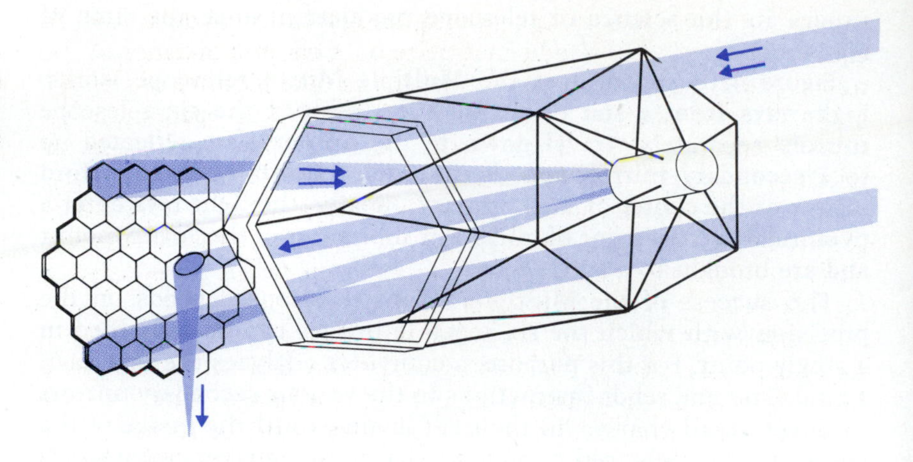

FIGURE 4.17
A possible telescope design based on a mosaic of 36 separate hexagonal mirrors, each ground to a slightly different shape. The telescope has the light-gathering power of a single 400-inch mirror.

puter, so as to bring the beams from the 36 separate mirrors continually to a single focus (Figure 4.17).

The Space Telescope

Unusual telescope designs, such as the multiple mirror, thin lens or mosaic mirror, are designed to increase the range of the instrument by increasing its light-gathering power. Another method for increasing the range of a telescope is to place it in a satellite orbiting above the earth's atmosphere. Because the light from distant objects enters an orbiting telescope without passing through the atmosphere, the image it forms is not blurred by shifting air masses; that is, the "seeing" is always perfect.

The image of an orbiting telescope is still blurred to some degree by diffraction effects, but will be considerably sharper than the image formed by a instrument of comparable size located on the ground. In addition, the orbiting telescope can view the Universe in all wavelengths of the electromagnetic spectrum, and not merely those wavelengths that pass freely through the atmosphere without absorption. These advantages of the orbiting telescope make it one of the most promising innovations in the history of astronomy.

NASA plans to place a large telescope in orbit in the late 1980s. The telescope will have an objective mirror with a diameter of 94 inches and will be carried into orbit in the cargo bay of the Space Shuttle. This instrument, called the Space Telescope, will form images about 10 times sharper than images produced at the focus of the largest telescopes on the ground; its resolution of detail will be equivalent to reading the license plate on a car at a distance of 100 miles. The Space Telescope will be able to take pictures of a planet comparable in quality to those obtained from NASA spacecraft. Pictures of the moon, for example, will be as sharp as in photographs taken by the Apollo astronauts 60 miles above the moon's surface. While the spacecraft can only secure a fleeting series of images as it passes by a planet, a space telescope will be able to map changes in surface conditions and track mammoth storms on the planet Jupiter, for example, over long periods of time.

In addition, the Space Telescope will be able to see farther out in space. The reason for this is again the lack of atmospheric blurring and the increased sharpness of the image, which means that the light from a distant star or galaxy will be focused on a smaller area. Therefore, it will form a brighter image which will stand out better against the background illumination. This background illumination, which tends to fog a photographic plate and swamp a faint image, is a problem for every telescope. In telescopes on the ground, the main sources of plate fogging are the lights of nearby cities and a faint radiation from atoms in the upper atmosphere, called the night air glow. These factors are not present for telescopes in space, but there is still some fogging due to light from the stars in the Milky Way, as well as the "zodiacal light," that is, sunlight scattered from grains of dust in the solar system.

The Space Telescope will also be able to collect and focus infrared and ultraviolet radiation. Most of this radiation is filtered out by the earth's atmosphere and does not reach telescopes on the ground. (Chapter 5)

Scientist astronauts will use the Space Shuttle to inspect the telescope regularly. Every five years they will return the mirror to the earth in the Shuttle for cleaning and recoating. The Space Telescope will be controlled by astronomers in a command post on the ground, who will issue radio commands to point it at astronomical objects of special interest. The radio commands cause small rotating wheels mounted on the satellite to speed up or slow down. If such a wheel is speeded up in one direction, the entire telescope will react by turning in the opposite direction. Through this system the telescope will be able to lock onto a given star, and hold its axis steady for as long as 10 hours at a time, with a pointing accuracy of one part in 20 million.

Since science specialists will be going up in the Shuttle to visit the Space Telescope periodically, it would be possible, in principle, to use photographic film for recording of the images. However, fast-moving particles trapped in the magnetic field of the earth would cause fogging of the photographic film or plates within two months. If a scientist or technician could be ferried up to the Space Telescope on the Shuttle every month, film would be practical, but a monthly Shuttle trip would be prohibitively expensive. Therefore, electronic devices—charged coupled devices (CCDs) and electronic image intensifiers—are used instead. These devices are discussed on pages 105–107.

Because the Space Telescope does not have to wait for the advent of darkness or for clear weather, it will be able to operate for twice as long each year as the best-located telescopes on the ground. The increased range and sharper images formed by this remarkable instrument are expected to contribute to the resolution of such major astronomical problems as the nature of quasars, the size of the Universe, and the presence of planets around other stars.

ELECTRONIC IMAGING DEVICES

New developments in the design of telescopes have been accompanied by vast improvements in the sensitivity of the devices used for recording the image. The eye is rarely used by astronomers for this purpose because it cannot store up light received over a long period of time. Until recently, photographic plates were usually employed instead, because they afford the possibility of long-time exposures, capturing images of faint, distant objects that the eye cannot detect. Astronomers are particularly interested in such faint objects because among them are exceedingly distant galaxies billions of light-years away. The light from these distant galaxies must have started out on its trip to the earth billions of years ago; thus, a study of the images and spectra formed from this light can give us a picture of the Universe at an earlier stage of its existence, answering questions of great interest related to the origin of the Cosmos.

Suppose a galaxy is very distant and faint, and even a full night's exposure does not suffice to record any information about it. This will happen if the light from the galaxy is so weak that less than 1000 photons from it are collected by the telescope during the course of the night's exposure. The reason for that limitation lies in the photographic plate itself; a minimum number of photons—approximately 1000—must strike a grain in the emulsion in order to blacken it. In principle the difficulty could be overcome by constructing a telescope so large that its mirror collects at least a thousand photons from the faint galaxy in one night, but that solution to the problem would be extremely expensive. Alternatively, one of the larger telescopes currently in operation could be dedicated to the study of that single galaxy for a long period of time. During this period the same plate would be exposed to the faint rays from the galaxy night after night. In practice, large telescopes are hardly ever used in this way, because only a few such instruments exist in the world and many astronomical investigations of great importance compete for time on them.

Image Intensifiers

An ingenious electronic device called the image intensifier, or image tube, provides astronomy with a more practical way of gathering more light from faint objects. The image intensifier is placed at the focus of the telescope and the photographic plate is mounted behind it. One out of roughly every 10 photons that enter the front of the image tube triggers a cascade of 1000 photons that leave the rear of the tube and strike the photographic plate. Thus the time required for a photographic exposure is reduced by a factor of 100. An image tube, mounted at the focus of the 200-inch telescope, converts it into the equivalent of a 2000-inch instrument in light-gathering power.[1]

The image intensifier is shaped like a tube or cylinder, usually a few inches in diameter and six or eight inches long. The front face of the tube is constructed like the surface of a light meter or a photoelectric cell, with a thin layer of material that gives off electrons when light falls on it. Light from the telescope strikes this layer, ejecting electrons from it. Because the layer is exceedingly thin, the electrons penetrate through it and emerge from the other side, entering the interior of the tube.

The other end of the tube contains a metallic foil charged with positive electricity. The positive charge exerts a strong force of attraction on the negatively charged electrons, pulling them down the tube with increasing speed. When the electrons reach the far end of the tube, they are traveling many times faster than they were when they started out.

A phosphor layer, similar to a TV screen, is located immediately behind the foil. The electrons strike the phosphor layer. Each electron, on hitting the phosphor, creates a flash of light, just as in a TV set. The light flash contains approximately one thousand photons. In this way, working through the intermediate stage of the fast-moving electron, the image tube converts the single photon that enters into a burst of many photons that leave. A photographic plate is located behind the phosphor. The burst of photons from the phosphor falls on the plate, blackening a grain in the emulsion and recording the image.

[1] The light-gathering power of a mirror is proportional to the area of the mirror, πr^2, where r is the radius of the mirror. A tenfold increase in radius produces a 100-fold increase in area.

Charge Coupled Devices

An even more recently developed type of image intensifier called the charge coupled device, or CCD, resembles an oversized silicon circuit chip. This device is a thin wafer of silicon whose surface is divided electrically into the squares of a grid. The light striking any particular square on the grid generates an electric charge proportional to the intensity of the light at that point. A mosaic pattern of electric charges builds up across the surface of the grid. This mosaic provides the image of the object (Figure 4.18). Thus, instead of the image being recorded by blackened grains of a silver compound in the photographic emulsion, it is recorded as a two-dimensional array of electric charges.

One advantage of this device lies in the fact that the patterns of electricity can be read directly into a computer memory. Thus, the results of many successive exposures can be stored on a computer for later analysis. Another advantage is the fact that the CCD is approximately 50 times more sensitive than the eye or a photographic plate. Still another and very important advantage relates to the fact that if a photographic plate is exposed to a distant galaxy,

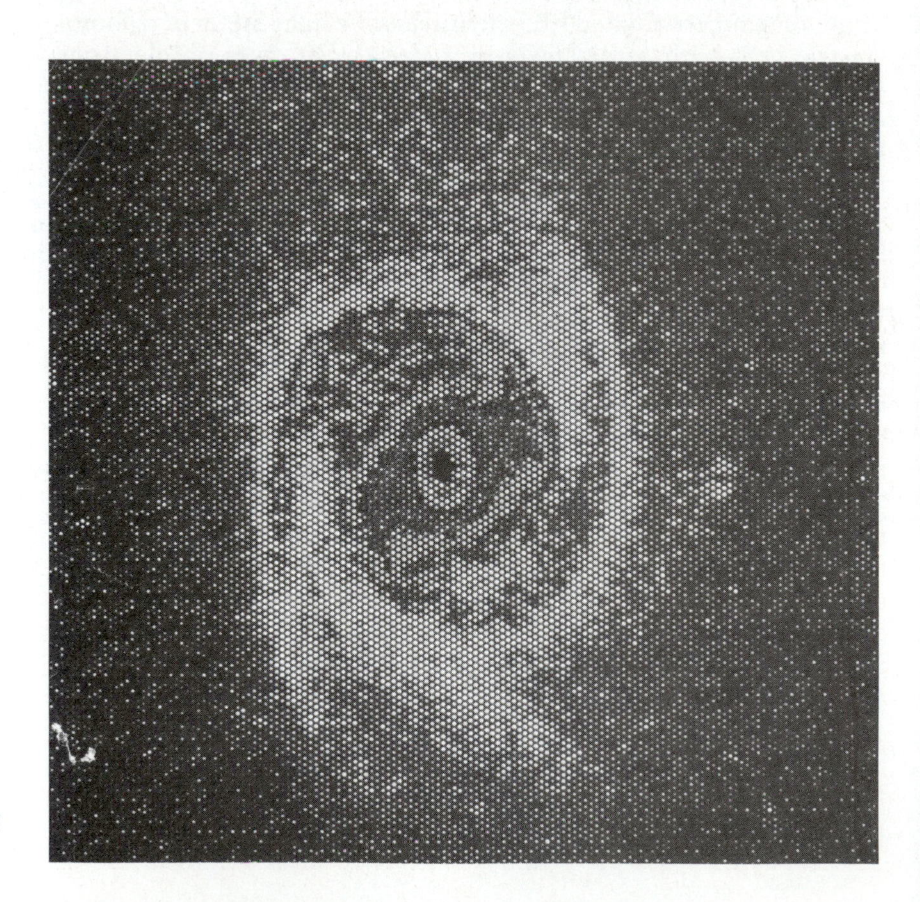

FIGURE 4.18
An image of a barred spiral galaxy formed with a CCD, showing the mosaic pattern.

for example, long enough to capture the faint outer regions of the galaxy, the details in the bright inner regions will be overexposed and details in it will be lost. CCDs, however, are not readily "overexposed." They can accurately record details in different areas of a galaxy over a much larger range of intensities than either the eye or a photographic plate.

THE SPECTROSCOPE

We will see in the next chapter that astronomers acquire much of their information by breaking up the light from a star or a galaxy into its separate wavelengths and measuring the intensity of light at each wavelength. Every element in a star radiates a characteristic pattern of wavelengths. By separating the light into its individual wavelengths, we can detect these patterns and determine which elements are in the star, how much of each the star contains, the temperature of the star, how fast it is moving through space, and a great deal of other important information about it.

The eye cannot be used for this purpose because it is unable to resolve a ray of light into its individual wavelengths. To separate the wavelengths in a beam of light, the eye must be aided by an instrument called the *spectroscope.* Spectroscopes separate the various wavelengths in a beam of light from one another so that each wavelength can be examined individually. If the separate wavelengths are photographed instead of being examined by eye, the instrument is called a *spectrograph.*

FIGURE 4.19
The bending of a ray of light as it enters glass.

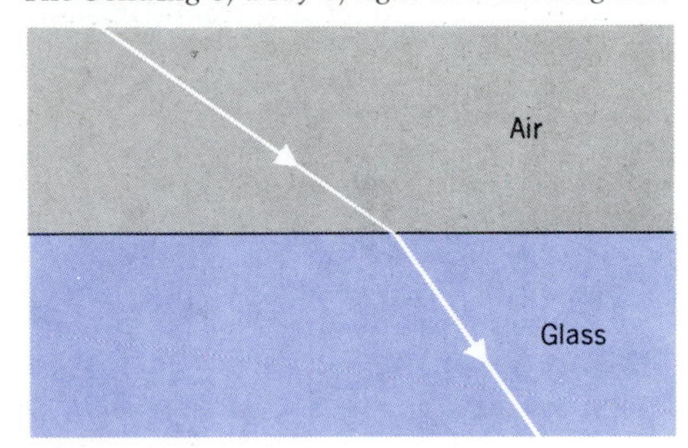

Prism Spectroscopes

The oldest type of spectroscope makes use of a property of light traveling through glass. As we point out in our explanation of the telescope, a ray of light is bent as it enters a piece of glass because light travels more slowly in glass than in air (Figure 4.19).

The speed of light in glass depends on its wavelengths; the speed of blue light in ordinary glass is approximately 121,000 miles per second, whereas the speed of red light is 122,000 miles per second. Because of the difference in speeds, blue light is bent by a larger angle than red light on entering a piece of glass obliquely.

When a ray of light composed by many wavelengths enters a piece of glass, the light breaks up into a number of different beams, each beam consisting of one wavelength (Figure 4.20). The beams travel through the glass in different directions, according to their wavelengths. In other words, the glass separates the wavelengths in the ray of light. When the light emerges from the glass into the

FIGURE 4.20
The separation of a ray of light into component colors as it enters the glass.

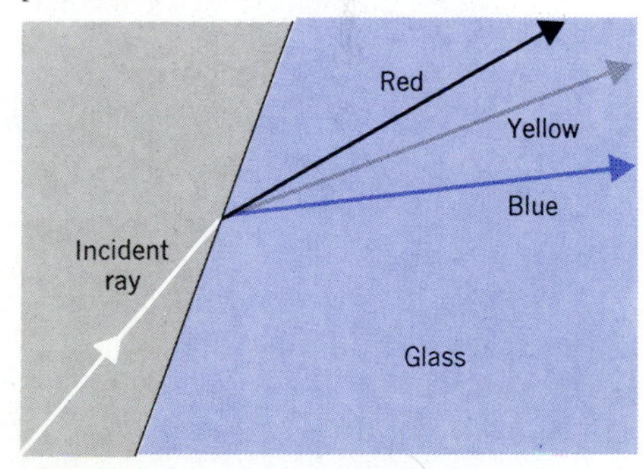

air again, it should be possible for the eye to see each of these wavelengths.

If the glass is cut in the shape of a prism, the rays will be bent twice—once when they enter the prism, and a second time when they leave it. On entering and leaving the prism, they will bend by angles that depend on their wavelengths. Each time, the blue wavelengths will bend more than the red. Thus, using a glass in the shape of a prism roughly doubles the spread between the blue and the red wavelengths in the beam. This is the principle of the prism spectroscope or spectrograph. If the ray of light contains two wavelengths that are very close together, they will still overlap after passing through the prism. In order to separate close-lying wavelengths, the light is usually passed through a narrow slit before it enters the prism. After leaving the prism, the light will consist of several narrow beams, each beam being an image of the slit in the one color or wavelength characteristic of the substance emitting the light. The eye or the camera sees the images of the slit as narrow lines of light. These are called *spectral lines* (Color Plate 1).

As noted above, the set of spectral lines makes up the *spectrum* of the light that is being analyzed. We will see that each element's atoms radiate their own characteristic pattern of wavelengths. These wavelengths, which pass through a spectroscope and are analyzed into a line spectrum, are a means of identifying the presence of that element. If the light from a star contains the spectral lines of a familiar chemical element, we know that the element exists in the star.

Figure 4.21 shows the components of a complete spectroscope. The gas discharge tube shown at the left is a typical light source.

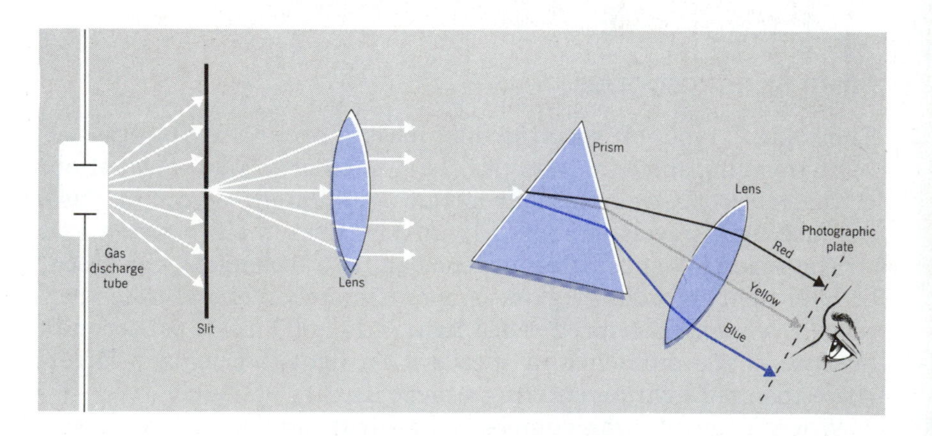

FIGURE 4.21
The principle of the prism spectroscope.

The end of the spectroscope facing the light source is sealed off from light except for a narrow slit, which admits a small amount of light from the source. The light from the slit passes through the prism after first passing through a lens placed between the slit and the prism. The lens makes the beams diverging from the slit run parallel. The parallel beam enters the prism and emerges from the other side as a number of distinct beams, one for every wavelength in the original light from the gas discharge tube. The beams travel in different directions according to their wavelengths. The light then passes through a second lens that focuses its components into a number of images of the slit. There is one slit image or spectral line for each wavelength in the original light.

A device that records a photographic image of a spectrum is called a spectroscope. Figure 4.22 shows a prism spectrograph attached to a telescope.

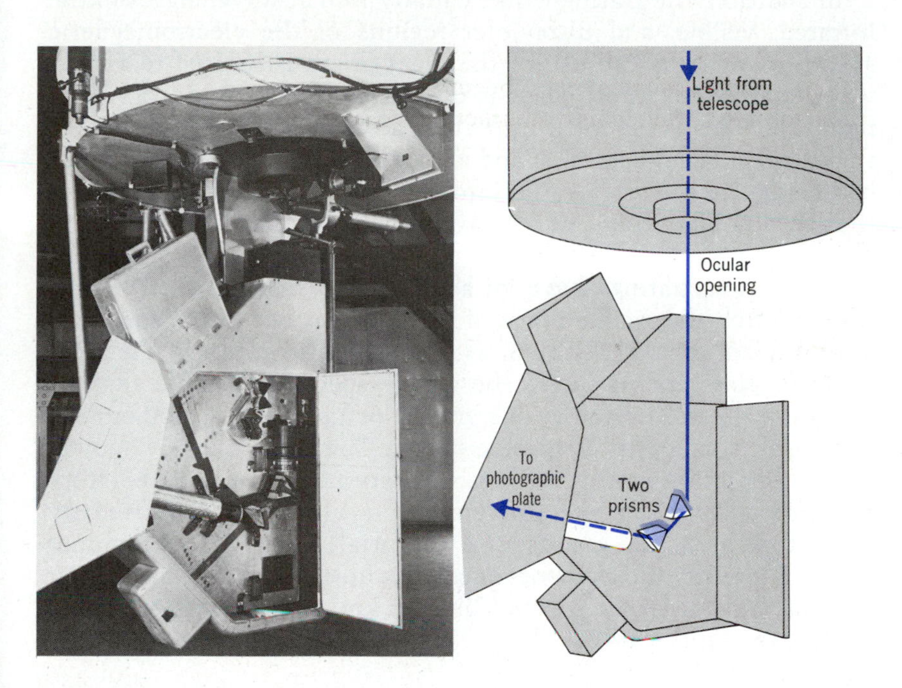

FIGURE 4.22
A prism spectrograph attached to a large reflecting telescope. The spectrograph contains two prisms arranged in tandem to increase the separation of the various wavelengths in the light beam. The diagram at right shows the path of the ray of light from the ocular opening through the two prisms to the photographic plate on which the spectrum will be recorded. (The ocular is removed when the spectrograph is attached, because its function is taken over by one of the lenses in the spectrograph.)

The Importance of the Spectrograph

A big telescope is a very impressive object, and even a small telescope has great appeal, because one can actually look through its eyepiece and see for oneself many of the interesting objects in the heavens. The spectroscope, on the other hand, is no more than an adjunct to the main instrument, mounted on the end of the telescope as a minor attachment. Although it is capable of producing a pretty pattern of colors, normally one does not look through it, but instead a photographic film or plate is placed at the exit end of the instrument to record the distribution of wavelengths as an uninteresting pattern of gray and black lines. Yet if this seemingly modest instrument is removed, the value of the telescope is enormously reduced; the telescope captures the appearance of celestial bodies, but the spectrograph analyzes their nature.

The Grating Spectrograph

The glass prism is not the only device that can spread light into a spectrum. A device called a diffraction grating can also be used. Astronomers prefer the diffraction grating because it spreads light into a broader spectrum than the prism and can, therefore, separate very closely spaced lines.

In addition, the grating works equally well at wavelengths in the infrared, visible, and ultraviolet regions of the electromagnetic spectrum whereas the use of the glass prism is limited to visible wavelengths only, since glass is opaque to infrared and ultraviolet radiation. This is the most important advantage of the grating over the prism, since the infrared and ultraviolet regions of the spectrum have supplied much valuable information about stars and galaxies that astronomers could never have gathered from visible radiation alone.

Diffraction gratings have an additional advantage over prisms. They stretch the visible range of wavelengths out into a spectrum spanning an angle of as much as 60 degrees. In contrast, the most effective glass prisms spread the visible spectrum over an angle of only 20 degrees. The ability of a grating or a prism to spread out the spectrum is called its dispersive power. The high dispersive power of the diffraction grating enables the astronomer or the laboratory spectroscopist to separate different spectral lines that lie extremely close together. This accuracy is essential when the astronomer seeks to identify the elements occurring in stars by comparing their spectral lines with the spectral lines of known elements measured in the laboratory.

Main Ideas

1. How the telescope works; how a lens forms an image.
2. Refracting and reflecting telescopes.
3. Types of reflecting telescopes.
4. Light-gathering power, resolution, magnification.
5. Atmospheric blurring vs. diffraction.
6. New telescope designs.
7. Advantages of the Space Telescope.
8. Image intensifiers.
9. Charge coupled devices.
10. Prism and grating spectroscopes.

Important Terms

Atmospheric blurring
Cassegrain reflector
charge coupled devices
concave mirror
convex mirror
diffraction
diffraction grating
focus
grating spectrograph
image intensifier
lens
multiple mirror telescope
Newtonian reflector
objective lens
ocular lens
plate (photographic)
primary mirror
prism spectrograph
reflecting telescope
refracting telescope
refraction
secondary mirror
seeing
space telescope
spectral lines
spectrum
wavelength

Questions

1. An observatory has two telescopes, one twice the diameter of the other. How much more light does the larger telescope gather compared with the smaller? How does the resolution of the two telescopes compare?
2. Discuss the relative advantages and disadvantages of refracting and reflecting telescopes.
3. A two-inch telescope can distinguish features on the moon larger than 10 miles in size. What is the smallest sunspot that can be seen with this instrument?
4. Consider two telescopes. The first is a conventional reflecting telescope whose primary mirror is 50 inches in diameter. The second is a multiple mirror telescope consisting of six 6-inch mirrors separated by 50 inches. Compare the resolution and light-gathering power of these telescopes. (The surface area of a circular mirror is approximately πR^2.)
5. What factors blur the sharpness of the image produced by a telescope? How can you reduce each of them?
6. Suppose an advertisement offers a telescope for sale with a magnification of 400. What important information should you seek before you buy the telescope? Explain your answer.
7. What factor limits the size of both refracting and reflecting telescopes? How do the new telescope designs circumvent this difficulty?
8. Explain the advantages that a telescope in space has over one on the surface of the earth.
9. Briefly describe three advantages that image intensifiers and charge coupled devices have over photographic emulsions in recording astronomical images.

NEW WINDOWS ON THE HEAVENS

5

Stars and galaxies give off many kinds of radiation in addition to visible light. Discoveries made in the twentieth century reveal that they also emit X-rays, gamma rays, ultraviolet rays, infrared radiation, and radio waves. Of these various kinds of radiation, only two —visible light and some types of radio waves—pass freely through the earth's atmosphere. The other types of radiation are either largely absorbed by the atmosphere or do not reach the ground at all. Astronomers describe this situation by saying that the astronomer has two "windows"—one in the visible region and the other in the radio region.

Until 1931, astronomical observations were confined to the visible window, but in that year an electrical engineer named Karl Jansky stumbled across a phenomenon that led to the discovery of the radio window, and to the birth of radio astronomy. Jansky was not thinking about astronomy at the time, but trying to solve a practical problem in telephone communications. Telephone conversations between London and New York were frequently interrupted by static and hissing noises, and Jansky had the assignment of finding out where the static came from. After he eliminated obvious causes of static, such as thunderstorms, one type of noise remained. This was a hissing sound that did not seem to be connected with earthly phenomenon. Probing into the nature of the mysterious hiss, he found that it came from a fixed direction in space, which later turned out to be the center of our galaxy. Jansky had stumbled on radio waves from space. His accidental discovery opened up the new field of radio astronomy.

The 210-foot radio telescope at Goobang Valley in Australia.

With the advent of the space age, astronomers have been able to open still more windows on the heavens by placing their instruments on satellites orbiting far above the absorbing effect of the earth's atmosphere. The new branches of astronomy created by satellites involve observations of infrared rays, ultraviolet rays, X-rays, and gamma rays. These recent developments have accelerated the pace of astronomical discovery to the point where astronomy is now one of the most rapidly advancing fields in modern science.

ELECTROMAGNETIC WAVES

Radio waves, X-rays, ultraviolet rays, and all other kinds of radiation collected by astronomers are manifestations of a single phenomenon. They are all trains of electric and magnetic vibrations traveling through space. These trains of vibrations or waves are created by the movements of electrically charged particles in the star or galaxy. When an electric charge is jiggled, a train of electric vibrations spreads out into the space around it, just as a pebble dropped into a pond will create a train of waves that spread across the water. According to the laws of physics, when such a train of electric waves is produced, it is accompanied by a train of magnetic waves that move with it through space at the same speed. The two kinds of waves are tied together inseparably; one cannot exist without the other. Together they are called an *electromagnetic wave.*

Light is one kind of electromagnetic wave. Gamma rays, X-rays, ultraviolet rays, infrared rays, radar, television signals, and radio waves are other types of electromagnetic waves. They differ from one another only in their wavelengths. In the list above, gamma rays have the shortest wavelengths—about a billionth of an inch or less—and radio waves have the longest—from hundreds of feet to miles. Together with light, these radiations of different wavelengths form the *electromagnetic spectrum.*

The first clue to the fact that light is an electromagnetic wave came from the work of an English physicist named J. C. Maxwell. Maxwell calculated the speed with which electromagnetic waves travel through space and found that they move at the extraordinary velocity of 186,000 miles per second. Since light was the only phenomenon known to travel at that speed, Maxwell concluded that light was a train of electromagnetic waves. Later, physicists realized that heat, or infrared radiation, was another kind of electromagnetic wave. The newly discovered X-rays also turned out to be electromagnetic waves. All these waves, which make up the electromagnetic spectrum, travel through space at the same speed, and differ only in their wavelengths.

The Electromagnetic Spectrum

Visible light is that particular part of the electromagnetic spectrum to which the retina of the human eye happens to be sensitive. The cells of the retina respond to electromagnetic waves with frequencies between 4.3×10^{14} vibrations per second (usually written as cycles per second and abbreviated cps) and 7.5×10^{14} cps. Hence this band of frequencies is called the *visible* region of the electromagnetic spectrum. When electromagnetic waves with a frequency of 7.5×10^{14} cps strike the retina of the eye, the signal sent to the brain registers this light as "blue-violet" in color. If electromagnetic waves with a somewhat lower frequency strike the retina as, for example, waves whose frequency is 6.5×10^{14} cps, the eye signals to the brain that "blue" light has arrived. If the frequency of the light striking the retina decreases further, the eye sends to the brain, in succession, signals indicating the sensation of "blue-green," "green," "yellow," "orange," and "red" light. Light whose frequency is 4.2×10^{14} cps provides the sensation of a "deep red" color (Color Plate 1).

If the electromagnetic waves have a frequency higher than 7.5×10^{14} cps, the eye does not respond to them. Such waves, lying beyond the violet edge of the spectrum, are called *ultraviolet* light. If the waves have a frequency lower than 4.3×10^{14} cps, the eye again does not respond to them. These waves, whose frequency is lower than the lowest frequency of visible light at the red end of the spectrum, are called *infrared light* or *infrared radiation.*

The human eye has evolved in response to the need for seeing objects on the earth's surface with the aid of sunlight. If it were sensitive to frequencies far outside the visible region, that is, far outside the band of electromagnetic radiation that gets through the atmosphere, there would be little difference between night and day, and your eyes would be of relatively little value to you. On another planet, with a different atmosphere that is, perhaps, strongly absorbing in what we call the visible region, but transparent in the infrared region, evolution might generate creatures with infrared-sensitive eyes; or, on still another planet, there might be creatures with ultraviolet-sensitive eyes, depending on the electromagnetic wavelength for which the atmosphere of the planet is transparent.

Electromagnetic waves are often described in terms of their wavelength rather than their frequency. An example will demonstrate the conversion from wavelength to frequency. Suppose an electron vibrates one million times per second, creating an electromagnetic wave with a frequency of 10^6 cps. That is, in one second a million waves are sent out. Since the velocity of light is 186,000 miles per second, this train of one million waves has moved 186,000 miles in that second. Clearly the length of each wave is 186,000/1,000,000 = 0.186 miles = 1230 feet.

This is an example of a general formula connecting frequency and wavelength:

$$\text{wavelength} = \frac{\text{speed of wave}}{\text{frequency}}$$

The formula holds for any type of wave and is not restricted to light waves.

Figure 5.1 shows the complete electromagnetic spectrum with frequencies and wavelengths indicated. The names given to the various parts of the spectrum also are shown. Wavelengths in Figure 5.1 are given in centimeters. Other units of wavelengths for electromagnetic radiation in common usage include: one angstrom (Å) = 10^{-8} centimeters (cm); one micron (μ) or micrometer (μm) = 10^{-6} meters (m). The wavelengths of the visible light is approximately 5000 angstroms or 0.5 microns.

Column 2 of the figure indicates how transparent the earth's atmosphere is to electromagnetic radiation of various wavelengths. This part of the diagram indicates the types of electromagnetic waves that are useful as tools in astronomy. It shows that the atmosphere blocks radiation from outside the earth everywhere except in the visible region, a part of the infrared, and a part of the radio region. These bands of electromagnetic radiation are three windows of the atmosphere through which earthbound astronomers can look out at the Universe.

The first window, in the radio region, covers the band of wavelengths extending from a few millimeters up to 100 meters. This is a true window, with very little atmospheric absorption of radio waves over that large range of wavelengths. The second window, in the visible region, has been the source of most astronomical knowledge until recent times. The third astronomical window, in the infrared region, extends from the long-wave end of the visible spectrum in the deep red, at a wavelength of approximately 7000 angstroms, to approximately one millimeter, on the border of the radio region. The infrared window, unlike the radio window, is murky or opaque in many places, but a few relatively transparent bands have been found in this region, and they have already yielded important information that will be discussed in later chapters.

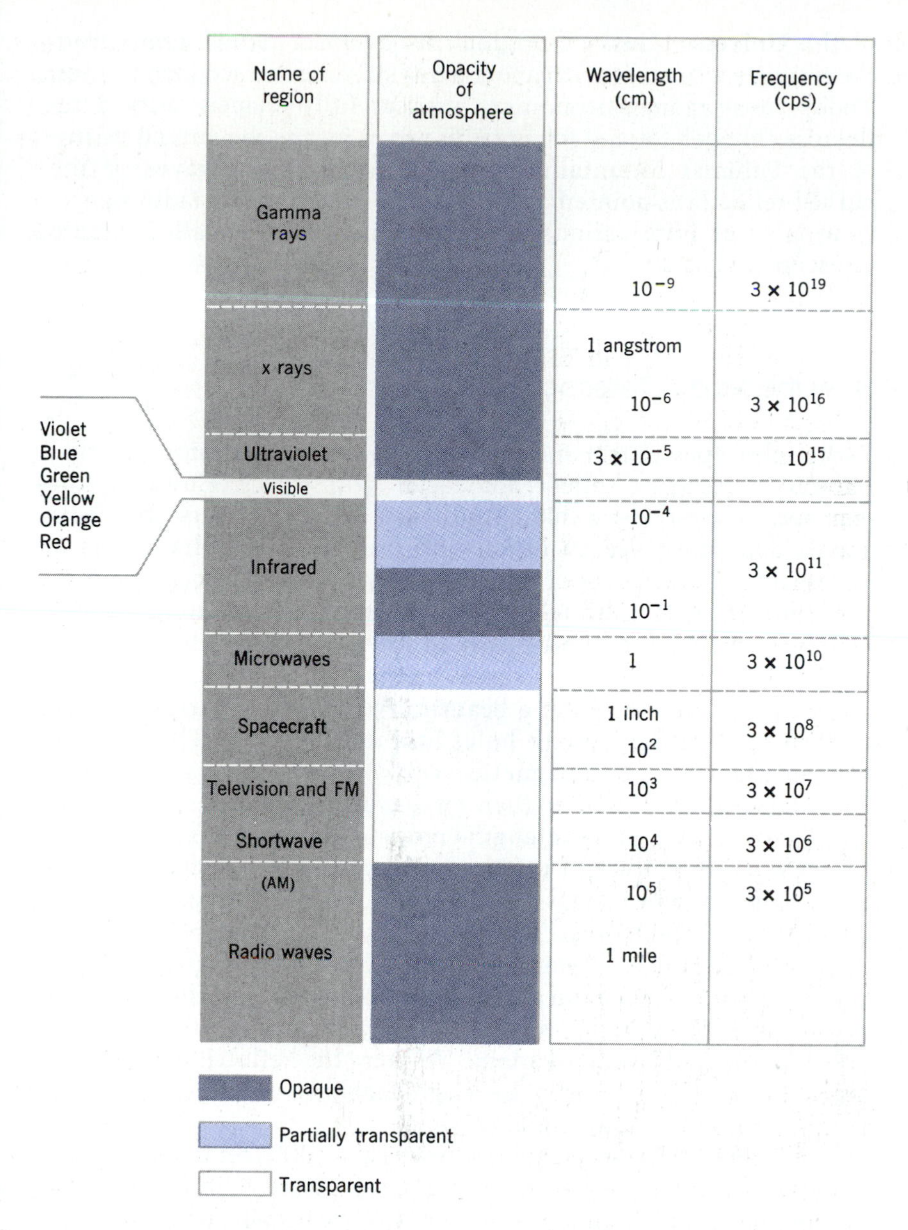

FIGURE 5.1
The electromagnetic spectrum.

RADIO TELESCOPES

The birth of radio astronomy was described on page 113. Although radio telescopes came into existence only fifty years ago, they have already added a large body of information to our knowledge of the Universe. Jansky's original discovery of radio signals from the heavens was made with a few wires stretched on wooden frames. Today most radio astronomers use carefully shaped, curved reflectors that collect and focus radio waves just as the curved mirror of the Palomar Mountain telescope focuses light waves. If the curved reflector is pointed at the sky and used to study radio signals coming to us from various directions in space, it is called a radio telescope.

How the Radio Telescope Works

Radio telescopes collect and focus radio waves in the same way that ordinary telescopes collect and focus light waves. However, the concave "mirror" in a radio telescope which collects the radio waves does not have to be smooth and brightly polished, as is necessary for the mirror of a high-quality optical telescope. In fact, the "mirror" of a radio telescope can have holes in it, or even be constructed out of wire mesh—as in the Australian radio telescope shown opposite page 113—instead of solid glass or metal, and yet work as well as if it were a beautifully finished spherical mirror.

How can a "mirror" with holes in it reflect waves? The explanation is that an electromagnetic wave cannot "feel" any hole or irregularity that is smaller than its wavelength in size. Most radio astronomy is done at wavelengths ranging from about one-half yard to a few yards. For use at these wavelengths, the reflecting surface need only be "true" to a spherical or parabolic surface to within a fraction of an inch or so. The surface itself can resemble a gravelly or pebbly pavement, or even have holes, but as long as the bumps or the holes are no more than a fraction of an inch in size, the telescope mirror will reflect radio waves just as well as if it were a finely polished, smooth piece of metal. In fact, the reflecting surface of a radio telescope is usually so rough that it is not called a mirror at all. Instead, it is called a dish.

The largest radio telescope in the world—1000 feet in diameter—is located in Puerto Rico, in the hills near Arecibo on the western end of the island. Figure 5.2 shows this great dish, which is constructed within a natural bowl in the hills. The telescope "looks" in different directions by changing the location of the radio receiver at its focus. The direction in which the telescope points also varies during the course of the day as the earth rotates. The combined variation is sufficiently great for the instrument to be "pointed" at all the planets in the solar system as well as a number of galaxies and nebulas.

FIGURE 5.2 (opposite)
The thousand-foot dish in Arecibo, Puerto Rico. The dish was made by placing wire mesh over a natural bowl in the mountains. The bowl is oriented vertically upward and is immovable, but receives signals from many directions as a result of the earth's rotation. Additional flexibility in direction is obtained by shifting the antenna, which is mounted on a trolley suspended from cables 500 feet above the valley floor.

Comparison with Optical Telescopes

The radio telescope differs from the optical telescope in three important respects. First, it uses an antenna plus a radio receiver as a receiving device, instead of the eye or a photographic film. It would do no good to "look" through a radio telescope, because the eye is completely insensitive to radio waves. Photographic film is also insensitive to radio waves; hence, you cannot use a camera to record the "images" that a radio telescope focuses. The only device that reacts sensitively to radio waves is an antenna—an electrically conducting wire along which electrons can move freely. To understand how the antenna works, remember that a radio wave is a kind of electromagnetic wave. As the train of electric waves moves past the antenna, it exerts a changing electric force on the electrons in the antenna. The electric force reverses its direction repeatedly as the waves move by. In response to this continually reversing force, the electrons surge up and down the length of the antenna. The antenna is connected to a radio receiver, which detects the tiny electrical changes produced by the surging electron currents in the antenna and magnifies their strength a million times or more so that they can be easily measured and recorded (Figure 5.3*a*).

The receiver is connected to a recording device, which may be a pen moving across a slowly revolving drum of paper. As the radio telescope scans the sky, it may happen to point in the direction of a strong source of radio waves such as a radio galaxy, which is the equivalent of a bright object for an optical telescope. The radio receiver will detect a sharp increase in the strength of the signal, and the pen will record a peak on the drum of paper. An example is shown in Figure 5.3*b*, which is a drawing of a recording made by a radio telescope pointed at a galaxy.

If the signal from the radio receiver were connected to a loudspeaker instead of to a recording pen, you would hear the distant

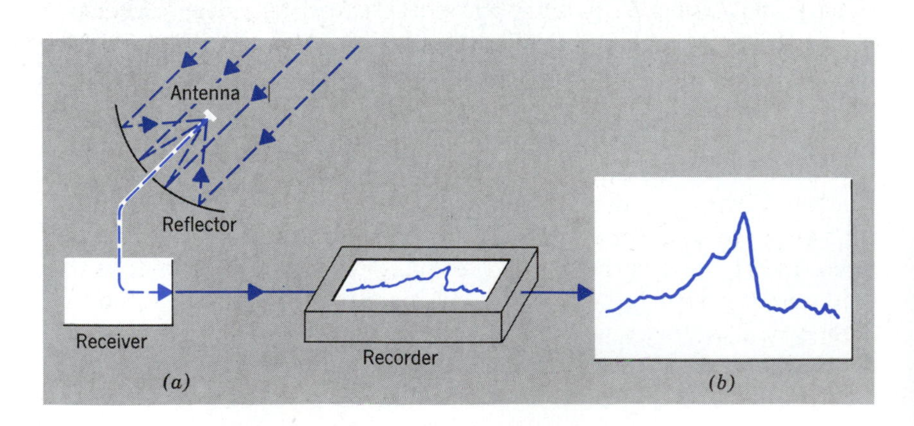

FIGURE 5.3(a)
Schematic diagram of a radio telescope.

FIGURE 5.3(b)
Recording of a signal from a radio galaxy.

galaxy as a crescendo of static emerging out of the general background of hissing noises, as the axis of the telescope swept past the direction of the radio galaxy. The effect would be similar to that produced on your automobile radio when you drive under a high-voltage wire.

Signals from Intelligent Beings

What is the origin of the radio signals coming to use from outer space? The term "radio signal" seems to mean a broadcast, or a Morse code sent out by intelligent beings. However, the signals radio astronomers record are not of this kind. In fact, no astronomer has yet received a signal that clearly indicates the presence of intelligent life in space. The radio signals picked up from stars and galaxies resemble a steady noise, or static. They are rapidly fluctuating radio waves, containing all frequencies jumbled together in a chaotic fashion, and they can be explained by natural causes, without resorting to a theory of radio transmitters constructed by intelligent beings.

Although radio signals received from space show no obvious signs of having been produced by intelligent life, many people have wondered whether a few messages produced by intelligent beings might be mixed into the static of natural radio waves. Radio astronomers in the United States and the Soviet Union have made several attempts to detect signals from intelligent beings, but thus far no successes have been reported. (See Chapter 20.)

Resolving Power: The Sharpness of a Radio Image

The ability of a telescope to produce sharp images is called its resolving power. The resolving power of a telescope usually is limited by the diffraction effect described in Chapter 4. According to the theory of that effect, the blurring depends on the wavelength of the light or other electromagnetic radiation entering the telescope; for a telescope of a given size, the longer the wavelength, the greater the degree of blurring. Light waves have wavelengths in the neighborhood of 5,000 angstroms or 1/20,000 of an inch. The wavelengths of the waves used in radio astronomy usually are, however, in the neighborhood of a few inches to a foot. This means that the blurring and loss of detail in an image obtained with a radio telescope will be many thousands of times greater than that obtained with a normal telescope focusing light waves.

The diameter of the dish of a radio telescope is generally much larger than that of even the largest optical telescope. Since, as was explained in Chapter 4, the blurring effect of diffraction becomes smaller as the size of the mirror or dish increases, this fact compensates to some degree for the blurring due to the longer wavelength of radio waves. Even so, the resolution of radio telescopes tends to be poorer than that of optical telescopes.[1]

Arrays of Radio Telescopes

The relatively poor definition of detail in images[2] formed with radio telescopes has limited the attempts of astronomers to exploit the full potential of an important new area of astronomy. In recent years several ingenious methods have been developed for circumventing this limitation without building an extremely large telescope, whose cost would be prohibitive.

All the methods depend on a simple idea. Imagine a radio telescope with a dish a thousand miles in diameter. This dish will, of course, produce a very sharp image because of its large diameter and correspondingly small diffraction effect. Now suppose that in your imagination you divide the dish up into many small segments. Let each segment be, say, 100 feet in diameter. Finally, suppose all the segments are removed except two, located at opposite ends of the original 1000-mile mirror. What is left? The answer is: Two separate radio telescopes, functioning as two segments of a 1000-mile dish, with a small diffraction effect and an excellent definition of detail. Of course, the collecting area of the two segments is much less than the area of the original 1000-mile dish. The important point, however, is that the resolving power of the two dishes, work-

[1] The theory of diffraction shows that the angular resolution—that is, the smallest separation (in angle) between two points in an object that can just be resolved—is given by the formula

$$1.2\frac{\lambda}{d}$$

where λ is the wavelength of the radiation and d is the diameter of the telescope mirror or lens. For the 200-inch telescope, this formula tells us that two objects can be barely distinguished if their separation is 1/40 of a second in angle, or one part in two million. (Blurring by atmospheric effects makes the angular resolution of the 200-inch telescope about 40 times poorer than this in practice.) For the 1000-foot Arecibo telescope, the formula gives an angular resolution of approximately one part in 3500, or one minute of arc, for radio waves with a wavelength of a few inches. This is as good as the resolution of the human eye, but much poorer than the resolution of a good optical telescope.

[2] What is a radio "image"? The radio telescope collects radio waves coming from one direction in the sky at one time; these form one small area in the radio image. Other points in the image are recorded by looking in other directions, one at a time. In this way the image is built up as a mosaic of separate measurements.

ing together, is equivalent to the resolving power of a 1000-mile dish.

This idea is the basis of a new method in radio astronomy called "Very Long Baseline Interferometry," or VLBI. Radio telescopes as far apart as South Africa and Australia, or Sweden and West Virginia—separated by approximately 6000 miles—have been linked in VLBI networks to obtain radio images with a definition of detail equal to that of a dish whose diameter is 6000 miles—nearly equal to the diameter of the earth. The resolution of such a combination of radio telescopes is approximately one thousandth of a second of arc. At this resolution, a person on the earth could see astronauts moving on the surface of the moon.

In order for the VLBI method to work, the signals from the separate radio telescopes must be combined with very precise timing—a millionth of a second or better—for sites hundreds or thousands of miles apart. This is accomplished by using atomic clocks, which gain or lose no more than a millionth of a second per year, located at the separate telescope sites. The time signals from each atomic clock are put on magnetic tape that also records the radio signals collected by the telescope (Figure 5.4). After the observations are

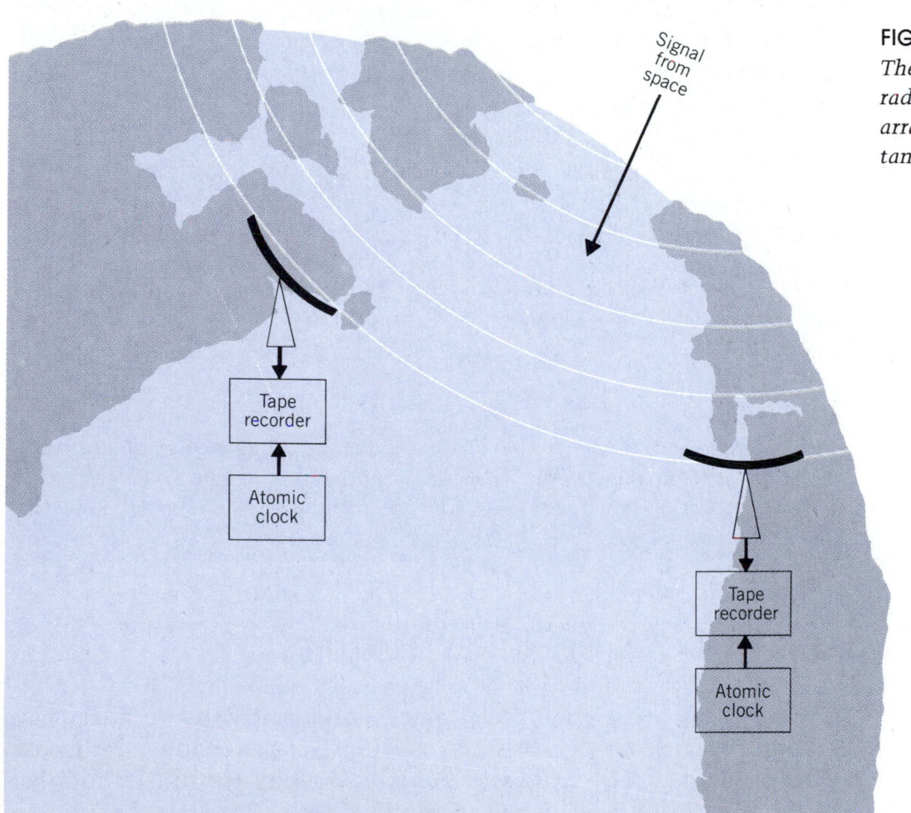

FIGURE 5.4
The Very Long Baseline Interferometer. Two radio antennas with recorders and clocks are arranged to combine signals received simultaneously on separate continents.

completed, the tapes are brought together and read into a computer, which synchronizes the timing marks and combines the separate records. If the timing is not synchronized in this way, the train of waves arriving at one telescope will not be precisely matched to the train of waves arriving at the other, and the two signals, when added, will produce a jumbled record rather than the equivalent of a single signal from different parts of one large dish.

The Very Large Array, or VLA, is a variation of the VLBI concept. The VLA is an array of radio telescopes located near Socorro, New Mexico. It consists of 27 radio dishes mounted on mobile platforms that roll along railway tracks, extending over several miles of desert (Figure 5.5). Since the dishes are only a few miles apart, precise synchronization of two signals can be obtained by connecting the dishes with cables. The tracks permit the dishes to be moved closer together or farther apart, providing the equivalent of a radio telescope with variable diameter and resolving power. The image of a cosmic jet shown in Color Plate 14 was obtained with instruments based on the VLA and VLBI concepts.

FIGURE 5.5
Some of the radio dishes in the Very Large Array, in Secorro, New Mexico.

NEW WINDOWS OPENED BY SATELLITES

Observations with radio telescopes are usually made from the ground, because the earth's atmosphere has a window that freely admits radio waves with wavelengths ranging from a fraction of an inch to hundreds of feet. Ground observations are limited to this radio window plus the visible window through which ordinary light penetrates the atmosphere. But the electromagnetic spectrum includes many other wavelengths which together can provide a rich source of information regarding the Universe. Unfortunately, these regions—which include gamma rays, X-rays, ultraviolet radiation, and infrared radiation—are excluded from ground observations by the screening effect of the atmosphere. However, observations in the excluded regions of the spectrum can be carried out above the atmosphere from orbiting satellites.

X-Ray Astronomy

X-ray astronomy, carried out from satellites, ranks only below optical astronomy and radio astronomy in the number and variety of interesting results it has yielded. However, X-rays present a problem for the astronomer because they cannot be collected and focused in an ordinary telescope. An X-ray, because of its great penetrating power, would pass straight through a mirror instead of being reflected. In that case, how can an X-ray telescope be constructed?

The answer lies in a special property of X-rays; if an X-ray is incident on a mirror surface at a grazing angle, nearly parallel to the mirror, it will bounce off and not penerate the surface. Working on this principle, X-ray astronomers have built special telescopes with "grazing-incidence" mirrors. Several mirrors are nested, one inside the other, to increase the collecting area of the telescope (Figure 5.6). The mirrors are aligned so that X-rays from a distant astronomical object are focused at a single point. The largest X-ray telescope placed in orbit thus far, called the Einstein Observatory, produced X-ray images of the sky with a degree of detail approximately equal to that of large optical telescopes. The Einstein Observatory has detected thousands of separate X-ray sources in

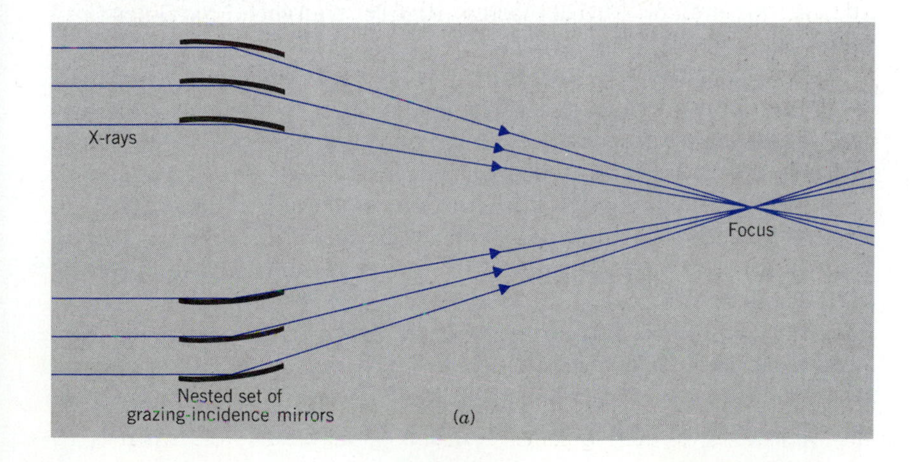

FIGURE 5.6
Basic design of a grazing-angle X-ray telescope.

the sky. It has obtained detailed X-ray images of supernova remnants—the shells of hot gas surrounding exploding stars (Chapter 9), of pulsars or neutron stars, and of galaxies and clusters of galaxies. It has also observed intense streams of X-rays coming from the remarkable objects known as quasars (Chapter 12).

In general, what kind of object would be expected to emit X-rays? Ordinary stars, whose surface temperatures are some thousands of degrees, emit most of their radiation in the visible region of the spectrum. As the temperature at the surface of the star rises, the emitted radiation shifts to shorter wavelengths. If the temperature at the surface reaches millions of degrees, the star radiates X-rays. In other words, stars that emit X-rays have extremely hot surfaces. Neutron stars are an important example, as will be seen in Chapter 9.

Black holes are another likely source of X-rays, as also will be seen in Chapter 9. When a black hole draws matter to its surface, the powerful force of the black hole's gravity accelerates the matter to very high speeds, and raises it to million-degree temperatures. As a result, the gas surrounding a black hole emits a large amount of X-radiation. In addition, such powerful energy sources as supernovas, quasars, and exploding galaxies may also be strong sources of X-rays.

Gamma-Ray Astronomy

Gamma rays are electromagnetic waves with wavelengths even shorter than those of X-rays—in the neighborhood of a billionth of an inch. Because of their great energies, gamma rays are very difficult to collect and record. Their penetrating power is so great that not even the grazing-incidence mirrors developed for X-ray telescopes will suffice to reflect and bring them to a focus. Instead gamma-ray telescopes are built around instruments developed by nuclear physicists for detecting highly energetic particles and radiation.

Although gamma-ray astronomy is in a less advanced stage than X-ray astronomy, American astronomers plan to put into orbit several gamma-ray telescopes of advanced design around 1989. These telescopes will be mounted on a satellite called the Gamma Ray Observatory. The new instruments are expected to shed light on some of the most important questions in astronomy. For example: Why is the Universe composed of matter rather than antimatter? Matter and antimatter are two similar forms of material substance, but they cannot coexist; when matter and antimatter meet, they annihilate each other in a shower of gamma rays. These gamma rays produced by the annihilation have characteristic energies; in other words, they have a very sharply defined spectrum. If a gamma-ray telescope detected gamma rays with these particular energies coming from distant parts of the Universe, that would indicate that antimatter is present in those regions and is being annihilated in collisions with matter.

A relatively uniform diffuse glow of gamma rays has been detected in the Universe. The gamma-ray glow has properties suggesting that it may be, in fact, a remnant of the gamma rays produced at an earlier stage in the Universe's evolution when large regions of matter and antimatter, may have existed in contact with matter being annihilated at their boundaries, and showers of gamma rays produced. This theory could mean that half the galaxies in the Universe today are composed of antimatter. There is no firm evidence against this suggestion, and very tentative evidence, from the gamma-ray glow, in its favor. The Gamma Ray Observatory is expected to shed light on this fascinating question.

Gamma-ray telescopes also have detected several regions in the sky that emit intense amounts of gamma radiation; yet the gamma rays do not seem to be coming from any object visible in other wavelengths; no stars, quasars or galaxies emitting light, X-rays, or radio waves can be seen in these regions. Any event so energetic as to produce copious amounts of gamma rays would be expected also to produce an abundance of X-rays and radio waves, if not visible radiation. What kind of object can emit only gamma rays, and nothing else? The answer is not known.

Another puzzle in gamma-ray astronomy relates to bursts of gamma rays, first detected about ten years ago by American satellites designed to monitor nuclear weapons tests. These intense bursts lasted from a fraction of a second to about a minute, and had intensities as much as thousands of times greater than the normal levels of gamma radiation received from the sky. The sources of the bursts appeared to be randomly distributed in the sky. The discovery of these gamma-ray bursts offers a tantalizing hint of the existence of some new kind of astronomical object, never before observed by astronomers or imagined by theorists.

The Gamma Ray Observatory satellite will carry an instrument especially designed to monitor the heavens for gamma-ray bursts, measuring their directions, changes with time, and energies. At the moment, gamma-ray bursts appear to be a curious but minor phenomenon. However, the absence of any explanation for these bursts indicates that they may be the harbingers of new phenomena, lying outside the boundaries of current scientific knowledge.

Ultraviolet Astronomy

The ultraviolet region lies immediately adjacent to the short wavelength end of the visible spectrum. The near ultraviolet region, from 4000 to 3000 angstroms, penetrates partially to the ground, but wavelengths shorter than 3000 angstroms are almost entirely absorbed in the atmosphere. This region—the so-called far ultraviolet—can only be explored from space.

As will be seen in the following chapter, each chemical element—that is, each of the 92 different kinds of atoms in the Universe—produces a characteristic pattern of wavelengths called the spectrum of that element. Often the spectrum of an element includes radiation both in the visible region and the ultraviolet region. In that case, the presence of the element can be detected from the earth, using a spectroscope placed at the focus of a telescope, as described in Chapter 4. But in other cases the spectrum is produced mainly in the ultraviolet region, and the observations must be made with spectroscopes and telescopes mounted in orbiting satellites.

The largest ultraviolet telescope orbited thus far, called Copernicus, was launched in 1972. It was 10 feet long, 7 feet in diameter, and contained a total of 11 telescopes, the largest having a mirror with a diameter of 32 inches (Figure 5.7). The mirrors were similar to those used in ground-based telescopes. As with the planned Space Telescope, Copernicus used photoelectric cells that converted light into electricity, as a means of measuring and recording spectra, rather than the photographic plates normally used for this purpose.

Among the major discoveries made by Copernicus was the observation of elements in the interstellar gas that had never been detected from the ground. Silicon, magnesium, and iron were observed in this way, and their abundances were found to be depleted by large factors below the amounts expected. Since these three elements are major constituents of rocks, the implication in this finding is that a great deal of such material is locked up in space in the form of small grains of rock.

FIGURE 5.7
The Copernicus satellite photographed during a prelaunch test.

Copernicus also observed a broad spectral band in the ultraviolet, at a wavelength of 2200 angstroms. This band is believed to be due to interstellar carbon, perhaps in the form of flakes of graphite. And Copernicus found evidence of heavily ionized atoms of oxygen—oxygen atoms stripped of five of their outer electrons. Presumably these electrons were removed in violent collisions occurring in the interstellar gas, implying that the temperature of that gas is very high. Estimates based on the Copernicus observations indicate that some of the interstellar gas is at a temperature of several hundred thousand degrees—much hotter than the surface of any star. This gaseous material apparently forms a hot halo that surrounds the entire Galaxy.

Finally, Copernicus searched for, and detected, deuterium or heavy hydrogen—the variety of hydrogen whose atom has a nucleus consisting of a neutron plus a proton, rather than a single proton. It was always believed that deuterium must exist in the interstellar gas, but its presence has been almost impossible to detect from the earth. The Copernicus measurements indicated that interstellar deuterium does in fact exist, and has an abundance of approximately one atom to every hundred thousand atoms of ordinary hydrogen.

The deuterium measurement may seem like a very fine detail in the study of the interstellar medium, but it has turned out to have great significance for the understanding of the evolution of the Universe. According to theoretical studies, most of the deuterium now present probably was made during the early moments of the history of the Universe, shortly after the Big Bang (Chapter 13). Calculations indicate that the amounts of deuterium left over from that early time depends, in turn, on the density of matter in the early Universe. If the density of matter is great, collisions are frequent and many nuclear reactions occur. In these reactions, deuterium is consumed in the process of building up helium and other elements. If the density of matter is small, fewer collisions and nuclear reactions occur, and more deuterium is left over. Thus, a measurement of the amount of deuterium in the Universe today yields information on the density of matter in the Universe when it was young. As will be seen in Chapter 13, the Copernicus observations on the abundance of deuterium, by limiting the density of matter in the early Universe, have illuminated one of the outstanding cosmological questions: Will the Universe expand forever?

Infrared Astronomy

As the temperature of an object decreases, its electromagnetic radiation shifts towards longer wavelengths. The sun, with a surface temperature of 6000°K, radiates its peak intensity in the visible range, but cooler and less massive stars like Barnard's star, for example, radiate most of their energy in the infrared. Many objects in the sky have temperatures too low to produce appreciable emission at visible wavelengths, but high enough to produce a detectable amount of infrared radiation. Since atmospheric molecules—molecules of water vapor, carbon dioxide and ozone—absorb infrared waves, it is desirable to carry a telescope and infrared detector above the atmosphere in order to study these infrared-emitting objects.

Initially the telescope and infrared detector were carried aloft in balloons and high altitude aircraft. In 1983 NASA launched the first Infrared Astronomy Satellite (IRAS) on behalf of an international team of astronomers. The IRAS contained a detector in a kind of thermos bottle cooled by liquid helium to a temperature of two degrees above absolute zero. The sensitivity of the detector at this extremely low temperature is sufficient to pick up the heat from a piece of warm toast at the distance of the moon.

The IRAS can detect dark asteroids and comets which are too

FIGURE 5.8
A map of the large Magellanic Cloud at a wavelength of 0.1 millimeters in the far infrared, obtained by the Infrared Astronomy Satellite. The map is made by repeated scans across the large Magellanic Cloud. Each line in the illustration is proportional to the infrared intensity measured during one scan.

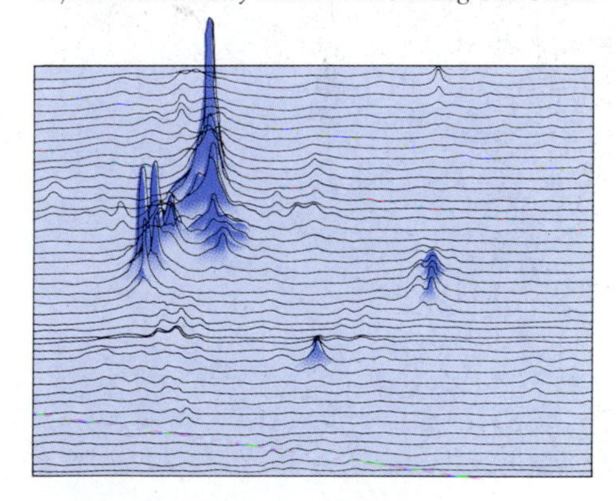

dim to be detectable by their visible light, but are warmed slightly by the sun's rays and radiate at infrared wavelengths. The satellite played an important role in the discovery of Comet Iras-Araki-Alcott, which passed by the earth in 1983 at a distance of only 3.1 million miles. The IRAS also should be able to detect a tenth planet in the solar system if one exists beyond Pluto's orbit (Chapter 15).

Infrared astronomy has led to the discovery of relatively cool red giants and also of cool red dwarfs. It is possible that this branch of astronomy will yield the discovery of still smaller and cooler brown dwarfs, lying on the dividing line between small stars and large planets. Infrared observations are also valuable in the study of certain galaxies and quasars that emit copious amounts of infrared energy, in some cases dwarfing the total emission of energy from our galaxy at all wavelengths.

Finally, whenever a hot object, such as a massive star, is embedded in a cloud of dust, the dust absorbs the energy radiated by the hot object at short wavelengths in the visible and the ultraviolet regions, and converts this energy to infrared radiation or heat. Figure 5.8 shows a map of infrared emission from the large Magellanic Cloud, a satellite galaxy of the Milky Way, obtained with the Infrared Astronomy Satellite. The conspicuous source of infrared at the upper left is the Tarantula nebula, a seed bed of new stars in the large Magellanic Cloud. The stars are embedded in a large cloud of dust which has absorbed their energy.

Another observation made with the Infrared Astronomy Satellite indicates that the star Vega is surrounded by a shell or ring of particles, probably about one millimeter in diameter, and possibly including even larger objects. The observations suggest that Vega may be surrounded by a solar system in the process of formation. This would be the first direct confirmation for the existence of solar systems other than our own in the Universe.

Main Ideas

1. Light as an electromagnetic wave.
2. Important regions in the electromagnetic spectrum.
3. Relationship between wavelength and frequency.
4. Transparency of the atmosphere to radiation of various wavelengths.
5. How the radio telescope works.
6. Comparison between radio telescopes and optical telescopes.
7. Limits on the resolving power of radio telescopes.
8. Arrays of radio telescopes: VLBI, VLA.
9. Telescopes above the atmosphere.
10. Major discoveries in X-ray, Gamma ray, ultraviolet and infrared astronomy.

Important Terms

angstrom	radio waves
dish antenna	resolution
electromagnetic spectrum	ultraviolet telescope
frequency	very long array (VLA)
gamma ray telescope	very long baseline interferometry (VLBI)
infrared telescope	visible light
light	X-ray telescope
microwave	wavelength
radio telescope	window

Questions

1. The FM radio band centers on a wavelength of about 100 megahertz. (One hertz = one cycle per second (cps); one megahertz = 10^6 cps.) What is the wavelength in feet of a radio wave with this frequency? What is the frequency (in cycles per second) of a radio wave whose wavelength is 1.86 miles? A light wave whose wavelength is 1000 Angstroms?
2. What is meant by a "window" in the atmosphere? What are the important atmospheric windows from the astronomer's point of view? What are the names of the respective branches of astronomy?
3. How do the following differ from visible light?
   (a) X-rays
   (b) Infrared radiation
   (c) Radio waves
   (d) Gamma rays
4. Why are astronomers so interested in sending telescopes up in satellites?
5. What portions of the electromagnetic spectrum would you select for viewing from an orbiting astronomical observatory? Why?
6. Why do most radio telescopes have poor spatial resolution compared to optical telescopes? Discuss the relative advantages and disadvantages of optical and radio telescopes.
7. What is the advantage of using the VLBI? Why are atomic clocks necessary for the VLBI?
8. Why is an infrared telescope in orbit superior to the same telescope in a balloon or aircraft? What kinds of objects can be observed with an orbiting IR telescope?
9. Use the formula for angular resolution to calculate the diameter of a radio telescope needed to "see" a black hole at the center of a galaxy if the black hole is 10 million miles in diameter and the galaxy is one billion light-years away. From the same formula, calculate the size of an optical telescope in earth orbit needed to read a newspaper headline on the moon.

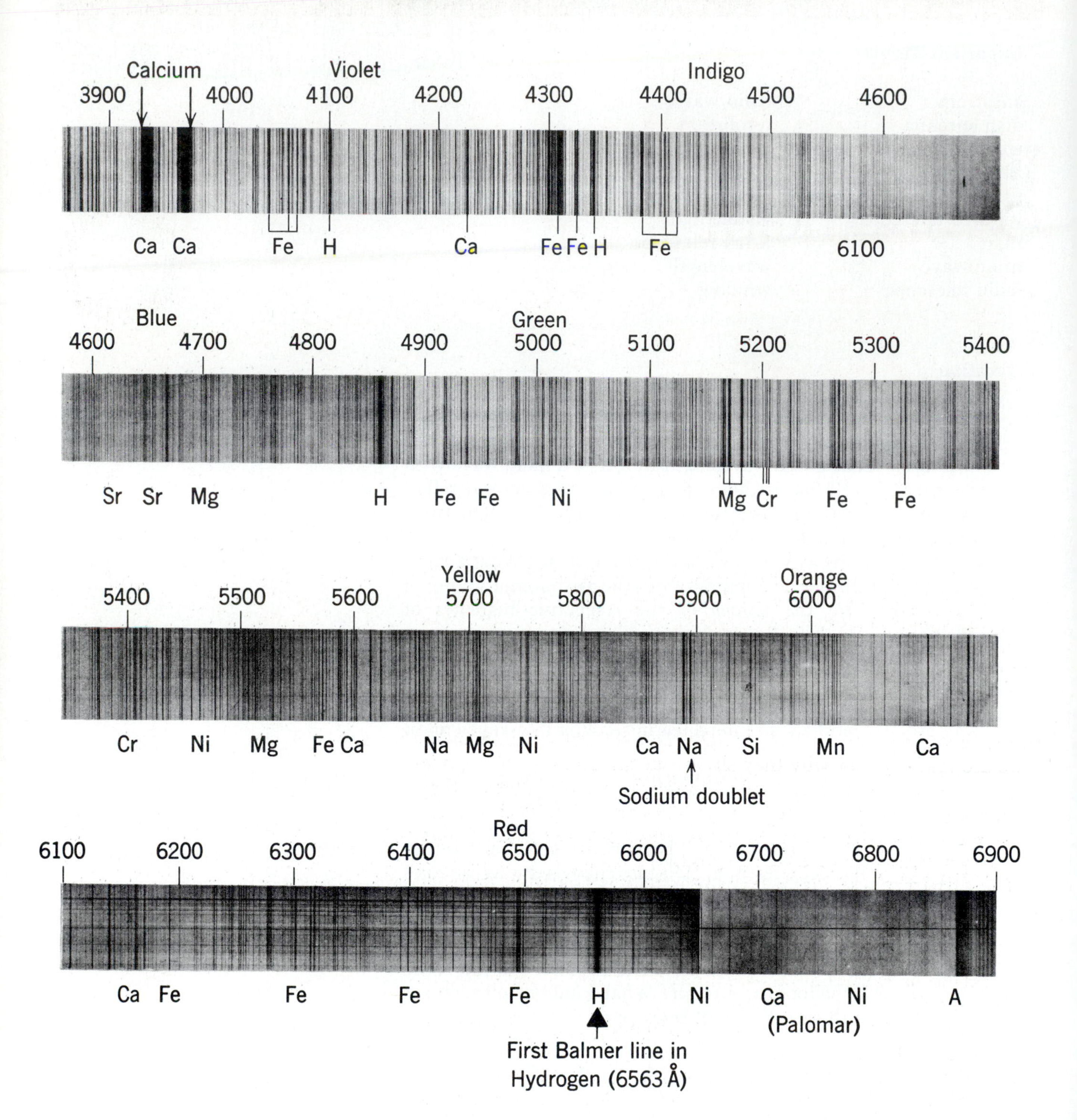

Calcium
Violet
Indigo
3900
4000
4100
4200
4300
4400
4500
4600
Ca
Ca
Fe
H
Ca
Fe
Fe
H
Fe
6100
Blue
Green
4600
4700
4800
4900
5000
5100
5200
5300
5400
Sr
Sr
Mg
H
Fe
Fe
Ni
Mg
Cr
Fe
Fe
Yellow
Orange
5400
5500
5600
5700
5800
5900
6000
Cr
Ni
Mg
Fe
Ca
Na
Mg
Ni
Ca
Na
Si
Mn
Ca
Sodium doublet
Red
6100
6200
6300
6400
6500
6600
6700
6800
6900
Ca
Fe
Fe
Fe
Fe
H
Ni
Ca
Ni
A
(Palomar)
First Balmer line in
Hydrogen (6563 Å)

THE MESSAGE OF STARLIGHT

6

Astronomy began to grind to a halt in the middle of the nineteenth century. The optical telescope had reached a high state of perfection, and nobody yet dreamed of other windows in the sky. Tens of thousands of stars had been catalogued and put away on the shelf. Two new planets unknown to the ancients had been discovered, but a thorough search had failed to reveal signs of any others. Astronomers seemed to know everything that could be learned about the stars as points of light. They did not know what the stars are made of, nor why they shine, but this knowledge seemed to be forever denied to man because of the enormous distances that separated our solar system from other stars. Astronomy—the oldest science—was ready to enter the graveyard of dead subjects. It was revitalized by a basic discovery in physics.

The discovery was that a hot atom radiates light at a series of wavelengths peculiar to that element. The chemical composition of a star could therefore be determined by analyzing the light coming from the atoms in that star. This is the message of starlight. It reveals to us what the stars are made of, and how they are born, evolve, and die.

A German physics professor named Kirchhoff, working with the chemist Bunsen, who invented the Bunsen burner, was the first to realize that flaming objects send out a coded message about themselves to all who have the wisdom to read it. Kirchhoff discovered how to break the code, and with his discovery astronomy was reborn.

Fraunhofer lines in the Solar Spectrum.

THE ATOM

The electrons in an atom circle around the nucleus, attracted to it by the electrical force, in the same way that satellites circle in their orbits around the earth under the attraction of the gravitational force. However, the orbiting electron and the orbiting satellite turn out to behave very differently. When we launch a satellite into an orbit, the orbit can be at any distance from the earth that we choose, depending on the power of the rocket. If a certain amount of rocket power will get a satellite into a given orbit, an increase in that rocket power will put it into a higher orbit. Any orbit is possible for the satellite if we put enough power into launching it. This is a simple fact that agrees with our everyday experience on the earth: If you throw a ball up into the air, it will rise a certain distance; if you throw it up a little harder, it will rise a little higher.

But the familiar laws of nature, as we know them in the everyday world, do not apply to the world of the atom. According to these laws, an electron can only circle the nucleus in certain allowed orbits of definite radius. In the hydrogen atom, for example, the smallest orbit possible to the electron has a radius of 0.53 Å. (The angstrom, equal to 10^{-8} cm, is a convenient unit for expressing distances between atoms; its symbol is Å.) No orbits smaller than 0.53 Å are allowed in the hydrogen atom. That is, no hydrogen atom has ever been found in which the electron's orbit is smaller than 0.53 Å.

Moving outward, the next allowed orbit in the hydrogen atom has a radius of 2.12 Å. No hydrogen atom exists in the world in which the electron orbit has a radius between 0.53 Å and 2.12 Å. The rules of the atomic world forbid these intermediate radii. Furthermore, the rules are very precise; the orbit radius has to be *exactly* 0.53 Å or 2.12 Å. An orbit with a very slightly different radius—0.52 Å, say, or 2.10 Å—never occurs.

The limitation on possible sizes of orbits is not the only peculiarity of the atom. The *number* of electrons in each orbit is also limited. When dealing with orbiting satellites, we can place as many satellites in an orbit of a given radius as we wish; if a satellite is placed in an orbit with an altitude of 110 miles, no law of nature prevents us from placing a second satellite in an orbit with precisely the same altitude. But in the world of the atom, the number of electrons in a given orbit is limited by a complicated set of rules, which have been worked out by atomic physicists over many years and are a part of the laws of the atom. If these rules specify that a given orbit may contain no more than two electrons, for example, we can be sure that we will never find an atom in which there are more than two electrons in that particular orbit.

The allowed orbits for the electrons in an atom are called *shells.* The laws of the atom specify the radii of the electron shells in each atom and the maximum number of electrons that may occupy each shell.

The carbon atom, for example, has precisely two electrons in a shell with a radius of 0.2 Å, and four electrons circling in a shell outside the inner group, at a radius of 0.5 Å. According to astronomers, this is true not only of all the carbon atoms examined on the earth but also of every carbon atom in the Universe.

Our experience with everyday events provides no precedent for such peculiar laws. These laws become important only when we investigate the properties of an object as small as an atom or smaller; on a larger scale of distances, such as an inch, a foot, or a mile, their effect is negligible. When the Danish physicist, Niels Bohr, proposed a special set of laws for the world of the atom around 1910, many other physicists objected, but within a few years it became clear that the properties of the atom could not be explained in any way except by the special rules proposed by Bohr, and now these laws of atomic physics are universally accepted. However, they are still just as hard to understand in terms of everyday experience as they were 70 years ago.

Electron Shells in the Atom

Every type of atom has its own special set of electron orbits or shells; the radii of these shells are peculiar to each element. The number of shells increases from small atoms to bigger ones. However, the maximum number of electrons in each shell remains the same for all elements. The innermost of the allowed orbits, called the first shell, can hold up to two electrons and no more. It is this shell that contains two electrons in the carbon atom. The second allowed orbit is called the second shell. Eight electrons are the maximum number permitted in the second shell. In carbon, as we noted, the second shell contains only four electrons. When four electrons are placed in the second shell of the carbon atom, they make up, with the two electrons in the first shell, a total of six electrons. Six electrons exactly cancel the positive charges on the six protons in the carbon nucleus. Thus an additional electron passing by feels no electrical attraction toward the carbon atom and cannot be drawn into an orbit in the second shell, even though there is a place for it there.

We have now laid the foundations for building up the entire periodic table of the elements in terms of the number of electrons in the various shells. We start with hydrogen, which contains one electron in the first shell (Figure 6.1*a*).

Next is helium, which contains two electrons in the first shell (Figure 6.1*b*). Beyond helium is lithium, with three electrons. In the world of the atom, it is not possible to place a third electron in

FIGURE 6.1
Atoms of (a) hydrogen and (b) helium (p = proton, n = neutron, e = electron).

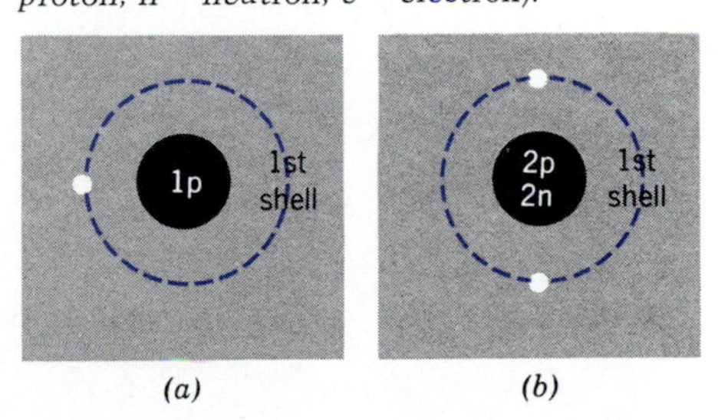

FIGURE 6.2
The lithium atom.

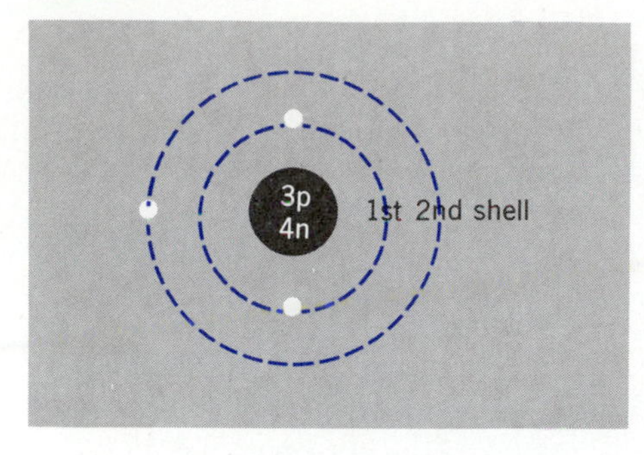

the inner shell. Therefore, the third electron must go into the second shell (Figure 6.2).

After lithium comes beryllium, with a total of four electrons. Once more, two go into the first shell, and the next two must go into the second shell. Beryllium is followed by boron with five electrons, of which two must be located in the first shell and three must be placed in the second. Then comes carbon, which we have already discussed. Carbon is followed by atoms with successively five, six, seven, and eight electrons in the second shell. These atoms are, respectively, nitrogen, oxygen, fluorine, and neon. At neon, the process of filling the second shell stops because the neon atom has eight electrons in this shell, and eight is the maximum number allowed by the laws of the atom.

By the time we come to the elements near the end of the periodic table, for example, uranium, we find the first four shells fully occupied, the fifth shell nearly full, and the sixth and seventh shells partly occupied (Figure 6.3).

FIGURE 6.3
An atom of uranium.

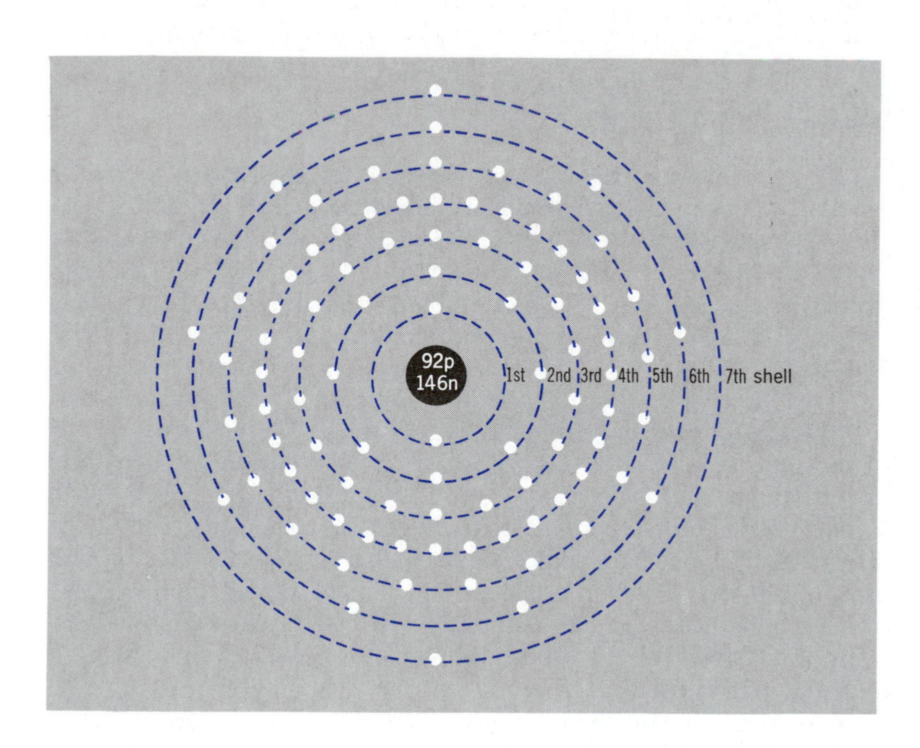

Notice, as we proceed in building up the table of the elements, that it is always the charge of the nucleus, or the number of protons in the nucleus, that sets a limit on the number of electrons we can place in the shells. The number of protons in the nucleus determines the electron structure of the atom.

Ground States and Excited States of the Atom

We have been describing the orbits of the electrons in the atom when it is undisturbed. This undisturbed state is called the ground state of the atom. If another particle collides with the atom, the electron may be forced out of its ground state. For example, in a gas, the atoms continually collide with one another; or, as another example, the atom may be hit by a fast-moving atomic projectile in a nuclear physics experiment. Occasionally, an electron will absorb enough energy in such a collision to break the bond that holds it in its orbit. This excited electron will then jump outward to a new orbit, in which it circles the nucleus at a greater distance than it did before. An atom in which an electron has absorbed energy and jumped into a larger orbit is said to be in an excited state.

If the force of the collision is very great, the electron may leave the atom entirely. An atom that has lost an electron is said to be ionized; the part of the atom left behind by the departing electron is called an atomic ion, or simply an ion. If the atom has lost one electron, it is sometimes referred to as being singly ionized; if it has lost two electrons, it is doubly ionized; and so on.

Usually it is the electrons of the outermost shell that are affected by collisions. The electrons of the inner shells are more tightly bound to the nucleus and less readily disturbed by outside forces; also, the electrons in the shells surrounding them act as an electrical screen against disturbances.

Emission of Light from an Excited Atom

Suppose an atom has been excited. What happens to this excited atom? Does the electron stay in the excited orbit, circling there indefinitely? If the orbiting electron were similar to an orbiting satellite, the answer would be yes, it could stay there almost indefinitely. However, the laws of the atom give a very different answer. When an electron has been kicked upward to an excited orbit, it circles in that orbit for a definite amount of time, characteristic of that particular orbit and that particular atom, and then collapses back down to the original orbit in a sudden transition. Typically, the electron stays in an excited orbit for one one-hundred-millionth of a second before collapsing. At the same time that the electron goes back to the ground state orbit, the atom emits a flash of light (Figure 6.4). The energy of the flash is equal to the difference between the energy of the electron in the ground state orbit and its energy in the excited orbit.

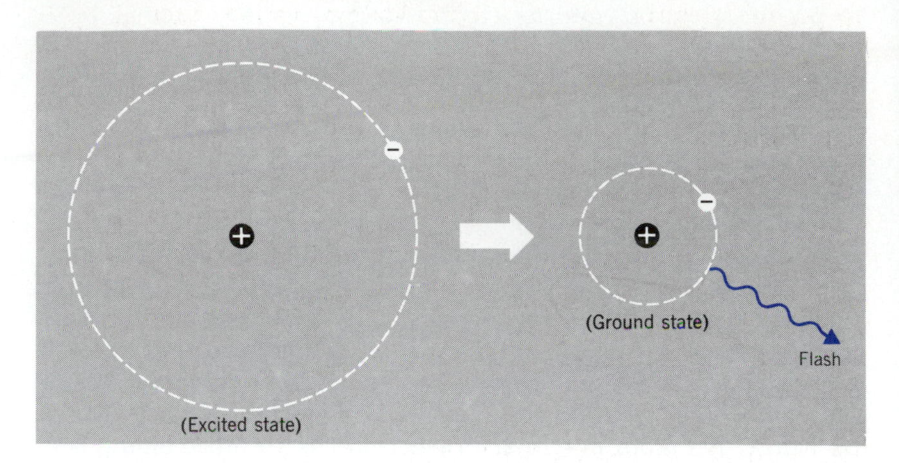

FIGURE 6.4
An excited hydrogen atom collapses to the ground state, emitting a photon.

The flashes of light are called *photons.* Photons travel at the speed of light and have energies that depend on their wavelengths; the shorter the wavelength of the photon, the greater its energy. To explain effects like refraction we must describe light as a train of waves; but to best understand the effects that occur when light is emitted by atoms, we must describe the light in terms of photons.

Collisions are not the only means by which the electrons in the atoms can be propelled from the ground state to a higher state. Suppose a beam of light shines through a gas composed of atoms. This beam of light is equivalent to a hail of photons. If the energy of the photons in the beam is precisely equal to the difference between the energy of the electron in its ground state orbit and the energy of the electron in one of its excited orbits, a photon can be absorbed by an atom in the gas, kicking the electron in the atom from its ground state to an excited orbit (Figure 6.5).

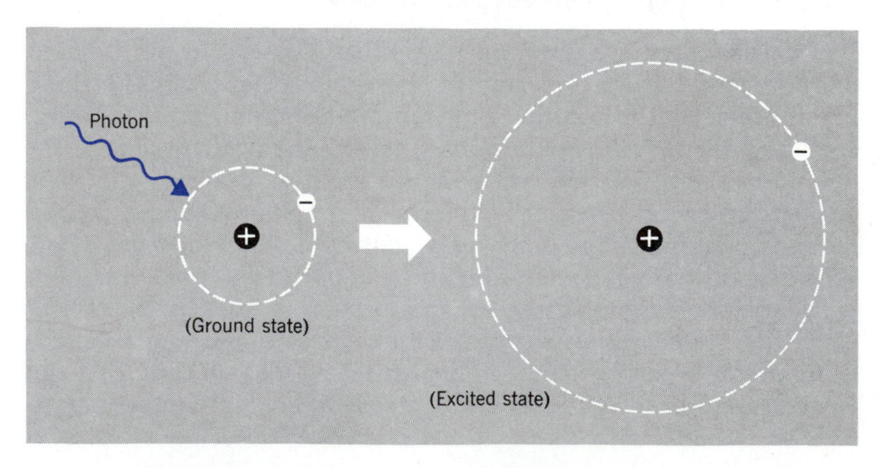

FIGURE 6.5
A hydrogen atom is raised to an excited state by the absorption of a photon.

ATOMIC SPECTRA

The size of an electron's orbit and the amount of energy required to get the electron from its ground state to various excited states depend on the strength of the force holding the electron to the nucleus. This force depends, in turn, on the number of protons in the nucleus. That is, it depends on the nature of the element involved. Thus, each element has its own set of electron orbits, which are the same for every atom of that element but are substantially different from the electron orbits of every other element. This set of facts—the uniqueness of electron orbits in an element—is the reason that underlies the use of atomic spectral lines as the means of identifying an element. Many branches of modern science depend on this property of the atom.

The Spectrum of a Heated Gas

Suppose we have a glass container filled with a pure gas consisting of atoms of one kind, such as hydrogen or helium. If the temperature of the gas is low, nearly all the atoms will be in the ground state. If the gas is now heated, collisions will become more violent, and an increasing fraction of the atoms will be raised to excited states. Collapsing from these excited states to the ground state, these atoms will emit photons with wavelengths characteristic of the kind of atom that makes up the gas in the container. Of course, photons of many different energies and wavelengths will be emitted by the atoms of the heated gas, because each atom possesses many excited states, and transitions can occur between any one of these excited states and the ground state or from one excited state to any other excited state. As a result of the emission of all these photons with various energies and wavelengths, the heated gas will glow with a light composed of many different colors.

An Example: The Spectrum of the Hydrogen Atom. Consider the hydrogen atom once again as an example. When a hydrogen atom undergoes a transition from its first excited state to the ground state, a photon is emitted with a wavelength of 1216 Å, deep in the ultraviolet portion of the spectrum (Figure 6.6).

FIGURE 6.6
Transition of a hydrogen atom from the first excited state to the ground state with emission of a photon.

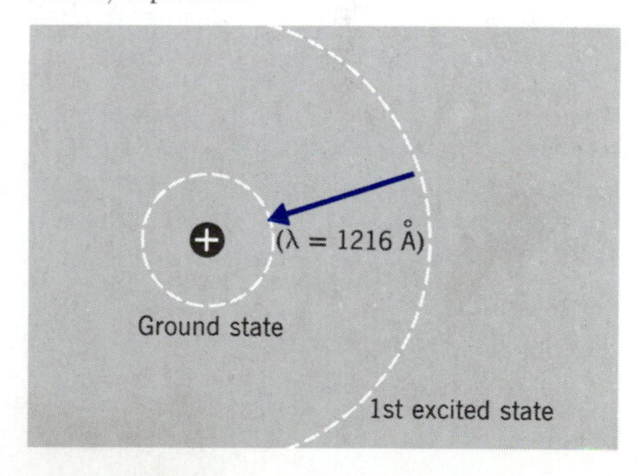

A line at this wavelength should be the first to appear if hydrogen gas is heated to a high temperature and examined through a spectroscope. If the temperature is raised still higher, or the hydrogen atoms are excited in some other way, as, for example, by passing an electric discharge through the gas, some of the atoms will be raised to the second excited state of hydrogen. A hydrogen atom raised to the second excited state can collapse down to the ground state in either of two ways: it can proceed directly from the second

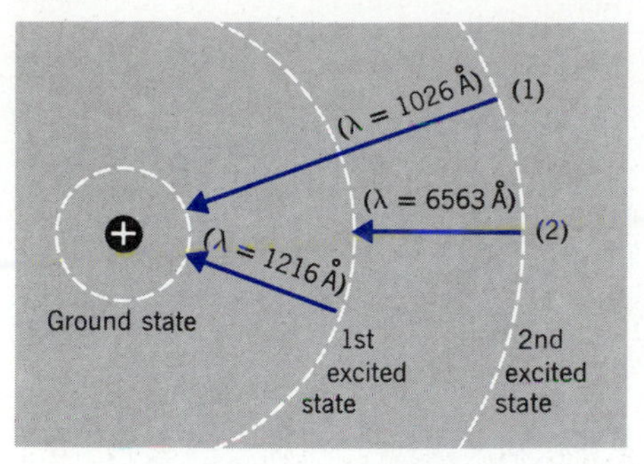

FIGURE 6.7
Transitions from the first and second excited states of the hydrogen atom to the ground state.

excited state to the ground state, or it can undergo a transition to the first excited state as an intermediate step, and from there go to the ground state. In the first case, a single photon is emitted at a wavelength of 1026 Å, corresponding to the energy difference between the second excited state and the ground state. In the second case, two photons are emitted, the first with a wavelength of 6562 Å, in the red region of the spectrum, and the second with a wavelength of 1216Å, in the ultraviolet region. The arrows in Figure 6.7 show three transitions between the first and second excited states and the ground state. Each transition produces a photon whose wavelength is shown in parentheses. The sum of the energies of the 1216 Å photon and the 6563 Å photon is equal to the energy of the single photon emitted in the direct transition to the ground.

As higher states of excitation are reached, the number of alternative ways in which the atom can return to the ground state also increases, and the series of lines emitted by the gas becomes very complex. However, we can introduce a degree of order into this complicated series of lines of different wavelengths by separating the lines into the following groups.

In the first group, place all the lines produced when the hydrogen atom undergoes a transition from any excited state to the ground state; this series of lines, of which the first one has a wavelength of 1216 Å, is known as the *Lyman series.*

In the second group, place all the lines that are produced when a hydrogen atom undergoes a transition from any excited state to the first excited state; this series, in which the first line has a wavelength of 6563 Å, is known as the *Balmer series.*

In the third group, place all the lines that are produced when the atom undergoes a transition from any excited state to the second excited state; this series, in which the first line has a wavelength of 18,756 Å, is known as the *Paschen series.*

Figure 6.8 shows the transitions, indicated by arrows, that give rise to the various lines in the above series, and Figure 6.9 shows the first 12 lines in the Balmer series of the hydrogen atom.

Figure 6.8 brings home the fact that each group, or series of lines, represents a transition to one particular final state of the hydrogen atom. This interpretation of the lines emitted by the hydrogen atom was first put forward by the great Danish physicist, Niels Bohr, more than 70 years ago, at the beginning of the twentieth century. Prior to that time, no one had any idea of the meaning of the spectra emitted by atoms. In spite of the failure of physicists to make any sense out of these singular sequences of spectral lines, an enormous amount of work went into measuring the wavelengths of the lines and classifying them, partly for their usefulness as the fingerprints of atoms and partly in the hope of learning something about the structure of the atom, which was at that time an entirely mysterious object.

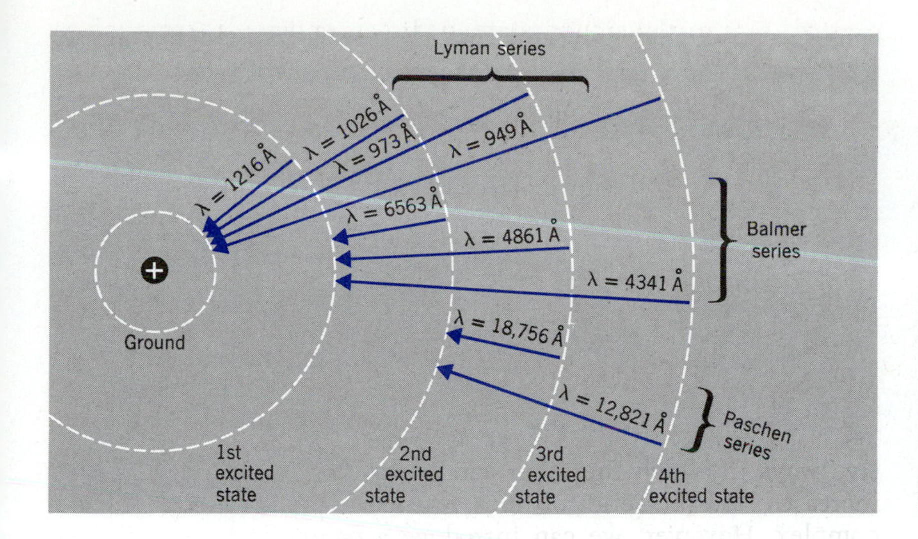

FIGURE 6.8
Formation of the Lyman, Balmer, and Paschen series of spectral lines in the hydrogen atom.

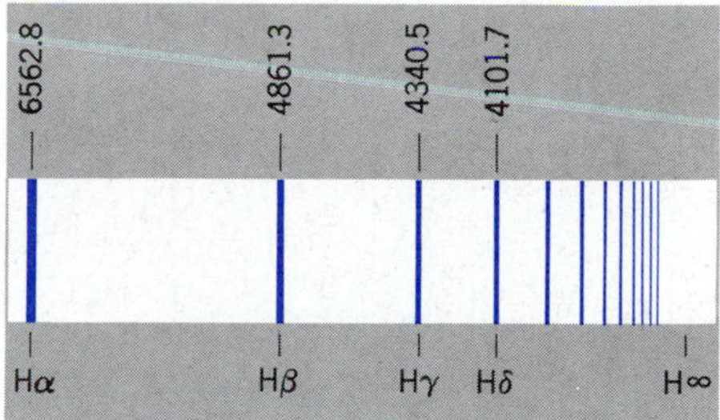

FIGURE 6.9
The Balmer series.

Emission Spectra

The sequences of lines emitted by atoms of a given kind, when heated to a high temperature, is called the emission spectrum, or sometimes the bright-line spectrum of that atom. An enormous amount of effort has been invested in spectroscopic examination of the light emitted by atoms. The wavelengths of the lines in the spectra of all the elements have been carefully measured, tabulated, and published. The wavelengths of most lines have been measured with an accuracy of a fraction of an angstrom; and, in the case of certain lines, the precision of the wavelength measurement is one ten-thousandth of an angstrom. The wavelengths of the lines in the spectrum of an element are as unique a characteristic of that element as your fingerprints are of your identity. The published tables of wavelengths of the lines in the spectra of the elements are equivalent to the fingerprint files of the FBI.

Absorption Spectra

Suppose we have a source of light that emits radiation of all visible wavelengths, such as, for example, a tungsten lamp. If the light from this lamp is examined through a spectroscope, we will see the full range of colors in the visible spectrum, starting with violet at the short-wavelength end, and shading imperceptibly into indigo, blue, green, yellow, orange, and finally to red in the long-wavelength end of the visible region. Now let us place a glass container, filled with a gas of unidentified atoms, between the spectroscope and the light source so that the light from the source must pass through the gas on its way to the spectroscope (Figure 6.10).

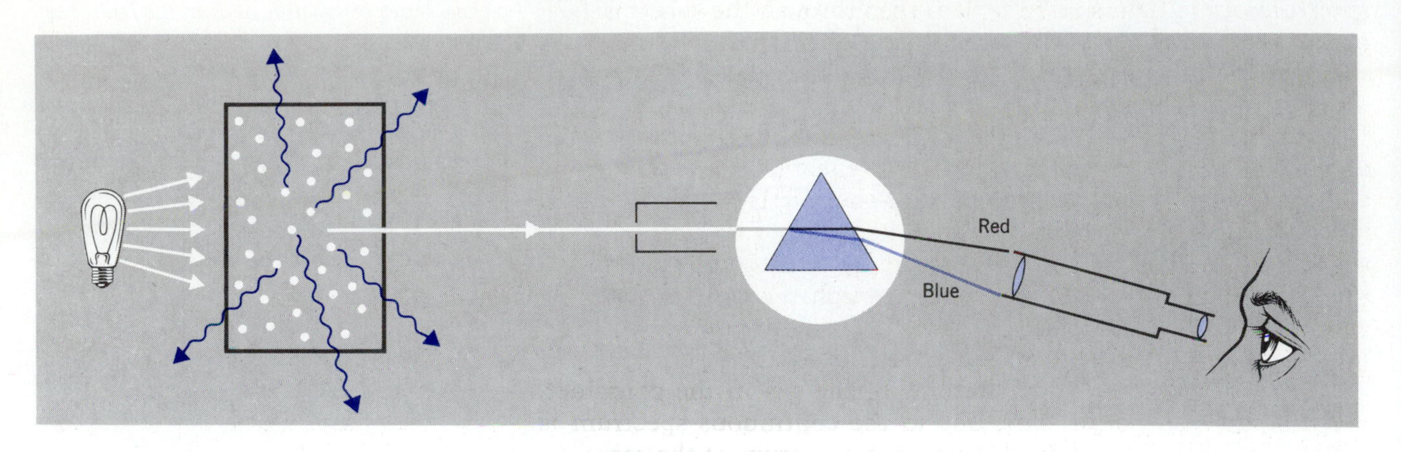

FIGURE 6.10
The removal of selected wavelengths from a beam of light passing through gas.

The original light contains photons of all wavelengths and energies; among these photons are some whose energy is precisely equal to the difference between the energies of the ground state and one of the excited states of the type of atom that fills the container. Such a photon can be absorbed by an atom in the container, raising the atom from its ground state to an excited state. This absorption process removes the photon of that particular wavelength from the beam of light. The excited atom collapses quickly to its ground state again, emitting another photon of precisely the same energy and wavelength as the photon it had just previously absorbed. However, the new photon need not be emitted *in the same direction* in which the old one was traveling; in fact, it is usually emitted in a different direction (Figure 6.11).

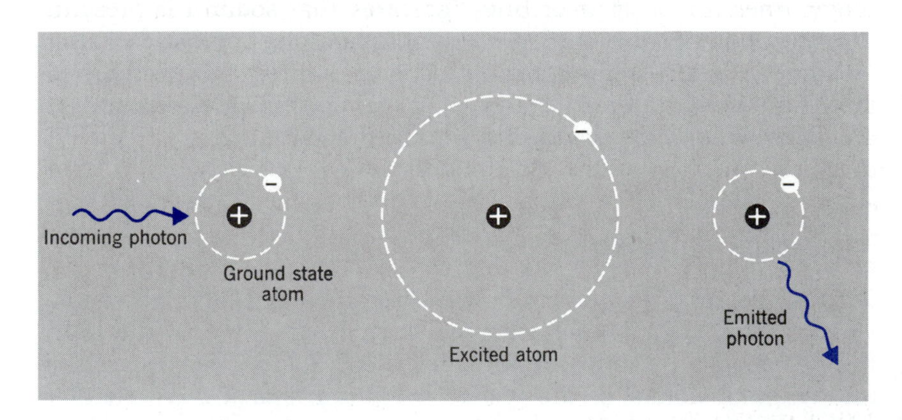

FIGURE 6.11
Atom absorbing a photon and emitting it in a different direction.

Thus, the reemitted photons do not enter the spectroscope because they are not traveling in the right direction to do so. They leave the container at the sides, the top, or the bottom, and they are removed from the beam as far as an observer looking through the spectroscope can determine (Figure 6.10). Therefore, the continuous spectrum of the light source will, as seen through the spectroscope, be marked by dark lines at certain wavelengths. The wavelengths that are "dark" are those whose photons have been removed by absorption in passing through the gas.

But what wavelengths are these? These are the wavelengths that correspond to the difference between the energy of the ground state and the energy of an excited state for the atoms of the gas. It is precisely these wavelengths at which light would be emitted if the gas were heated to a high temperature, so as to produce a bright-line or emission spectrum. If we photograph the continuous spectrum in the above experiment, marked by dark lines at certain wavelengths as we have described, and place it alongside a photograph of the emission spectrum of the atoms of the gas in the container, we will find that every dark line in the continuous spectrum is matched by a bright line in the emission spectrum of the atom.

As an example, the sodium atom has two very intense emission lines located close to one another in the yellow region of the spectrum, at 5890 Å and 5896 Å. The states of the sodium atom that produce these lines are easily excited when the sodium atoms are placed in a flame; it is because of this that the color of a Bunsen burner flame turns yellow if salt is thrown on it. Color Plate 3 shows the two yellow lines in the emission spectrum of sodium, as seen through a spectroscope that is pointed at a flame to which sodium has been added. Above the emission spectrum of sodium is an illustration of an absorption spectrum of sodium, produced when a glass container of sodium vapor is placed between the spectroscope and a light source. The two dark lines in the yellow region of the spectrum occur at precisely the same wavelengths as the two bright lines. Either pair of lines indicates that sodium is present.

That is, the pattern of the dark lines in the continuous spectrum, produced when a beam of light passes through the gas and some of the photons are absorbed from this beam, is the fingerprint of the type of atom that makes up that gas, in just the same sense that the emission spectrum produced by that atom when heated also is its fingerprint. The dark lines in the continuous spectrum of the gas are called its *absorption spectrum.* Either the absorption spectrum or the emission spectrum can be used to identify an element.

THE SPECTRUM OF A STAR

Scientists have observed thousands of lines in the sun's spectrum, and by comparing their wavelengths with the wavelengths of lines emitted by elements in laboratory experiments on the earth, they have detected the presence of 67 elements in the sun. The particularly fine photograph of the sun's spectrum shown opposite page 133 stretches across the full range of the visible spectrum, from the violet end at 3900 Å to the red end at 6900 Å. The wavelengths and colors are noted above the spectrum, and some lines are labeled below the spectrum with the name of the element responsible. Approximately 1000 lines are visible in this photograph.

Stellar Absorption Spectra

Stars are flaming balls of gas whose surfaces are at high temperatures ranging up to tens of thousands of degrees. Violent collisions occur at these temperatures, raising many atoms to excited states, and producing a rich emission spectrum with a great number of lines. These lines may be so many in number as to nearly fill in the spectrum of colors. Moreover, because of the high density of the gas in the star, collisions are very frequent, so that an electron in an excited state often is disturbed by a second collision before it has a chance to collapse down to its ground state orbit. Such disturbances blur the sharpness of the transition to the ground state and spread each line in the spectrum out into a broad band of color. Neighboring lines in the spectrum overlap one another as a result, so by the time the light leaves the surface of the star the separate spectral lines have been blurred into a *continuous spectrum,* that is, a rainbow of light including all wavelengths.

Stars have atmospheres just as planets do. The atmosphere of a star lies above its surface, and consists of the same elements that are contained in the body of the star, but at very low density. The light radiated from the surface of a star must pass through this atmosphere on its way out. The atoms in the star's atmosphere will absorb some of the wavelengths in the continuous spectrum of radiation coming up from the surface beneath. As a result, when the spectrum of the star is examined in a spectroscope by an astronomer on the earth, he will find that it is crossed by dark lines that are wavelengths of the atoms composing the star's atmosphere. These atoms have left their fingerprints on the star's light. They have created what is called an absorptive spectrum. This is the way in which astronomers determine the materials out of which stars are made: they collect the starlight in telescopes, spread it out into its component wavelengths, and photograph the resultant absorption spectrum.

Relationship between Spectral Lines and Surface Temperature of a Star

When you compare the absorption spectrum of the sun with the spectra of other stars, you find that some stars have sunlike spectra but the spectra of other stars look quite different. For example, consider the blue-violet region, running from 3900 Å to 4600 Å. In this region, the sun's spectrum shows two lines for hydrogen (at 4101 Å and 4340 Å) plus hundreds of lines produced by iron and other elements (Figure 6.12*a*). The spectrum of the star Vega shows the same two hydrogen lines in the blue-violet (Figure 6.12*b*) but they are much blacker and more intense than the sun's spectrum and, apart from the two hydrogen lines, the spectrum of Vega shows very little else. The many lines that are present in the sun's spectrum in the blue-violet region are absent.

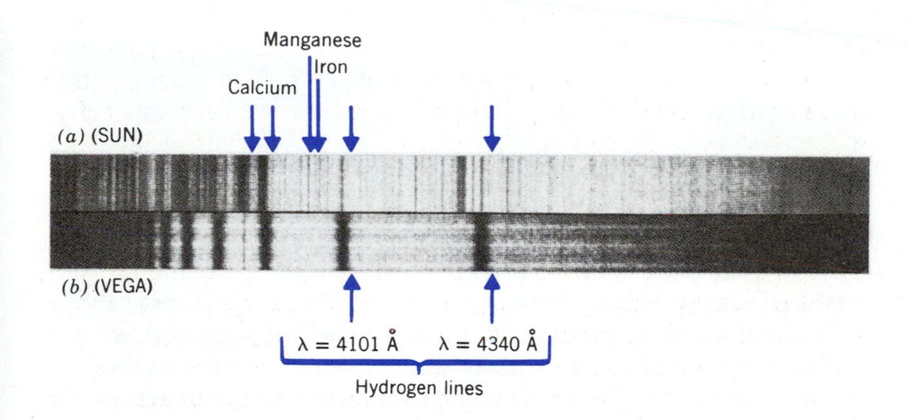

FIGURE 6.12
The absorption spectra of (a) the sun and (b) Vega.

At first you might conclude that Vega has a much higher percentage of hydrogen than the sun, because its spectrum shows stronger absorption lines for that element. You might also think that Vega has no iron, because the lines of this element are missing from its spectrum, and that the sun, whose spectrum is crowded with iron lines, must have a rich abundance of this metal.

Actually, the differences in the spectra of the stars are deceptive. All stars are made of a similar mixture of materials. There are some differences in composition from one star to another, but these differences are far less than the variations in the spectra of stars would seem to indicate.

To understand why the spectrum of one star can differ greatly from the spectrum of another, even though their compositions are the same or very nearly the same, consider the two hydrogen lines in the spectra of the sun and Vega (Figure 6.12). One of these lines is produced when the electron in the hydrogen atom absorbs a photon and is kicked upward from the first excited orbit to the fourth excited orbit. The other line is produced when the electron is kicked upward from the first excited orbit to the fifth excited orbit (Figure 6.13).

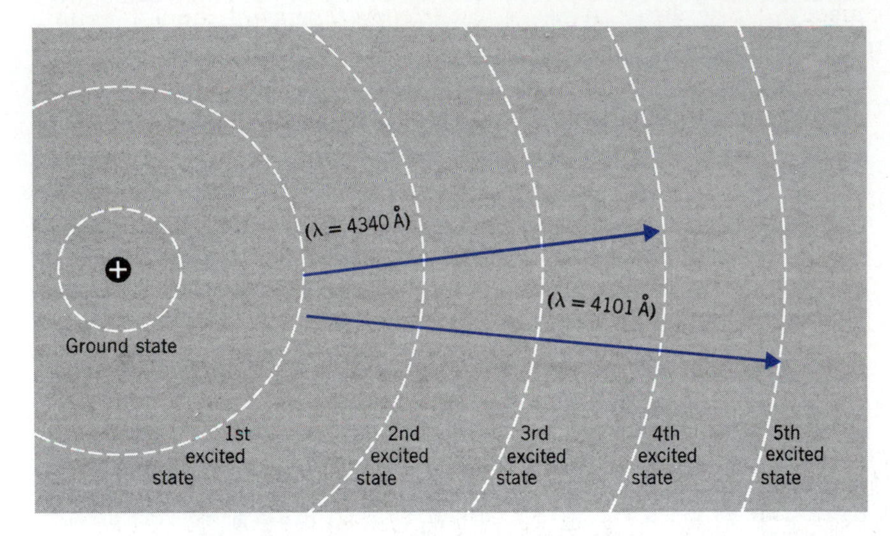

FIGURE 6.13
Transitions producing hydrogen lines in the solar spectrum.

In both cases, the atom must be in the first excited state above the ground state to produce this line. Now suppose that we are dealing with a star whose surface temperature is relatively low. In this star, because of the low temperature, nearly all hydrogen atoms at the surface will be in the ground state. Few atoms will be in excited states. Therefore, in the spectrum of such a star the Lyman series of hydrogen lines in the ultraviolet, corresponding to transitions from the ground state to excited states, will appear prominently. However, the Balmer series, including the two lines referred to above, will be relatively weak because the Balmer lines can be formed only when the hydrogen atom is initially in the first excited state, and at the low temperature that we are assuming for this star very few atoms will be in any excited state.

Now consider a star whose surface temperature is somewhat higher. A greater number of hydrogen atoms on the surface of this star will be in excited states than in the case of the cooler star; therefore, transitions from the first excited state upward to higher states will occur frequently. The spectrum of this star should show the hydrogen lines at 4101 Å and 4340 Å with greater intensity than the cooler star.

Suppose we consider a star of still higher temperature. The fraction of its hydrogen atoms in the first excited state will be still greater, and the intensity of the absorption lines corresponding to the Balmer series will also be greater.

Looking at the spectra of these stars of increasing temperature, one after the other, we will see that the hydrogen absorption lines grow stronger from one star to the next. We will be tempted to conclude that the amount of hydrogen increases from star to star. But the conclusion would be false. The variations in spectra are caused by differences in the surface *temperatures* of the stars and not by differences in the *amount* of hydrogen.

Figure 6.14 shows this effect clearly. The figure presents the spectra of seven different stars, from Vega to the sun, placed under one another and lined up so that the wavelengths match from spectrum to spectrum. The positions of the third and fourth lines in the Balmer series of hydrogen are marked. The surface temperature of each star is indicated next to its spectrum at the right. We see that the intensity of the hydrogen lines increases from star to star as the surface temperatures go up from 5800°K to 10,000°K.

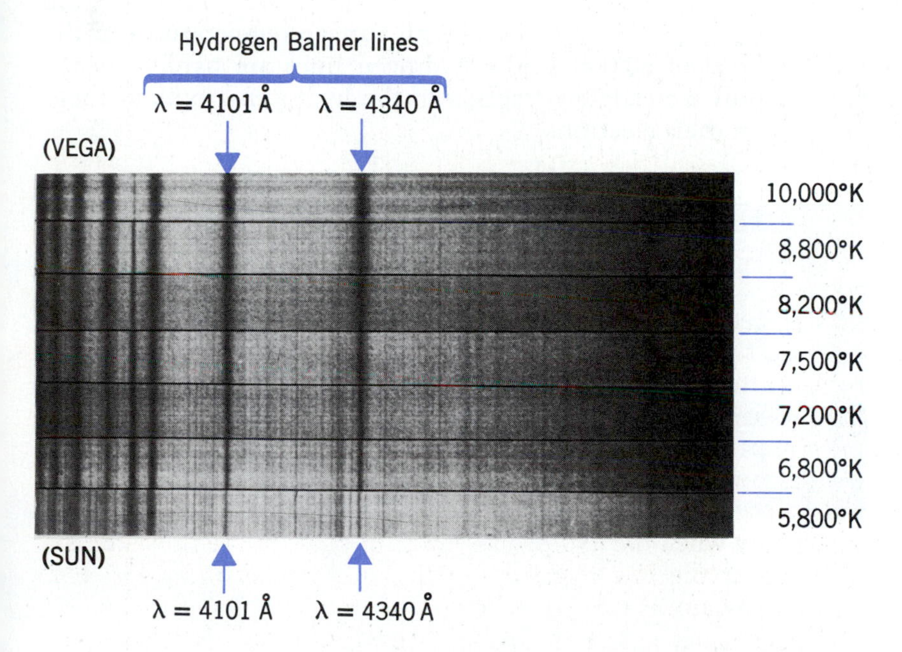

FIGURE 6.14
Effect of temperature on the intensity of hydrogen lines.

The higher temperature of Vega also explains the absence of the iron lines from its spectrum, mentioned on page 145. These lines, which appear prominently in the solar spectrum, are produced by *neutral* iron atoms, possessing their full complement of electrons. Because of the relatively high temperature of Vega's surface, however, nearly all iron atoms in Vega are *ionized;* that is, they have lost at least one electron in collisions. An ionized iron atom also possesses an absorption spectrum, but it lies in the ultraviolet rather than the visible region.

Suppose a star is even hotter than Vega. Will the hydrogen lines be still stronger than they are in Vega's spectrum? It would seem so according to our explanation. However, another factor entering into the case of extremely hot stars cuts down the intensity of the hydrogen lines. At an extremely high temperature most of the hydrogen atoms are stripped of their electrons—that is, ionized—by the violent collisions that take place in the hot gas. An ionized hydrogen atom is a proton, which cannot absorb light. Thus, in extremely hot stars, the hydrogen lines become fainter again.

You can see this effect clearly in Figure 6.15. The figure shows the spectrum of a star like Vega and the spectra of five other stars, once again all lined up under one another so that the wavelength scales match, and arranged in a sequence of surface temperatures with the Vega-like star on the bottom and the hotter stars above it. The sequence of surface temperature marked at the right runs from 10,000°K at the bottom to approximately 35,000°K at the top. The hydrogen lines are discernibly fainter in the hotter stars at the top of this figure. In very hot stars, with surface temperatures in the neighborhood of 50,000°K, the hydrogen lines are hardly visible, because only a negligible fraction of the hydrogen atoms in these stars retain their electrons.

FIGURE 6.15
Effect of increasing temperature on the intensity of hydrogen lines.

Classification of Stellar Spectra

Suppose we arrange all the stars in the sky in a sequence, with the hottest stars at the top of the list and the coolest at the bottom. What will their spectra look like? Astronomers use precisely this scheme to classify the complicated varieties of stellar spectra. The sequence has been arbitrarily divided into classes called spectral classes and to each class a letter has been assigned. The spectral classes are arranged in order of decreasing temperatures, with the hottest stars coming first and the coolest last. The letters that designate class are, from hottest to coolest, O B A F G K M. The sequence is easy to remember if you associate it with the first letter of each word in the sentence: *"Oh, Be A Fine Girl, Kiss Me."* This sentence has been burned into the memories of many generations of astronomy students.

Why is this peculiar sequence of letters used? Originally, the stars were classified according to the strength of the hydrogen lines in their spectra. They were divided into 16 groups, with the stars possessing the strongest hydrogen lines listed first and the stars with the faintest hydrogen lines listed last. The groups were lettered in alphabetical order from A through Q (with J missing for some reason). Later, astronomers realized that a classification by surface temperature made more sense, and the stars were rearranged, but the force of tradition prevailed, and the letter designations of the original listing were retained, although they no longer followed alphabetical order.

The stars with the most intense hydrogen lines originally were designated by the letter A, and they are still designated by this letter although, because they are not the hottest stars, they are no longer at the beginning of the sequence. Stars with quite faint hydrogen lines, placed near the end of the original sequence and designated with the letter O, are the hottest class of stars, and they are therefore at the head of the modern sequence.

Subclasses of Stars. Each spectral class has a range of temperatures. For example, B stars are defined as stars with surface temperatures in the range from 10,500°K to 30,000°K. However, astronomers can measure finer differences in the surface temperatures of stars within such large ranges. They describe these differences by subdividing each type of star into a maximum of 10 parts. The B stars, for example, are divided into 10 groups ranging from B0 to B9. The surface temperatures of the stars in each of these subdivisions are listed in Table 6.1.

TABLE 6.1
Surface Temperatures of B Stars

Type	Temperature (°K)	Type	Temperature (°K)
B0	30,000	B5	15,500
B1	23,000	B6	13,500
B2	21,000	B7	14,000
B3	18,000	B8	12,000
B4	17,000	B9	10,500

An Album of Stars. The photographs of the spectra of the principal types of stars are collected in Figure 6.16. Each spectral type or class is divided into ten subclasses running from 0 for the hottest to 9 for the coolest subclass. Two subdivisions are included for each type in order to display the smoothness of the transition in stellar spectra from the hottest to the coolest stars. The most significant lines in the spectra are indicated at the top and bottom of the figure. At the left of each spectrum appears the name of a familiar star of this type, if a familiar example exists. At the right is shown the temperature corresponding to the spectral type and the color of the star.

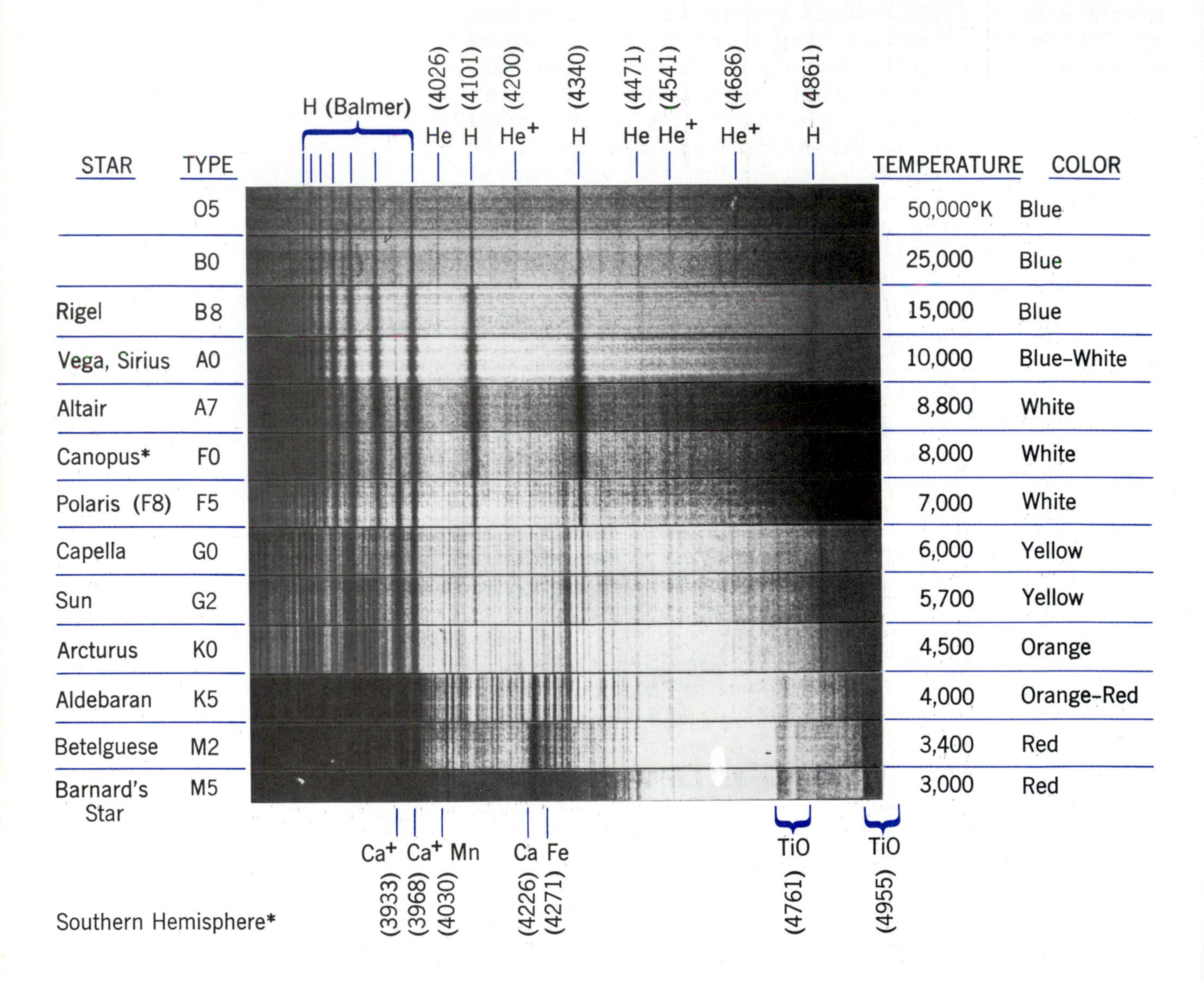

FIGURE 6.16
Temperature versus spectral type.

DETERMINATION OF THE ABUNDANCES OF THE ELEMENTS FROM STELLAR SPECTRA

Since every element leaves its fingerprints on the spectrum of a star in the form of a unique set of spectral lines, in principle it should be possible to determine the abundances of all the chemical elements in a star[1] directly from these spectral lines. How can we measure the relative abundances of the elements in a star accurately from the analysis of the star's spectrum?

As we saw in the case of hydrogen, the question does not have a simple answer. A faint set of absorption lines for a particular element does not necessarily mean that the abundance of that element is low. In the photograph of the sun's spectrum on page 132, for example, the two intense calcium lines at 3934 Å and 3968 Å are much more intense than the Balmer line of hydrogen at 6563 Å, but the abundance of hydrogen in the sun is several hundred thousand to a million times greater than the abundance of calcium. The explanation is that the intensity of an absorption line of an element depends not only on the number of atoms of the element that are present, but also on the fraction of these atoms that are in the correct initial state to produce that absorption. The 6563 Å line in the hydrogen spectrum is produced when an electron moves up from the first excited state of hydrogen to the second excited state. At the temperature of the sun's surface, most hydrogen atoms are in the ground state, and relatively few are in the first excited state. Consequently, the 6563 Å line is faint.

If the fraction of the atoms of each element in the ground state, the first excited state, the second excited state, and so on, could be determined, the difficulty would be resolved. The astronomer obtains this information with the aid of a calculation, using accurate formulas that predict how many atoms will be in the ground state and in each excited state for a given temperature. In applying these formulas to a star with a given mass and radius, the astronomer first guesses at the value of the temperature on its surface and the abundance of the elements in its atmosphere, then calculates the intensities of the lines in the spectrum with the aid of the formula, and finally compares the theoretical spectrum with the observed spectrum. He adjusts and readjusts the values of the temperature

[1] Actually, the spectral lines yield the abundances in the star's *atmosphere*, where the absorption takes place, and not the *interior* of the star. However, it is generally assumed that all stars are formed out of a well-mixed, homogeneous cloud of materials, and therefore that the composition of a stellar atmosphere is representative of the star as a whole. The core of the star is, of course, never homogeneous, since new elements are always accumulating there as a result of nuclear fusion reactions (Chapters 8 and 9).

and the relative abundances that he had assumed in order to secure the best possible agreement between theory and observation, finally arriving at a value for both the abundances of the elements in the star *and* the temperature at its surface.

The graph of the cosmic abundances of the elements on page 229 is partly based on computations of this kind. The temperatures listed for various spectral types on page 149 are derived from similar computations.

SHIFTS IN THE WAVELENGTHS OF SPECTRAL LINES; THE DOPPLER EFFECT

Sometimes, examination of a stellar spectrum reveals that the familiar spectral lines are not at their expected wavelengths. For example, the Balmer pattern may be there, but in the wrong place; that is, the Balmer lines are shifted to longer or shorter wavelengths. What is the meaning of this shift in wavelength? Does it signify that there are several kinds of hydrogen atoms in the Universe?

The answer is that the shift in the spectral lines in these cases is not caused by a change in the nature of the atoms in the star, but by a motion of the star relative to the observer. An Austrian physicist named Christian Doppler discovered in 1842 that the wavelength of a train of waves is changed if the source of the waves is in motion relative to the observer. This change is called the *Doppler effect* or the *Doppler shift.* The Doppler shift can occur in any kind of wave motion, and is not limited to electromagnetic waves. It arises whenever the observer and the source of the waves are moving toward or away from one another and it does not matter whether one is moving and the other is at rest or both are moving; the observed Doppler shift is the same.

The explanation of the Doppler shift is not complicated. Suppose an oscillating electric charge—an electron in an atom or a star—is sending waves into space in the direction of the earth (Figure 6.17). Suppose also that the star is moving backward in the direction away from the earth. Waves continue to be added to the wave train, one after the other, as the source oscillates up and down. Each of these waves moves out into the space at the same speed as before. However, during the time in which the oscillating charge creates one wave, the charge itself moves backward, so that the length of the wave is stretched (Figure 6.18).

The amount of the increase in the wavelength depends on the velocity of the star and the velocity of the wave train. If the source moves slowly in comparison to the wave train, each wave will be stretched by a correspondingly small amount. If the source moves backward nearly as fast as the wave train moves forward, the waves

FIGURE 6.17
A source of light generates an electromagnetic wave.

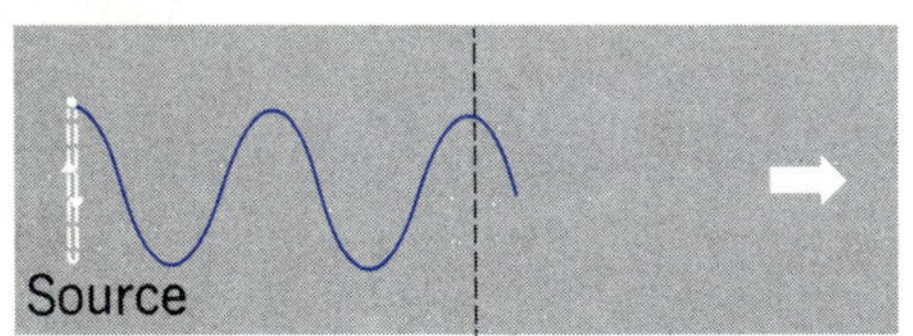

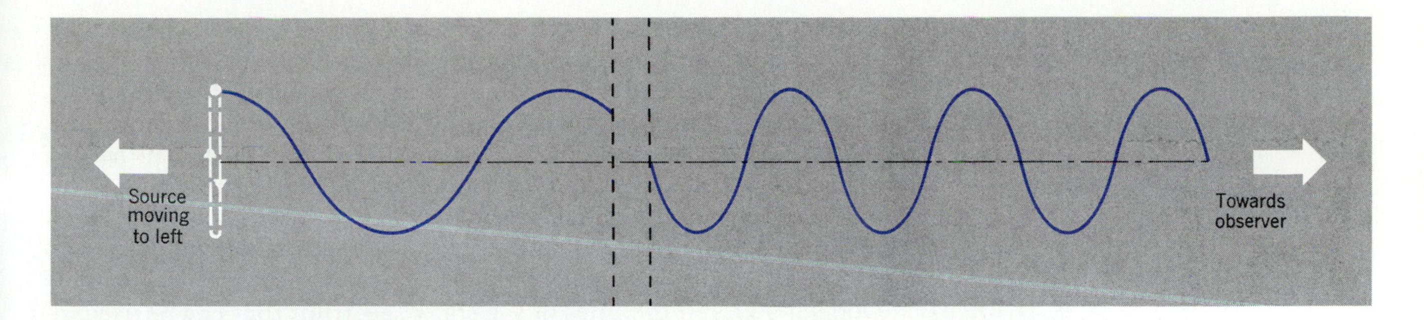

FIGURE 6.18
The stretching of waves by the receding motion of the source.

will be stretched enormously. These ideas can be expressed in a formula giving the change in wavelength $\Delta\lambda$ produced by a motion of a velocity v, when the speed of the waves is V and their original, unstretched wavelength is λ:

$$\frac{\Delta\lambda}{\lambda} = \frac{v}{V}.$$

The same type of reasoning applies when the star moves toward the earth instead of away from it. In this case, the length of a wave is decreased because the oscillating charge moves forward during the time in which the wave is produced. Again the amount of the effect depends on the speed of the star and the speed of the wave train: the larger the speed of the source in comparison to the speed of the waves, the greater the effect. The same formula can be used for this case as for the case of backward motion, if we agree to attach a plus sign to the velocity of the star when it is moving away from the earth and a minus sign to it when it is moving toward the earth. This rule is reasonable because the distance from the source to the observer increases in the first case and decreases in the second.

The formula also applies to the case in which the star is fixed and the earth is moving, or the case in which both are moving. The formula can be used in any of these cases if v is defined as the net speed at which the star and the earth approach or recede from one another.[2]

Information Yielded by the Doppler Effect.

The shift in the wavelength of spectral lines may, at first, seem like a hindrance to the analysis of stellar spectra. In fact, it often turns out to be a rich source of information. For example, if the lines in a

[2] This formula is not precise because it does not include effects of relativity theory. The exact formula derived from relativity theory is $\lambda(\text{new}) = \lambda(\text{old})\sqrt{(1 \pm v/c)/(1 \mp v/c)}$, in which c is the speed of light. The upper sign is valid for a receding source and the lower sign for an approaching source. When v/c is small the approximate and exact formulas are in close agreement.

star's spectrum form a familiar pattern, but every line in the pattern is shifted by an amount proportional to its wavelength, it can be inferred that the star is moving relative to the earth through space. The velocity with which the star moves—at least, the component of the star's velocity toward or away from the observer—can be determined from the formula for the Doppler shift. As another example, if a star is rotating on its axis, a part of the star's surface will be moving toward the earth and another part will be moving away from the earth, as a result of the rotation. These motions will produce Doppler shifts in the lines of the star's spectrum that can be used to infer the rate of rotation.

OTHER INFORMATION YIELDED BY STELLAR SPECTRA

Stellar spectra are a gold mine of facts regarding the structure of stars. In addition to revealing the ingredients of a star, the temperature on its surface and the properties connected with the Doppler effect, the star's spectrum can reveal whether the star is a small object, like Barnard's star, or a distended red giant, like Betelgeuse; and it can determine the strength of the star's magnetic field, if any. The spectrum of a star can even reveal that the star is shedding its outer layers in a massive ejection of material to space. Appendix B contains an explanation of the methods used to obtain this information.

Main Ideas

1. The structure of the atom; the buildup of the table of the elements.
2. Ground states and excited states of the atom.
3. The uniqueness of atomic spectral lines.
4. Three types of spectra: emission, absorption, and continuous.
5. The similarity between laboratory spectra and stellar spectra.
6. A comparison of the spectra of different stars.
7. The effect of temperature on stellar spectra.
8. The division of stellar spectra into spectral types.
9. The abundance of the elements.
10. The Doppler effect in stellar spectra.

Important Terms

absorption spectum
atom
atomic spectral lines
Balmer series
continuous spectra
Doppler broadening
Doppler effect (shift)
electron
emission spectra
excited state
Fraunhofer lines
ground state

ionization
Lyman series
nucleus (atomic)
photon
periodic table
shells (electron)
spectral class (type)
stellar rotation
stellar spectra
surface temperature
transition

Questions

1. What are the differences between the everyday world and the world of the atom.
2. What is the most fundamental property that distinguishes one chemical element from another?
3. Why does an atom radiate photons with discrete wavelengths and not a continuous range of wavelengths? Explain why the wavelengths of light radiated from the atoms of an element are unique to that element.
4. Draw an energy-level diagram of an atom with three states labelled 1, 2 and 3, corresponding to electron orbits of successively greater radius. Draw an arrow representing the transition from state 1 to state 3. Does this transition represent *emission* or *absorption?* Is the wavelength of the photon longer or shorter than it would have been for a transition from state 2 to state 3?
5. Consider the atom of hydrogen, the once-ionized atom of helium, the 5-times-ionized atom of carbon. What are the similarities in these spectra? What are the differences?
6. Considering the fact that when excited atoms collapse to lower levels, they emit photons with the same wavelengths as those absorbed, why do absorption lines appear in the spectra of gases?
7. Consider a cloud of gas exposed to radiation from a nearby hot star. Describe the appearance of the spectrum of the cloud observed at right angles to the direction of the star. Describe the appearance of the spectrum of the star when observing in the direction of the star.
8. What would happen to the sun if nuclear reactions were suddenly extinguished?
9. List the spectral classes of stars in order from higher to lower temperatures, and the approximate surface temperatures of these classes. For each class, mention one property of the star's spectrum.
10. Indicate the spectral class and probable temperature of stars whose spectra have the following characteristics:
    (a) Lines of ionized helium weak; exceedingly weak lines of neutral helium and hydrogen.
    (b) Strong lines of neutral helium; weak lines of ionized helium.
    (c) Strong hydrogen lines; no helium lines.

THE HERTZSPRUNG-RUSSELL DIAGRAM

7

A person has two basic properties that stand at the top of any list of his physical characteristics. They are easily estimated by looking at him, and are always used as the basic means of identifying or classifying the individual in a group of people. The two properties are *height* and *weight*. Suppose you make a graph of height versus weight and plot a point for each member in a group. Such a plot is shown in Figure 7.1.

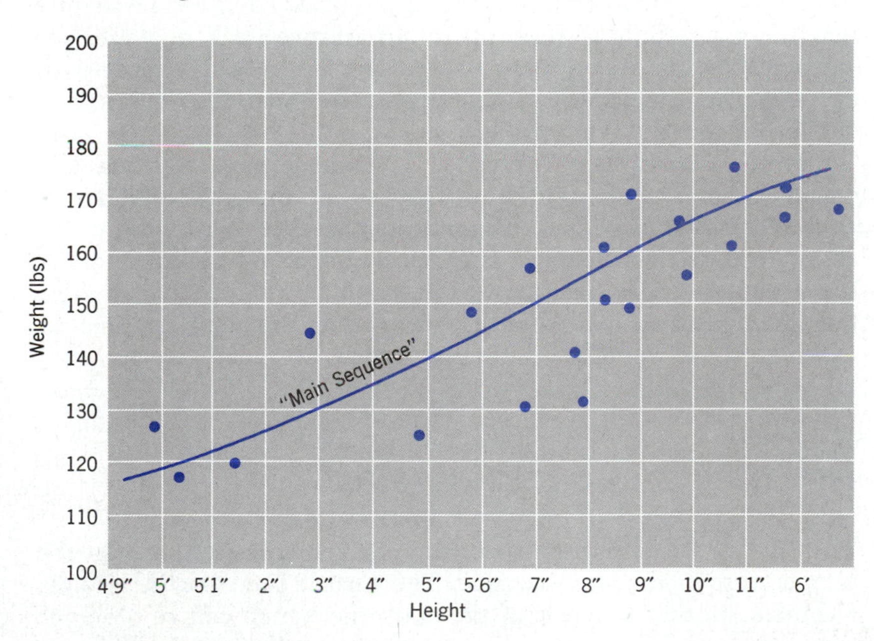

FIGURE 7.1
Plot of height vs. weight for a random sample of 21 adults.

E. Hertzsprung (left) and H. N. Russell (right).

Unless the group is very unusual, you will find that most of the points lie close to a line. You might call this the "Main Sequence" of physical properties of adults in the United States. The "Main Sequence" drawn through the middle of these points probably is typical of what would be obtained by plotting the height and weight of any other similar group in the United States. The significance of the "Main Sequence" line is that it gives a connection between height and weight in the population; this connection does not hold true, in general, for any *individual* member of the population, it only holds true for the *average* person.

Does a star possess properties, analogous to height and weight in a person, that may be considered to be its fundamental observable attributes? Does it have two properties that would enable us to classify stars in the sky in the same way that we classify people? A partial answer comes to us immediately: the *temperature* of a star, as evidenced by its spectral type or its color, is one basic property of the star. Is there another? A moment's thought indicates that the intrinsic *brightness* of the star is a second basic property. Is there a third? The answer is no; at least, no other property comes to mind that is as fundamental as the temperature and the intrinsic brightness of a star.

Color and brightness—these are the two properties that every star presents to the naked eye, and they are the only two. Presumably with these thoughts in mind, a Danish astronomer named Hertzsprung and an American astronomer named Russell separately decided to plot data equivalent to color and brightness of all stars for which this information was available. Hertzsprung and Russell had no prior thought of what they might find. Their hope was only that the population of the skies might arrange itself in some way when plotted on such a chart, so as to seem meaningful to their eyes, and that thereby they might acquire a clue to the inner nature of this distant object known as a star.

The two astronomers succeeded beyond their expectations. The diagram that grew out of their investigations, universally referred to as the *Hertzsprung-Russell* diagram, has become as essential to the astronomer as the calculator is to the engineer. As a preliminary to the interpretation of this invaluable graph, we must develop more precise concepts of *color* and *brightness* than the qualitative senses in which these words are normally used in everyday speech.

THE COLOR OF A GLOWING OBJECT

The lines in the absorption spectrum of a star are determined by the temperature of the star's surface. The surface temperature also determines another property of the star, one which can be observed without the aid of a special instrument such as the spectroscope. This property is the color of the star.

Color, Wavelength, and Temperature

What is color? Color means wavelength; the quality perceived by the eye as color is the dominant wavelength in the visible radiation received from an object. A star, or any other hot object, emits electromagnetic radiation stretching across the entire spectrum from infinitely short wavelengths to infinitely long ones. However, the waves emitted at various wavelengths do not have equal intensities. The intensity is always weak at very short and very long wavelengths, and it is strongest at some wavelength in between. What determines the dominant wavelength and, therefore, the color of the radiation?

A moment's reflection will tell you the answer. Suppose that you place a bar of iron in a furnace and steadily raise its temperature. At first, the iron feels hot and sends out heat but does not glow visibly; at this point the peak of its radiation is in the infrared region. As the temperature of the iron increases, it begins to emit radiation in the visible region, first turning dull red in color, then cherry-red, then yellow, and finally, at the highest furnace temperature, becoming white-hot.

This simple thought-experiment—which has its counterpart in many experiences of everyday life involving kitchen stoves, soldering irons, and so on—shows that the *temperature* of an object determines the wavelength at which it radiates most of its energy. As the temperature of the object increases, the radiation emitted by it moves toward shorter wavelengths—from the infrared to the red, then to the yellow, to the blue, and eventually, if the temperature mounts high enough, presumably past the blue and the violet into the ultraviolet.

As a parenthetical question, you may ask why an object becomes white-hot at furnace temperatures. This seems difficult to understand because white is not one of the colors of the spectrum. The answer is that when an object is heated to the point where its peak intensity is in the visible spectrum—in the yellow, let us say—it will also emit radiation at surrounding wavelengths in the blue, green, and red. That is, it will radiate a mixture of all wavelengths in the visible region. Such a mixture of all colors is recorded by the eye and brain as the sensation of *white* light.

Wien's Law. There is a formula connecting the temperature of a glowing object and the wavelength at which it radiates with the greatest intensity. This formula, known as Wien's law, is

$$T = \frac{2.89 \times 10^7}{\lambda_{\mathrm{max}}}$$

where T is the temperature in degrees Kelvin and λ_{max} is the wavelength at which the intensity has its peak. (The wavelength is expressed in angstroms.) Notice that the formula agrees with your everyday experience on the way in which the color of a heated object changes as its temperature increases: the hotter the object, the shorter the wavelength of its most intense radiation.

Connection Between Color and Spectral Type

FIGURE 7.2
Scattering of sunlight by the atmosphere.

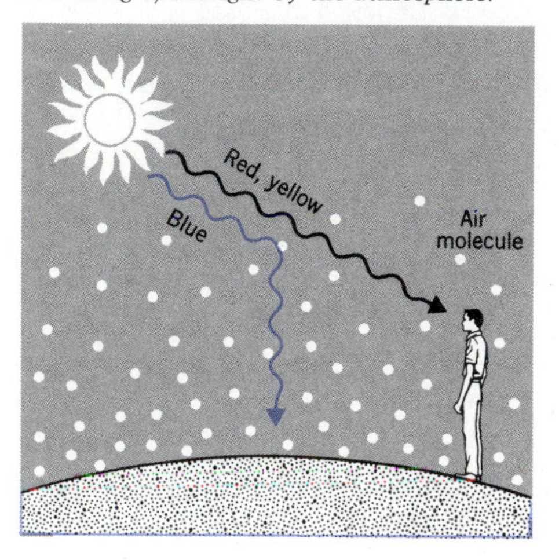

Stars, like all glowing objects, obey Wien's law. The hottest stars—the ones of the O and B types—radiate most of their energy in the far ultraviolet region. A relatively small amount of their energy is emitted in the visible region, with emphasis on the blue end of the visible spectrum, giving these stars a bluish-white color. A-type stars radiate a more uniform distribution of intensity in the visible region and, therefore, are white in color. As the surface temperature decreases, and the peak intensity moves to longer wavelengths, the color of the star changes from blue-white to white to yellow-white, then to yellow, to orange, and finally to red. F stars are yellow-white, G stars like the sun are yellow, K stars are orange, and M stars are red.

Yet the color of a star does not correspond exactly to its peak of radiation. For example, the sun radiates most intensely in the blue-green part of the visible spectrum, but it appears yellow-white rather than turquoise. Why? The answer has two parts. First, an object radiating at peak intensity in the blue-green region will also radiate appreciably at surrounding wavelengths, producing a sensation of white light with a bluish or blue-green cast. Second, the earth's atmosphere affects the rays of light from the sun in a way that tends to remove the blue-green component. Molecules in the air scatter the sunlight as it traverses the atmosphere. If the atmosphere scattered all wavelengths to the same degree, there would be no effect on the color of the light from the sun, but actually the shorter wavelengths are scattered more strongly than the red. Thus, looking at the sun in the sky, you see light from which a large amount of blue has been removed. This scattering effect eliminates the bluish cast that we might otherwise expect in sunlight, shifting the peak of the spectrum toward the yellow (Figure 7.2).

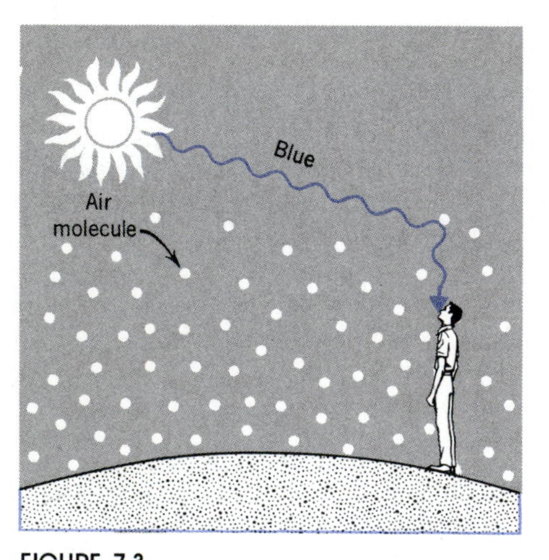

FIGURE 7.3
Why the sky is blue.

Incidentally, the scattering effect explains several other phenomena. The sky is blue because the rays of light that enter your eye are the rays from the sun that were scattered downward toward you by molecules of air. Because the blue wavelengths are scattered more strongly than the yellow or the red, the light you see in the sky is primarily blue (Figure 7.3). The sun turns orange and then red as it sets because its rays pass through the maximum thickness of air at sunset. The atmosphere scatters all of the blue and green light, leaving only a mixture of yellow, orange, and red in the spectrum of the sun (Figure 7.4).

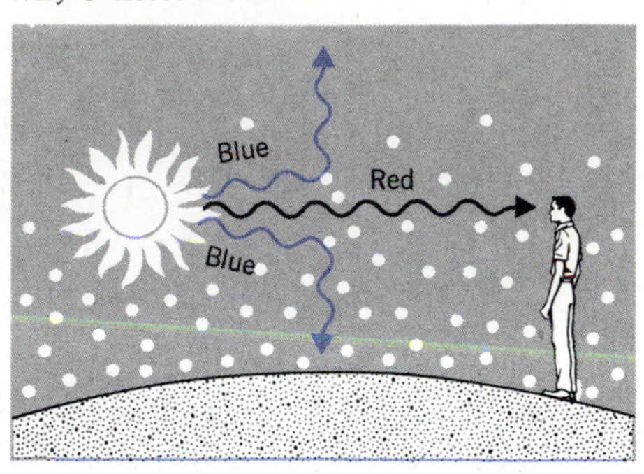

FIGURE 7.4
Why sunsets are red.

A Plot of Intensity of Radiation Versus Wavelength

Scientists have measured in the laboratory the amount of energy radiated at different wavelengths by objects at various temperatures. Theoretical physicists have also calculated the distribution of radiated energy. Both measurement and theory predict a bell curve, with a sharp decrease of energy on the short wavelength side of the peak and a slow falloff on the long wavelength side. The circles on the graph in Figure 7.5 represent measurements made in the laboratory of the intensity of radiation emitted by an object at a temperature of 1600°K. The solid line in this graph shows a theoretical calculation of the energy radiated by an object at the same temperature. The agreement between the measurements and the theoretical curve is nearly perfect.

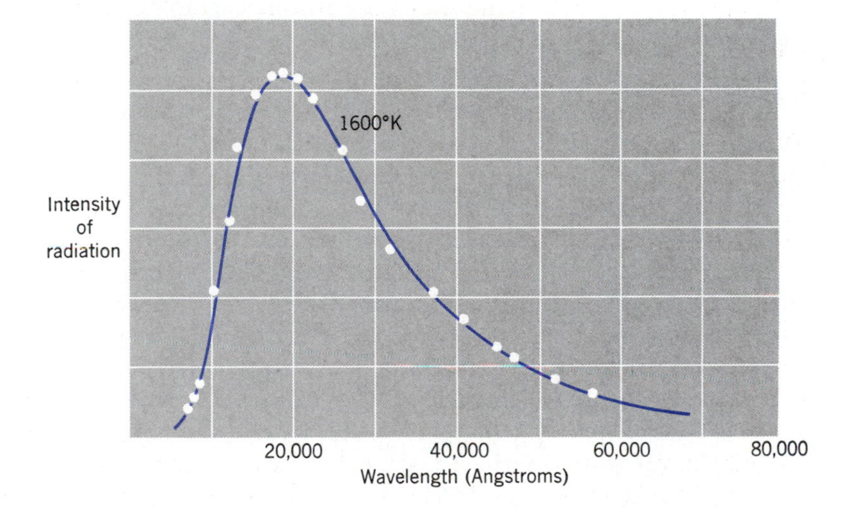

FIGURE 7.5
Measured radiation from an object at 1600°K.

The sun has a temperature of roughly 6000°K. Figure 7.6 (p. 162) shows the energy radiated at different wavelengths by the sun,[1] or by any other star or object with a surface temperature of 6000°K. The shaded portions of the spectrum lie outside the visible region; the visible part of the spectrum is shown without shading.

Stars much hotter than the sun emit most of their radiation in the ultraviolet. Vega, with a surface temperature of 10,000°K is

[1] Neglecting the absorption lines and bands in the sun's spectrum.

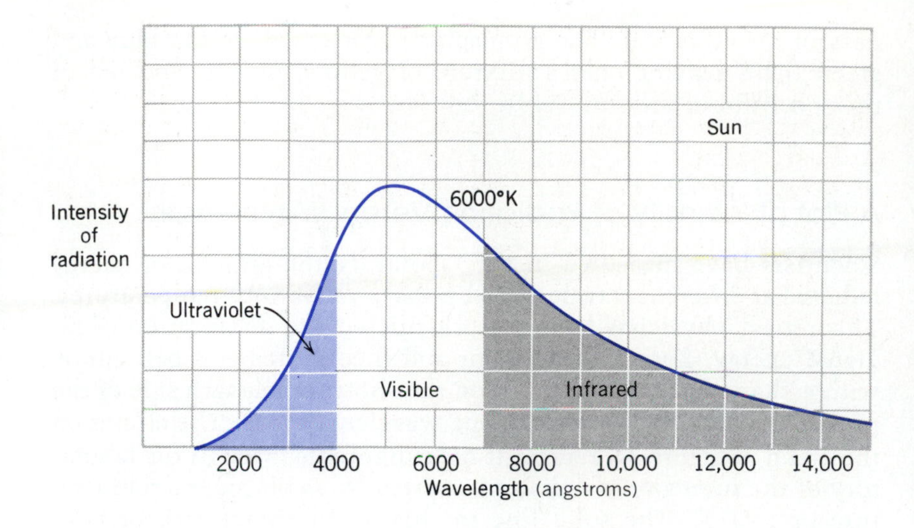

FIGURE 7.6
Radiation from an object at 6000°K.

an example of a hot star. The graph in Figure 7.7*a* shows approximately the relative amount of energy radiated by Vega at different wavelengths. One-half of the energy radiated lies in the ultraviolet region, about one-third is radiated in the visible region, and the remaining part lies in the infrared region.

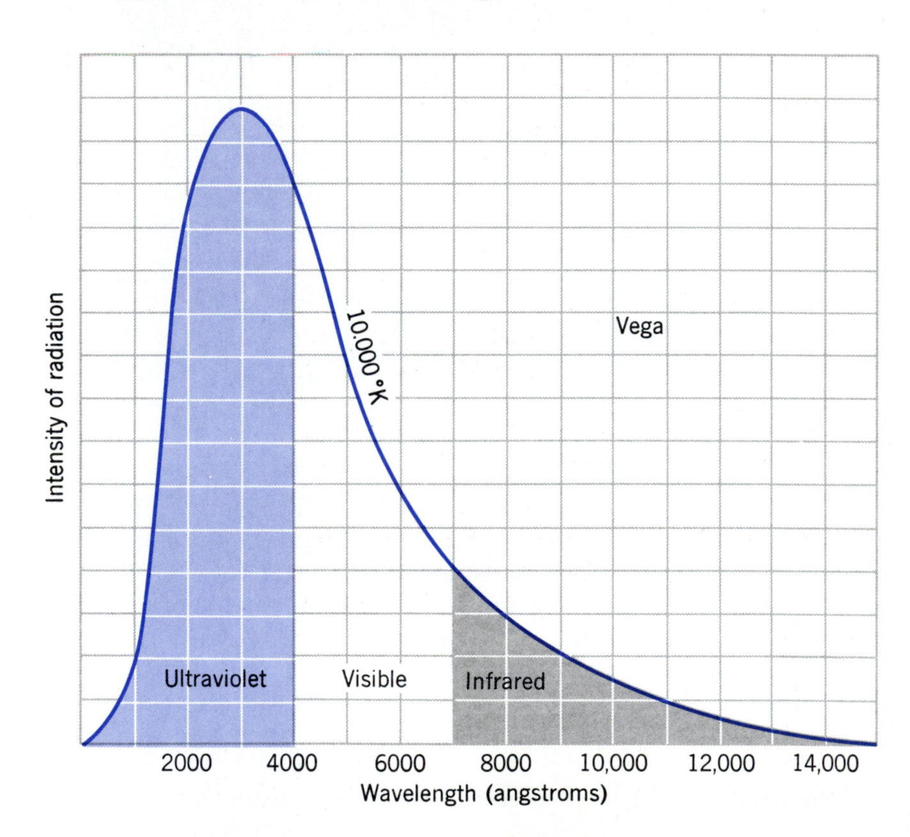

FIGURE 7.7(a)
Radiation from an object at 10,000°K.

Few stars have surface temperatures exceeding 50,000°K. A star at this temperature emits its peak radiation at a wavelength of 560 Å, deep in the ultraviolet region. Such stars shine mostly in the ultraviolet; they emit a relatively small part of their radiation as visible light.

Many stars are cooler than the sun and emit most of their energy in the red and infrared, radiating little energy in the visible region and practically no energy in the ultraviolet. As an example, consider Barnard's Star with a temperature of 3000°K (Figure 7.7*b*). This cool star will emit its peak of radiation at a wavelength of 9600 Å, in the infrared region.

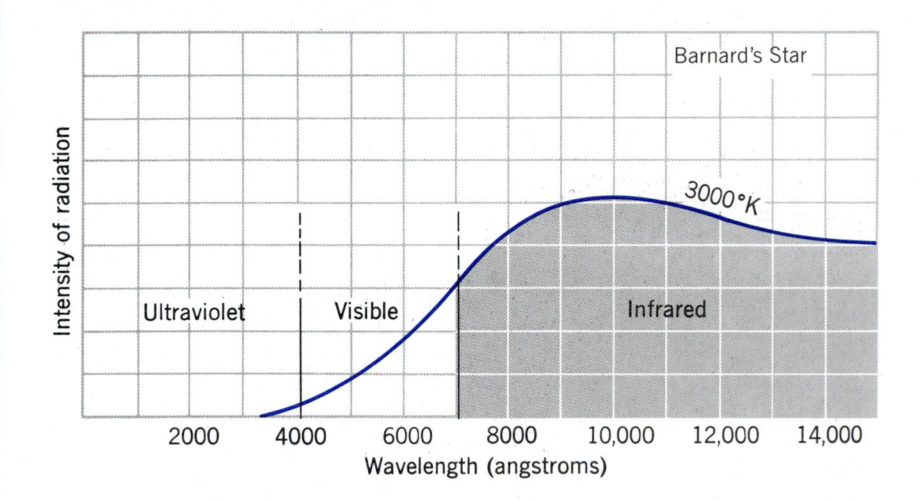

FIGURE 7.7(b)
Radiation from an object at 3000°K.

The intensity of the radiation in Figure 7.7*b* has been exaggerated relative to the intensity in Figure 7.7*a* in order to provide a clearer picture of the amounts of energy radiated by a cool star in different wavelength regions. However, if the two intensity curves are plotted on one graph using the same vertical scale, they have the appearance shown in Figure 7.8 (p. 164). The radiation from a star at a temperature of 6000°K, roughly that of the sun, is also shown in Figure 7.8.

Cool, red stars are more numerous than any other kind of star in the sky. They are considerably more numerous than stars resembling the sun. However, very few red stars are visible, partly because red stars tend to be exceedingly dim and partly because they radiate primarily in the infrared region.

Some red stars are conspicuous exceptions to the rule that red stars tend to be very dim. These stars, known as red giants, are relatively cool, emitting most of their energy in the red and the infrared; nonetheless, they are very luminous. Antares is an example of a red giant. Although its surface temperature is only

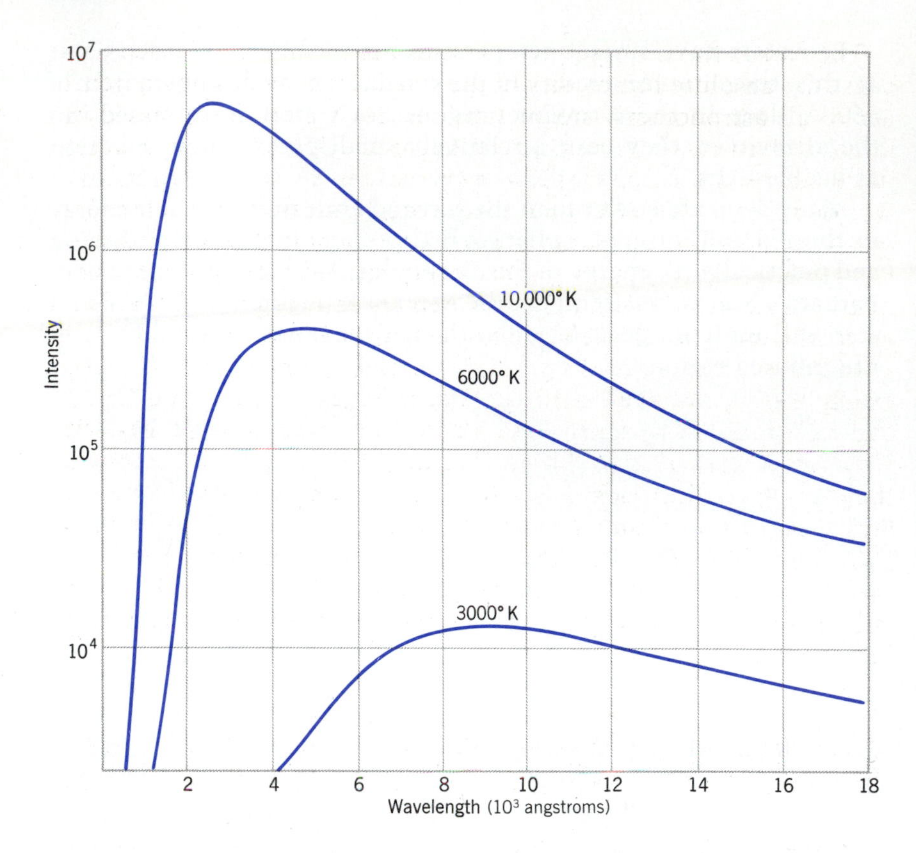

FIGURE 7.8
Comparison of radiation from objects at several temperatures. A logarithmic scale is used because of the large difference in intensity at different temperatures.

3600°K, it is thirty thousand times more luminous than the sun; that is, it radiates thirty thousand times as much energy into space every second.

LUMINOSITY AND BRIGHTNESS

The radiant energy emitted by a star is referred to as its luminosity by astronomers. The absolute luminosity of a star is the amount of energy radiated into space each second from the star's surface. This energy is partly in the form of visible light and partly in the form of radiation at other wavelengths, such as the infrared or ultraviolet.

Astronomers and physicists usually measure absolute luminosity in ergs per second. An erg is a unit of energy in the centimeter-gram-second system. It is a very small amount of energy; when a fly collides with a window screen it imparts about one erg to the screen on impact. A 100-watt bulb radiates about one billion ergs per second into space.

The sun radiates about 4×10^{33} ergs per second[2] into space. Because the absolute luminosity of the sun is a very well-known number to all astronomers, stellar luminosities are often expressed in units of the sun's luminosity, written as $L_\odot$. The circle with a dot in its center is generally used as a symbol for the sun in astronomy (and astrology). Alpha Centauri—the star nearest to our solar system—has an absolute luminosity of $2L_\odot$, meaning that it radiates 8×10^{33} ergs of energy into space every second.[3] If the earth were in the same orbit around Alpha Centauri as it is around the sun, its temperature would be close to the boiling point of water.

Massive stars have luminosities ranging up to $10^6\ L_\odot$. An example of a massive, luminous star is Rigel, a star in the Orion constellation. If the earth were placed in an orbit around Rigel it would vaporize instantly. Barnard's Star—the star second nearest to the sun—has a luminosity of $10^{-2}\ L_\odot$. A planet circling Barnard's Star at our distance from the sun would have a temperature of minus 200° Fahrenheit. The dimmest stars observed up to now, called red dwarfs, have luminosities extending down to $10^{-6}\ L_\odot$. The sun falls in the middle of the range between the most luminous and least luminous stars.

Apparent Luminosity or Brightness and Apparent Magnitude

When you look at the stars in the sky, some seem considerably brighter than others, but this impression does not necessarily correspond to their absolute luminosity. For example, Vega seems brighter than Deneb. Actually it is 1600 times fainter, but Deneb is roughly 55 times farther away. Because the intensity of light falls off as the square of the distance, the apparent luminosity of Deneb is diminished by a factor of about 3000, relative to Vega. Consequently, Vega appears brighter than Deneb, although Deneb is intrinsically more luminous than Vega.

Astronomers use a unit called apparent magnitude to measure apparent luminosity. This unit dates back to the second century B.C., when the Greek astronomer Hipparchus drew up a list of stars visible to the naked eye and assigned a number to each one to indicate its relative brightness. He assigned the number "1" to the brightest stars, which came to be known as "first-magnitude stars." In Hipparchus' list, Sirius, Vega, and Deneb were first-magnitude stars. All the stars in the Big Dipper were second-magnitude stars.

[2] If you were to express this number in words, it would be 4 billion trillion trillion.

[3] Alpha Centauri is a triple star. This figure refers to the luminosity of the largest member of the triplet.

Today, the faintest stars that the naked eye can see are classified as sixth magnitude.

The system devised by Hipparchus is still in use today, but the units of first magnitude, second magnitude, and so on have acquired a more precise meaning. Instead of judging the brightness of a star with the naked eye, astronomers use a photometer to measure the visible radiation from each star more accurately. The photometer readings reveal that, on the average, first-magnitude stars in the Hipparchus scale are 100 times brighter than sixth-magnitude stars. Physiologists find that when the eye observes a linear increase in brightness in a series of light sources, the measured increase in brightness turns out to be geometrical. That is, if the eye records an impression of a uniform *step*-increase, the actual increase is a constant *multiple* from one source to the next. In the case of Hipparchus' scale, the eye observes five uniform steps of increasing brightness from sixth- to first-magnitude stars. Therefore, the true increase in brightness from the sixth to the first magnitude must be $R \times R \times R \times R \times R = R^5$, where R is the ratio of brightness of a star of one magnitude to a star of the next magnitude. Since the photometric observations show that the ratio of brightness of first- to sixth-magnitude stars is 100, we have

$$R^5 = 100 \qquad \text{and} \qquad R = (100)^{1/5} = 2.51$$

In other words, a first-magnitude star is 2.51 times brighter than a second-magnitude star and $(2.51)^2 = 6.31$ times brighter than a third-magnitude star. The table (Figure 7.9) tabulates the relative brightness of stars of various magnitudes compared to the brightness of a first-magnitude star.

This table gives the relative brightness of the *average* for each class of stars. *Individual* stars will differ from the average for their class to some degree. For example, Deneb, a first-magnitude star, appears a little fainter than the average for a first-magnitude star, although it is definitely brighter than a second-magnitude star. The photometer readings place it between the two, with a relative magnitude of 1.3.

Some first-magnitude stars are brighter than the average for their class. For example, Vega is about 60 percent brighter than the average for a first-magnitude star. Because brighter stars have smaller numbers on the relative magnitude scale, the apparent magnitude of Vega is 0. Sirius—the brightest star—is four times brighter than Vega. The apparent magnitude of Sirius must be *negative.* It turns out to be −1.4. The sun is 100 billion times greater in apparent brightness than the average first-magnitude star. Its apparent magnitude is −26.7.

FIGURE 7.9
The magnitude scale.

Apparent Magnitude	Brightness Relative to First-Magnitude Stars
1	1
2	1 : 2.51
3	1 : 6.31
4	1 : 15.85
5	1 : 39.82
6	1 : 100

With the aid of time exposure, the 200-inch telescope on Mount Palomar can photograph stars down to the twenty-third or twenty-fourth magnitude. A twenty-third-magnitude star has the apparent brightness of one candle viewed from a distance of 10,000 miles.

Absolute Magnitude. Reference books on astronomy often list the "absolute magnitude" of a star in addition to its apparent magnitude. The absolute magnitude is another way of expressing the absolute luminosity of a star. The absolute magnitude of a star is defined in terms of a unit of distance called the parsec, equal to 3.26 light-years. The explanation for the use of this unit is given in Appendix A. A star's absolute magnitude is the apparent magnitude that this star would have if moved from its actual position to a new position at a distance of 10 parsecs from the sun. According to this definition, a star with an absolute magnitude of "-1" radiates 1.2×10^{35} ergs per second into space. The absolute magnitude of the sun is 4.7. The absolute magnitude of Deneb—a considerably brighter star than the sun—is -7.2. The absolute magnitude of Barnard's Star is $+10.7$.[4]

Distance. The distance to a star determines the relation between its absolute and apparent magnitudes. If a star is farther away than 10 parsecs or 32.6 light-years, its apparent magnitude is greater than its absolute magnitude. If the star is closer than 10 parsecs, its apparent magnitude is smaller than its absolute magnitude. If the star is at a distance of 10 parsecs, the two quantities are identical. These relationships are contained in the following formula connecting a star's absolute magnitude M, apparent magnitude m, and distance d measured in parsecs

$$M = m + 5 - 5 \log d \text{ (parsecs)}$$

The apparent magnitude of a star can always be measured. If the star's absolute magnitude or luminosity can be estimated, its distance can be determined from the formula by solving for d:

$$d = 10^{(1/5)(m-M+5)}$$

Appendix A discusses the use of this method to determine the distances to stars and galaxies.

[4] These numbers are bolometric magnitudes, which are based on the total energy radiated by a star at all wavelengths. The numbers in the previous paragraph are called visual magnitudes and are based only on the energy radiated in the visible part of the spectrum.

THE POPULATION OF STARS

Let us plot the luminosity and surface temperatures of the 12 stars within 10 light-years of the sun (Figure 7.10). Inspection of the graph shows that for stars, just as for people, nearly all the points fall on or near a line running diagonally across the graph; they are not scattered randomly all over the area of the graph. We can immediately draw the following conclusion: there is a connection, on the average, between the intrinsic luminosity and the surface temperature of a star. Furthermore, we can see that the connection is such that the hottest stars are the brightest ones and the coolest stars

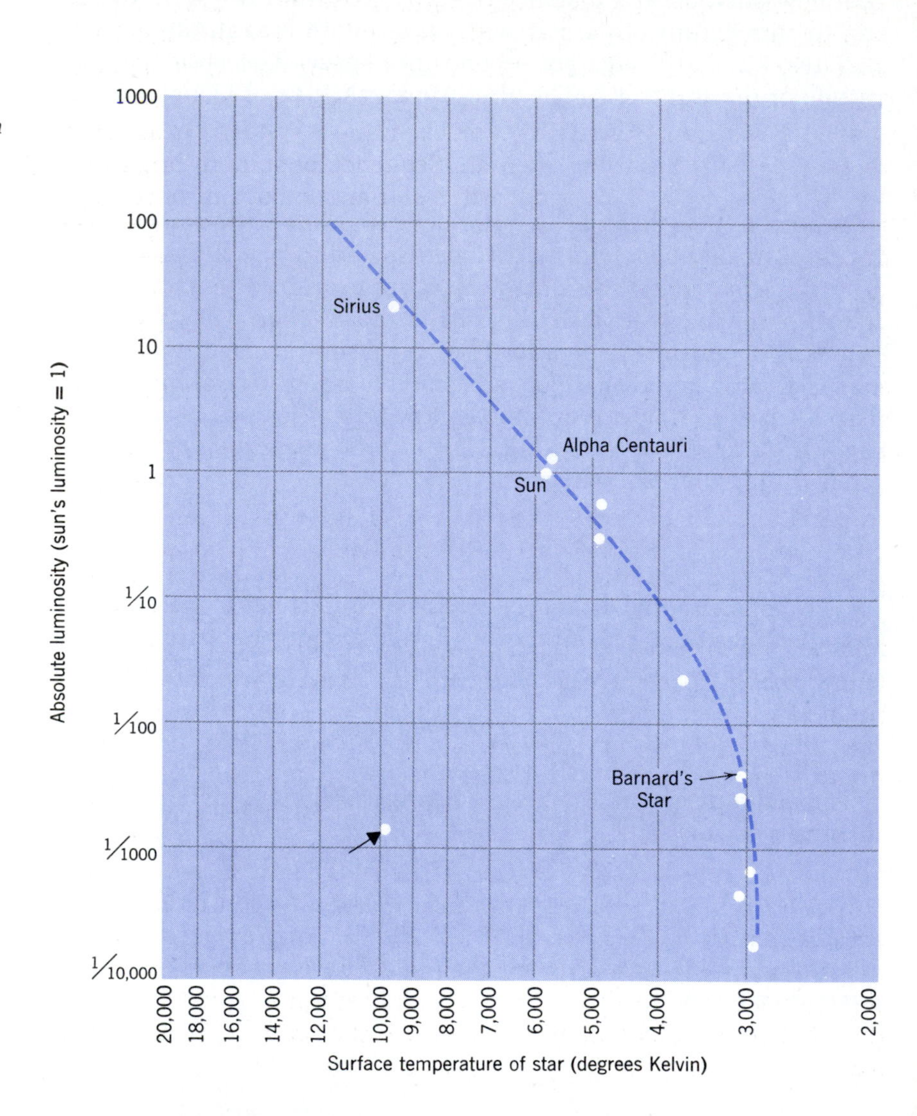

FIGURE 7.10
Luminosity vs. temperature for stars within a distance of 10 light-years from the sun.

are the faintest. This result agrees with our everyday experience regarding glowing objects: The hotter a glowing object, the more energy it radiates.

The Main Sequence

The dashed line drawn on the graph represents a connection between brightness and temperature. This line can be called the Main Sequence of basic properties of the population of stars or, at least, of the stars that are located near us in our galaxy.

In one respect, the star plot differs from the people plot. Although most of the stars cluster along the main sequence line, one out of the 12, marked by an arrow, lies quite far from the line, near the bottom of the plot. Inspection of the graph shows this star has a somewhat higher temperature than the sun but is intrinsically 300 times fainter. No person could possibly deviate from the main sequence line of height and weight in the human population as much as this star deviates from the Main Sequence of stars in brightness and temperature. Discovering such a star at the bottom of the plot is as much a surprise as it would be to discover a thriving person who was six feet tall and weighed only a few pounds, or one who was four feet tall and weighed 1000 pounds.

What can be the explanation for this extremely peculiar star? Can there be more than one Main Sequence? Let's enlarge the volume around the sun to a sphere with a radius of 20 light-years containing 90 stars, including the sun. The graph in Figure 7.11 shows the results of a plot of luminosity versus surface temperature for the 90 stars. We see that most of the stars continue to lie very close to the Main Sequence line, just as in the case of the smaller sample whose plot is shown in Figure 7.10. However, a substantial number of stars, although still a minority, lie quite far from the Main Sequence. The peculiar star lying at the bottom of the earlier graph is now joined by several others also located near the bottom, well below the Main Sequence. They are indicated by arrows. One of them has a surface temperature of 14,000°K. If this were a normal star following the Main Sequence, it would be 200 times brighter than the sun; yet, the graph shows it to be far fainter!

Figure 7.11 (p. 170) shows that the majority of the stars in our neighborhood are exceedingly faint, only a few being as bright as the sun. We would expect the brightest stars in the sky to be our neighbors, but this turns out not to be the case, for only a dozen of the 90 stars in the list can be seen with the naked eye under normal conditions. The bright stars that make up the familiar constellations of the night sky are for the most part a different group, relatively distant, but so luminous that they outshine the relatively dim nearby stars.

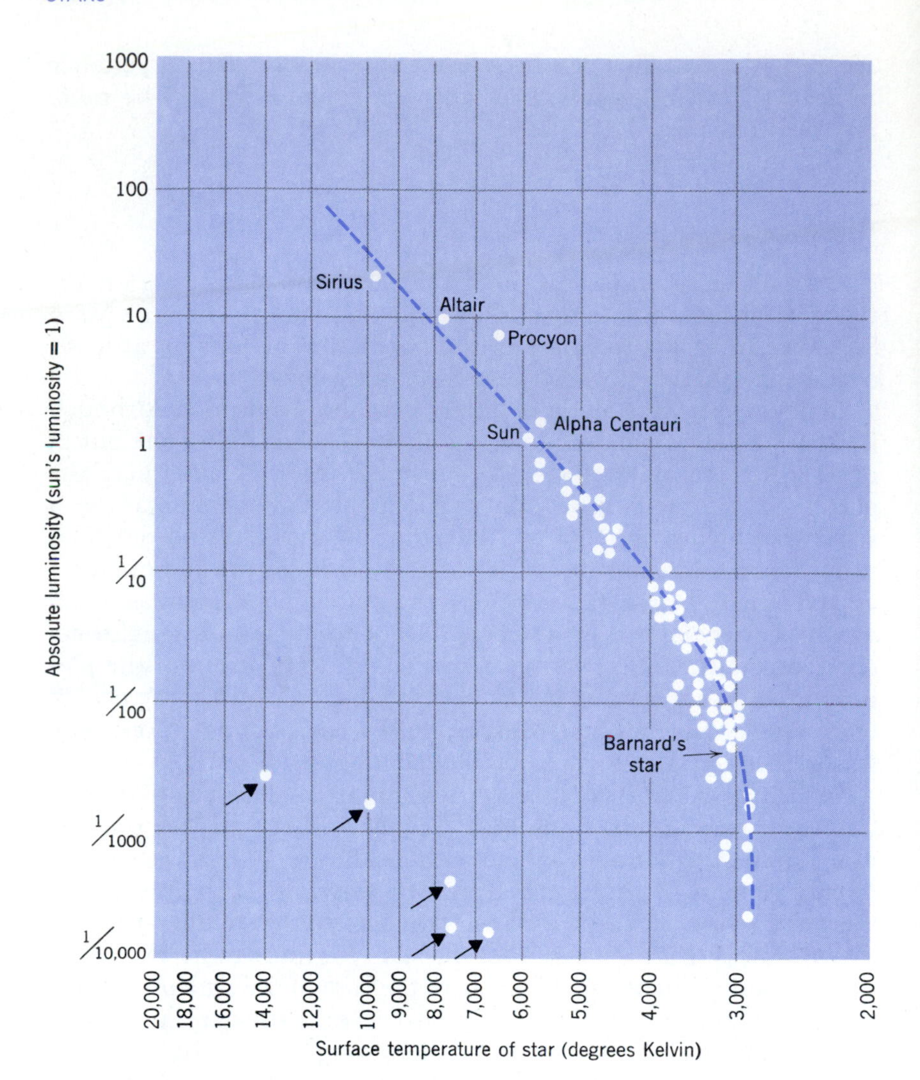

FIGURE 7.11
Luminosity vs. temperature for stars within a distance of 20 light-years from the sun.

Although stars brighter than the sun are relatively rare, they are exceedingly interesting because they are the giants of the sky in size and luminosity. Will these luminous stars follow the Main-Sequence line on the H-R diagram? The answer appears in Figure 7.12, showing the H-R diagram for the 100 brightest stars in the sky, plotted together with the 90 nearest stars for comparison. Again the majority of the stars are clustered around a line which is an extension of the Main-Sequence line for the faint stars, but, as in the case of the faint stars, one group of stars departs from the general pattern. The unusual stars form a trail leading upward and to the right, into a region of stars that have very low surface temperatures, and therefore are red in color, but nonetheless are enormously bright. Normally a low-temperature, red-color star is very faint in

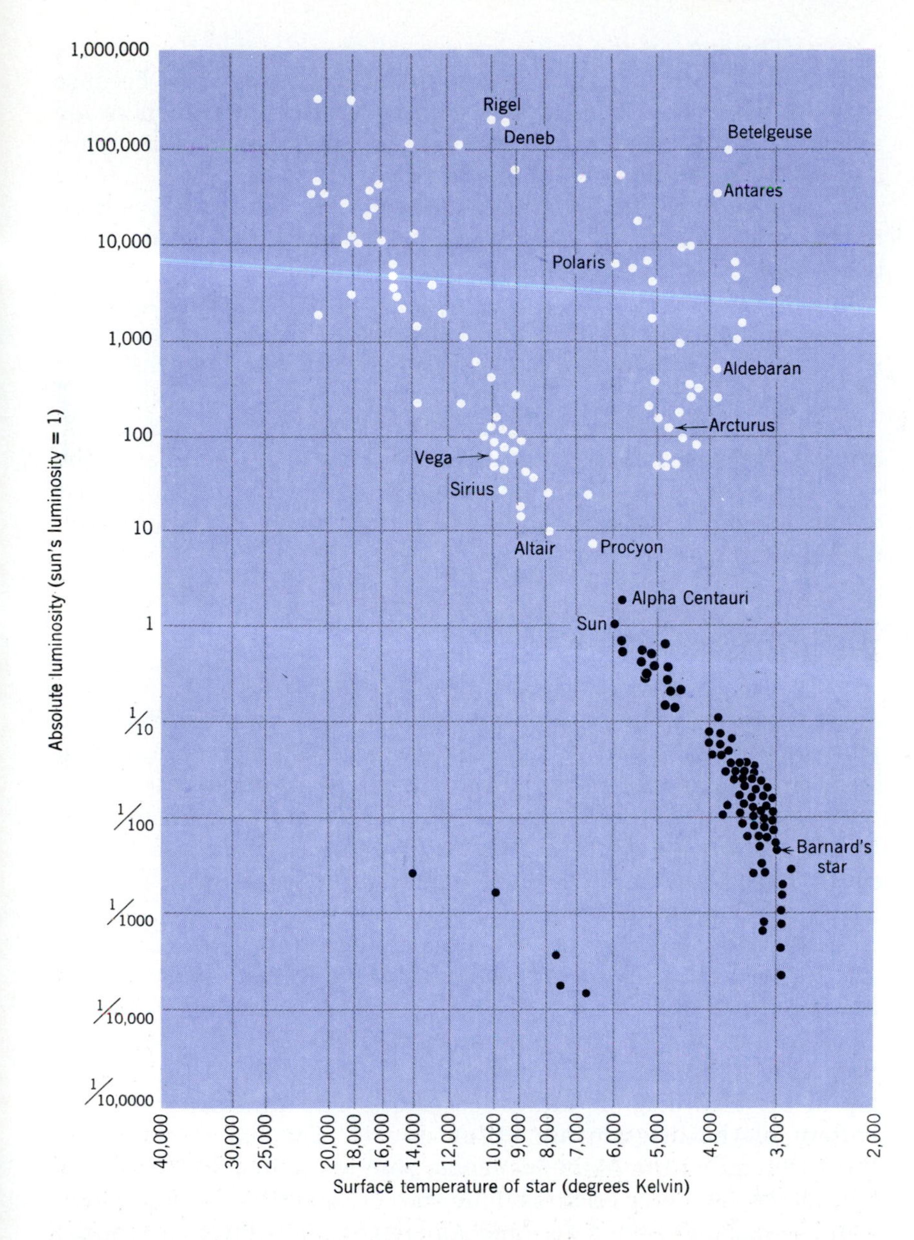

FIGURE 7.12
Luminosity versus temperature for the 100 brightest stars (light circles) and 90 nearest stars (dark circles).

comparison to the sun. Betelgeuse, however, is a star at the upper right that is deep red in color, with a surface temperature of only 3600°K, and yet it is 100,000 times brighter than the sun!

Dwarfs and Giants

The stars in the lower-left region of the graph and those in the upper-right region are so remarkable that they have been given special names. The stars at the lower left are called *white dwarfs;* they are named dwarfs because they are small and faint, and white because they are white-hot. The stars at the upper right are known as *red giants;* they are called giants because they are large and luminous, and red because they are red in color.

It turns out that the dwarf stars actually are very small, as their name implies. A typical white dwarf has a diameter of about 20,000 miles, only twice the diameter of the earth, although its mass is approximately the same as that of the sun, which has a diameter of almost one million miles. White dwarfs are dense, compact stars, so dense that a teaspoon of matter from a white dwarf weighs about 10 tons.

Red giants, on the other hand, are very large in size, typically 100 million miles in diameter, or about 100 times the diameter of the sun. These huge stars have, however, approximately the same mass as the sun in most cases. Thus, the average density of the matter within a red giant is very low.

THE HERTZSPRUNG-RUSSELL DIAGRAM

A great deal of interesting information can be obtained from star plots of the kind that we have discussed above. Hertzsprung first discussed a star plot of variables equivalent to brightness versus temperature in 1911, and Russell discussed a similar plot in 1913. In the years since then, this type of graph has come to occupy a central position in astrophysics because of the valuable results that have been obtained from it.

In honor of the two pioneers, plots of brightness versus temperature for stars are known today as *Hertzsprung-Russell diagrams.* For brevity, they often are referred to simply as *H-R diagrams.* The line along which most of the points are clustered is called the Main Sequence in the H-R diagram. The luminous stars that lie near the top of the H-R diagram are called giants. The faint stars near the bottom of the diagram are called dwarfs. The bright stars near the upper end of the Main Sequence, blue-white in color, are called blue giants. The very brightest blue stars are classified as blue supergiants; Rigel and Deneb are blue supergiants. The faint stars at the lower end of the Main Sequence, orange or red in color, are called orange dwarfs or red dwarfs; Barnard's Star is a red dwarf. The sun, yellow-white in color, occasionally is referred to as a yellow dwarf.

The Numbers of Stars of Different Types. About 90 percent of the stars in the sky are Main-Sequence stars. The remaining 10 percent are divided between white dwarfs, red giants, and a few odd minor varieties. Among the Main-Sequence stars, the red dwarfs are the commonest variety. The blue giants are the rarest.

The Significance of the Main Sequence

The stars along the Main Sequence are the "normal" stars in the sky. They are the family of stars to which the sun belongs. The red giants and the white dwarfs, on the other hand, appear to be "abnormal" stars, with a completely different connection between brightness and temperature. Let us put aside these puzzling, peculiar stars for the moment; their nature will be explained in full later in the chapter. In this section we wish to concentrate on the Main-Sequence stars, and ask ourselves this question: What is the difference between one Main-Sequence star and another? That is, can we discover one single property or characteristic of stars that is responsible for the distribution of stars along the Main Sequence? Is there a property of stars such that, for example, the cool, red, faint stars at one end of the Main Sequence have a small amount of this property, and the hot, blue, luminous stars at the other end of the Main Sequence have a great deal of it?

The answer is affirmative; the stars along the Main Sequence differ from one another only with respect to their *mass.* The red, faint stars at the lower end of the Main Sequence have small masses; that is, each of these stars is made up of a relatively small amount of matter. The blue, luminous stars at the upper end of the Main Sequence are very massive; that is, each of these stars contains a great deal of material. Of course, even the stars at the lower end of the Main Sequence, which we described as having "small" masses, still are enormously massive. The reddest, faintest stars that appear in the H-R diagram on page 171 contain 30,000 times as much mass as the earth.

The sun lies in the middle of the range between the smallest, least massive stars and the largest, most massive stars. Its mass, which is 2×10^{33} grams, is 300,000 times greater than the mass of the earth.

The faintest and reddest star ever observed thus far has a mass about one twenty-fifth that of the sun. The brightest, bluest and most massive stars on the Main Sequence have masses approximately 60 times greater than the mass of the sun.

That is the significance of the Main Sequence: it is a sequence of stars arranged in order of increasing mass, from the smallest stars at the lower end of the sequence to the most massive stars at the upper end. The *mass* of a Main Sequence star, and nothing else, determines its temperature, its brightness, and its place on the H-R diagram.

The graph in Figure 7.13 is the H-R diagram on page 171 with the values of stellar mass marked off along the line. All masses are expressed in units of the sun's mass, which is 2×10^{33} grams or 2×10^{27} tons. Deneb and Rigel—two of the brightest stars on the diagram—are 20 times as massive as the sun.

The Ends of the Main Sequence

Does the Main Sequence extend indefinitely far in both directions? Is it cut off at either end or both ends? It seems likely that there is a lower limit to the masses of stars; that is, if an object is extremely small, it cannot be a star and is more likely to resemble a planet. This guess turns out to be correct; the lower limit for the mass of a star is not known precisely, but is believed to be approximately one one-thirtieth the mass of the sun. This is roughly 100 times the mass of the planet Jupiter and 30,000 times the mass of the earth.

It is not surprising that there is a *lower* limit to the mass of a star, since a great deal of mass is needed to create the enormous pressure and temperatures that cause a star to burn nuclear fuel at its center. It is a surprise, however, that there also appears to be an *upper* limit to the mass of a star. The explanation for an upper limit to stellar masses appears to be connected with the fact that massive stars are also highly luminous (Figure 7.13) (p. 175). In fact, the light coming from the interior of a massive star is so intense that the outward-moving photons, colliding with the gaseous matter, exert a strong pressure on the outer layers of the star. If the luminosity of the star is greater than $10^6\ L_\odot$ — which corresponds, according to theoretical estimates, to a mass of about 60 $M_\odot$ — the pressure created by the photons is great enough to blow off part of the outer layer. The photons continue to remove significant amounts of mass from the star's outer layers until its mass has been reduced to 60 $M_\odot$ and its luminosity to $10^6\ L_\odot$. In agreement with this prediction, the most massive stars that have ever been observed are about 60 times as massive as the sun.

The Meaning of the Red Giants and the White Dwarfs

What are the peculiar red giants and white dwarfs? They are the giant stars that lie above the Main Sequence in the upper-right corner of the H-R diagram, and the dwarfs that lie below, in the lower-left corner of the H-R diagram. Are they special stars that were born abnormal? Or are they transient stages through which every star passes in its lifetime? Astronomers were puzzled by these questions for many years, but today we know the answers as a result of laboratory experiments combined with theoretical investigations carried out on high-speed electronic computers. The experiments have revealed how a star derives its energy by burning nuclear fuel, while the computations show the changes that occur in a star as its fuel is used up.

The combination of laboratory experiments and theoretical studies has provided a complete picture of the life story of a star,

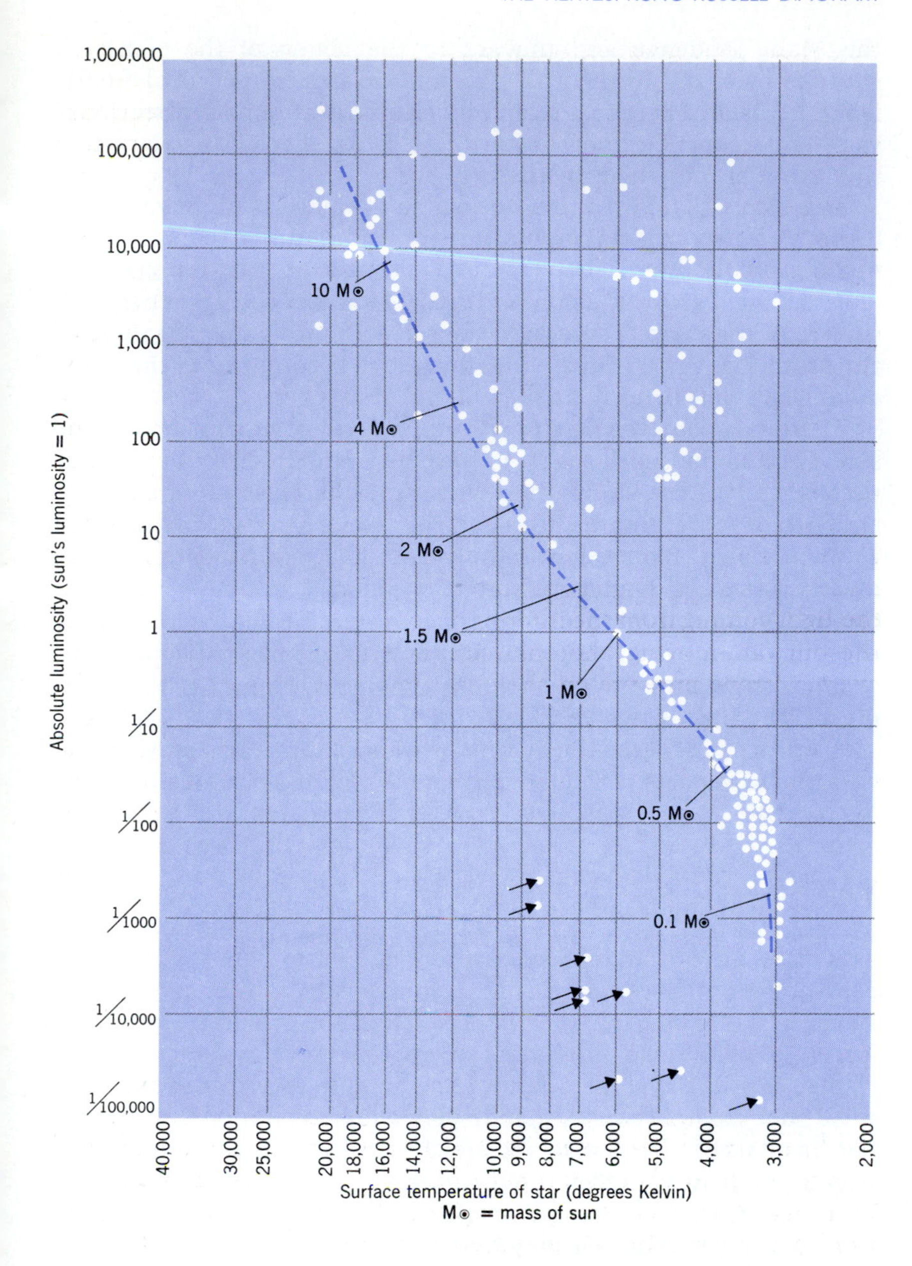

FIGURE 7.13
Masses of the Main-Sequence stars.

which will be discussed in Chapters 8 and 9. According to this picture, red giants and white dwarfs are ageing stars that have partly or largely exhausted their fuel and are headed toward extinction. With the aid of the latest theoretical studies, we can trace the path of a star on the H-R diagram as it passes through all the stages of its existence. We are now certain that the giant and the dwarf are stages in the lifetime of a normal star. When a star has burned up a large fraction of its fuel, it begins to feel its age and moves off the Main Sequence and upward to the region of the giants. It shuttles back and forth in this region for some time, until finally its end is near. Then, in a relatively short time, it crosses back over the Main Sequence line toward the left and dives downward into the region of the white dwarfs.

The sun will follow this course some billions of years in the future, first increasing in luminosity and turning red, and then moving off the Main Sequence and upward into the red giant region. As a red giant, the sun will radiate enough energy to melt the surface of the earth. Thereafter, the sun will move down again, this time into the region of the white dwarfs, to become one of the many dying stars that litter the Galaxy.

There is a complication to the story. Stars of modest or average size, such as the sun, usually follow the route described; but a star that is, let us say, more than four times as massive as the sun, ends its days in a more spectacular fashion. After its life as a red giant is ended, it does not simply fade away to become a white dwarf; instead, the massive star may collapse on itself in a cataclysmic implosion. Rebounding from the explosion, it blows some of its materials out into space in the event known as the supernova; or if it is sufficiently massive, it may contract without limit, becoming a black hole in space (Chapter 9).

A full account of the life history of a star requires an acquaintance with the basic facts of nuclear physics, which is a part of Chapter 8.

VARIABLE STARS

The stars on the Main Sequence change very slowly during the course of their lives, increasing in luminosity as they grow older. When a star moves off the Main Sequence to become a red giant (pages 203–207), its brightness increases more rapidly, but the change is still too slow to become detectable in a year or in a century. Stars exist, however, that display a rapid variation in light output, in marked contrast to the steady, unflickering output of ordinary stars. In some cases these variable stars change in brightness by as much as a factor of two over a period of a day or so. The variations may be irregular or they may follow a regular cycle, with the cycle repeated exactly from one time to the next.

Cepheid Variables

One type of variable star that undergoes regular variations in brightness has come to assume a special importance in astronomy. This type of star, called a cepheid variable, is found in a region of the H-R diagram located approximately halfway between the red giant region and the upper end of the Main Sequence.[5] Examples of the changes observed in the light output for two cepheid variables are shown in Figure 7.14. The first (a) is a cepheid variable with a period of approximately 10 days, and the second (b) is a cepheid with a period of one month.

The most familiar cepheid variable in the sky is the North Star, Polaris. Its brightness varies by a factor of 9 percent every four days.

The location of the cepheid variables in the H-R diagram is shown in Figure 7.15 (p. 178). This diagram also shows the positions of the red giants and of other important members of the family of stars.

As will be seen in Chapters 8 and 9, the stars found in the region between the upper Main Sequence and the red giants are in a relatively advanced stage of their lives. Theoretical studies show that for a brief period in this stage of its life, a star may begin to oscillate, collapsing on itself and expanding outward repeatedly. Its size can change by as much as 30 percent during each oscillation. These rhythmic changes in size are accompanied by large changes in luminosity.

The surface temperature of a cepheid variable also changes during the course of each cycle of variations in brightness. As might be expected, the surface temperature is highest when the brightness is at a peak, and lowest when the brightness is at a minimum. For example, a cepheid variable with a period of six days falls in surface temperature from 7500 °K at peak brightness to 6000 °K at minimum brightness. The corresponding change in its spectral type is from F2 to G2. Figure 7.16 (p. 178) shows the light curve for a cepheid variable with a period of about six days, with the corresponding changes in spectral type indicated along the light curve.

The period during which a star is a cepheid variable may last as

[5] The name is derived from the fact that the first star of this type to be observed was called Delta Cephei.

FIGURE 7.14 (a) and (b)
Light curves for two cepheid variables.

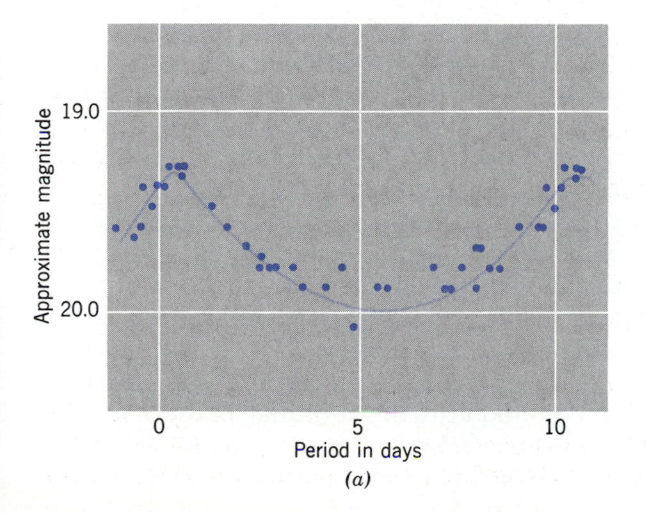

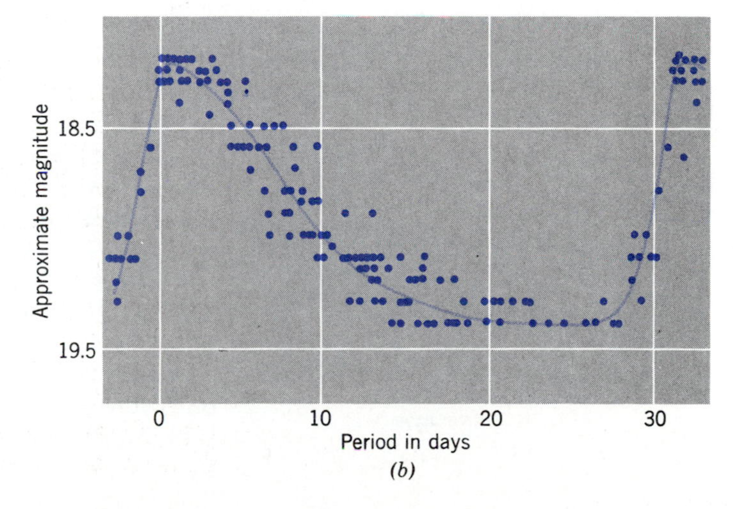

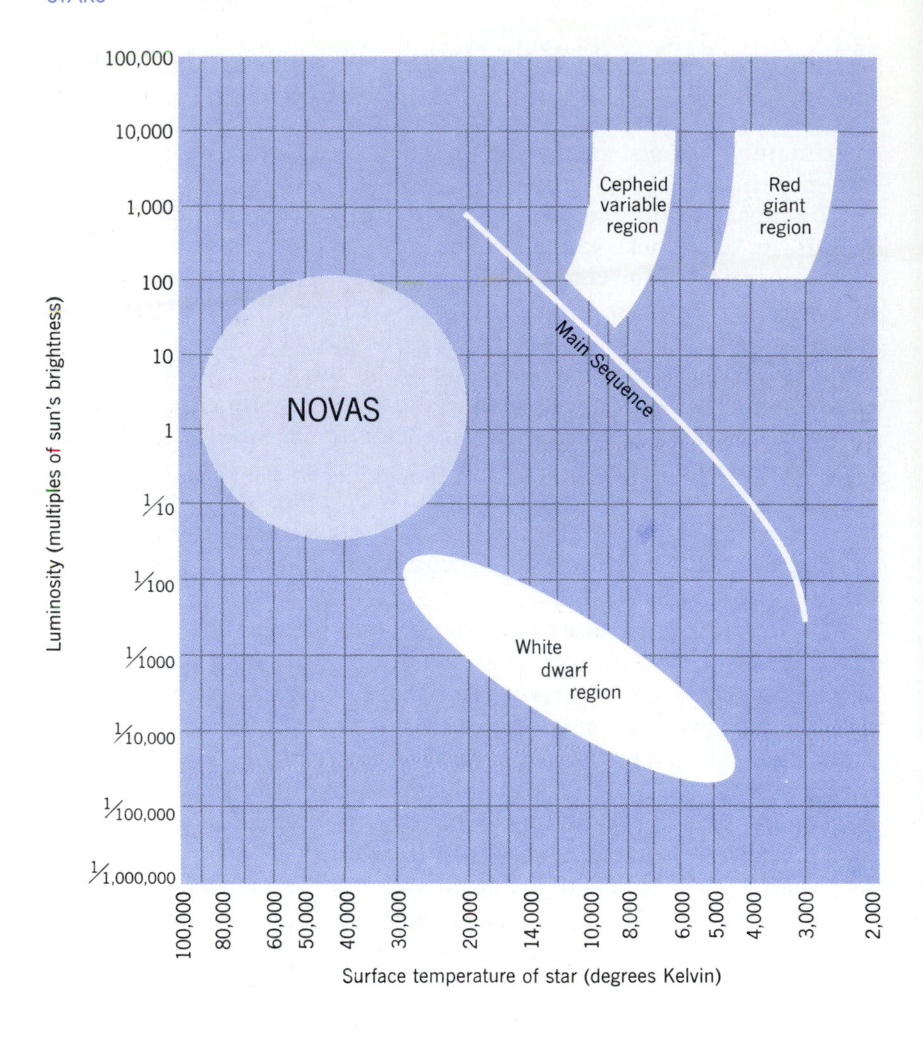

FIGURE 7.15
The position of variable stars in the H-R diagram.

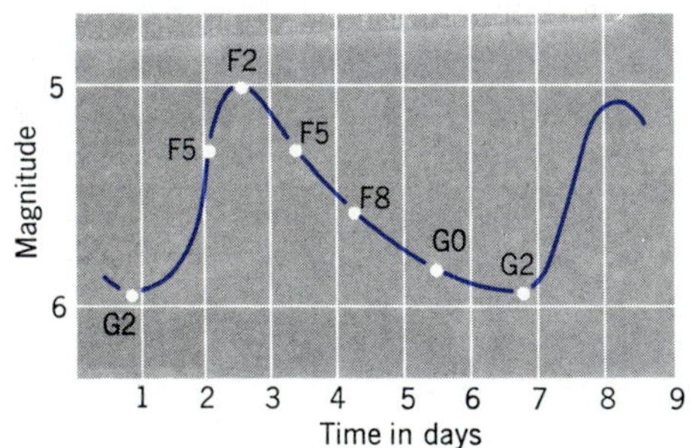

FIGURE 7.16
Change in spectral type during the light cycle of a cepheid variable.

long as one million years. Although this seems a very long time, it is a small fraction of the billion-year lifetime of the typical star. That is why only a few of the stars in the sky at a given time are cepheid variables, in spite of the fact that most stars pass through this stage at least once during their lives.

The cepheid variables have an importance in astronomy that is far greater than their small numbers would otherwise indicate. They have become the most important yardstick of the astronomer for measuring distances to other galaxies. Their usefulness in this connection depends on the fact that the period of oscillation of the cepheid variable—that is, the interval of time from the beginning of one cycle of changing light output to the beginning of the next cycle—is uniquely related to the average luminosity of the star. Thus, by measuring the period of the light variations in a cepheid, we can deduce its absolute luminosity. Since we can also measure its apparent luminosity, its distance immediately follows. This method gave one of the first measures of the distance to nearby galaxies.

Novas

The family of stars contains another group whose light output varies with time. This group of stars, unlike the cepheid variables, may be found anywhere within a large region on the H-R diagram (Figure 7.15). Also unlike the cepheid variables, which fluctuate in intensity rhythmically according to a fixed pattern, this second group of stars flares up suddenly and unpredictably at intervals of time. On the average, the flare-ups occur every 30 to 50 years for a given star of this type. The flare-ups also differ from the changes of light output displayed by cepheids in the sense that they are very violent, with the total energy output of the star increasing by as much as a factor of a million over a short period of time.

The irregularly flaring stars are called *novas.* They are found on the H-R diagram to the left of the Main Sequence, and above the white dwarf region. Novas are believed to be, in fact, stars that have lived out most of their lives and are on the way to becoming white dwarfs. As we mentioned above, the white dwarf is the final stage in the life of all stars of modest size.

Why is the nova different from other ageing stars that are about to become white dwarfs? Do all ageing stars flare up as novas for a time, as they approach this stage? Or do only a small number do so, as the result of some special property? From the small number of novas in the sky at any time—no more than one thousand in our galaxy at this moment—it is clear that the nova must be a very special kind of star. A clue to the property of novas that makes them special may lie in the fact that many of the novas observed thus far are found to belong to binary stars; that is, they are one of two stars circling closely around one another under the attraction of their mutual gravity. Why the presence of a white dwarf in a binary should lead to a violent flare-up from time to time is not clearly understood, but it is believed by the theoretical astronomers that the outbursts occur when matter is pulled off the partner star and onto the surface of the white dwarf by gravity. This material, deposited on the surface of the white dwarf, produces an instability in the white dwarf that leads to a violent outburst of radiation, or nova outburst. The newly acquired matter is blown out into space by the nova outburst and, as a consequence, the nova subsides into a normal white dwarf state again, until some later time when enough matter has been drawn from its companion star to produce another unstable condition, and trigger another flare-up.

Although novas are a spectacular phenomenon for the astronomer to observe, their occurrence does not contribute any basic new knowledge to our picture of stellar evolution. Novas are more of a curiosity than a central feature in the life story of the stars. The contrary is true for supernovas, which are flare-ups of a star that are even more violent than nova outbursts. Supernova outbursts, as will be seen in Chapter 9, play a critical role in stellar evolution and in the history of the Universe. In fact, it can be said that without the supernova outbursts that terminate the lives of some of the more massive stars in the sky, we would not be here today.

BINARY STARS: DETERMINATION OF STELLAR MASSES

The mass of a star determines its place on the Main Sequence in the H-R diagram. Unlike temperature and luminosity, the mass of a star is not directly observable. The best method for determining stellar masses depends on the properties of binary stars.

Mass Determination from Binaries

A large fraction of the stars in the sky are *binary stars,* that is, two stars revolving around a common center and bound to each other by their mutual force of gravitational attraction. Binaries are not concentrated in any particular part of the H-R diagram. In fact, the two individual stars making up a binary may appear in widely separated portions on the diagram.

Binaries are particularly interesting members of the family of stars because they lead to a method of determining stellar masses. Other methods are explained in the Appendix. The method based on observations of binaries is more important than any other because it is applicable to many stars, and covers a large range of masses. The general idea behind the method can be illustrated by considering a special case in which one star in the binary is much more massive than the other. The larger, more massive star controls the movement of the smaller star by its gravitational pull, just as the sun controls the motion of the earth. For an orbit of a given size, the mass of the larger star determines the rate at which the smaller one revolves around it. The more massive the central star is, the greater the pull of its gravity will be, and the faster the small star must revolve around it in order not to be drawn in. Therefore, if the size of the orbit and the period of revolution can be measured for the small star, it should be possible to determine the large star's mass. The mass of the sun is determined in this way from the size of the earth's orbit and the length of the earth's year.

A similar relationship holds if the two stars in the binary have comparable masses. In that case, the theory of motion of objects under gravity shows that both stars revolve in ellipses around their common center of mass. If the sizes and shapes of the two ellipses and the period of revolution can be measured, the mass of each star can be calculated by an extension of the idea described above.

Visual Binaries. The period of revolution for a visual binary is determined by direct observation of the orbits of the two stars over an extended period of time. The shapes of the orbits and the apparent size or angular size of each orbit are also determined by direct observation (Figure 7.17). If the distance to the binary is known, the true sizes of the orbits can be calculated from their apparent sizes. The mass of each star is calculated from the period of revolution and the orbital sizes and shapes with the aid of formulas derived from Newton's law of gravity.

The calculation is complicated by the fact that generally the plane of the orbits is tilted at an unknown angle to the observer's line of sight, making it impossible to determine the shape of the orbit accurately. Methods for dealing with this complication are described below.

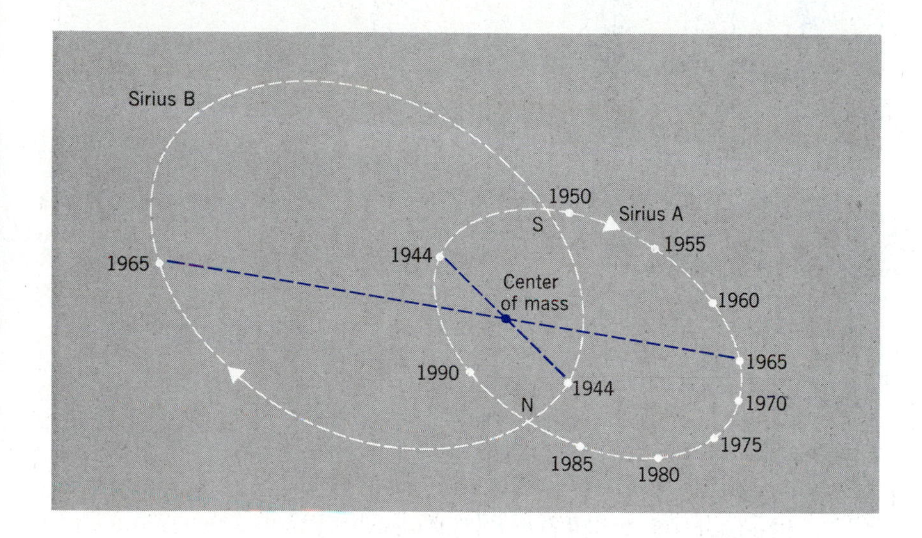

FIGURE 7.17
The observed and predicted positions of Sirius A and B between 1944 and 1990. The center of mass around which both stars revolve is the intersection of lines connecting their positions at two different times (e.g., 1944 and 1965).

Spectroscopic Binaries. For spectroscopic binaries, the effect of the orbital motion can only be seen indirectly, as a Doppler shift in the spectrum of each star. The magnitude of the Doppler shift gives the velocity of the star along the line of sight of the observer. The spectra of the two stars are superimposed, and since the orbital velocities along the line of sight are different for the two members of the binary, each line is split into two components (Figure 7.18*a*). Twice in each orbit, the velocities of both stars are tangential and the Doppler shift for each is zero. At this time, the double lines in the spectrum coalesce into single lines (Figure 7.18*b*). The period of the cycle of changes in the wavelengths of the lines is the period of the orbital motion.

The magnitudes of the Doppler shifts provide information about the component of the orbital velocity of each star along the line of sight to the observer. With the aid of the relationships provided by

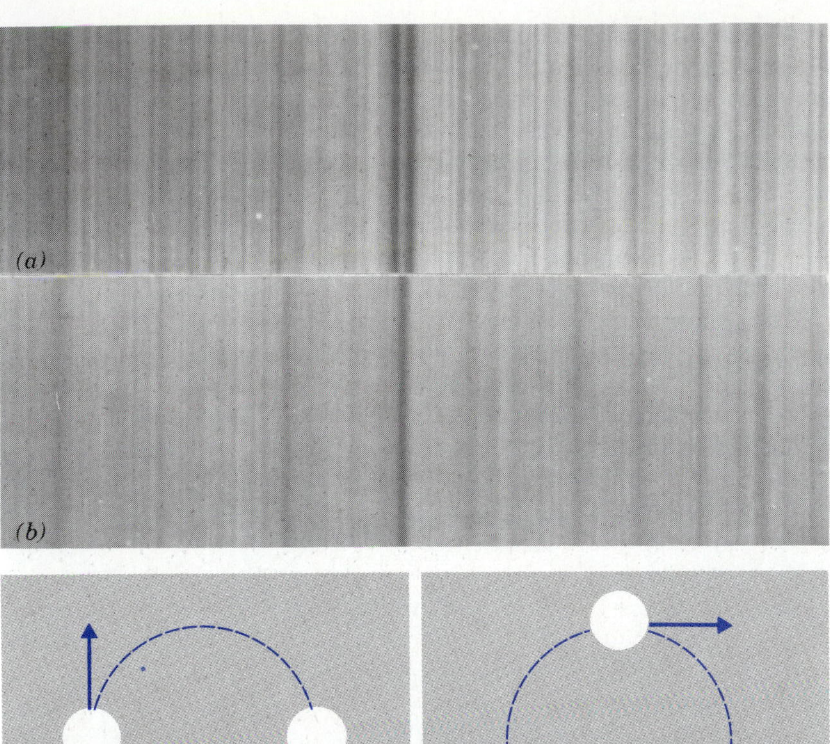

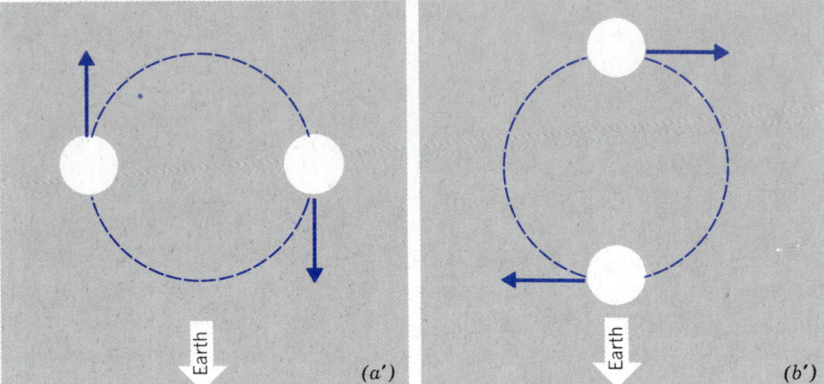

FIGURE 7.18
Two spectra of the spectroscopic binary Mizar, one of the stars in the handle of the Big Dipper. The two stars in the binary revolve around a common center every 20.5 days. In the upper spectrum (a) each line is split into two components by the Doppler shift because one member of the binary is moving toward the earth and the other is moving away (a′). In the lower spectrum (b) the lines are not split because at the time this spectrum was photographed the two stars happened to be moving tangentially across our line of sight, with no motion toward or away from the earth (b′). The cycle of splitting and coalescence of the lines is repeated every 20.5 days.

Newton's law of gravity, the information regarding the velocities can be converted into information on the sizes of the two orbits. In the special case of a circular orbit, for example, the orbital velocity, v, radius, R, and period of revolution, P, are connected by the formula,

$$v = \frac{2\pi R}{P}$$

For an elliptical orbit the relation is more complicated. In this case, the observed velocity is plotted as a function of time, and an ellipse is determined that gives the best fit to the observations.

Tilt of the Orbit Plane. For both types of binaries, the inclination of the plane of the orbit to the observer's line of sight presents a problem. The problem is readily solved for visual binaries. Suppose for clarity that one star has a considerably smaller mass than the other, and, therefore, describes a large and more accurately measurable orbit. This orbit is an ellipse with the more massive star at one focus. The problem is to find out how the plane of the ellipse is tilted with respect to the observer's line of sight. Figure 7.19 suggests the solution. On the left is the elliptical orbit as it would appear if viewed perpendicular to the plane of the orbit. The focus is indicated by the letter F. The position of the focus is known, being the location of the more massive star. On the right, the same ellipse is shown as viewed by the earth observer, with its plane tilted.

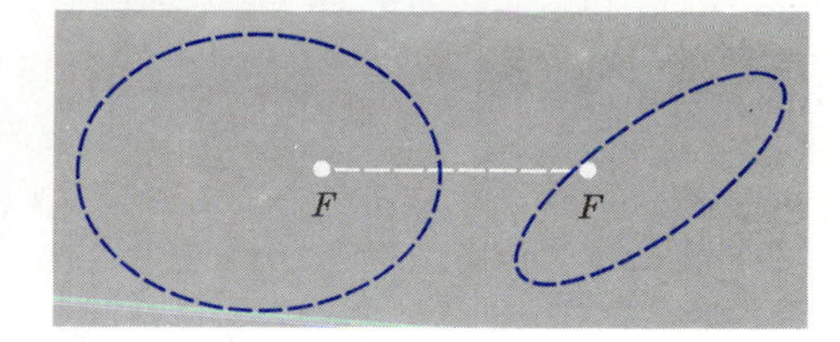

FIGURE 7.19
Effect of a tilted orbit plane on the shape and position of the focus (F) of an elliptical orbit.

The tilted, foreshortened orbit is still an ellipse, although of greater eccentricity than the original one. However, the focus of the foreshortened ellipse no longer coincides with the position of the massive star. In general, the massive star will not even be on the axis of symmetry of the foreshortened ellipse. This asymmetry betrays the tilt. An example is the path of 70 Ophiuchi B around A, shown in Figure 7.20. The curve is the best-fitting ellipse to positions of 70 Ophiuchi B observed between 1850 and 1940. It is clear to the eye that the position of 70 Ophiuchi A is not on the line of symmetry for this ellipse, indicating that the plane of the true ellipse is tilted relative to the observer.

The angle of the tilt can be determined by adjusting the orientation of the orbit until the focus of the ellipse coincides with the position of the massive star.

For spectroscopic binaries, the problem of determining the tilt of the orbit plane is generally insoluble. This is because the Doppler shift only yields partial information about the velocities, namely, their radial components. The best that can be done in this case is to make an assumption regarding the inclination of the orbit plane, and deduce the mass corresponding to this assumption.

If, for example, the orbit is assumed to be precisely edge-on to the observer's line of sight, the analysis can be carried out, and yields a lower limit to the mass of each of the two stars. One can go a little further by assuming that the orbit planes of binaries are randomly oriented in space. By averaging the mass over this random distribution of orbit planes, we obtain a correction to the lower limit. When this is done separately for binary components of different spectral types, the result is a rough value for the mass of each spectral type.

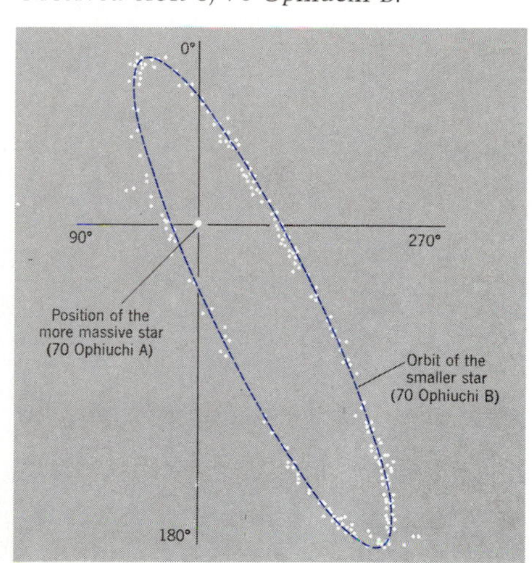

FIGURE 7.20
Observed orbit of 70 Ophiuchi B.

Eclipsing Binaries. For one special type of spectroscopic binary, known as the *eclipsing binary,* the problem of the unknown tilt of the orbit plane disappears. The eclipsing binary is a double star whose orbit plane happens to be nearly edge-on to the observer, so that one star periodically moves behind the other and is repeatedly eclipsed. As a result, the intensity of the light from the binary varies periodically. A periodic variation in the light from a binary usually indicates that it is eclipsing. Thus, the plane of its orbit is known, and the masses of its components can be determined unambiguously.

The Light Curve

Eclipsing binaries can yield other interesting information, in addition to stellar masses. The information comes out of a study of the details of the light variation from the binary as one star passes behind the other during the course of an orbit.

Suppose, for example, that the more massive star is a red giant and the less massive star is still on the Main Sequence. This is a plausible situation, since the two stars were formed at the same time, but the more massive star evolves more quickly. Then the less massive star will be smaller in radius and higher in surface temperature. Label the more massive star "A" and the less massive one "B." Star B will be totally eclipsed by Star A once in each orbit. That is, the radiation from B will be completely blocked out for a finite length of time as it passes behind A. In this case the "light curve" for the binary—that is, the plot of measured light intensity as a function of time—has the following characteristic appearance: the light curve is at a maximum when the two stars are separated; once in each orbit, the light curve falls as the smaller star begins to pass behind the larger one; when the smaller star is totally eclipsed, the light curve reaches a minimum, and remains constant at this minimum value until the smaller star starts to reappear on the other side; at that point the light curve begins to rise again.

Later in the same orbit, B passes in front of A and blocks a part of the radiation from A, causing another dip in the light curve. Since B

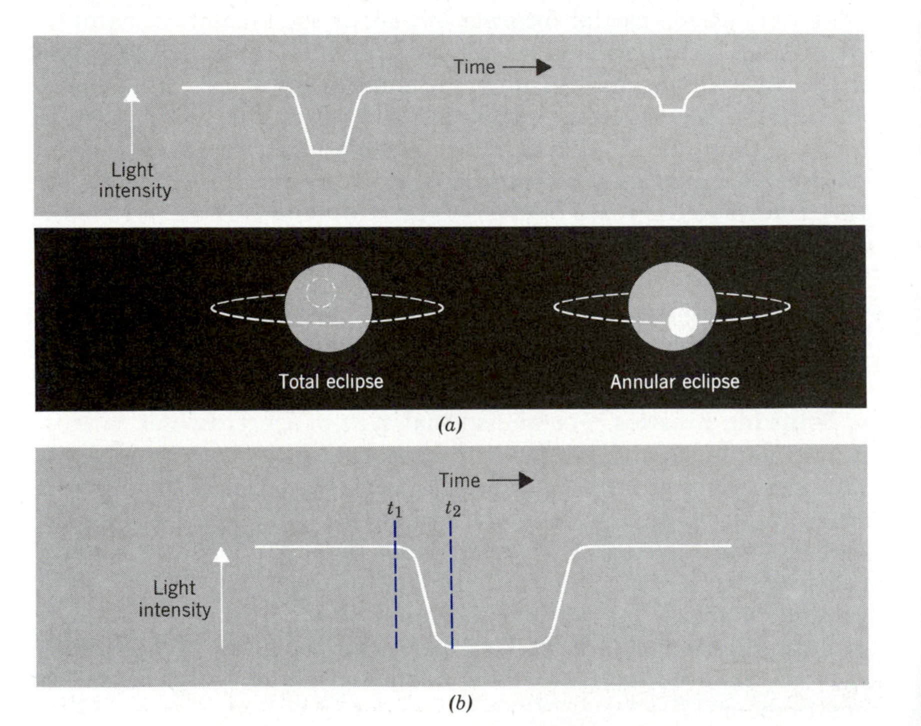

FIGURE 7.21
Light curve for a completely eclipsing binary.

is smaller than A, this eclipse is annular. The dip in the light curve during the annular eclipse has the same shape as during the total eclipse. However, the dip is shallower because, while the same area is blocked off in both cases—that is, the area of the smaller star—the surface radiation *per unit area* is less in the case of the red giant than for its companion.

A light curve for a binary of this type, called a completely eclipsing binary, is shown in Figure 7.21*a*.

The unmistakable feature of the light curve of this particular type of eclipsing binary is the *flatness* at the bottom of each trough. This flatness indicates that the main eclipse is total; hence one star is considerably smaller than the other.

The radius of the smaller star can be determined from the light curve as follows: the velocities of the stars are known from their Doppler shifts; if the sloping portion of the trough in the light curve lasts from time t_1 to t_2, and the relative velocity of the two stars in this portion of the orbits is V, the diameter of the smaller star is $V \times (t_1 - t_2)$ (Figure 7.21*b*).

If the orbit is tilted slightly to the observer's line of sight, the light curve will display one deep and one shallow trough per orbit, as before, but the troughs will not be flat-bottomed, because the finite tilt of the orbit plane prevents the disk of star B from lying completely within the disk of star A at any point during the orbit. Two *partial* eclipses occur in each orbit. Because the eclipses are partial, the troughs are not as deep as in the previous case. Figure 7.22 shows the light curve for a partially eclipsing binary.

It is clear that the light curve of an eclipsing binary depends on (1) the relative luminosities of the two stars, (2) their radii, (3) the surface brightness of each, that is, the surface temperature, and (4) the precise angle of inclination of the orbit of the binary. The details of the light curve also are influenced by such features as the way in which density and temperature fall off with increasing height in the atmosphere of each star, or the presence of a stream of gas

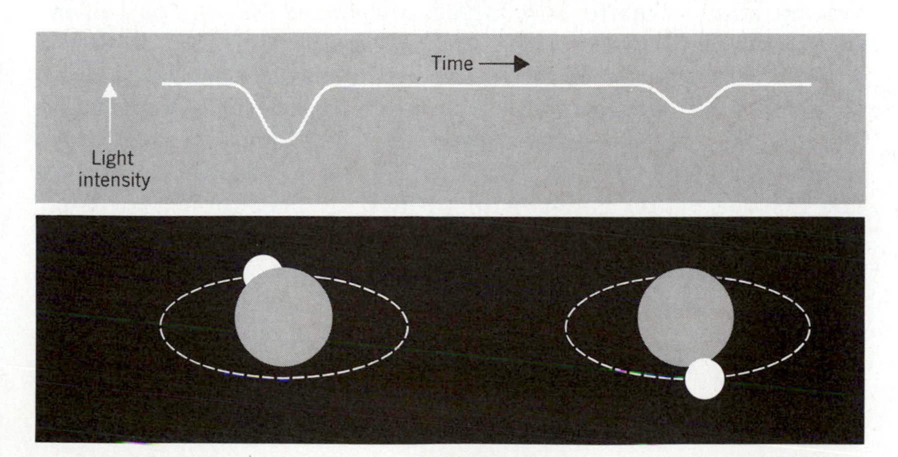

FIGURE 7.22
Light curve for a partially eclipsing binary.

flowing from one star to the other. Information regarding all these properties has been obtained from careful studies of the light curves of eclipsing binaries.

Close Binaries

If the stars in a binary are sufficiently close together, the gravitational pull of one star can capture matter from the other. In some cases, a star can lose a large fraction of its mass to its companion in this way.

In every binary an imaginary boundary lies between the two companion stars, separating their gravitational spheres of influence. A transfer of mass occurs when some of the material in the outer envelope of one star crosses the boundary and enters the gravitational domain of its companion. This is likely to occur when a star that is a member of a close binary moves off the Main Sequence and its outer layer expands as it enters the red giant phase of its life. Blue giants or blue supergiants also have distended outer envelopes and will lose mass to their companion star if they are in a close binary. In typical cases, a red giant or a blue giant in a close binary can lose more than one-half of its mass to its companion in the course of time.

The transfer of matter between the stars in a close binary can have striking consequences. The transferred material is drawn to the surface of the receiving star by that star's gravity; it picks up speed as it approaches; when it collides with the surface it liberates heat, raising the temperature of the surface. According to one theory, if the receiving star is a white dwarf this increase in temperature produces a nuclear detonation on the surface of the white dwarf, a nova outburst.

A similar sequence of events could also provide observational evidence for the existence of black holes in space. A black hole is a concentration of matter so compact and dense that the pull of its gravity prevents rays of light from escaping (page 233). A direct proof of the existence of black holes is impossible since they emit no light at all and absorb all light that crosses their boundaries. But suppose a black hole is a member of a close binary and its companion is a giant. The giant will lose mass to the black hole; as this gaseous matter is drawn in toward the black hole, it will be accelerated to extremely high speeds by the intense gravitational pull of the black hole and will become very hot, reaching a temperature of many millions of degrees. At this temperature, the hot, luminous material converging on the black hole will emit radiation in the X-ray region of the electromagnetic spectrum. Rockets and satellites have identified a source of X-rays called Cygnus X-1 that fits this description of a black hole very closely (page 236).

Main Ideas

1. Relationships involving temperature of an object, color and wavelengths of emitted radiation; Wien's law.
2. Variation of intensity of radiation with wavelength.
3. Definitions of luminosity, apparent magnitudes and absolute magnitude.
4. Plots of luminosity vs. surface temperature for populations of stars.
5. Definitions of Main Sequence stars, giants and dwarfs.
6. Definition of the H-R diagram.
7. Interpretation of the Main Sequence in the H-R diagram; relation between mass and position on the H-R diagram.
8. Determination of stellar masses from the orbits of binary stars.
9. Information obtainable from the light curve of an eclipsing binary.

Important Terms

absolute magnitude
apparent magnitude
binary star
blue giant
Cepheid variable
eclipsing binary
luminosity
Main Sequence
mass of a star
nova
orbit plane
photometer
red dwarf
red giant
solar mass
spectroscopic binary
surface temperature
variable star
white dwarf
Wien's law

Questions

1. Calculate the wavelength of maximum intensity of radiation for glowing objects with temperatures of 3000, 6000 and 9000 degrees Kelvin.
2. Why is the sky blue? Why is the sun red at sunrise and sunset?
3. Explain the difference between apparent magnitude and absolute magnitude.
4. What is the luminosity of the sun? What does luminosity specify about a star?
5. Figure 7.8 shows that the hotter a glowing object, the more radiation it emits at all wavelengths. According to the Stefan-Boltzmann law, the total amount of energy emitted at all wavelengths, per unit area of the emitting surface, is proportional to the fourth power of the temperature of an object. Using this

law, compute the ratio of the energy emitted per unit area for objects 3000 and 10,000 degrees Kelvin.

6. Sketch or trace a blank H-R diagram and plot the following list of stars on it. ($L_\odot$ = sun's luminosity.)
9500°K, 19 $L_\odot$ (Sirius A); 6400°K, 6 $L_\odot$ (Procyon); 12,200°K, 250 $L_\odot$ (Regulus); 3100°K, 0.004 $L_\odot$ (Barnard's Star); 26,500°K, 30,000 $L_\odot$ (Mimosa); 5800°K, 1.0 $L_\odot$ (Sun); 2950°K, 2600 $L_\odot$ (Mira); 4600°K, 0.6 $L_\odot$ (Beta Centauri); 5100°K, 0.7 $L_\odot$ (36 Ophiuchi); 9000°K, 0.008 $L_\odot$ (Sirius B). From the H-R plot of the stars, which stars do you immediately recognize as peculiar? How?
7. Look at the H-R diagram in Figure 7.13. The mass increases along the Main Sequence, but the luminosity increases much faster. From these facts, what can you deduce about the relative lifetime of massive stars compared to low mass stars? Explain.
8. Compute the average density of a white dwarf with the sun's mass and 1/100th of the sun's radius, and compare with the density of lead. (Density equals mass divided by volume.)
9. Compute the average density of a red giant with the sun's mass and 100 times the sun's radius, and compare with the density of air at room temperature and pressure.
10. Dark spherical globules of interstellar matter, believed to be clouds of gas condensing to form stars, emit radiation peaked at about 30,000 angstroms. Using Wien's law, calculate the surface temperature of such a cloud.
11. Why are there so many more Main Sequence stars than red giants?
12. H-R diagrams often are plotted using spectral type instead of surface temperature, and absolute magnitude instead of luminosity. Make a new plot of the H-R diagram in Figure 7.11 in terms of these quantities. Use the information in this chapter and in Chapter 6.
13. Why is the observed number of visual binaries fewer than the observed number of spectroscopic binaries?
14. Referring to Figure 7.21, how would you use the light curve to determine the diameter of the larger star in the binary?

STELLAR EVOLUTION

8

A star's life begins in the swirling mists of hydrogen that surge and eddy in the space between the stars. The photograph on the facing page shows one of these clouds, located about 4000 light-years from the solar system in the direction of the constellation Monoceros. In the random motions of such clouds, atoms sometimes come together by accident to form small, condensed pockets of gas. Stars are born in these accidents.

GRAVITATIONAL CONDENSATION

How does a condensed pocket of gas form in space? For simplicity let us concentrate our attention on three atoms of hydrogen that happen to be neighbors of one another in an interstellar cloud. These atoms are labeled by the numbers 1, 2, and 3 in Figure 8.1. Let us suppose that at a given moment the three atoms happen to be moving toward one another, as shown by the arrows in Figure 8.1*a*. A short time later, the three atoms will have come together as a consequence of their motions, as shown in Figure 8.1*b*. They now form a small pocket of condensed gas in space.

Because the atoms are very small, they rarely collide. Usually they pass by one another in the course of the same motions that brought them together, and then separate again, as shown in Figure 8.1*c*. As a result, the condensation disperses.

As the atoms pass one another, however, each atom exerts a

Stars in formation.

FIGURE 8.1
Random motions of atoms in space.

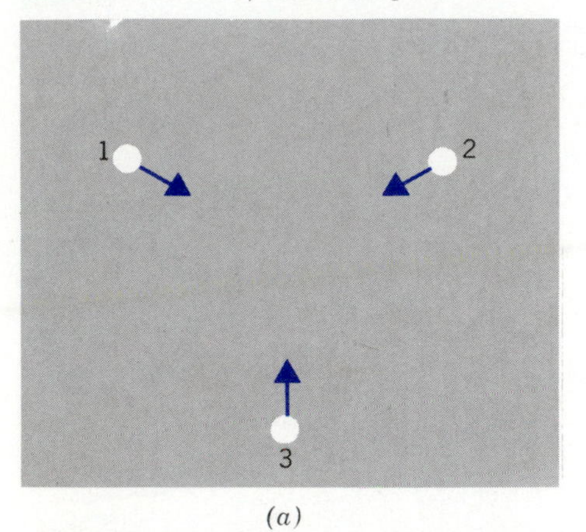

(a)

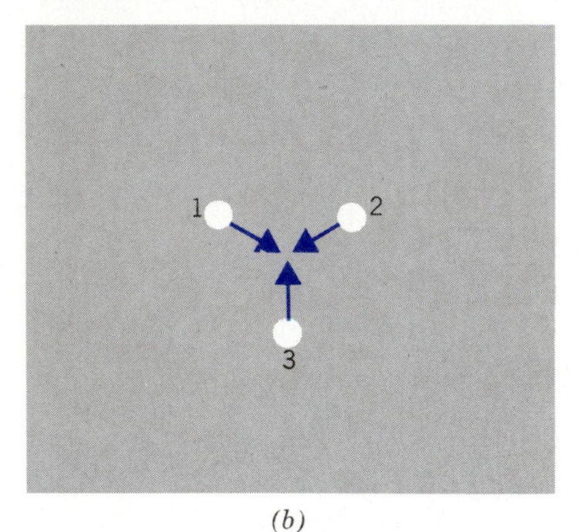

(b)

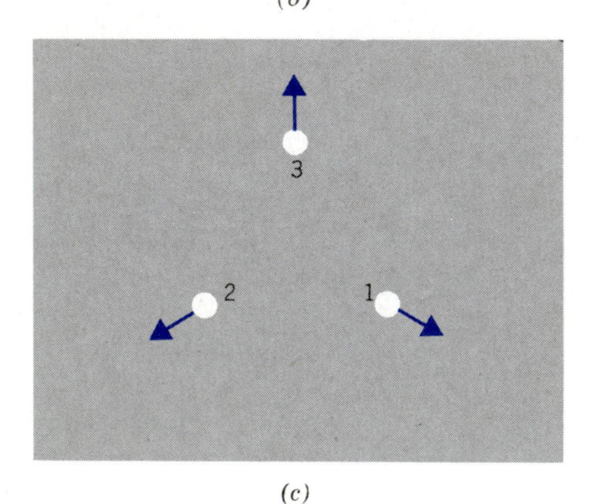

(c)

small gravitational attraction on its neighbors, which counters the tendency of the atoms to separate. If the gravitational attraction were sufficiently strong, it could hold the three particles together and prevent them from dispersing again to space. (The atoms would have to get rid of some of the energy they picked up during their approach, by radiating it away.) The effect of gravity would have been to convert the pocket of gas from a temporary condensation to a permanent one.

Gravity, however, is not a very strong force; it is the weakest natural force known to man. As a consequence, under normal conditions the force of gravity that a few atoms exert on each other is not strong enough to tie them permanently together.

But suppose that there are not merely three, four, or five atoms in the pocket of gas, but a very large number. Because the force of gravity extends over great distances, each atom feels the gravitational pull of all the other atoms in the pocket. If the numbers of atoms is sufficiently large, the combined effect of all these minute pulls of gravity will be powerful enough to prevent any of the atoms in the pocket of gas from leaving the pocket and flying out into space again. The pocket becomes a permanent entity, held together by the mutual attraction of all the atoms within it.

This is the heart of the theoretical explanation of the birth of stars. According to the theory, a star is conceived when a condensed pocket of gas forms in outer space, and when the number of atoms in the pocket of gas is so great that their own gravity holds them together permanently. The pocket is not yet a star, but it will become one a little later. This tight cluster of atoms, held in the grip of its own gravity, is called a protostar.

Protostars

How large must a cluster of atoms be before its own gravity is strong enough to hold it together permanently? If three or four atoms are not enough, will a million atoms suffice? Or a trillion? Theoretical astronomers, using pencil and paper and the laws of physics, have calculated the number of atoms that are necessary. The results show that the answer depends strongly on the temperature of the gas, which controls the speed at which the atoms move about. Clearly, the higher the temperature and the higher the speed of the atoms, the more difficult it is for gravity to hold them together.

Under normal conditions in interstellar space, the required number of atoms turns out to be about 10^{60} (1 followed by 60 zeros). If we translate this staggeringly large number of atoms into grams, using the fact that most of the atoms in space are hydrogen and one hydrogen atom weighs approximately 10^{-24} grams, we arrive at the result that the condensed cloud has a mass of 10^{36} grams. A single star typically has a mass of approximately 10^{33} grams. Thus, the condensed cloud contains the mass of about a thousand individual stars.

On the basis of this result, astronomers believe that star formation occurs in two stages. First a large cloud condenses under its gravity. This cloud may contain the mass of hundreds or thousands of stars. Then, the large cloud fragments into many smaller condensations, each with approximately the mass of one star. This two-step process typically leads to the formation of a group or cluster of many stars. There is support for the idea in the fact that such clusters are observed to be quite common. The Pleiades, shown in Color Plate 10, are an example of a cluster containing about 120 stars.

Not all stars occur in well-defined clusters held together by gravity. Many are found in looser clusters called *associations.* According to a theory that has been advanced to explain the existence of these associations, after a cluster has formed the most massive and hottest stars in it, of O and B types, emit intense ultraviolet radiation that drives away the gas in the dense cloud out of which the cluster of stars has been forming. Before that happened, the sum of the masses of these stars, plus the mass of the gas and dust in which they are embedded, had produced enough gravitational force to hold the cluster together. Now, with much of the gas gone, the gravitational force is diminished; the stars are no longer bound together, and they start to disperse; they are an association rather than a cluster.

The same effect can be produced by the massive stars in a cluster in a different way, when they come to the ends of their lives and explode as supernovas. The supernova explosions create an expanding wave of pressure that sweeps out into the interstellar gas, driving much of the gas away and diminishing the gravitational pull that had been holding the stars together. The result, again, can be an association rather than a cluster.

There is recent evidence that supernova explosions, produced by the demise of massive stars in a cluster, also play an important role in star formation. The pressure wave resulting from the supernova explosion compresses the surrounding gas in the interstellar medium, triggering the formation of new stars. Star formation may be the result of this process combined with, or augmented by the process of random gravitational condensation described above.

Photographs taken in certain regions of the sky show small, dark globules of matter that look just like stars, or clusters of stars, in the process of formation. An example is the large cloud of gas and dust in the direction of the constellation Monoceros (the Unicorn) known to astronomers as Nebula NCG 2237. Figure 8.2 shows a part of this nebula. The luminous clouds in the photo are collec-

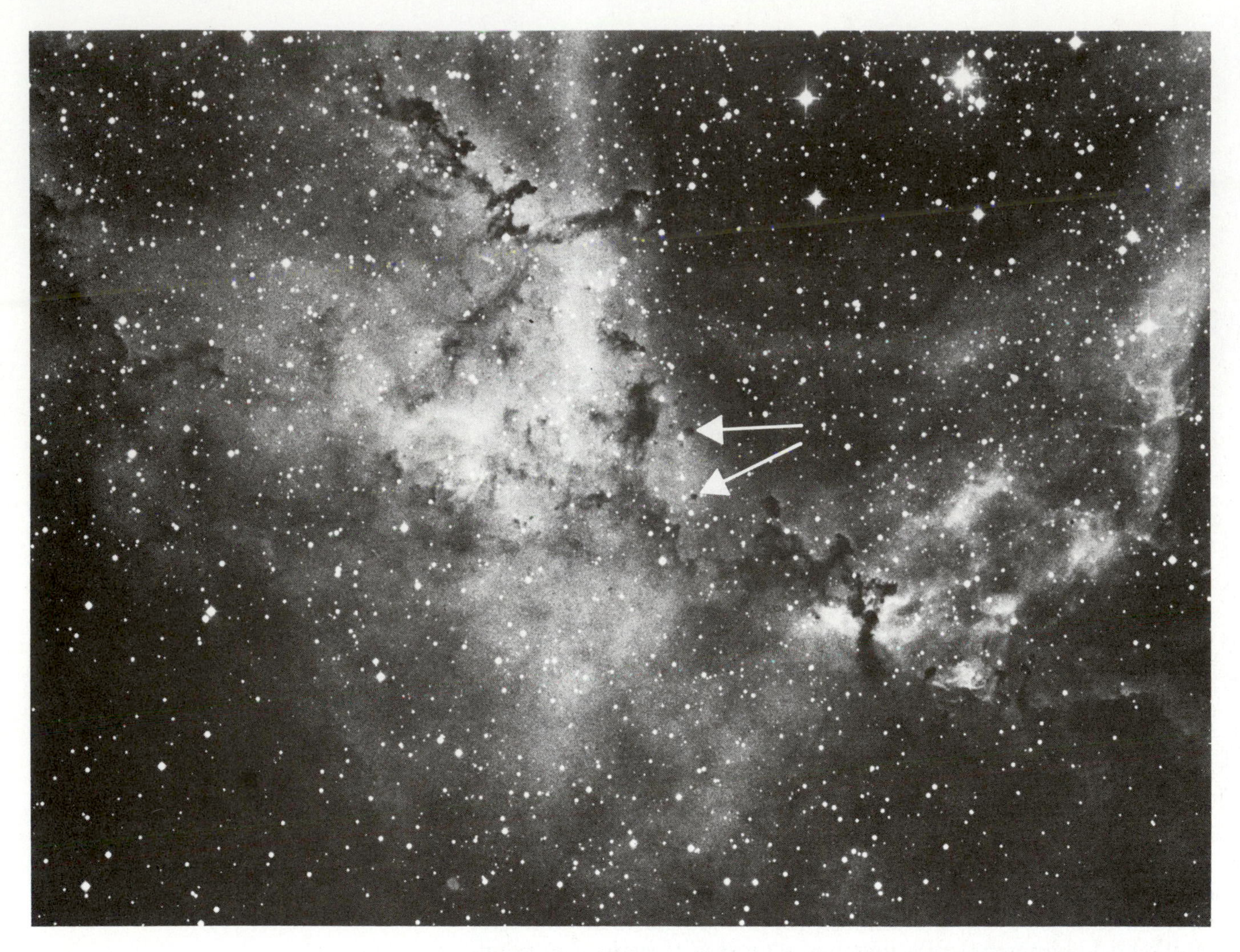

FIGURE 8.2
A part of the nebula NGC 2237. The dark spots (arrows) are concentrations of gas and dust in which stars are likely to form. The smallest spots are a few trillion miles in diameter.

tions of hydrogen atoms that have been heated by the absorption of radiation from the newly formed hot, young stars embedded in their midst. The arrows point to a number of small, dark spots called globules,[1] which contain clusters of atoms that are probably in the process of forming stars. The smallest spots are a few trillion miles in diameter.

The Orion Nebula

The famous and beautiful Orion Nebula, shown in Color Plate 5, is another region that astronomers consider to be a breeding ground

[1] The globules are dark because they contain substantial amounts of dust in addition to gaseous material. The gas is relatively transparent, but the grains of dust absorb light very effectively.

for new stars. Embedded in the relatively dense gas of the Orion Nebula are enormous numbers of stars, including some that are known to have formed very recently.

The Orion Nebula has about 10,000 stars in a volume of 20 light years across, compared to about 10 stars in the same volume of space in ordinary regions of the Milky Way Galaxy. Many of the thousands of stars in the Orion Nebula are small and relatively dim, resembling the sun, but four stars in one group, called the Trapezium, are very massive and extremely hot. Because a massive, hot star lives only a short time, these stars of the Trapezium must have been formed quite recently. They are probably not more than a few hundred thousand years old.

Stars as massive and hot as those in the Trapezium emit a large part of their energy in the form of ultraviolet radiation. This ultraviolet radiation ejects electrons from the atoms of gas in the cloud; that is, it ionizes them. The ionized atoms subsequently capture stray electrons from the surrounding matter of the cloud. If the captured electron returned to the ground state of the atom in one jump, it would emit an ultraviolet photon with about the same energy that had ionized the atom in the first place. However, this does not happen in most instances. Instead, the electron is captured into a higher orbit and then tumbles down to the ground state in one or more separate jumps, landing in intermediate orbits on the way. The photons emitted in these intermediate transitions have less energy than the ultraviolet photon that was originally absorbed, and therefore emit light at a longer wavelength. The wavelength is frequently such as to place the emitted light in the visible part of the electromagnetic spectrum. This visible light causes the beautiful red, blue, and mauve colors of the nebula.

The Orion Nebula is barely visible to the naked eye as a relatively small patch of faint luminosity, in the midst of the stars that form the sword hanging from the belt of Orion the Hunter (see Chapter 2, Figure 2.36. The densest regions of the cloud contain a million or more atoms per cubic centimeter. Normal regions of the interstellar medium in the Milky Way Galaxy, in contrast, have only about one atom per cubic centimeter.

Because the density of atoms in the Orion cloud is very great, collisions between atoms are frequent. Sometimes the atoms stick together in these collisions to form molecules. Two atoms of hydrogen can combine to form a molecule of hydrogen; atoms of carbon and hydrogen can combine to form a molecule of methane; carbon and oxygen combine to form carbon monoxide or carbon dioxide; and so on.

One of the most abundant molecules in the Orion cloud, and similar interstellar clouds, is carbon monoxide. This molecule has become a tracer for astronomers seeking to study the nature of the clouds in which stars form (Figure 8.3 (p. 196) and Color Plate 5).

FIGURE 8.3
The Orion Nebula. The shaded areas are particularly dense concentrations of carbon monoxide, revealed by their radio emission. The carbon monoxide measurements reveal that the entire cloud of molecules is at least 10 times larger than the visible nebula near its center.

Chapter 11 describes other molecules that have been found in clouds in which stars are forming.

The Rising Temperature of a Protostar

The pockets of gas shown in Figure 8.2 are not yet stars; they are only condensations of gaseous matter on the way to becoming stars. Astronomers call these condensations protostars. How does a protostar—a tenuous collection of atoms drawn together out of the cold gas of space—become the flaming sphere of gas we call a star? The answer depends again on the force of gravity, which draws every atom in the protostar toward the center of the cloud (Figure 8.4*a*). As gravity pulls the atom in the cloud toward the center, the protostar shrinks in size and its density increases (Figures 8.4*b* and *c.*).

At the same time that the cloud becomes denser, it also becomes hotter because the atoms in the protostar, "falling" toward the center, pick up speed like any falling object. Since the average speed of the atoms in a gas determines the temperature of that gas, this means that the temperature of the contracting protostar rises.

At the start of this process, the temperature of the protostar is the same as the temperature of the relatively cold interstellar gas out of which it formed. However, as the gas cloud contracts under the attraction of its own gravity, the temperature at the center mounts, eventually reaching 50,000°K.

At a temperature of 50,000°K, the hydrogen and helium atoms at the center of the protostar collide with sufficient violence to dislodge all electrons from their orbits around the nuclei. The original gas of atoms, each consisting of an electron circling around a nucleus, becomes a mixture of two gases, one composed of electrons and the other of nuclei.

At this stage the globe of gas has contracted from its original size, which was trillions of miles in diameter, to a diameter of 100 million miles. To understand the extent of the contraction, imagine the Goodyear blimp shrinking to the size of a grain of sand.

After still more time passes, the protostar shrinks to 50 million miles, its internal temperature rises to 150,000°K, and its surface temperature rises to 3500°K. By this time, the protostar is a highly luminous object, hundreds of times more luminous than the sun, even though its surface is considerably cooler and redder than the surface of the sun. (There is no paradox in the fact that a relatively cool object can be highly luminous. The explanation is that the protostar is 50 times larger in diameter than the sun, and therefore has 2500 times more surface area. Although each square centimeter of the surface radiates less energy because the surface temperature is lower, the total amount of energy radiated from the protostar is enormous because its area is so great.)

At this point the protostar makes its debut on the Hertzsprung-Russell diagram. The combination of a low surface temperature and a high luminosity places the protostar in the upper right-hand corner of the H-R diagram, in approximately the same part of the diagram as the red giants. However, the protostar has arrived in that region by a path very different from that of the red giants, which are not newly forming stars, but stars that are growing old.

Protostars should emit enough radiation in the visible region of the spectrum to be seen by astronomers on earth with a telescope. However, because a protostar stays in this visible stage for such a relatively short time, very few of these objects will be in the heavens at any given time. Therefore, astronomers see them very rarely. In addition, they are difficult to detect because they are imbedded in dense clouds of gas and dust which absorb their visible radiation and convert it to infrared.

FIGURE 8.4
Three stages in the collapse of a protostar under gravity. The arrows indicate the force of the protostar's gravity pulling each atom toward the center.

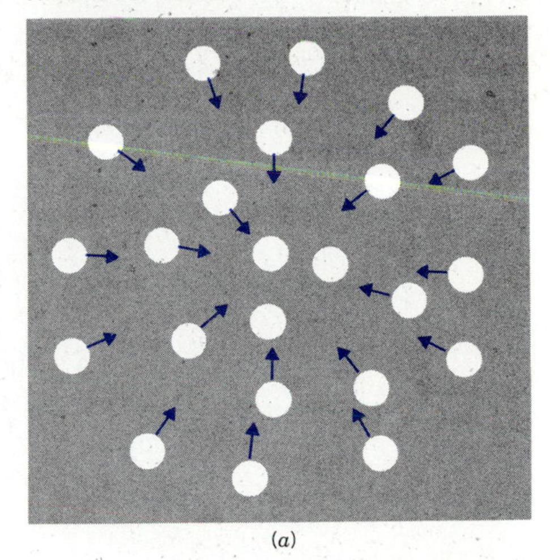

(a)

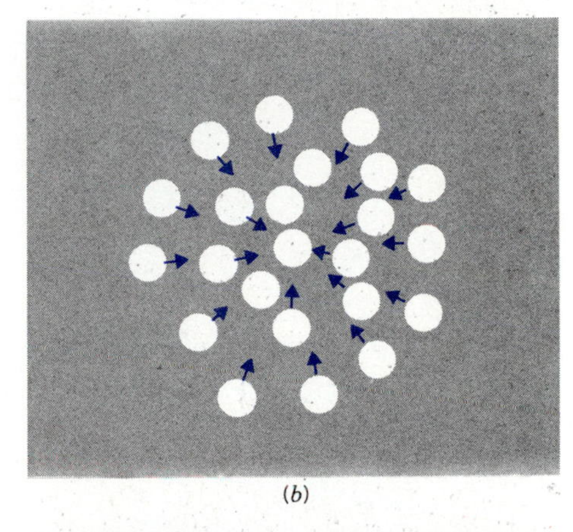

(b)

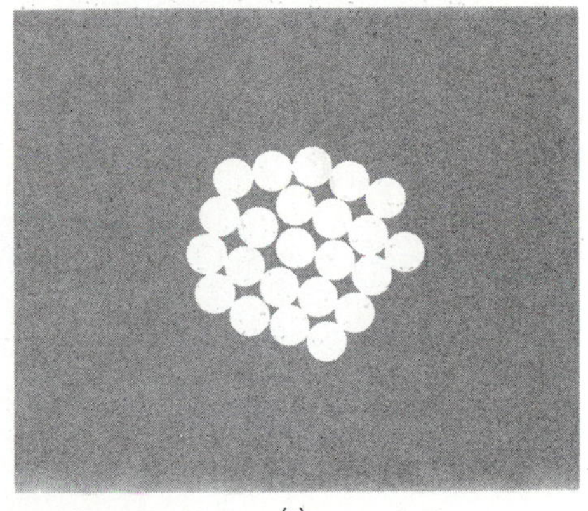

(c)

STARS OUT OF PROTOSTARS

The evolutionary path of a contracting protostar is shown in Figure 8.5 (p. 198) as it appears on the Hertzsprung-Russell diagram. This is the path of a star with the same mass as the sun. The age of

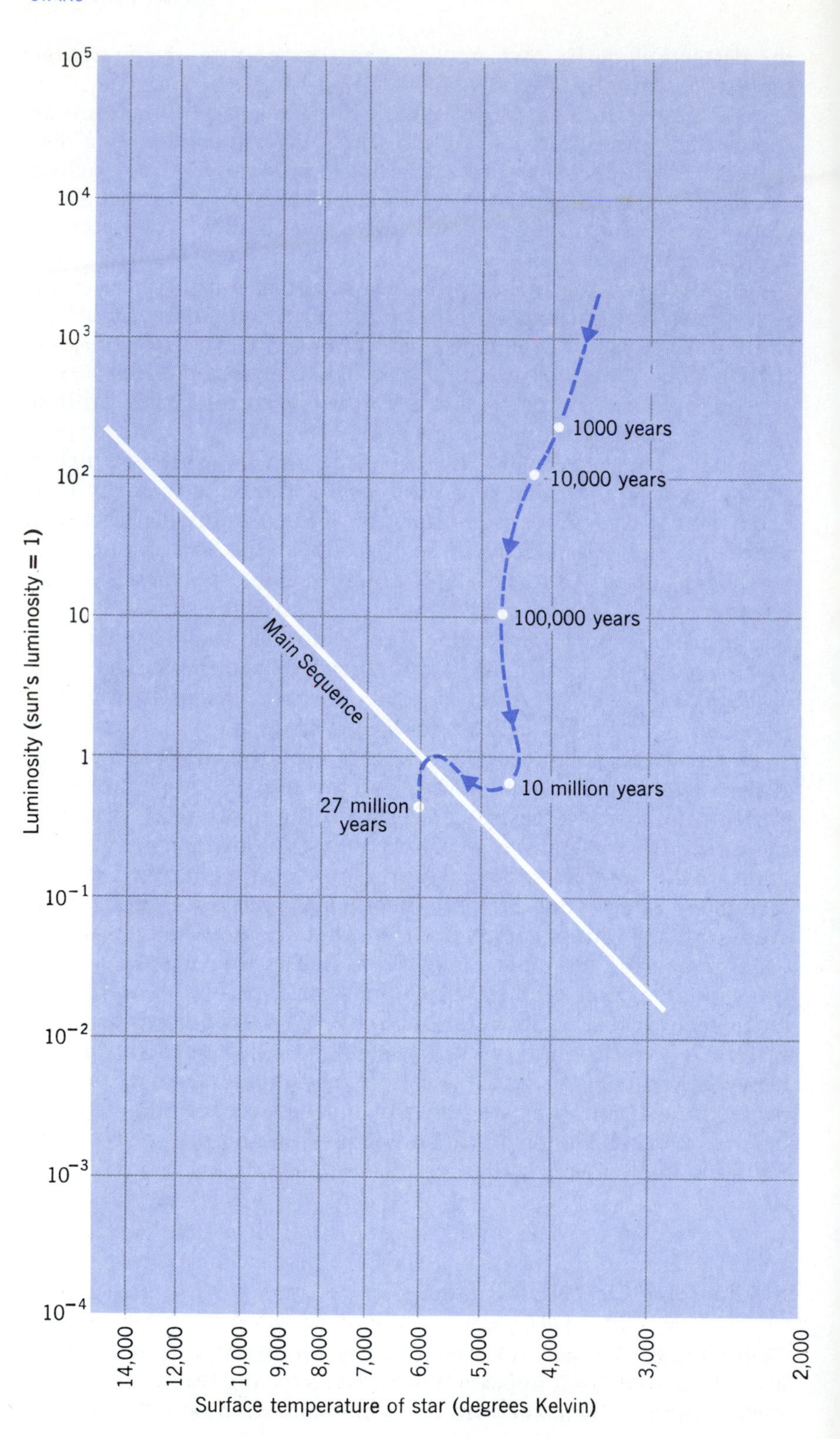

FIGURE 8.5
The track of a newly forming star across the H-R diagram.

the protostar is marked at successive points along the diagram. You can see that the protostar moves down the H-R diagram very quickly at first, and more slowly later on.

Ignition of Hydrogen

When the protostar has been contracting for about 10 million years, a critical event occurs. The protostar has been collapsing since its formation, and has now shrunk from its original size of trillions of miles to a diameter of about 1.5 million miles. This diameter is close to, but slightly greater than, the size of the sun. At the same time, the temperature at the center of the protostar has risen to 10 million degrees Kelvin.

Ten million degrees Kelvin marks a critical threshold in the life of the collapsing sphere of gas. At this temperature, for the first time, the protons at the center of the protostar collide and stick together in a nuclear fusion reaction. The nuclear reaction releases an enormous amount of energy to the surroundings. The release of nuclear energy marks the birth of the star.

The Electrical Barrier

Why are such high temperatures, ranging up to millions of degrees, needed to produce a significant amount of energy by nuclear fusion? The explanation is connected with the electrical forces between protons. Two protons repel one another electrically because each proton carries a positive electric charge. If the protons come very close to each other, the electrical repulsion gives way to the even stronger force of nuclear attraction. However, the range of the nuclear attraction is only one ten-trillionth of an inch; outside that distance, it is not effective.

Under ordinary circumstances, the electrical repulsion serves as a barrier to prevent as close an approach as this. In a collision of exceptional violence, however, the protons may pierce the electrical barrier which separates them, and come within the range of their nuclear attraction. Collisions of the required degree of violence occur with significant frequency in a gas only when the temperature of the gas climbs to about 10 million degrees Kelvin.

Suppose the temperature in the contracting protostar has reached 10 million degrees. Now the protons in the center are moving and colliding at speeds great enough to penetrate the electrical barrier. They come within the range of the nuclear force of attraction; immediately they are seized by this very strong force of attraction; they rush violently toward one another and collide. In the collision, because of the great strength of the nuclear attraction, the two particles fuse together, forming a heavier nucleus called a deuteron, in place of the two separate protons that existed before. At the same time, the energy of their collision passes to the surroundings, and sets up a counterpressure to the star's gravity. This counterpressure slows down the star's collapse. The star now proceeds to live out its life in a balance between the inward force of gravity and the outward pressures generated by the nuclear energy released at its center.

NUCLEAR REACTIONS IN A STAR

The collision and fusion of two protons into a nucleus is an example of a *nuclear reaction.* Physicists have studied many such reactions in nuclear accelerators, and their findings have come to play a very important role in astronomy. The results of the physicists' research show that the collision of two protons is the first step in a series of reactions that govern the life of a star.

According to the accelerator experiments, as well as theoretical studies, when two protons collide and fuse, at the very moment of the collision one proton sheds its positive charge of electricity in the form of a positive electron, also called a positron. The removal of the positive charge from the proton leaves behind a neutron. The neutron is locked to the other proton by the nuclear force of attraction to form a deuteron (the nucleus of deuterium, or heavy hydrogen). Figure 8.6 shows a collision between two protons to form a dueteron.

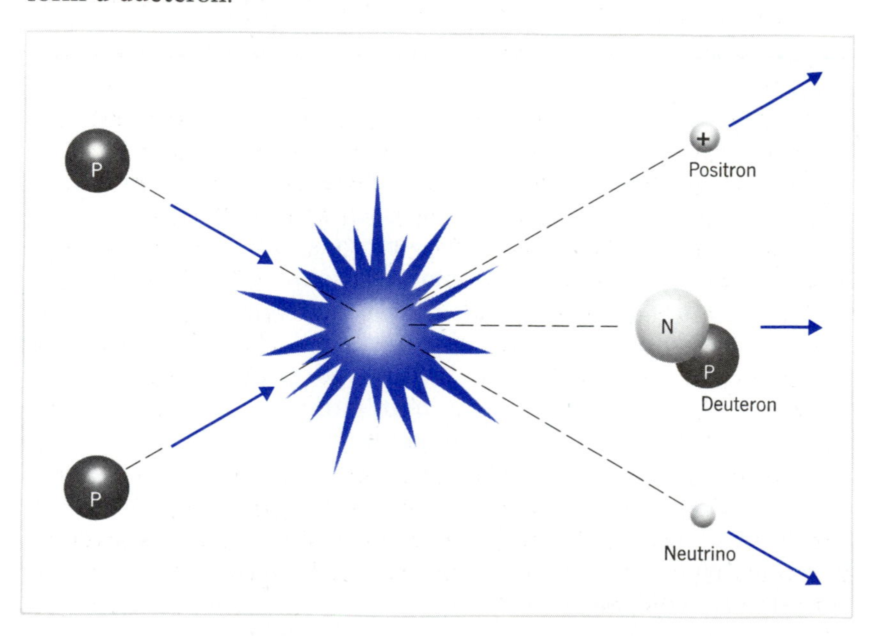

FIGURE 8.6
Fusion of two protons to form a deuteron.

TABLE 8.1
Presents a Glossary of Some Basic Particles Appearing in These Reactions.

p (proton)	A positively charged particle, relatively massive; it is the nucleus of the hydrogen atom and one of the two basic building blocks of heavier nuclei.
n (neutron)	Electrically neutral particle about the same mass as the proton; it is the other basic component of atomic nuclei.
d (deuteron)	A particle composed of a proton and neutron bound together, containing the same charge as the proton but double its mass; it is the nucleus of heavy hydrogen or deuterium.
e or e^- (electron)	A negatively charged, relatively light particle 1/1840 times the mass of the proton.
e^+ (positron)	Similar to the electron, with identical mass, but positively charged.
ν (neutrino)	A chargeless and massless or nearly massless particle; it is produced in some nuclear reactions, generally with a positron or electron, and carries off some of the energy released in the reaction; because it has no electric charge and little or no mass, the neutrino can pass through large amounts of matter, such as the entire body of a star.
γ (gamma ray)	A photon or packet of electromagnetic radiation similar to an ordinary photon, but having extremely high energy and correspondingly short wavelength (less than 10^{-8} cm).

At the same time that the positron is released, an electrically neutral, ghostlike particle of small or zero mass, known as the neutrino, also appears. The positron and neutrino carry away most of the nuclear energy released in the fusion. (See Table 8.1.)

Nuclear physicists have developed a useful shorthand for nuclear reactions like the one shown above. Suppose we let the symbols d, p, n, e^+, and ν represent, respectively, the deuteron, proton, neutron, positive electron or positron, and neutrino. Then the reaction in which two protons fuse to form a deuteron can be written:

$$p + p \longrightarrow d + e^+ + \nu$$

The fusion of protons into deuterons is the first in a series of steps in which heavier elements are built up from lighter ones in the interior of stars. The continuing succession of nuclear reactions manufactures all the other elements of the Universe out of the basic ingredient, hydrogen.

Formation of Helium

Once a large number of protons has fused together within a star to form deuterons, a second set of reactions begins. A deuteron produced by fusion can collide and fuse again with another proton, forming the three-particle nucleus, He^3, the light isotope of helium (Figure 8.7). At the same time, the energy released by the fusion escapes in the form of an energetic photon, called a gamma ray.

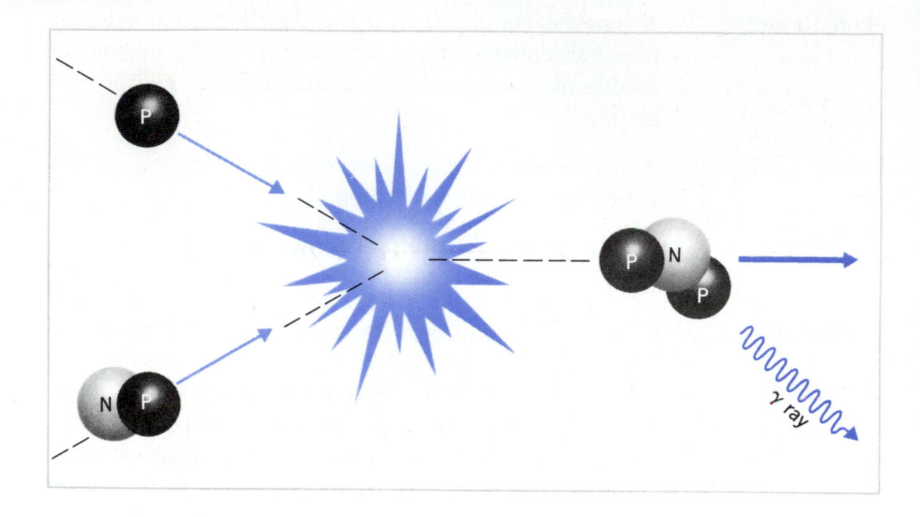

FIGURE 8.7
Fusion of a proton and a deuteron to form a He^3 nucleus.

Since protons are so abundant in stars, this reaction is quite likely to occur once an appreciable number of deuterons has been formed.

The shorthand form for the proton-deuteron reaction is:

$$d + p \longrightarrow He^3 + \gamma$$

The Greek letter γ (gamma) on the right-hand side of the equation is the symbol usually employed to represent the gamma ray. The gamma ray in the above reaction carries off most of the energy released in this particular fusion process, whereas the positron and neutrino carried off the energy in the proton-proton reaction. The lightest particles emerging from a reaction always carry off most of the energy.

Many experiments have been performed in nuclear laboratories in which a target containing deuterium, such as heavy water, is bombarded with fast-moving protons. The experiments show that a He^3 nucleus is frequently formed as a result of the bombardment, verifying that the fusion of deuterons and protons does actually occur, and is not just a theoretical possibility.

The experiments also show that two He^3 nuclei may fuse together to form a He^4 nucleus, with two protons emitted in the process (Figure 8.8):

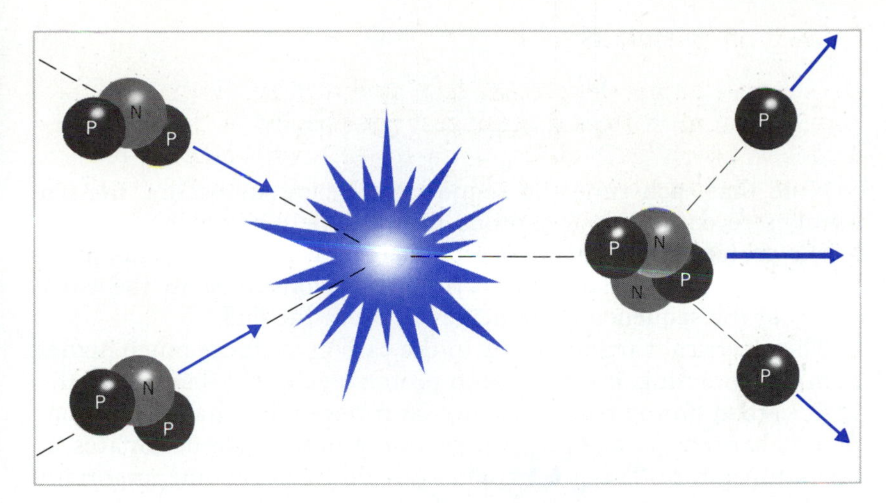

FIGURE 8.8
Fusion of two He^3 nuclei to form He^4.

$$He^3 + He^3 \longrightarrow He^4 + 2p$$

It turns out that an exceptionally large amount of energy is released by this reaction. Most of the released energy is carried off by the two protons.

The sequence of three reactions by which protons combine to form helium nuclei is called the proton-proton cycle. It is the main source of energy release in stars of one solar mass or smaller. The following equation summarizes the proton-proton cycle:

$$4p \longrightarrow He^4 + 2e^+ + 2\nu + 2\gamma$$

The Carbon Cycle

The sequence of reactions that starts with two protons fusing into a deuteron was undoubtedly the primary energy source when the Universe was young and made up largely or entirely of hydrogen. Other elements built up steadily in the Universe during the course of time, as a result of stellar evolution. Carbon, which is the first substance to be formed in a star after helium, was particularly abundant in the elements. We will see in Chapter 9 that some stars explode at the end of their lives, distributing their materials, including carbon, to space. When new stars form out of the enriched interstellar gases, an alternate way of transmuting hydrogen into helium becomes possible, in which the carbon nuclei play a critical role. This sequence of reactions, called the carbon cycle, is listed below.

$$C^{12} + p \longrightarrow N^{13} + \gamma$$

$$N^{13} \longrightarrow C^{13} + e^+ + \nu$$

$$C^{13} + p \longrightarrow N^{14} + \gamma$$

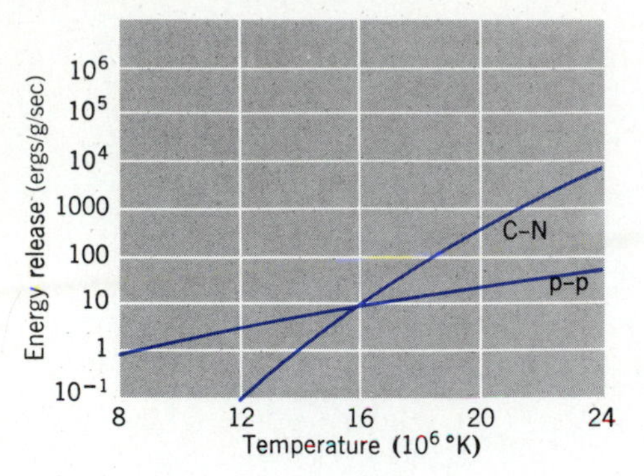

FIGURE 8.9
Comparison of temperature dependence for proton-proton and carbon cycles.

$$N^{14} + p \longrightarrow O^{15} + \gamma$$
$$O^{15} \longrightarrow N^{15} + e^{+} + \nu$$
$$N^{15} + p \longrightarrow C^{12} + He^{4}$$

Note that each time the sequence of reactions occurs, the net effect is to convert four protons into one helium nucleus, just as in the proton-proton reaction sequence. The carbon nucleus plays the role of a catalyst in the cycle; it is consumed in the early stages of the sequence but emerges again at the end.

The electrical barriers acting in the carbon cycle are much higher than those acting in the proton-proton cycle, the barrier in the basic carbon-proton reaction being six times higher than the proton-proton barrier. As a result, the proton-proton cycle dominates at low temperatures, and the carbon cycle does not become important until higher temperatures are reached. Figure 8.9 indicates the separate nuclear energy release from both cycles in ergs/g/sec for various temperatures. The proton-proton cycle is dominant up to a temperature of 16,000,000°K – slightly higher than the sun's central temperature – and the carbon cycle takes over above that level. About 10 percent of the sun's energy is contributed by the carbon cycle and the remainder by the proton-proton cycle.

The Main Sequence is divided into two segments, depending on whether the proton-proton or carbon cycle is the dominant energy source. The segment in which the proton-proton dominates corresponds to the *lower Main Sequence*; the other segment corresponds to the *upper Main Sequence*. The sun lies near the dividing point between the two segments of the Main Sequence.

THE MAIN SEQUENCE

The onset of hydrogen fusion – whether by proton-proton collisions or the carbon cycle – releases energy that halts the gravitational collapse of the star. For a star of one solar mass, the temperature at the surface at this time is 4500°K and the luminosity is half the present luminosity of the sun. The star is not yet on the Main Sequence, but is approaching it. As the release of nuclear energy at its center continues, the star becomes somewhat hotter and more luminous. It also shrinks slightly, to a diameter of one million miles. As these events proceed, the star moves upward and to the left on the Hertzsprung-Russell diagram. Finally, 17 million years after the onset of nuclear fusion, and about 50 million years after the first collapse of the protostar, the star comes to its resting place on the Main Sequence. Here it lives out most of the remainder of its life in a balance between the inward pressure created by the force of gravity and the outward pressure generated by the release of nuclear energy.

The fusion of hydrogen nuclei into helium nuclei dominates the longest single stage in the history of a star, about 90 percent of its lifetime in the case of a star of solar mass. The sun is in the middle of this stage; it came onto the Main Sequence 4.6 billion years ago, and will remain there for another 4 to 5 billion years before it moves off to die.

It is fortunate for us that the sun burns its hydrogen so slowly, for the evolution of life on the earth has also been a very slow process. According to the fossil record, simple forms of life, such as bacteria, appeared on the earth sometime during the first billion years of the solar system's existence; advanced forms of life did not emerge until several billion years thereafter. If the sun's lifetime had been, let us say, 100 million years or less, it is doubtful whether intelligent creatures would ever have populated the earth.

Relation Between Lifetime and Mass for a Star

Although the sun will live for about 10 billion years, other stars live for as short a time as a million years, and still others may live for as long as a trillion years or more.

Surprisingly, the largest stars live for the shortest time. They have more fuel to burn, but they burn it much more rapidly than the smaller stars. For example, a star 10 times as massive as the sun has 10 times as much fuel to burn, but it burns it at 10,000 times the rate of the sun, so that its lifetime is 1000 times shorter; that is, it lives for only 10 million years. Why does a massive star burn its fuel so much more rapidly than a smaller star? The great weight of such a star generates higher temperatures at its center, causing the protons to collide more violently than they do in a lighter star. Under the violence of these collisions, the electrical barrier between protons is more readily penetrated and the nuclear reaction rate goes up. In the more massive stars, the reaction rate climbs so rapidly with increasing temperature that a doubling of the temperature multiplies the reaction rate by a factor of 30,000.

On the other hand, a star with a mass one-tenth that of the sun should live for a trillion years. Barnard's Star is an example. It should still be shining long after the sun has gone out. Stars smaller than Barnard's Star are known to exist and should live for even longer times. These small stars live for such a long time that not a single one has died since the Galaxy itself came into existence, an event that occurred, according to the latest evidence, roughly 12 to 15 billion years ago.

Red Giants. During the hydrogen burning stage of a star's life, the helium produced by fusion of hydrogen atoms accumulates at the center of a star, where most of the reactions take place. When an appreciable amount of the hydrogen within the star has been converted into helium, and its center is filled with a core of pure helium, the star begins to show pronounced signs of age. It has taken the first step toward the *red giant* stage of its life.

Increase in the Temperature of the Helium Core

The first major change involves the conditions in the helium core. Because the temperature of the core is not high enough to fuse helium into heavier elements no energy is released there, and the central region of the star, which had been supported against gravitational collapse by the release of nuclear energy, no longer possesses the means for sustaining itself. Under the inward force of gravity, the helium core shrinks, and its temperature rises, just as the temperature of the entire star rose when it was collapsing at the beginning of its life. The center of the star is now steadily heating up. As the core gets hotter, it heats the hydrogen shell immediately surrounding it; this hydrogen commences to burn vigorously to form helium.

As time passes, the helium core continues to collapse and, as a consequence, its temperature continues to rise. Eventually, the rate of hydrogen burning in the shell becomes so great, because of the steadily mounting temperature, that the brightness of the star is noticeably affected. The envelope of the star absorbs some of this energy and expands even more rapidly than before, but this outward expansion can no longer absorb the enormous outpouring of radiation that comes from the brightly blazing hydrogen shell. The rate of energy release within the star is now hundreds of times greater than it was when the star was in the prime of its life. Only a small part of this energy can be taken up in the expansion of the star's envelope; most of the energy reaches the surface of the star and escapes as radiation. The luminosity of the star, accordingly, soars upward. The star, still red in surface color, has become brilliantly luminous; it has become a red giant.

The Structure of a Red Giant

Red giants are very odd stars. The core of helium at the center of the red giant is enormously compressed, with a density equal to one ton per cubic inch. One-quarter of the mass of the entire star is packed into the core, although its radius is only one one-thousandth of the radius of the star. This core has a diameter of 20,000 miles—about twice the size of the earth—but it weighs nearly 100,000 times as much. Around the core lies a thin shell of burning hydrogen a few thousand miles in thickness. Around the shell of burning hydrogen is an enormously distended and very tenuous envelope of hydrogen gas, 100 million miles across.

To bring out the peculiar structure of red giants, suppose the star is reduced to a sphere the size of a basketball. Then the helium core, containing one-quarter of the mass of the star, is a dot at the center no larger than the period at the end of this sentence.

From Young Stars to Red Giants on the H-R Diagram. What happens to the plot of the star on the H-R diagram during these changes? We last left the star on the H-R diagram as it reached the Main Sequence, just after its transition from protostar to star. Figure 8.10 (p. 208) shows how the position of the star changes on the diagram subsequently. During the first 9 billion years of the star's life, it stays in the vicinity of the Main Sequence (points 1–3). In this stage its luminosity and diameter slowly increase, while the surface temperature remains approximately constant. During the next billion years, it first moves rapidly to the right to point 4; then it shoots vertically upward to the red giant region (point 5). The changes are very rapid towards the end; between points 4 and 5 the star mushrooms in size, and its diameter increases by a factor of 50 in the relatively brief period of 100 million years.

ONSET OF HELIUM BURNING

What happens after the star reaches point 5? The helium core becomes more and more compressed with the passage of time, and its temperature continues to rise. Therefore, the red giant becomes more and more luminous. Finally, the helium core reaches the critical threshold temperature of 100 million degrees Kelvin. When the core reaches 100 million degrees, a completely new episode begins in the life of the star.

The new episode is triggered by the fusion of helium nuclei in the previously inert core of the star. At a temperature of 100 million degrees Kelvin, the helium nuclei, for the first time, collide with sufficient violence to pierce the electrical barrier that normally keeps them outside the range of their nuclear attraction. Nuclear reactions take place in these collisions, in which carbon and oxygen are formed by the fusion of helium nuclei.

Formation of Carbon-Oxygen Core. The reactions proceed in several steps. First, two He^4 nuclei collide and fuse. Since the resultant nucleus has four positive charges, it is the nucleus of the element beryllium, or Be^8, containing four protons and four neutrons:

$$He^4 + He^4 \longrightarrow Be^8$$

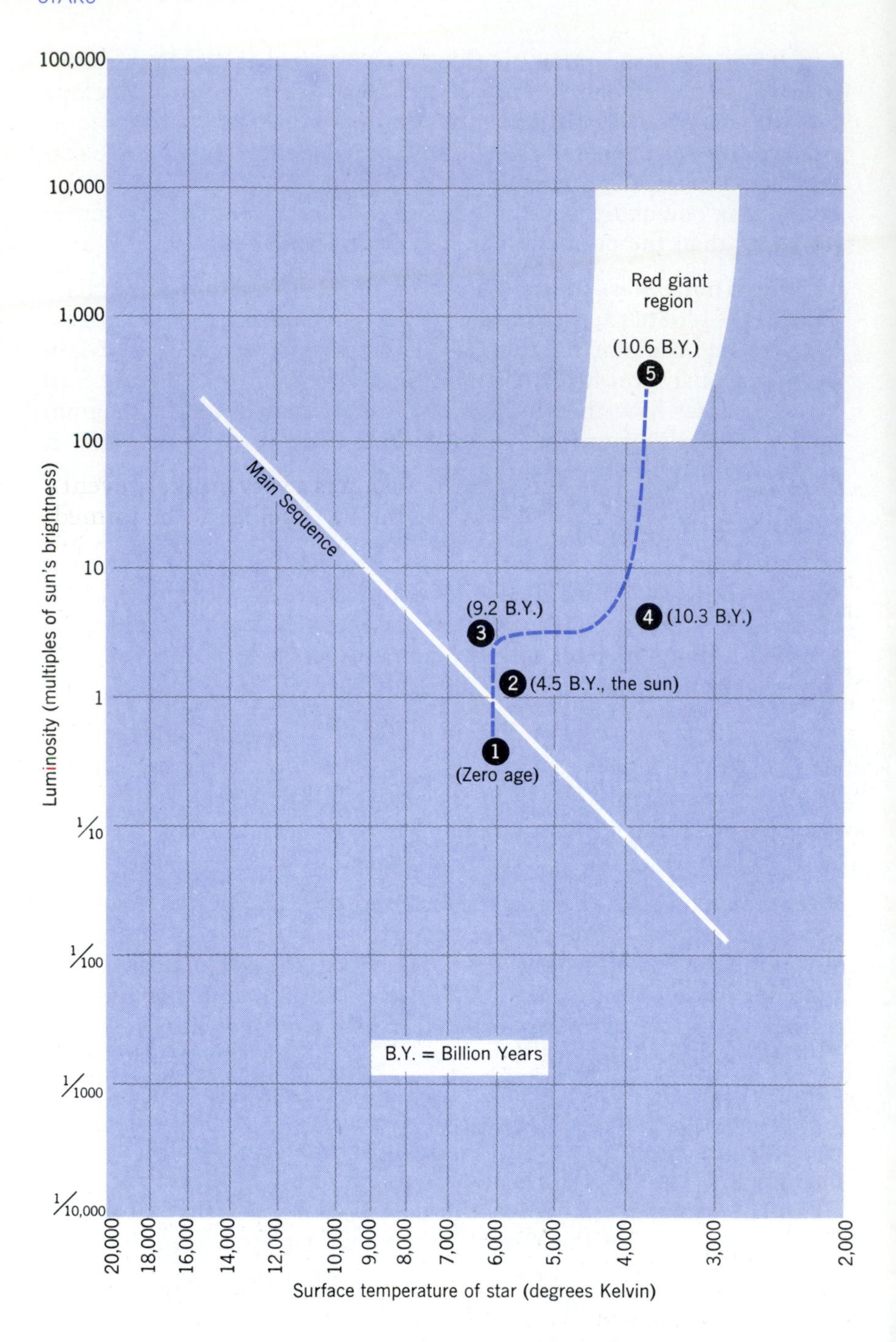

FIGURE 8.10
The track of a star from the Main Sequence to the red giant region.

But the Be^8 nucleus is not stable; although it can exist for a short time, the particles from which it is made do not attract one another strongly enough, and in less than one-trillionth of a second, it breaks up into two separate helium nuclei again.

However, if other He^4 nuclei are present, one of them may collide with the beryllium nucleus before it breaks apart, and fuse with it to form a nucleus with six neutrons and six protons, that is, a carbon-12 nucleus:

$$He^4 + Be^8 \longrightarrow C^{12} + \gamma$$

The carbon nucleus is tightly bound and very stable, and lasts forever once it is formed (unless, of course, it is disturbed by another nuclear collision).

If many He^4 nuclei are present, it also is possible for the C^{12} nucleus to be formed in one step, by the fusion of three helium nuclei:

$$3He^4 \longrightarrow C^{12} + \gamma$$

However, a simultaneous triple collision is a very unlikely event. Actually, it is much more likely for the Be^8 nucleus to be formed first, and the C^{12} nucleus to be formed immediately afterward by a collision between beryllium and helium nuclei, in spite of the fact that the beryllium nucleus stays together for such a short time.

The fusion of three helium nuclei to form carbon, like the fusion reactions that form helium itself, releases very large amounts of energy. When carbon is produced in stars by this process, the carbon nuclei frequently collide with helium nuclei and fuse again. This fusion creates an oxygen nucleus plus a substantial amount of nuclear energy which is carried off into the immediate surroundings by a gamma ray.

$$He^4 + C^{12} \longrightarrow O^{16} + \gamma$$

The Helium Flash

Once the fusion of helium into carbon and oxygen begins, you would expect the resultant release of nuclear energy to *add* to the luminosity of the star, driving it still further into the red giant region. But instead, a peculiar event occurs that checks the growth of the star's luminosity and sends it *down* the red giant branch.

At this point the helium core is packed in with a very high density, equal to many tons per cubic inch. Because the core is also very hot, the helium atoms are entirely stripped of their electrons. About 50 years ago a great theoretical physicist named Wolfgang Pauli discovered that when large numbers of electrons are packed into a small space, they become virtually incompressible. They behave as if they were a rigid solid, like steel, rather than a gas of small particles moving freely about. Because of the incompressibility of the electrons in the helium core, the core acts very much as if it were a sphere of solid steel.

Now consider what will happen to the core when the helium nuclei within it starts to burn. First of all, the temperature of the core rises. The rise in temperature should cause an expansion of the core. The expansion should cause a drop in the temperature and, therefore, in the nuclear reaction rate—that is, the rate of fusion of helium nuclei into carbon and oxygen. The drop in the nuclear reaction rate should stop the expansion of the helium. This is the safety-valve feature that is built into the star's structure. The star then lives on, burning helium at its center at a rate just sufficient to balance the inward attraction of gravity and the outward pressure produced by the helium burning. This is, after all, the way in which the star lived on the Main Sequence, although then it was the burning of hydrogen, rather than helium, that balanced the inward force of gravity.

But nothing of the sort happens to the core of the red giant. It has the properties of a sphere of solid steel and, like all solids, it expands only very slightly when heated. The core expands far less than it would if it were behaving like a normal gas. Because the helium core does not expand appreciably when heated, the star loses the safety-valve feature that used to be built into it. Imagine what must happen in this core without the safety-valve feature of expansion: helium nuclei burn, raising the temperature; the core does not expand; therefore, the temperature stays high; at the higher temperature, helium nuclei burn still faster; the core gets still hotter; it burns still faster; and so on. The reaction runs away with itself. After a time, the core becomes so hot that it literally explodes, just as a steel ball would vaporize and explode if a powerful enough charge of energy were put into it. The center of the star has changed from a controlled nuclear reactor to an uncontrolled nuclear bomb.

It takes a few hours from the onset of helium fusion for the core of the star to reach the explosion point. A few hours may seem like a long time when compared to the time it takes to trigger a hydrogen bomb, but remember that a few hours is the blink of an eye for a star. Because the time is very brief on this scale, the events from the commencement of helium fusion to the explosion of the core are called the *helium flash.*

After the Helium Flash

As soon as the core explodes, its density and temperature drop. The temperature of the surrounding shell also drops. Therefore, the rate of hydrogen burning in the shell decreases. As a result, for the first time since the star arrived on the Main Sequence 10 billion years previously, its brilliance diminishes substantially. The star moves vertically downward on the H-R diagram, in the general direction of the Main Sequence.

The diameter of the star decreases at the same time, for, with its nuclear energy sources greatly decreased, the red giant lacks the resources needed to keep its envelope distended. Relieved of the enormous pressures created by intense hydrogen burning in the shell, the envelope commences to collapse under the attraction of gravity, and the star begins to lose its swollen appearance.

The H-R diagram in Figure 8.11 (p. 212) shows the evolutionary track of the star in the period of its life that we are now describing. At point 5 of the previous H-R diagram the star becomes a fully developed red giant. At some time shortly after becoming a red giant, the star experiences the helium flash. Point 6 on the H-R diagram marks the helium flash. The helium flash sends the star down toward the Main Sequence again. This slide continues for about 10,000 years, with the luminosity and size of the star continually decreasing.

During the course of the star's descent from the red giant region, the helium in the core is again slowly but steadily compressed by gravity, because there are no nuclear energy sources within it to sustain it against this inward force. As always, the compression heats the helium core, and its temperature rises. After several thousand years, the temperature gets high enough to start a very small amount of helium burning at the center of the core. By the end of 10,000 years, the temperature of the core has risen enough, and the rate of helium burning has accordingly become great enough, for helium burning to constitute a major source of energy for the star. At this point the temperature in the core is roughly 200 million degrees Kelvin. Its structure is shown in Figure 8.12.

The release of substantial amounts of nuclear energy at the center of the star, through the burning of its helium, halts the star's descent toward the Main Sequence. The star now lies at point 7 on the H-R diagram (Figure 8.11). This is the second time that the temperature in the core has reached the level required for helium burning. The first time was the helium flash at point 6 on the H-R diagram. But this time no explosion occurs, because the helium core is far less dense than it was before the helium flash occurred. As a result, the helium core now behaves as a proper gas should. It expands in response to the renewed release of nuclear energy, dropping the nuclear reaction rate and reducing the rate of nuclear heating of the star. The star has recovered its safety-valve properties.

From point 7 the star moves horizontally to the left, approaching point 8 on the H-R diagram in Figure 8.11. It is coming into balance between the inward pressure of its gravity and the outward pressure generated by nuclear energy released in its interior. Carbon and oxygen are building up in the center of the helium core, just as helium built up in the center of the star when the star was in the first stage of its life. During this time carbon and oxygen steadily accumulate at the center of the star as the products of the

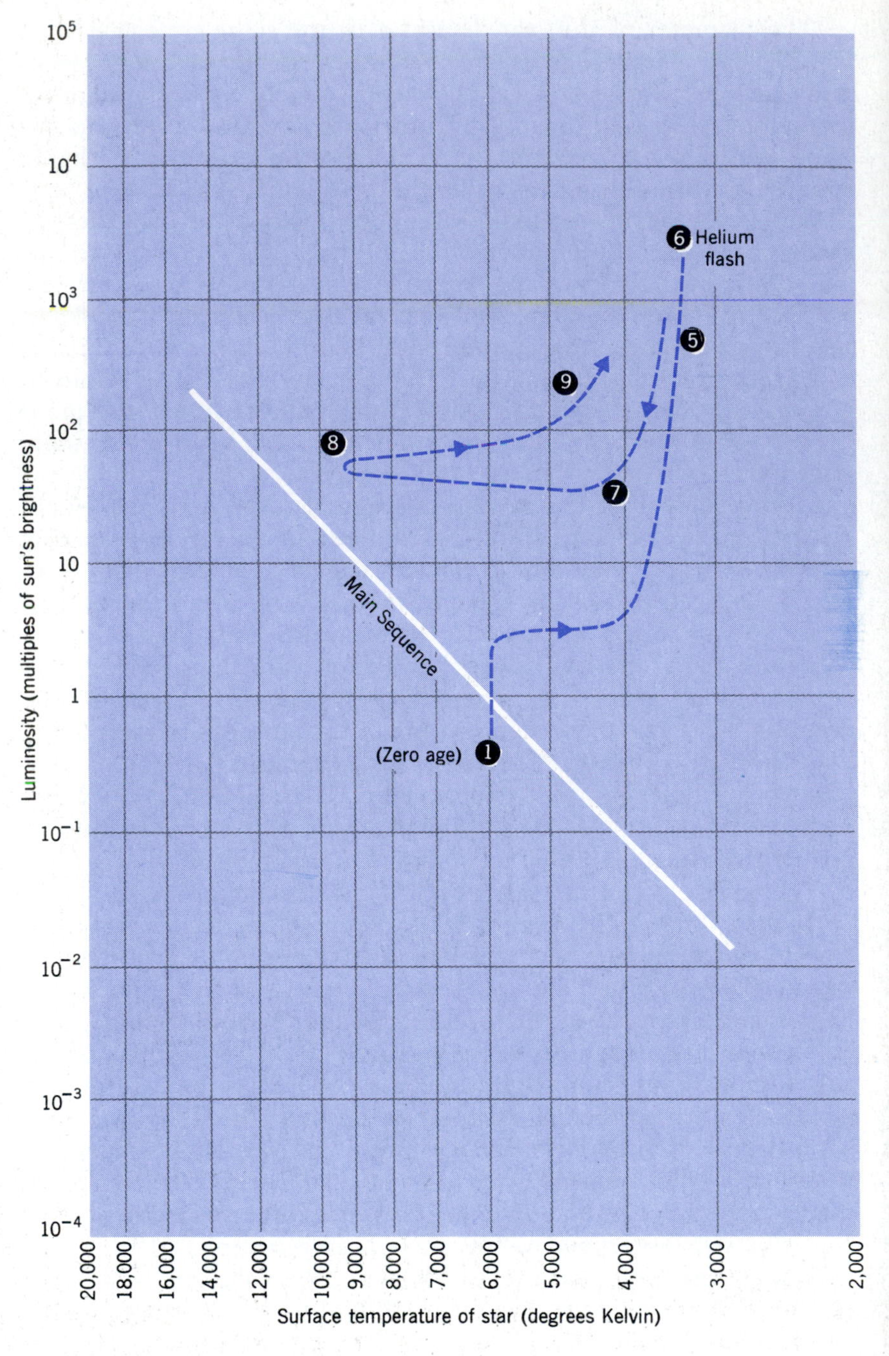

FIGURE 8.11
The late evolution of a star of solar mass.

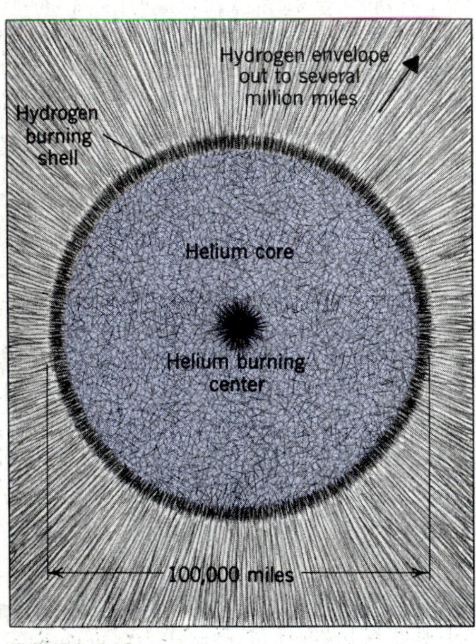

FIGURE 8.12
The structure of the core of a star shortly after occurrence of the helium flash and descent from the red giant branch. This stage corresponds to point 7 in Figure 8.11.

helium burning (Figure 8.12). As the carbon and oxygen build up, they begin to form an inner core within the core of helium. Following the pattern of the previous stage, the presence of the carbon-oxygen core halts the helium-burning reactions at the center. Helium burning now occurs in a shell surrounding the carbon-oxygen

I. STARS AND GALAXIES

PLATE 1. DISPERSION OF WHITE LIGHT BY A PRISM. Short wavelengths (blue) are bent through a larger angle than long wavelengths (red).

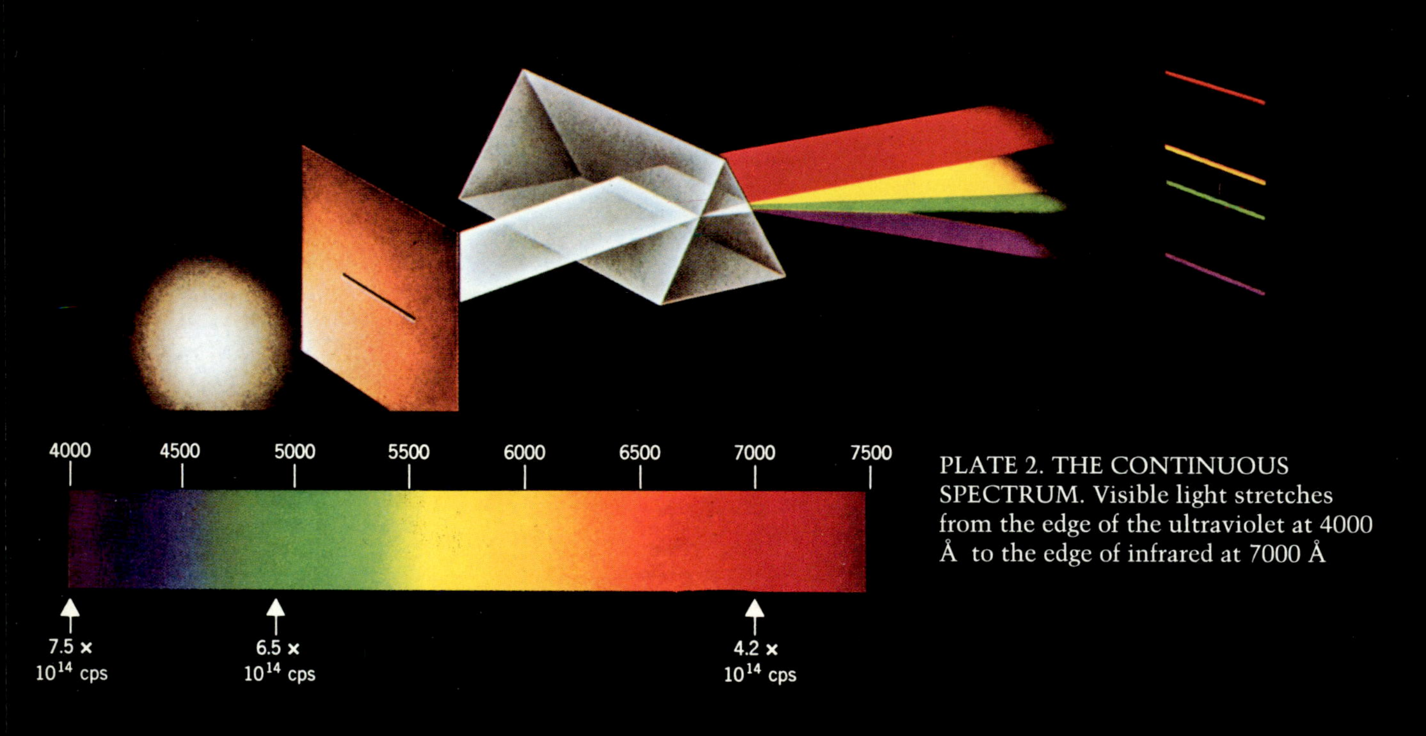

PLATE 2. THE CONTINUOUS SPECTRUM. Visible light stretches from the edge of the ultraviolet at 4000 Å to the edge of infrared at 7000 Å

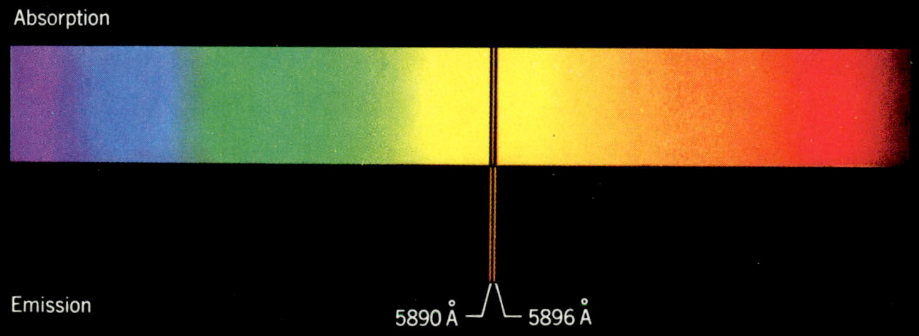

Absorption

Emission

5890 Å

5896 Å

PLATE 3. EMISSION AND ABSORPTION SPECTRA OF SODIUM. Every bright line in the emission spectrum is matched by a dark line in the absorption spectrum at the same wavelength, suggesting that an element can be identified by either its emission or its absorption lines.

PLATE 4. SERPENS. This nebula in the constellation Serpens shows star formation in the Galactic disk. The small, dark regions are believed to be protostars–pockets of condensed gas on the way to forming new stars.

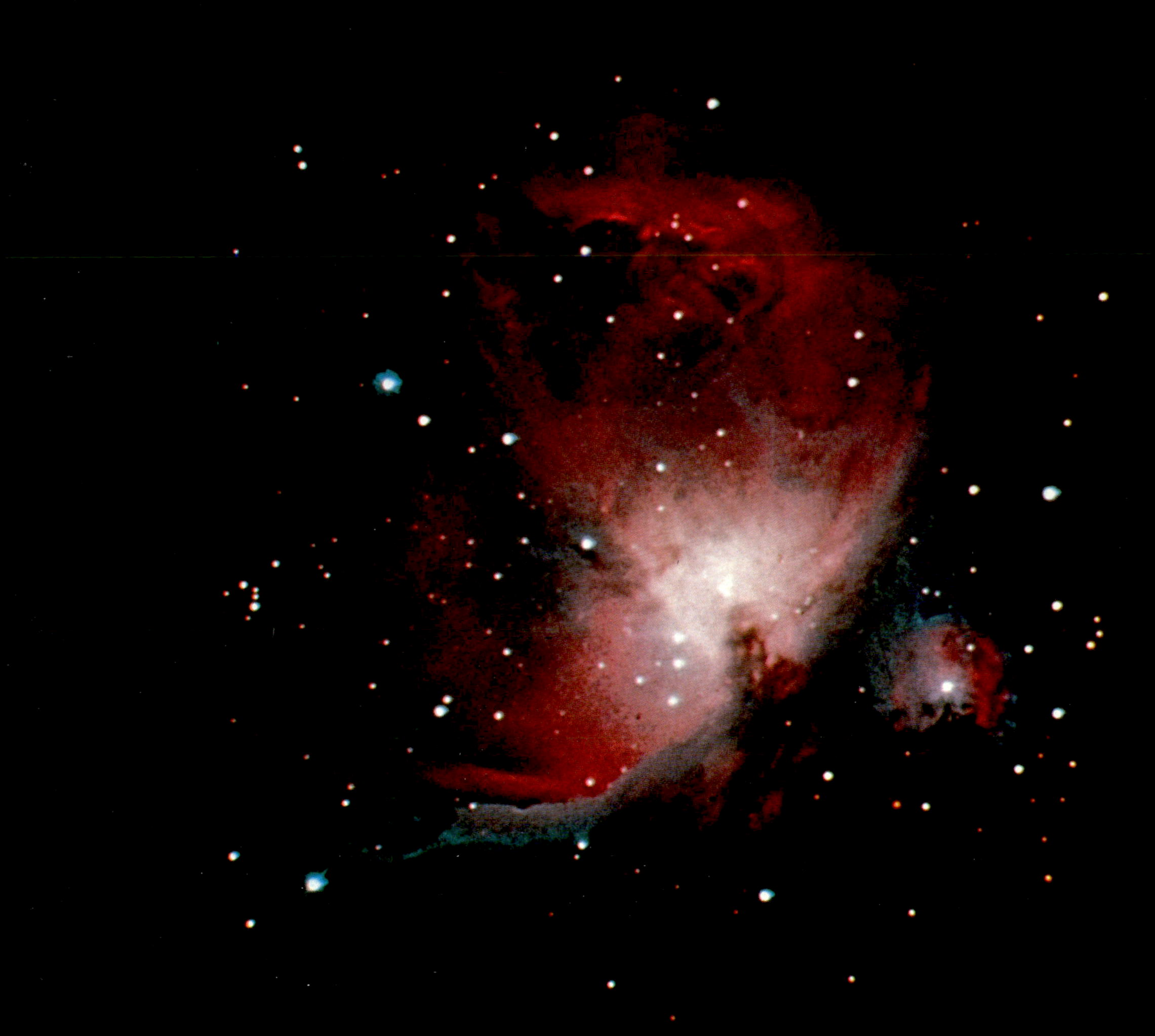

PLATE 5. THE NEBULA IN THE CONSTELLATION ORION. This is also one of the dense clouds of matter in the Milky Way that are breeding grounds of new stars. Most of the stars in the Orion Nebula were formed within the last 6 to 10 million years.

PLATE 6. THE TRIFID NEBULA. A luminous region, similar to the Orion Nebula, divided into three parts by dark lanes of obscuring dust in the Galactic disk.

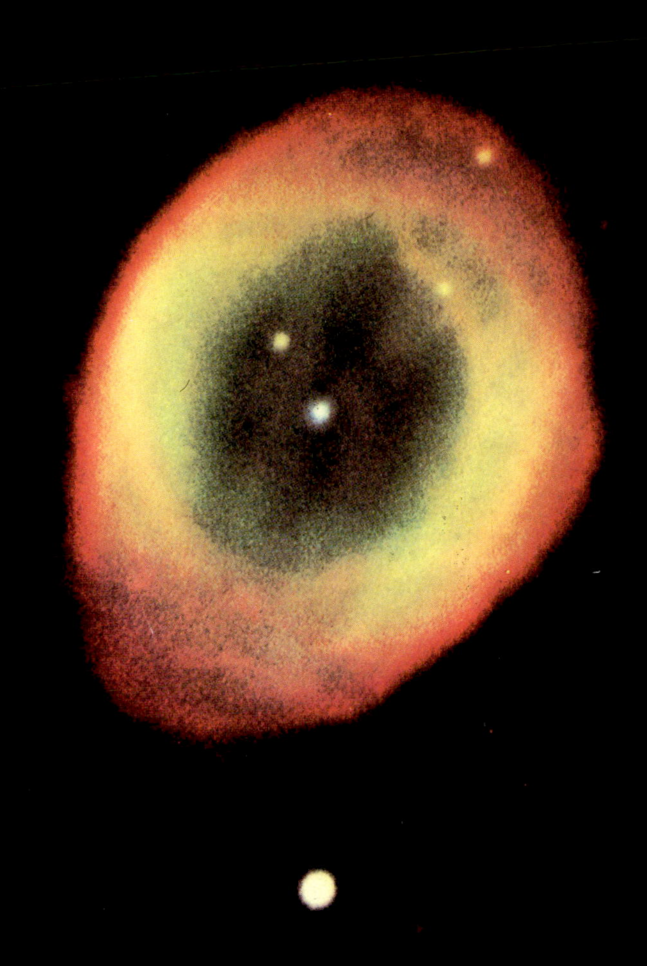

PLATE 7. AN AGEING STAR. The Ring nebula in the Constellation Lyra is a planetary nebula, or star undergoing transformation to the white dwarf stage. The former envelope of the star has become a shell of gas expanding rapidly outward. The core of the star–now visible at the center of the nebula–will become a white dwarf.

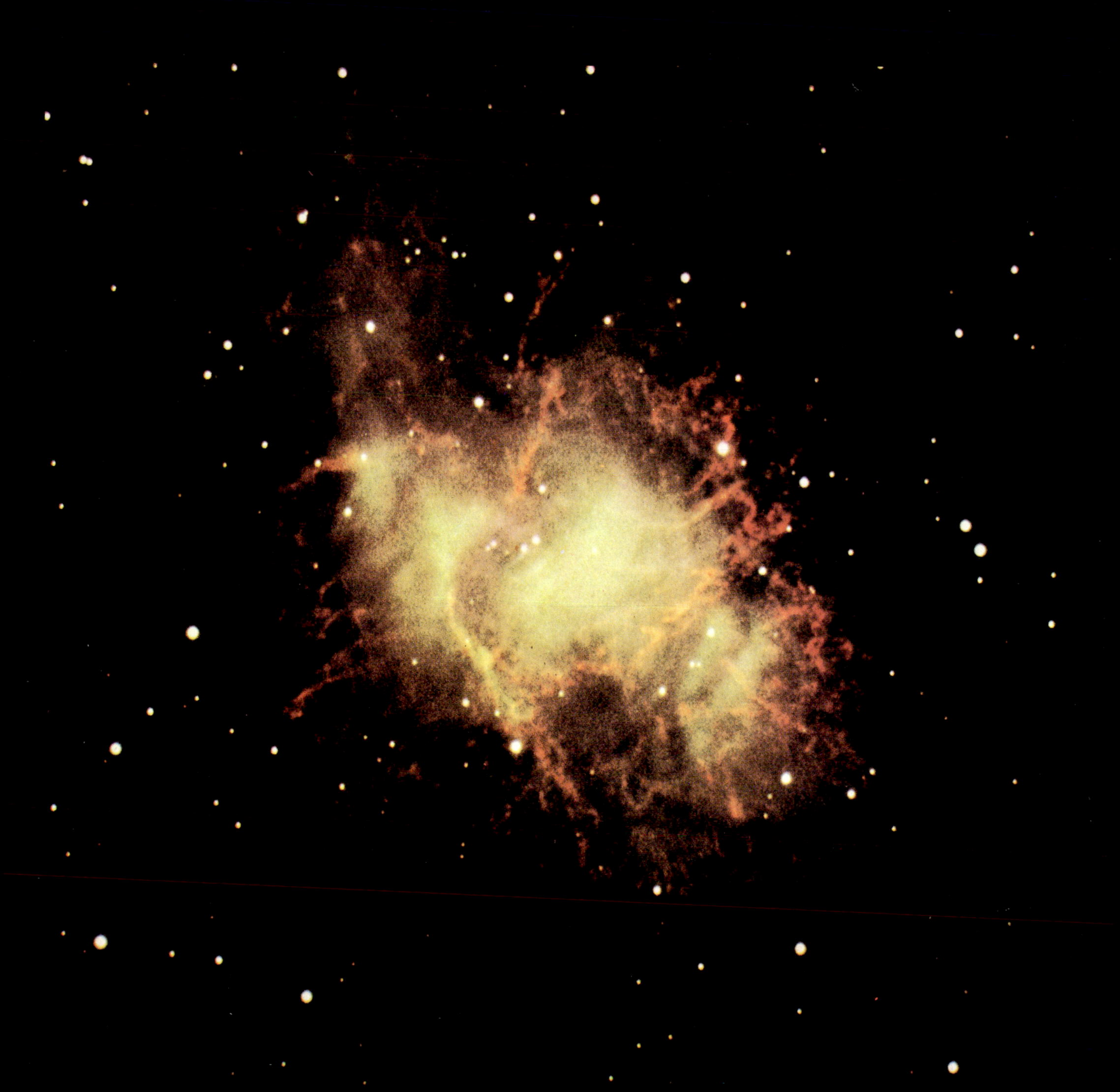

PLATE 8. A DYING STAR. The Crab Nebula is a remnant of supernova explosion of A.D. 1054. The compressed core of the supernova is a pulsar, barely visible in the photograph as a faint star in the center of the nebula.

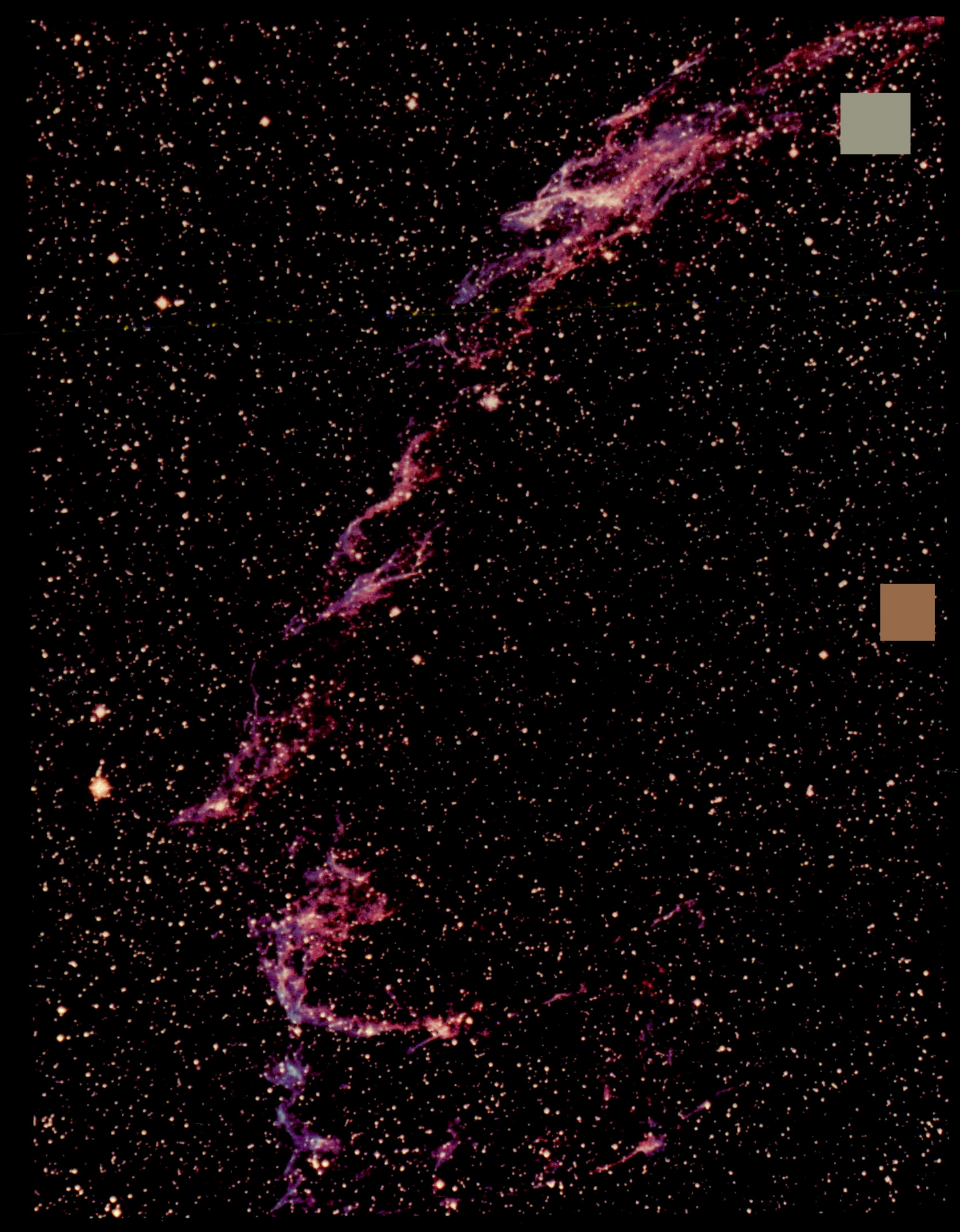

PLATE 9. THE VEIL NEBULA. This expanding spherical shell of gas is believed to be the remnant of an ancient supernova. It is about 70 light-years in diameter and is at a distance of 1500 light-years.

PLATE 10. A GALACTIC CLUSTER. The Pleiades are a young galactic cluster containing about 100 stars, formed 60 million years ago. Each of the hottest stars in the cluster is surrounded by the blue glow of its own light reflected from the surrounding interstellar dust.

PLATE 11. A GLOBULAR CLUSTER. M13 is a globular cluster containing approximately 300,000 stars, about 10 billion years old.

PLATE 12. THE LARGE MAGELLANIC CLOUD. This cluster of some 20 billion stars is a satellite galaxy of the Milky Way, bound to our galaxy by gravity. The luminous area at lower right is the Tarantula nebula, a region in which new stars are forming. The Large Magellanic Cloud orbits the Milky Way Galaxy at a distance of 170,000 light-years. The speed with which it moves in its orbit around our galaxy is one item of evidence for the existence of a massive galactic halo of dark matter extending beyond the visible edge of the Galaxy.

PLATE 13. THE WHIRLPOOL GALAXY (NGC 5194 or M51). (opposite) This spiral galaxy of Type Sc is at a distance of 12 million light-years. The irregular galaxy (NGC 5195) below appears to be a satellite of the Whir[illegible] Galaxy. The blue color in the spiral arms is proc[illegible] by hot,

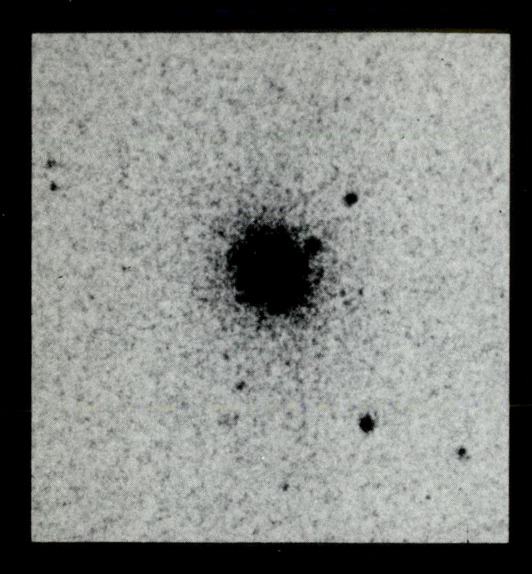

PLATE 14. A JET EMERGING FROM AN ELLIPTICAL GALAXY. NGC 6251 (left) is an elliptical galaxy 400 million light-years away. It is a strong source of radio waves. The radio image of NGC 4651 is shown below in three stages of increasing resolution. The higher-resolution images reveal an intense jet coming from a very small source at the center of the elliptical galaxy.

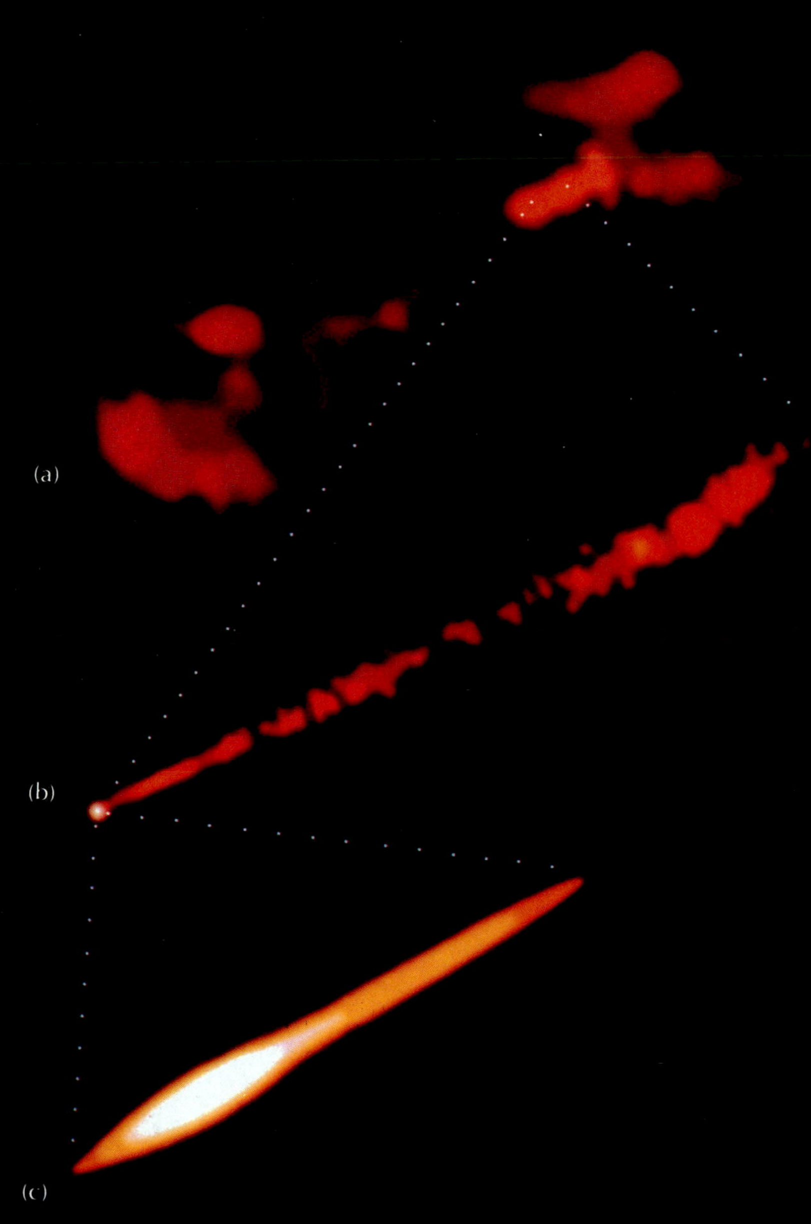

(a) [illegible]s image spans a distance of abou[illegible] million light-years. The posi[illegible] of the visible galaxy is marked by a "+". The red regions are radio-emitting clouds several million light years in size. Their energy appears to come from the galaxy.

(b) This image, about 700,000 light-years across, shows the region around the elliptical galaxy in greater detail. In this false-color image, shades of orange, yellow and white represent successively higher intensities of radiation. The image reveals a jet coming from a small and very intense source of radiation at left.

(c) The region near the intense source, shown in still greater detail wit[illegible]e aid of VLBI techniques, con[illegible]ms the impression of a cosmic jet of radiation. This image is only 10 light-years across. The jet comes from an extremely small object at left, considerably less than one light-year in size. Matter heated in the vicinity of a black hole at the center of the galaxy may be the origin of the jet.

core (Figure 8.13). At this stage the star has reached point 8 on the H-R diagram. The inner structure of the star is now analogous to its structure at point 3 in Figure 8.10, when hydrogen was burning in a shell around the inert helium core. The carbon-oxygen core acts as the helium core acted earlier; it contracts, and its temperature rises; the higher temperature heats the surrounding shell of helium, and the rate of helium burning increases. The heat from the increased helium burning is absorbed by the cool outer envelope, which keeps expanding. The temperature drops and the luminosity increases. The star starts up toward the red giant region once again. That is, it swells enormously and becomes very bright, although its surface is relatively cool and red in color.

The speed of the transition to the red giant region is now, however, much faster than it was when the star first moved up this path. Everything proceeds about 100 times more rapidly. The entire move toward the red giant region is completed in a few million years for a star with the mass of the sun, as compared to several hundred million years required to reach the red giant region for the first time.

When a core of carbon accumulates at the center of a star, and it enters the red giant region for the second time, the star is close to the end of its life.

FIGURE 8.13
The complete structure of the core of a star when an inner core of carbon and oxygen has accumulated, and the star has commenced its second ascent of the red giant branch, corresponding to point 9 in Figure 8.11.

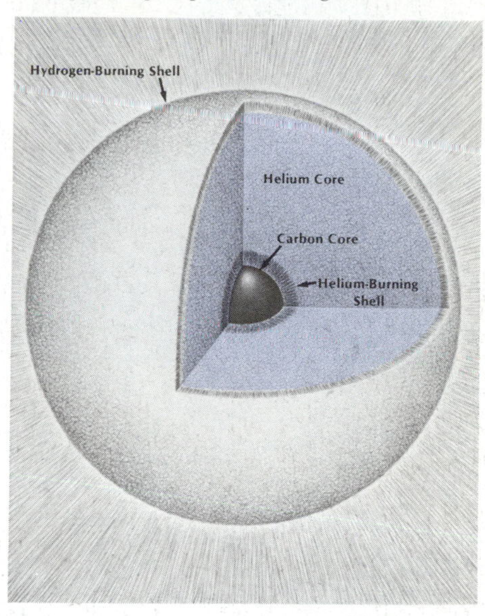

DEPENDENCE OF STELLAR EVOLUTION ON MASS

All stars have followed the same pattern up to this point. What happens next depends very much on the mass of the star. Stars with masses in the neighborhood of the sun's mass follow a pattern of evolution similar to that of the sun. These stars are formed by gravity, spend most of their lives burning hydrogen on the Main Sequence, and move upward and to the right into the red giant region when their reserves of hydrogen fuel are diminished (Figure 8.14*a*). They differ from the sun mainly in the pace of their evolution; stars smaller than the sun evolve more slowly; bigger stars evolve more rapidly.

A difference in the *pattern* of stellar evolution appears for the first time in stars with masses greater than roughly four times the mass of the sun. In fact, approximately four solar masses is a critical size for a star in the determination of its fate.[2] When a star with a mass greater than $4M_\odot$ depletes its reserves of hydrogen fuel and leaves the Main Sequence, its evolutionary track on the H-R diagram is nearly horizontal. It moves across the diagram to the right and directly into the red giant region, instead of moving upward

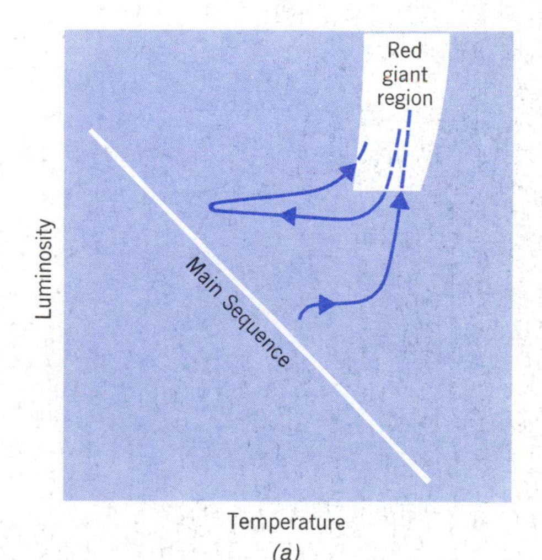

FIGURE 8.14(a)
Pattern of evolution for stars with masses less than $\sim 4M_\odot$.

[2] Theoretical studies of stellar evolution do not indicate the precise mass at which the transition occurs.

FIGURE 8.14(b)
Pattern of evolution for stars with masses greater than $\sim 4M_{\odot}$.

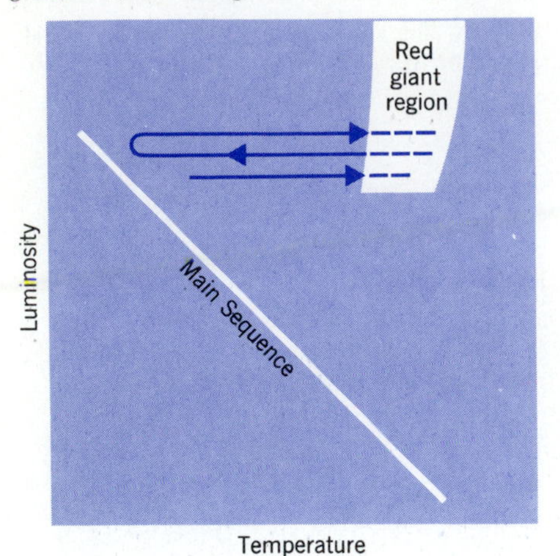

as in the case of the less massive stars. Later, when the star evolves back toward the Main Sequence and then returns to the red giant region for the second time, its track on the H-R diagram remains approximately horizontal (Figure 8.14*b*).

These differences in the tracks of stars of various masses do not change the general character of the story of stellar evolution. That is why the evolution of a star of one solar mass can be presented as a general model for the evolution of all stars. However, the single account of stellar evolution, unified for stars of all masses, breaks down at the end of the star's life. At that point, phenomena occur in the lives of very massive stars that have no counterpart in the lives of smaller stars. These phenomena, which involve the formation of supernovas, neutron stars, and black holes, are among the most interesting aspects of contemporary astronomy.

Main Ideas

1. The mechanics of gravitational contraction.
2. Birth of stars; rising temperature within a protostar.
3. Nuclear fusion reactions; the proton-proton reaction; the carbon cycle.
4. Transition from protostar to star on the Main Sequence; life on the Main Sequence; first signs of ageing; departure from the Main Sequence.
5. The red giant stage; structure of a red giant.
6. The helium flash; helium burning.
7. The track of an evolving star on the H-R diagram.
8. Relation between lifetime and mass.
9. Dependence of stellar evolution on mass.

Important Terms

cluster of stars
density
deuteron
electrical barrier
envelope (stellar)
gamma ray
gravitational condensation
helium core
helium flash
hydrogen burning
nebula
neutrino
nuclear fusion
positron
protostar
red giant

Questions

1. Why must a large number of particles come together at the same time for a star to condense from a gaseous nebula?

2. Why does the initiation of nuclear reactions at the center of a newly forming star halt the collapse of the star?
3. Are stars still forming today? How do we know? Where would they be likely formed?
4. Consider the reactions in the proton-proton cycle and the carbon cycle. Which set of reactions is more important in lower Main-Sequence stars? Why? Which set of reactions was more important when stars first formed in the Milky Way Galaxy?
5. Describe and contrast fission of heavy elements versus fusion of light elements. Why would stars obtain enormous amounts of energy from fusion but only minimal energy from fission?
6. Approximately what temperature is required to convert helium into carbon in a star? Compare with the temperature required to convert hydrogen into helium. Why are they different?
7. Outline the evolution of a one-solar-mass star from the contraction phase to the helium flash. Point out physically significant changes in the internal structure and observable surface properties of a star during hydrogen depletion, the formation of the hydrogen-burning shell, the helium-burning core and the helium flash.
8. Where on the H-R diagram does a star spend most of its lifetime? What happens to it when a helium core forms?
9. Why does a star leave the Main Sequence? Describe the sequence of events as the star moves away from the Main Sequence.
10. Describe the sequence of events that occur in the evolution of a one-solar-mass star during its second evolution to the red giant branch.
11. Why do very massive stars have much shorter lives than stars of relatively low mass? About how long is the sun expected to live? A star with 10 times the sun's mass? One-tenth the sun's mass? Considering this, where would you expect the greatest concentration of Main-Sequence stars on the H-R diagram? Explain.

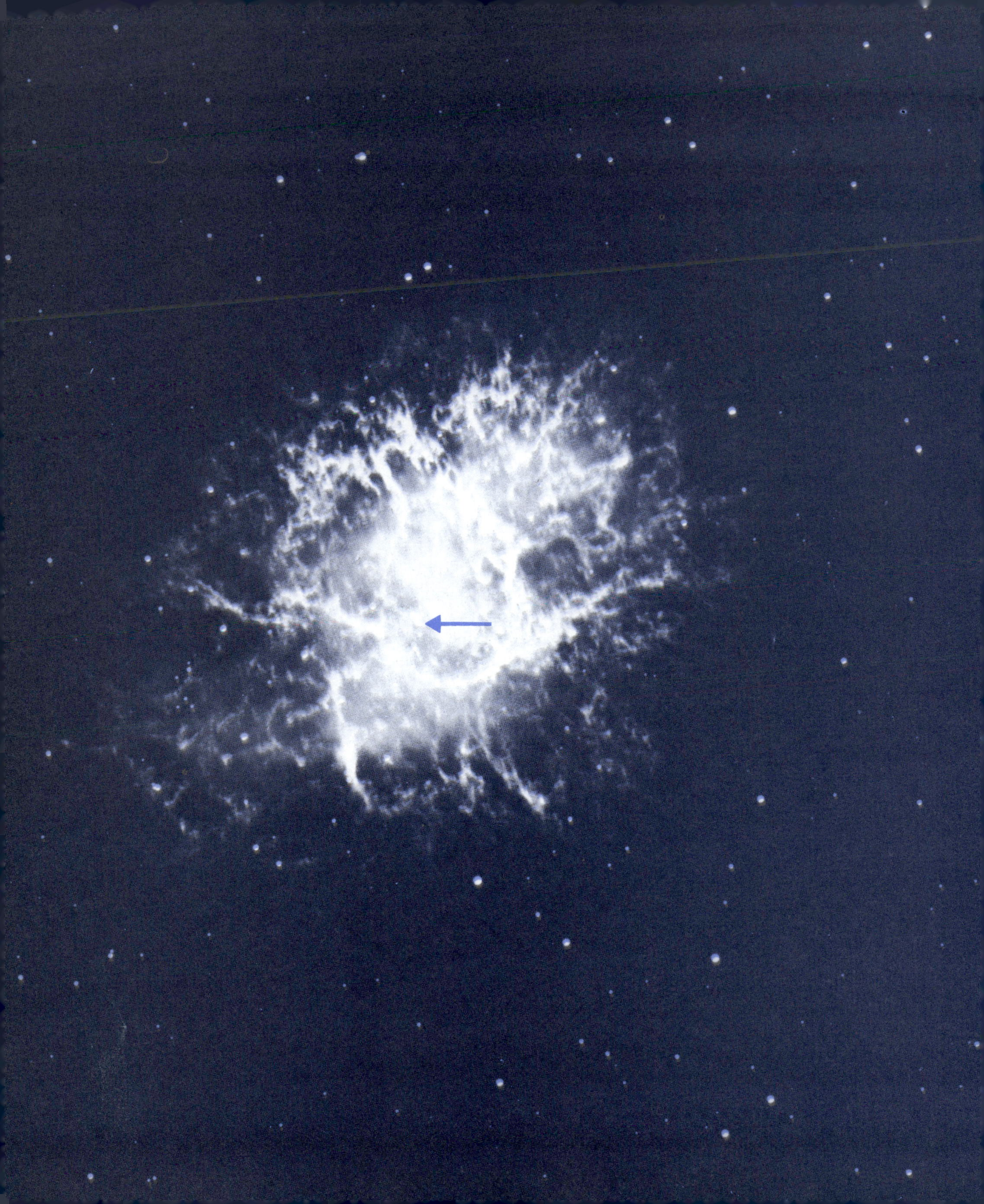

WHITE DWARFS, NEUTRON STARS, AND BLACK HOLES

Every star must exhaust its reserves of nuclear fuel eventually. When the reserves are sufficiently depleted, the star succumbs to the force of gravity and collapses. In the case of a small or medium-sized star, the collapse is relatively gentle; the star leaves the red giant region, moves to the left on the H-R diagram, and then moves downward, fading out slowly in the lingering death of the white dwarf. This will almost certainly be the fate of the sun.

A very large star has a different fate. It shuttles back and forth across the H-R diagram, from the neighborhood of the Main Sequence to the red giant region and back again. When this stage is over, the massive star moves toward the red giant region for the last time. Now its life is nearly over, but it does not fade away; instead, it blows up suddenly in the cataclysmic event known as the supernova. A glowing cloud of debris expands from the site of the supernova at speeds of thousands of miles per second. Buried in the cloud is the severely compressed remnant of the star's core—a neutron star or a black hole.

DEATH OF A SMALL STAR: THE WHITE DWARF

The dividing line between small and large stars is about four solar masses. Consider a star whose mass is less than $4M_{\odot}$, as it climbs the red giant branch on the H-R diagram for the second time. In the star's previous approach to the red giant region, its ascent was

The Crab Nebula; remnant of a supernova. The arrow points to the pulsar.

terminated by the onset of the helium flash (page 209). We would expect the second ascent to be terminated in a similar manner by the onset of a carbon flash, that is, an explosively rapid onset of carbon burning; however, a carbon flash cannot take place in this star because it is not massive enough to attain the temperature required for burning carbon.[1] According to cyclotron experiments, the carbon in the core must reach a temperature of 600 million degrees Kelvin before it can burn. Calculations indicate that if the star's mass is less than $4M_{\odot}$, compression by gravity does not produce enough heat at its center to raise the temperature to 600 million degrees. Thus, carbon cannot burn. Instead the star continues its climb up the red giant branch; its diameter increases, its surface temperature drops, and its color continues to redden.

The Planetary Nebula

Eventually, the outer layers of the star become so red—that is, so cool—that the nuclei in these layers begin to capture electrons to become neutral atoms. The formation of neutral atoms continues unchecked until a substantial part of the star's mass is in the form of neutral atoms rather than separate electrons and nuclei.

What happens when a neutral atom is formed by the recombination of an electron and a nucleus? The most important consequence is that a photon is emitted, carrying away energy with it. Usually the photon is absorbed by another atom or particle before it escapes from the star. Countless photons are created by the formation of neutral atoms, and then absorbed shortly thereafter on their way out of the star. Their absorption heats the outer layer.

The heat produced in the star's outer layers by the absorption of photons is modest in amount compared to the heat released by nuclear reaction at its center. Nonetheless, according to one theory this heat triggers profound changes in the star's appearance. The envelope, heated by the absorption of photons, expands. The outward expansion lowers the temperature of the envelope. At the lower temperature, more neutral atoms form from the separate nuclei and electrons in the envelope, and they in turn release still more energy in the form of photons. Most of the photons are, again, absorbed by nearby atoms in the star. They heat the star's outer layer, expanding it still further.

In other words, the theory predicts a runaway process in which the capture of electrons by nuclei heats the envelope and causes expansion, which cools the envelope, causing more electron capture, which leads to more expansion. The envelope of the star ex-

[1] The core contains oxygen as well as carbon. However, the carbon burns before the oxygen because carbon nuclei have a lower electrical barrier.

pands outward faster and faster, until it leaves the star entirely. In effect, the envelope of the star blows off into space and becomes a tenuous, nearly transparent shell of atoms that continues to move rapidly outward.

The core, which was formerly concealed by the envelope, now stands exposed to view. If someone were observing the star during the course of this entire process, he would see an amazing change in its appearance. At the start the star would seem normal. Then, when the envelope had started to expand but was still dense enough to conceal the core, the observer would see the surface of the expanding, relatively cool envelope, and the star would present the appearance of a large, luminous, red object. When the envelope had expanded out far enough to become more or less transparent, so that the core was exposed, the observer would see a small, white-hot object—the core—surrounded by a softly glowing diffuse shell of gas—the blown-off envelope.

Such objects have been observed, and are called planetary nebulas. The name "planetary nebula" came into use because the astronomers who first observed these nebulas through small telescopes thought the images resembled those of planets. We now know that planetary nebulas have no connection with planets or solar systems, but the name has persisted.

Figure 9.1 shows the structure of a planetary nebula clearly. This photograph was taken with the 100-inch telescope at Mount Wilson.

FIGURE 9.1
A planetary nebula (NGC 7293).

A striking color photograph of another planetary nebula is included in the color inset.

What happens to the core of the star after the envelope blows off? The core is more or less unaffected by the departure of the envelope, and it continues to burn helium in the helium-burning shell at the same rate as before. Therefore, the luminosity of the star, which is controlled entirely by the burning of the helium in the shell, remains constant.

However, the plot of the star's position in the H-R diagram changes dramatically when the envelope blows off, because initially we are plotting the position of the cool envelope of the star (around 3500°K), but after the envelope blows off, we plot the position of the hot core (around 50,000°K) that remains. Thus, on the temperature axis there is a shift from 3500°K to 50,000°K. Because there is no change in luminosity during this increase in surface temperature, the star's evolutionary track shoots horizontally across the H-R diagram to the left.

These changes are depicted in the H-R diagram in Figure 9.2. The star's envelope starts to expand at point 10. At point 11 the hot core of the star is fully exposed. At this point the star, if photographed, would look like the Ring Nebula in the constellation Lyra. (See Color Plate 7.)

The White Dwarf

At point 11 on the H-R diagram, the star begins its transition from the core of the planetary nebula to a white dwarf. The star is now composed of a carbon-oxygen core surrounded by a helium-burning shell (Figure 9.3). At this point the temperature of the core is still not high enough for fusion to occur, and there is therefore no source of nuclear energy at the center of the star to offset the attraction of gravity and keep the star from collapsing. The star's core continues to contract slowly.

If it were not for the electrons in the star, the contraction would continue and the core would get hotter and hotter, until finally, at 600 million degrees Kelvin, the carbon nuclei would begin to burn. Before this can happen, however, the peculiar incompressibility of closely packed electrons comes into play, just as it did at an earlier stage in the star's life, immediately before the helium flash occurred. As before, the "solid-steel" incompressibility of the electrons brings the contraction to a halt. This event occurs when the star has a radius of about 5000 miles and a density of about 10 tons per cubic inch.

No one knows what happens in detail between points 11 and 12 on the H-R diagram. The theoretical calculations for that stage indicate that any one of a number of different possibilities could take

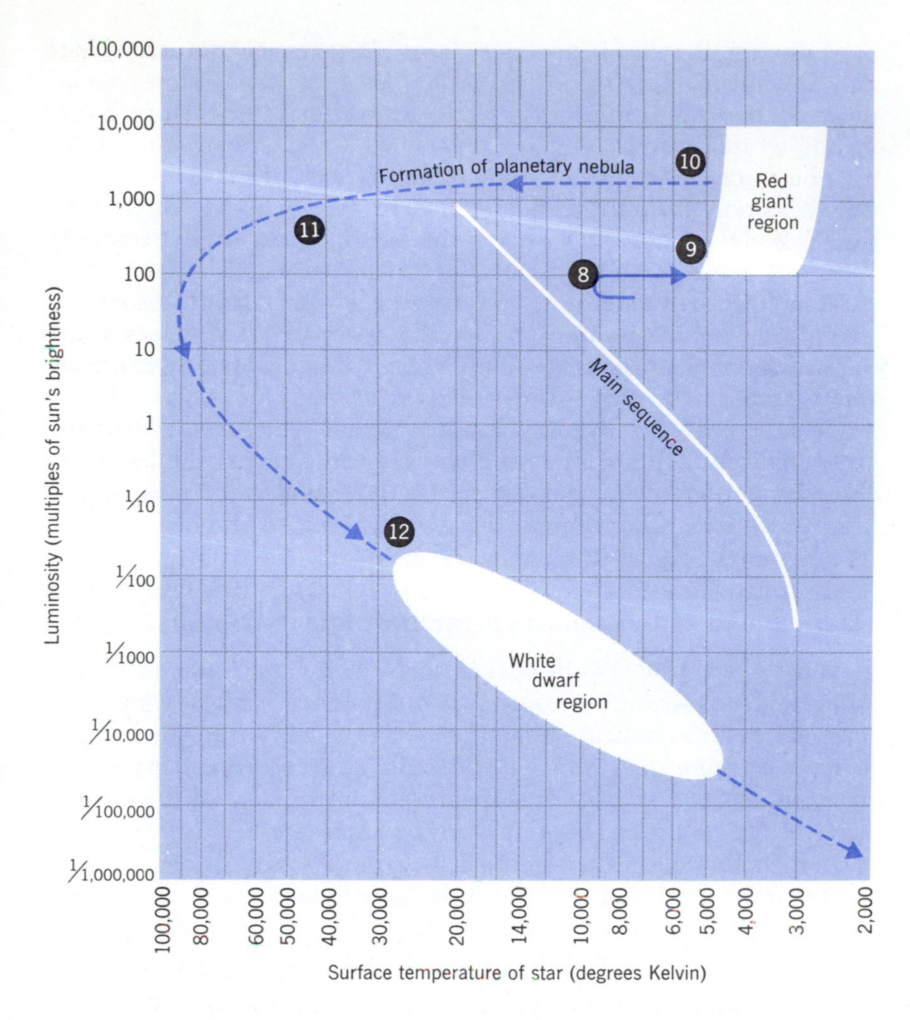

FIGURE 9.2
The track of a small star in the final stages of evolution.

place. Observations of stars do not provide a clear indication of what actually occurs, because few stars have been found between 11 and 12.

Once the star reaches point 12, however, its course becomes clear. In the region of point 12 the star is exceedingly dim in comparison to its luminosity in the earlier stage of its life; a star with the mass of the sun, for example, would be 100 times fainter at point 12 than the sun is at the present time. The diameter of the star is very much smaller at this stage than when the star was in its prime. A star originally the size of the sun would be about 20,000 miles in diameter, which is about twice the size of the earth. The shrunken star is exceedingly dense. Into its relatively small volume, no more than that of a good-sized planet, is packed an enormous mass, hundreds of thousands of times greater than the mass of the earth. A matchbox filled with material from this dense star would weigh 10 tons.

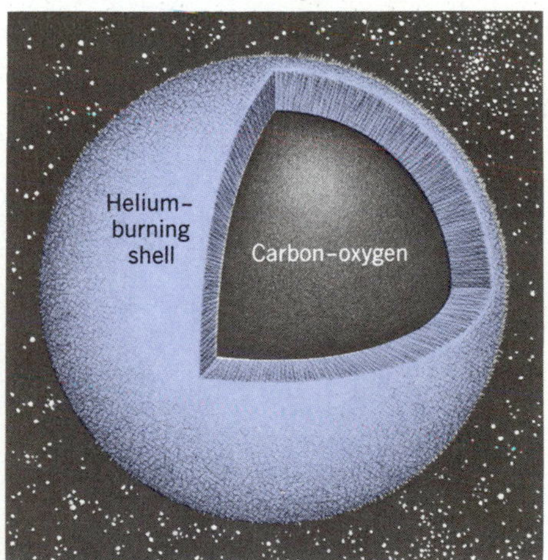

FIGURE 9.3
Structure of the exposed core of a planetary nebula during the transition to a white dwarf.

Although the star is now very faint, its surface is quite hot, with a temperature ranging up to 30,000 degrees. Such stars—small, dense, and exceedingly faint, but white-hot at the surface—are called white dwarfs.

The force of gravity on the surface of a white dwarf can be as much as one million times greater than gravity on the earth. Even if we should ever come across a white dwarf whose surface temperature has declined to a comfortable level, we would never be able to land people or even remote-controlled spacecraft on this strange world. A person attempting to land on a white dwarf would weigh 150 million pounds, and would literally be flattened by the enormous force of the white dwarf's gravity.

From point 12 onward, the star—now a white dwarf—shrinks very little in radius. Slowly the white dwarf radiates the last of its heat into space, moving downward in luminosity and temperature as it does so, and following the path leading to the dead stars at the bottom of the H-R diagram. Progressively the dwarf changes in color from white to yellow and then to red, until it fades to a cold, dark lump of matter and enters the graveyard of the stars.

Mass of a White Dwarf: Loss of Mass by Stellar Winds. Although stars with masses up to roughly $4M_{\odot}$ produce white dwarfs, theoretical studies indicate that the mass of the white dwarf itself cannot be more than $1.4M_{\odot}$. The explanation for this apparent inconsistency is that the white dwarf is only the core of the original star. Much of the mass of the original star has been blown away earlier in the star's life, before it became a white dwarf. Some of this mass is shed by the star during its red giant phase, in the form of a *stellar wind* that blows matter steadily off the surface. The remainder is blown off during the planetary nebula stage.

DEATH OF A MASSIVE STAR: THE SUPERNOVA EXPLOSION

A different fate awaits a star whose initial mass is greater than four solar masses. Because the weight of the star is so great, its collapse generates an enormous amount of heat. Now, according to theoretical studies of stellar evolution, the temperature at the center of the star can reach 600 million degrees. The attainment of that critical temperature sets in motion a train of events that eventually leads to the destruction of the star in a titanic explosion called a supernova.

Onset of Carbon Burning Within a Massive Star

In massive stars a core of carbon and oxygen forms, surrounded by a helium-burning shell, just as in smaller stars. As the carbon accumulates, the core begins to contract under its own weight, again as in small stars. In a small star the contraction continues until the star becomes a white dwarf, because the temperature of the core never gets high enough to ignite nuclear reactions in the carbon. But in a massive star, well before the core contracts to the size of a white dwarf, the temperature in the core reaches the 600-million-degree level at which carbon begins to burn.

With the onset of carbon burning, a sequence of nuclear reactions occurs in which progressively heavier elements are built up out of carbon. Experiments with nuclear accelerators have revealed many of the reactions that probably take place within a carbon-burning star. In a collision between two carbon nuclei, for example, the nuclei can fuse to form magnesium, with the energy of the fusion carried off by a gamma ray:

$$C^{12} + C^{12} \longrightarrow Mg^{24} + \gamma$$

Sometimes in these fusion reactions, especially when the reaction involves a collision between two good-sized nuclei, nuclear fragments, such as one or two protons, neutrons, or helium nuclei, fly off. In a collision between two carbon nuclei, a proton or a helium nucleus is apt to fly off, leaving behind sodium or neon, respectively:

$$C^{12} + C^{12} \longrightarrow Na^{23} + p$$

$$C^{12} + C^{12} \longrightarrow Ne^{20} + He^{4}$$

In reactions involving a collision between two oxygen nuclei, the results shown by the equations below are frequently observed in the laboratory.

$$O^{16} + O^{16} \longrightarrow S^{32} + \gamma$$

$$O^{16} + O^{16} \longrightarrow P^{31} + p$$

$$O^{16} + O^{16} \longrightarrow S^{31} + n$$

$$O^{16} + O^{16} \longrightarrow Si^{28} + He^{4}$$

Reactions between heavy nuclei can also take place if the nuclei are of two different types. For example, carbon and oxygen can collide to produce silicon:

$$C^{12} + O^{16} \longrightarrow Si^{28} + \gamma$$

As the star ages, more elements accumulate in its interior, and an even greater variety of reactions takes place. Experiments show that protons fuse with chlorine, for example, to produce argon, simultaneously emitting a gamma ray in the process:

$$p + Cl^{35} \longrightarrow A^{36} + \gamma,$$

Or a helium nucleus may fuse with a silicon nucleus to produce sulfur:

$$He^4 + Si^{28} \longrightarrow S^{32} + \gamma.$$

A chain of reactions occurs by which increasingly heavy nuclei are built up from the original nucleus by the successive addition, one after the other, of "small" particles, such as protons, neutrons, deuterons, and helium nuclei. At each stage, the light elements left over from early stages of burning are in the outer regions of the star, and the heavier elements formed from the later stages of burning are in the inner regions close to the center. The result is that the star has a layered or onionskin structure, with hydrogen in the outermost layer. Underneath the hydrogen layer are layers of helium, carbon, and so on, proceeding to the center (Figure 9.4).

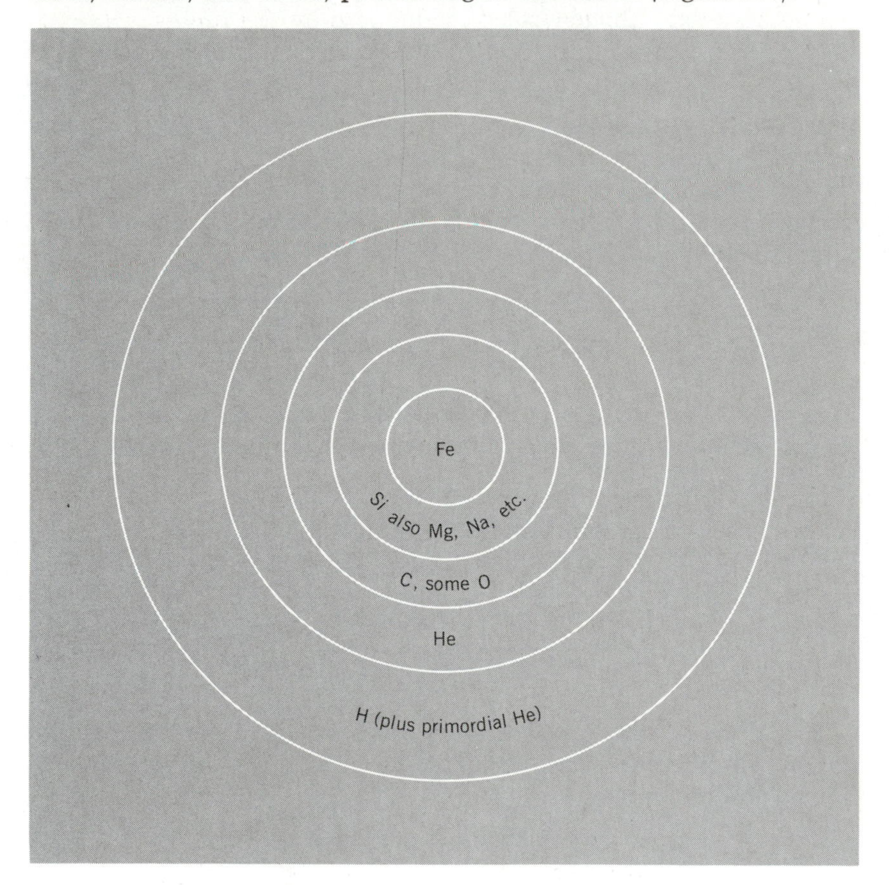

FIGURE 9.4
Layered structure of a massive star. The star's life starts with the burning of hydrogen to form helium. As the star ages and grows hotter at the center, helium begins to burn to form carbon and oxygen. Later, the carbon and oxygen burn to form successively heavier elements, from silicon to iron. When a substantial amount of iron forms, the star is near the end of its life.

Evolution of Moderately Massive Stars: Detonation of the Carbon Core

Theoretical studies suggest that for stars between four and ten solar masses, the buildup of heavier elements never proceeds very far, because a violent detonation occurs in the core almost as soon as the burning of carbon begins. This detonation is similar to the helium flash described on page 209, but is much more violent. As in the case of the helium flash, it is the result of the "solid-steel" incompressibility of the electrons in the core. As soon as the temperature in the core reaches 600 million degrees, the carbon starts to burn but, because of the peculiar unresponsiveness of the core, it does not expand when the carbon starts to burn. Expansion of the core would tend to drop the temperature, but since it does not occur, the temperature of the core rises rapidly. This causes the carbon to burn faster, and drives the temperature to a still higher level. A runaway effect develops in which the temperature and pressure mount rapidly, detonating the core and shattering all or a large part of it. Enormous pressures, ranging up to a trillion trillion tons per square inch, are generated in the core by the runaway detonation of the carbon. The pressures generated by the detonation of the carbon core cause the entire star to explode.

The exploding star is called a *supernova.* In the aftermath of the supernova explosion, a hot cloud of debris expands into space, carrying with it the elements the star has manufactured in the interior during its lifetime. All or a large fraction of the material within the star disperses to space in this explosion. At the original location of the star there now remains at most a compressed remnant of the core, containing a small fraction of the star's original mass.

TABLE 9.1
Threshold Temperatures for Onset of Nuclear Burning for Various Elements

Element	Temperature (°K)
Hydrogen	10 million
Helium	100 million
Carbon	600 million
Silicon	3 billion

Evolution of Very Massive Stars: Continued Burning of the Carbon Core

If the mass of the star is very great—more than about eight to ten solar masses—a carbon flash does not develop, because the density at the center of the star never becomes great enough to produce the electron "solid-steel" incompressibility that leads to these flashes.[2] As a consequence, the core expands and its density drops when the carbon starts to burn; the carbon flash does not occur; and the carbon in the core continues to burn without detonating. As the temperature of the core rises, the oxygen begins to burn as well.

When most of the carbon and oxygen in the core has been burned and converted to heavier elements, the reactions slow down and the star contracts again. The contraction heats the materials of the

[2] It is surprising that the most massive stars should have cores that are less dense. The explanation is connected with the great intensity of the radiation in the centers of these very massive and therefore highly luminous stars. So many photons are present, and the photons are so energetic, that as they move toward the surface they drive some of the gaseous matter of the star away from the center.

core to greater temperatures, causing other nuclear reactions to set in, and producing still heavier elements. The energy released in these reactions halts the contraction once more. In this way, through the alternation of collapse, heating, and renewed nuclear burning, many elements are produced inside the star. The table indicates the successively higher threshold temperatures at which burning starts for the heavier elements.

Buildup of Iron in the Core of a Very Massive Star

The successive stages of burning continue, and nuclear reactions produce successively heavier elements, until finally the element iron is reached. At this point the process stops, for iron is a very special element. Nuclear reactions involving iron do not yield energy; instead they *absorb* energy. Iron and a few neighboring elements are the only substances for which this is true. Because of the energy-absorbing properties of the iron nucleus, when a large amount of iron accumulates in the core of the star, nuclear energy can no longer be generated there; the iron nuclei, instead of adding fuel to the fire, tend to put the fire out. As a result, the pressure within the star drops sharply, and the star contracts. As before, the contraction yields heat. Normally this heat would slow down the contraction. However, the iron nuclei, because of their energy-absorbing property, absorb the heat produced by the contraction, and the star contracts further. The contraction continues; its pace accelerates; it turns into a catastrophic collapse.

The energy-absorbing effect of the iron is enhanced by the appearance of large numbers of neutrinos at this time. Because the neutrinos do not interact strongly with matter, most of them leave the core of the star, draining additional energy from the interior and greatly accelerating the collapse. In the last 24 hours of the star's life, its neutrino luminosity may be a trillion times greater than the luminosity of the sun. In fact, 10 to 100 times more energy is emitted in the form of neutrinos than in the energetic particles and radiation produced by the supernova explosion itself.

The consequences of the collapse are spectacular. The materials of the collapsing star pile up at the center, creating very high pressures and densities, and temperatures in the neighborhood of a trillion degrees. When the density at the center of the star is so great that neighboring nuclei touch one another, the star can be compressed no further, and the collapse comes to a halt. The collapsed star, compressed like a giant spring, now rebounds, and its materials begin to move outward.

A Second Kind of Supernova

The very high temperatures and densities produced in the star during its collapse cause a wave of pressure to travel outward from its center. When this wave of pressure reaches the star's surface, it tends to blow the outer layers of the star into space. The intensity of the outward-moving pressure wave is increased by two other factors: First, some of the neutrinos in the star's core are absorbed in the star's outer layers on their way out, heating those regions and also tending to blow them off the star. Second, sprinkled throughout the interior of the star are small amounts of helium, carbon, oxygen, and other nuclei. These elements were not used up in the earlier reactions, and would still be capable of yielding large amounts of nuclear energy if the temperature within the star were high enough to burn them. A substantial fraction of the unburned nuclei now undergo nuclear reactions because of the very high temperatures generated by the collapse. The reactions release nuclear energy which adds to the energy of the rebound.

As a result of these effects, the outer layers of the star do not merely return to their original position after the collapse, but move outward past that point at a high speed, traveling into space. In other words, the star explodes. It has become a supernova.

The supernova explosion that occurs in a very massive star in the aftermath of iron core formation is as violent as the supernova explosion that followed the detonation of the carbon core in somewhat less massive stars. The exploding star blazes up to a brilliance billions of times greater than its normal luminosity. For a short time it can be as bright as an entire galaxy.

Observation of Supernovas

If a supernova happens to occur nearby in our galaxy, it appears suddenly as a new star in the sky, brighter than any other and easily visible with the naked eye in the daytime. The last supernova that is known to have exploded in our galaxy was seen in Europe in 1604, and caused a sensation. One of the earliest reported supernovas was a brilliant explosion recorded by Chinese astronomers in A.D. 1054. At the position of this supernova there is today a great cloud of gas known as the Crab Nebula, expanding outward at a speed of 1000 miles per second, which contains the remains of the star that exploded 900 years ago (Color Plate 8).

When a supernova explodes in another galaxy, it cannot usually be seen with the naked eye, but it can be photographed through a

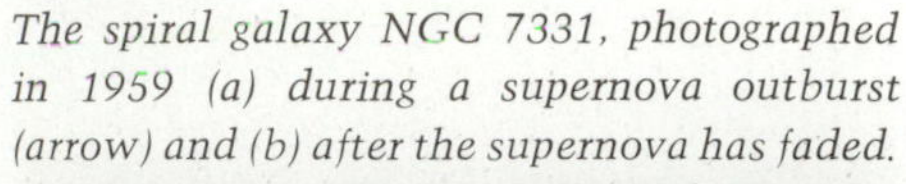

Figure 9.5
The spiral galaxy NGC 7331, photographed in 1959 (a) during a supernova outburst (arrow) and (b) after the supernova has faded.

(a)

(b)

large telescope. Several hundred supernovas have been photographed since the introduction of the camera to astronomy. A supernova in the galaxy NGC 7331, photographed in 1959, is shown in Figure 9.5*a* (arrow). Months later the supernova had faded to invisibility (Figure 9.5*b*).

The Aftermath of the Supernova; Buildup of Elements Beyond Iron

A star explodes when the buildup of elements inside it reaches the element iron. In that case, where do the elements in the Universe heavier than iron—silver, gold, lead, and so on—come from? Some heavy elements are produced earlier in the life of the star, as a result of thermal flashes occurring in the helium-burning shell. However, for a large number of elements, especially the heaviest ones, the answer has to do with the supernova collapse, and the very high temperatures generated in this collapse. As a result of these temperatures, many nuclei in the star are broken up, and neutrons and protons are freed. The neutrons and protons are captured by other nuclei, building up heavy elements. In this way the remaining elements of the periodic table, extending beyond iron, are manufactured in the final moments of the star's life. Because the time available for making these heavy elements is so brief, they never become as abundant as the elements up to and including iron. This explains the fact that in the graph of cosmic abundances of the elements (Figure 9.6), there is a steep drop in abundance, by a factor of 100,000, for elements beyond iron.

PULSARS AND NEUTRON STARS

The theory of exploding stars indicates that in some cases the entire star is shattered in the explosion; in other cases a compressed remnant of the star's core remains behind. In either case, a cloud of debris flies outward from the scene of the event. We know the subsequent fate of the cloud; it disperses to space and mixes with the primordial gases until its identity is lost. But what happens to that severely squeezed lump of matter that may be left behind at the center of the supernova after the star's outer layers are blown off? The answer to this question was unknown until 1967. In that year, pulsars—the most interesting objects to be found in the sky in many years—were discovered.

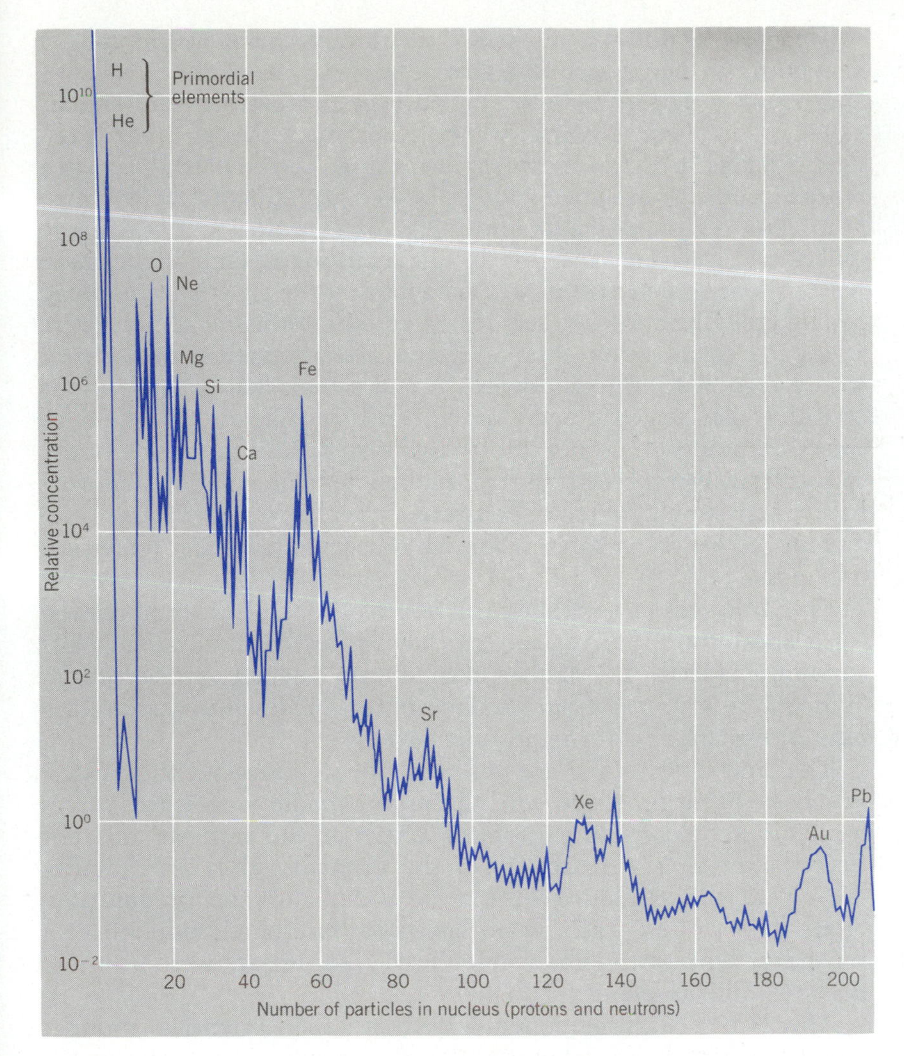

FIGURE 9.6
Cosmic abundances of the elements. The primordial substances, hydrogen and helium, are the most abundant. Next most abundant are carbon, nitrogen, oxygen and neon, which are made at relatively early stages in the life of a star. As the nuclear size and mass increase, the abundances of the elements continue to fall off until iron is reached. A peak occurs at iron because the iron nucleus is very stable—that is, tightly bound. Beyond iron the abundance falls off very sharply, because the elements heavier than iron are created only in brief moments of the final supernova explosion. The vertical axis of the graph is plotted in powers of 10 because of the enormous difference in abundance—a factor of more than a trillion—between the most and least abundant elements. According to a convention usually followed in plotting this type of chart, the element silicon (Si) is arbitrarily assigned an abundance of 10^6, and the abundances of all other elements are expressed relative to it.

Pulsars

The discovery came about by pure chance. Jocelyn Bell, an astronomy student at Cambridge University, received the task of investigating fluctuations in the strength of radio waves from distant galaxies. She found unexpectedly that certain places in the heavens were emitting short, rapid bursts of radio waves at regular intervals. Each burst lasted no more than one one-hundredth of a second. The rapid succession of bursts seemed like a speeded-up, celestial Morse code. The interval between successive bursts was about one second and extraordinarily constant. In fact, it did not change by more than one part in 10 million. A clock with this precision would gain or lose no more than a second a year.

No star or galaxy had ever before been observed to emit signals as bizarre as these. At first, some astronomers thought that intelligent beings might be beaming a message to the earth, and they referred to the Morse-code stars as LGM's, standing for Little Green Men. But the scientific community soon decided that the radio pulses had a natural and not an artificial origin. One of the main reasons for this conclusion was the fact that the signals were spread over a broad band of frequencies. If an extraterrestial society were trying to signal other solar systems, its interstellar transmitter would require enormous power to send signals across the trillions of miles that separate every star from its neighbors. It would be wasteful, purposeless, and unintelligent to diffuse the power of the transmitter over a broad band of frequencies. The only feasible way to transmit would be to concentrate all available power at one frequency, as we do on earth when we broadcast radio and television programs.

This cold reasoning dashed the hopes of romantics who believed for a short time that man might have received his first message from outer space. "LGM" disappeared from scientific conversation, "pulsar" took its place, and scientists settled down to search for a natural explanation of the peculiar signals.

The first clue to the answer was the sharpness of the pulses. When an object in space emits a burst of radio waves, the waves from different parts of the object arrive at the earth at different times, blurring the sharpness of the original pulse. The smaller the object, the less blurred the pulse and the shorter its duration. From the fact that each pulse lasted for one one-hundredth of a second or less, astronomers calculated that pulsars were no more than 10 miles in radius.

This was a startling conclusion. Until then scientists thought that the white dwarf—about 5000 miles in radius—was the smallest, densest star in the Universe. How could an object as massive as the sun be only 10 miles in radius? The matter in this compressed object would be one billion times denser than the matter in a white dwarf; a matchbox filled with material from it would weigh ten billion tons.

Neutron Stars

The answer goes back to a prediction made several decades ago. At that time, several theoretical astronomers pointed out that when a large star collapses and explodes as a supernova, the pressure on the core of the star compresses it so severely that individual electrons in the star's interior combine with neighboring protons and other nuclei. The negative and positive electric charges on the electrons and nuclei exactly cancel, and a ball of pure neutrons forms. According to the theory, this ball of neutrons is only 10 miles in radius, but has a large part of the star's original mass packed into that relatively tiny volume.[3]

Scientists dubbed the hypothetical ball of neutrons a neutron star. Starting in the 1940s, they searched for neutron stars assiduously, investigating with particular care the region at the center of the Crab Nebula, where the squeezed-down core of the supernova explosion of A.D. 1054 should have been located.[4] But no neutron stars were discovered and interest in them faded.

In 1968 a wave of excitement spread through the astronomical community when a pulsar was discovered at the center of the Crab Nebula, where astronomers had previously searched for a neutron star (Photograph facing page 217). Suddenly, many items of evidence fitted together like pieces of a jigsaw puzzle. A neutron star was predicted to exist at the center of the Crab Nebula; a pulsar was found there; and the neutron star and the pulsar are the only objects known that have the mass of a star packed into a sphere with a radius of 10 miles. Clearly, neutron star and pulsar are two names for the same thing—a fantastically compressed, superdense ball of matter, created when a massive star collapses at the end of its life.

Explanation of Pulses from Neutron Stars. One mystery remains to be explained. Why do neutron stars emit the sharp, regularly repeated bursts of radiation from which they derive their other name of pulsar? Scientists believe that a neutron star, like the sun and most other stars, is subject to violent surface storms that spray particles and radiation out into space. Each storm occurs in a localized area on the surface of the neutron star and sprays its radiation into space in a narrowly defined direction. When the earth lies in the path of one of these streams of radiation, our radio telescopes pick up the signals that indicate to us the presence of a pulsar, which is, in fact, the neutron star.

But if the pulsar sprays radiation steadily into space, why do we observe the radiation as a succession of isolated sharp bursts? The reason is probably that neutron stars, or pulsars, as most stars do, spin on their axes. In fact, it is entirely possible that they spin as

[3] Why does the compression stop at a radius of 10 miles? The answer is that within the nuclear force of attraction there is a short-range but extremely powerful force of repulsion that keeps two neutrons from coming arbitrarily close together. As will be seen in the next section, if a star is exceedingly massive, the pressure generated by its collapse is sufficient to overcome even this extremely strong repulsive force between neutrons, and the star collapses to less than a 10-mile radius. The result is the formation of a black hole.

[4] A faint star in the Crab Nebula had been tentatively identified as the supernova remnant in the 1940s, but no proof could be found that it was a neutron star.

rapidly as several times a second. As the neutron star spins, the stream of radiation from its surface sweeps through space like the light from a revolving lighthouse beacon. If the earth happens to lie in the path of the rotating beam, it will receive a sharp burst of radiation once in every turn of the neutron star.

This theory can be checked, because a spinning object must slow down gradually. Thus, the interval of time between successive bursts of radiation from a pulsar should increase. This prediction was confirmed by the discovery that the time between successive pulses from the Crab Nebula pulsar was getting longer, at the tiny but measurable rate of one one-billionth of a second per day.

More than 100 pulsars have now been identified. Presumably each one is a neutron star. The pulsar in the Crab Nebula is one of the most rapid; its bursts of energy reach us at the rate of 30 pulses per second, suggesting that this neutron star rotates 30 times a second. An even faster pulsar, discovered in 1982 and sometimes called the Millisecond Pulsar, rotates 642 times per second. This pulsar is slowing down at a remarkably small rate, with the time between its pulses increasing only a 100-trillionth of a second per day. Other pulsars have rates ranging down to one pulse every two seconds, indicating that these neutron stars are rotating more slowly. Since pulsars seem to slow down as they get older, the implication is that the Crab pulsar is one of the younger pulsars observed thus far. A major puzzle in the understanding of pulsars is the fact that the Millisecond Pulsar, with its rapid rate of rotation, should be very young, but the fact that its rate of spin is decreasing very slowly, and its surface is relatively cool, suggest that it is many millions of years old.

The Vela Pulsar. In one case in addition to the Crab, a pulsar has been located within a cloud of expanding gas that appears to be the remnant of a supernova explosion. This pulsar, called the Vela pulsar because it lies in the direction of the constellation Vela, is at the center of the Gum Nebula, a tenuous veil of interstellar matter about 1500 light-years away.

The pulses from the Vela pulsar reach us at the rate of 12 per second. Since the Vela pulsar is slower than the Crab pulsar, the supernova that produced this pulsar must have occurred considerably earlier than the Crab supernova. According to a rough estimate, the Vela supernova exploded between 5000 and 10,000 years ago. Since the Vela pulsar is very close to the earth—about one-tenth the distance to the Crab—the corresponding supernova explosion must have appeared as an extremely bright new star in the heavens, much brighter than the full moon for several weeks, and a fiery red color in its early stages. A supernova as close as this would be an awesome sight today, and surely was a terrifying spectacle to our forebears.

BLACK HOLES IN SPACE

With the realization of the connection between neutron stars, pulsars, and supernovas, many astronomers felt that the final pages had been written in the life story of the stars. But recent evidence has generated a suspicion that the neutron star or pulsar is not the ultimate state of compression of stellar matter. If the mass of a collapsing star is great enough, the core of the star may be squeezed beyond the 10-mile limit of the neutron star, until its radius has diminished to about 2 miles. At this point, the theory of relativity predicts the sudden occurrence of an extraordinary phenomenon.

Formation of a Black Hole

According to Einstein's theory, a ray of light should be affected by gravity, just as a mass would be. If Einstein is right, a ray of light emitted from a star will be pulled back by the star's gravity, as a ball thrown up from the surface of the earth is pulled back by the earth's gravity. When the star is normal in size—about one million miles in diameter—the force of gravity on its surface is not strong enough to keep the light rays from escaping, and they leave the star, although with somewhat less energy.

But if the matter of the star is squeezed into a very small volume, the force of gravity on its surface is very great. This can happen to the core of a star as a result of a supernova explosion. Suppose the core—which may be several times as massive as the sun—is squeezed down to a radius of a few miles. At that point, the force of gravity at the surface of this compact mass is billions of times stronger than the force of gravity at the surface of the sun. The tug of that enormous force prevents the rays of light from leaving the surface of the star; like the ball thrown upward from the earth, they are pulled back and cannot escape to space. All the light within the star is now trapped by gravity; no radiation can emerge. From this moment on, the star is invisible. It is a black hole in space.[5]

How Massive Must a Star Be to Create a Black Hole? While theoretical studies of stellar evolution indicate that extremely massive stars will become black holes when they collapse, these studies do not indicate precisely how massive the star must be in order for this to happen. Some studies indicate that the mass necessary to create a black hole in a supernova explosion may be as low as ten solar masses; in that case, approximately one star in 10,000

[5] If the mass of a star is exceptionally great calculations indicate that the entire star, and not merely the core, may become a black hole when the star collapses at the end of its life.

is a black hole, and there are 10 billion black holes in our galaxy. Other calculations indicate that the mass necessary to create a black hole is as much as 30 solar masses; in that case, about one star in a million becomes a black hole, and there are only about 100,000 black holes in our galaxy.

All the studies indicate, however, that when the mass of a star is close to four solar masses, a neutron star is produced at the end of the star's life; and when the mass of the star greatly exceeds four solar masses, a black hole is produced.

Properties of a Black Hole

The force of gravity within a black hole not only prevents light from escaping; it also prevents all physical objects from getting out of the hole. This property of black holes is another prediction of Einstein's theory, which asserts that no object can travel faster than light. If the black hole's gravity is so powerful that light cannot break its grip and escape to space, material objects cannot escape either. Everything inside the black hole is trapped there forever.

Any ray of light or physical object that enters the black hole from the outside is also trapped; it can never emerge again. The interior of the black hole is completely isolated from the outside world; it can take in objects and radiation, but cannot send anything back.

Once a black hole forms, gravity continues to draw everything inside it toward the center. According to current knowledge in theoretical physics, the star's volume contracts steadily, piling up material at the center in a dense lump. First the star shrinks to the size of a pinhead; then to the size of a microbe; then, still shrinking, it passes into the realm of distances smaller than any ever probed by man. At all times a mass of ten thousand trillion trillion tons remains packed into the shrinking volume.

The Fixed Boundary of a Black Hole. A black hole does not decrease in size as the matter within it shrinks toward the center. The radius of the black hole is not the radius of a tangible sphere of matter; it is the distance from the center of the black hole at which the force of gravity is sufficiently strong to keep light from escaping. Although the material within the black hole may be concentrated in a lump at the center, the force of gravity at the boundary of the black hole is the same as though this matter still filled the entire two-mile sphere.

Since black holes capture any material they encounter, the mass of a black hole will always tend to increase in time. A black hole is, in a sense, insatiable. As more matter enters it, its gravitational pull increases, and, therefore, its boundary expands.

This property does not imply that black holes act as gravitational vacuum cleaners, drawing in matter from the space around them. A ray of light, a star or a spaceship can pass by a black hole safely as long as it does not come too close. However, if the object is on a collision course with a black hole, it will enter the black hole and vanish. Even if its course carries it within a mile or two of the boundary, the gravitational pull of the black hole will curve the path of the object so that it enters and disappears.

The boundary of the black hole is called its *event horizon.* An observer outside the black hole can never discover anything about events that occur inside this boundary, because the rays of light and material particles emitted from events inside the boundary cannot penetrate to the outside world.

Size of a Black Hole. The formula for the radius of a black hole is

$$R = \frac{2G}{c^2} M$$

where M is the black hole's mass, G is the universal constant of gravitation, and c is the velocity of light. If R is measured in kilometers and M is measured in grams, the formula becomes

$$R = 1.5 \times 10^{-33} M$$

For a black hole with the mass of the sun or 2×10^{33} grams, the formula gives $R = 3$ kilometers or 1.8 miles. A black hole with the mass of a galaxy would have a radius of about 180 billion miles or one light-week. This is less than a millionth the size of our galaxy. If the mass of the earth were compressed into a black hole, its radius would be

$$\begin{aligned} R &= 1.5 \times 10^{28} \times 6 \times 10^{-27} \text{ cm} \\ &\simeq 0.9 \text{ cm} \end{aligned}$$

That is, the earth would be smaller than a golf ball.

Conditions Inside a Black Hole. What would happen to an astronaut who entered a black hole? The properties of black holes seem to suggest that he would be crushed by gravity. In actual fact he would be torn apart, because the part of his body closest to the center of the black hole would be pulled by a stronger gravitational force than any other part. Suppose, for example, the astronaut entered feet first; then his feet would be pulled more strongly than his head, and feet and head would tend to separate. The astronaut would feel as though he were stretched on a rack; a few thousandths of a second after entering the black hole, he would be dismembered; after a few more thousandths of a second, the individual atoms of his body would be broken into their separate neutrons, protons, and electrons; finally the elementary particles themselves must be torn into fragments whose nature is not yet known to the physicists.

Observational Evidence for Black Holes

Intuition tells us that an object as bizarre as a black hole cannot exist; yet theoretical studies of the evolution of massive stars, combined with the theory of relativity, assure us that whenever a very massive star undergoes a supernova explosion, a black hole must be left behind. At first this prediction may seem impossible to test, since the black hole by its nature is unobservable. However, recent results in X-ray astronomy (page 126) provide a tentative indication that black holes actually do exist. X-ray observations of the heavens made by rockets and satellites show a powerful source of X-rays, with unusual properties, in the constellation Cygnus. The source, named Cygnus X-1, is in the vicinity of a blue supergiant, whose spectrum shows that the supergiant is a member of a spectroscopic binary. Apparently, the X-ray source is the other member of the binary.

When these properties of Cygnus X-1 were announced, they seemed to be the evidence for black holes the theorists had been looking for. If the stars in a binary are relatively close together, the pull of each star's gravity draws matter away from its companion, and streams of gas flow back and forth between the two stars. If one star becomes a black hole, the gas drawn out of the other star will continue to stream toward it, but now, as this gas approaches the boundary of the black hole, it will be accelerated to extremely high velocities by the black hole's gravitational force. The rapidly moving particles, converging on the black hole, will collide with one another and produce an intense stream of X-rays, making the black hole an X-ray source of the kind observed by the X-ray satellites (Figure 9.7).

FIGURE 9.7
Artist's sketch of the hypothetical black hole in the constellation Cygnus. Matter drawn from the companion star (lower left) spirals in toward the black hole, becoming accelerated and heated. The high temperature of the in-falling gas causes the radiation of X-rays.

Is Cygnus X-1 a black hole? According to another theory, the invisible member of the binary could be a neutron star instead, since these stars also emit X-rays. However, from the properties of the binary in which Cygnus X-1 is located, its mass is estimated to be at least $6M_{\odot}$. Calculations on the structure of neutron stars show that the mass of a neutron star cannot be greater than approximately two or three solar masses. With neutron stars excluded, the black hole seems to be the most acceptable explanation for Cygnus X-1 remaining.

Another candidate for a stellar black hole was discovered in 1983 in the large Magellanic Cloud. The unusual object, called LMC X-3, is an intense source of X-rays. Like Cygnus X-1, it appears to be a member of a binary star, whose other member in this case is a blue giant. LMC X-3 and the blue giant circle around one another at a distance of 11 million miles with a period of 1.7 days. Because LMC X-3 and the blue giant are located in the Large Magellanic Cloud, the distance to them is known with relatively high accuracy to be 180,000 light-years. The accurate determination of the distance to the blue giant, combined with the measurement of its spectral type, leads to an accurate determination of its absolute luminosity and therefore its mass. This leads, in turn, to a more accurate determination of the mass of the unseen companion than was possible in the case of Cygnus X-1. The results indicate that the mass of the invisible companion is approximately $10M_{\odot}$, which is far above the limiting mass of a neutron star, strongly suggesting that LMC X-3 is a black hole. Initially there was a general reluctance to accept the existence of black holes because they are such peculiar objects, but LMC X-3 and Cygnus X-1 appear difficult to interpret in any other way.

EPILOGUE

The life story of the stars has an epilogue. When a supernova explosion occurs and the outer layers of the stars are sprayed out to space, they mingle with fresh hydrogen to form a gaseous mixture containing all the chemical elements. Later in the history of the Galaxy, other stars are formed out of clouds of hydrogen that have been enriched by the products of these explosions. The sun is one of these stars; it contains the debris of countless supernova explosions dating back to the earliest years of the Galaxy. The planets also contain the debris; and the earth, in particular, is composed almost entirely of it. We owe our corporeal existence to events that took place billions of years ago, in stars that lived and died long before the solar system came into being.

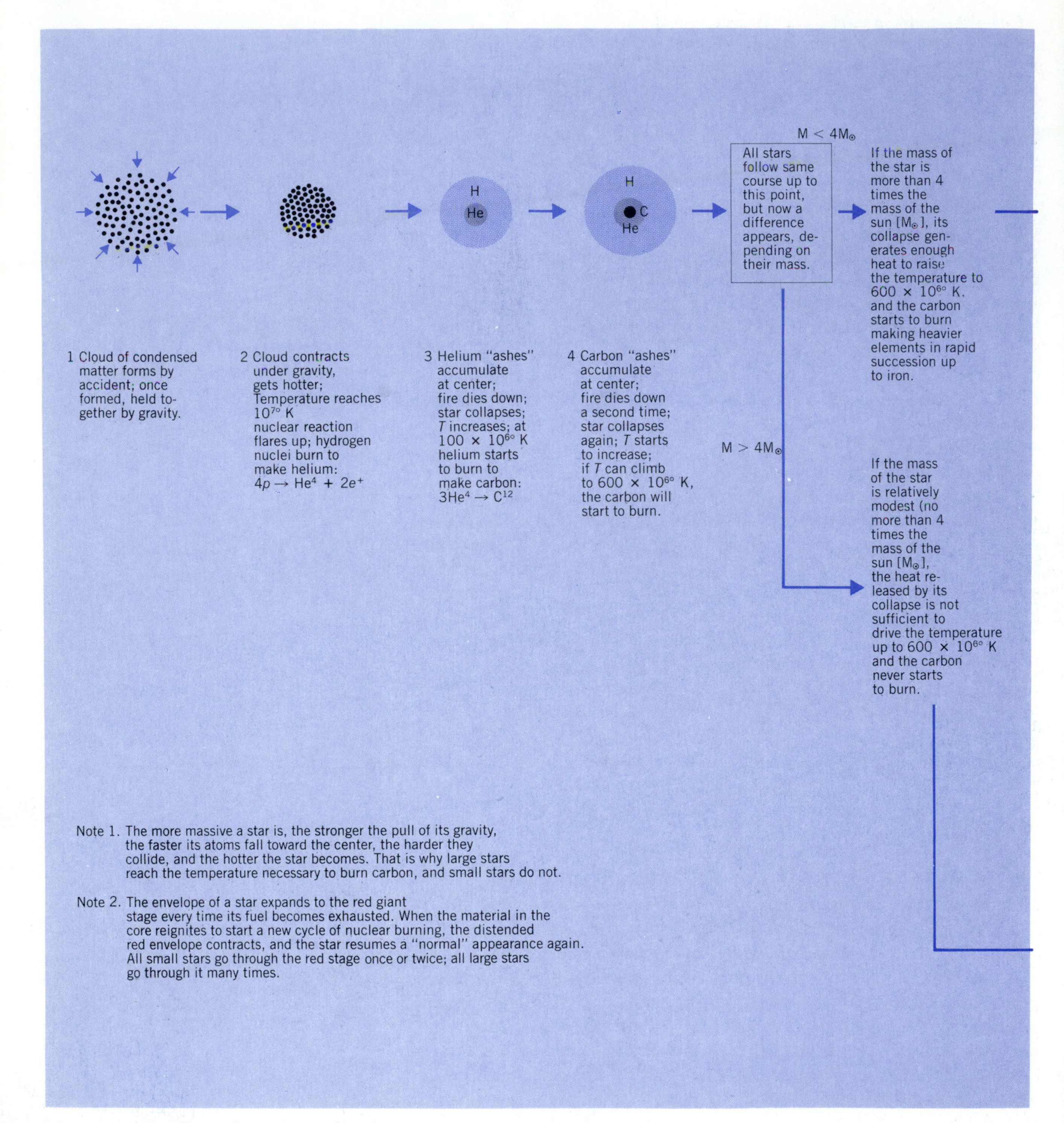

Main stages of stellar evolution for small and massive stars.

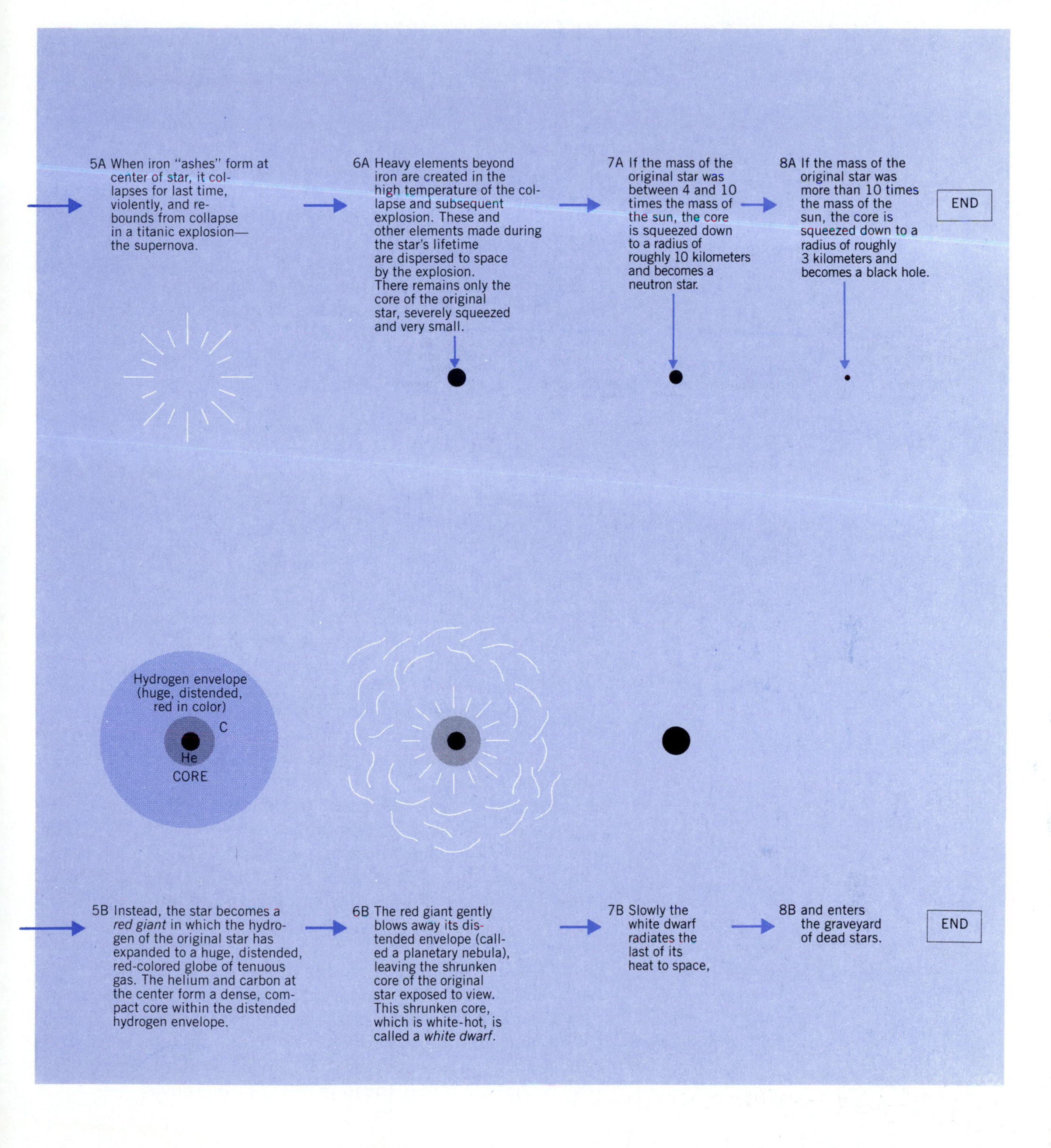
5A When iron "ashes" form at center of star, it collapses for last time, violently, and rebounds from collapse in a titanic explosion—the supernova.
6A Heavy elements beyond iron are created in the high temperature of the collapse and subsequent explosion. These and other elements made during the star's lifetime are dispersed to space by the explosion. There remains only the core of the original star, severely squeezed and very small.
7A If the mass of the original star was between 4 and 10 times the mass of the sun, the core is squeezed down to a radius of roughly 10 kilometers and becomes a neutron star.
8A If the mass of the original star was more than 10 times the mass of the sun, the core is squeezed down to a radius of roughly 3 kilometers and becomes a black hole.
END
Hydrogen envelope (huge, distended, red in color)
C
He
CORE
5B Instead, the star becomes a *red giant* in which the hydrogen of the original star has expanded to a huge, distended, red-colored globe of tenuous gas. The helium and carbon at the center form a dense, compact core within the distended hydrogen envelope.
6B The red giant gently blows away its distended envelope (called a planetary nebula), leaving the shrunken core of the original star exposed to view. This shrunken core, which is white-hot, is called a *white dwarf*.
7B Slowly the white dwarf radiates the last of its heat to space,
8B and enters the graveyard of dead stars.
END

Main Ideas

1. Advanced stages of evolution for small stars; planetary nebulas and white dwarfs.
2. Advanced stages of evolution for massive stars; onset of carbon-burning; synthesis of heavier elements up to iron.
3. The supernova explosion; synthesis of elements beyond iron; explanation of the cosmic abundance of the elements.
4. The core of the exploded star: pulsars, neutron stars and black holes.
5. Properties of a black hole: mass, size, conditions inside the black hole; observational evidence for black holes.
6. General implications of stellar evolution; genesis of the chemical elements.

Important Terms

black hole
cosmic abundance
neutron star
planetary nebula
pulsar
spectroscopic binary
stellar wind
supernova
white dwarf

Questions

1. Why do planetary nebulas not show up in substantial numbers on the H-R diagram, if a considerable fraction of all stars go through this phase?
2. What is the source of the energy radiated by white dwarfs? What kind of spectrum should a white dwarf have? Explain.
3. Why is gravity very strong on the surface of a white dwarf?
4. The Stefan-Boltzmann law states that if the temperature at the surface of an object is T, each square centimeter of the surface emits an amount of energy proportional to the fourth power of the temperature: Energy per unit area = constant $\times$ T^4. The total surface area of a sphere of radius R is $4\pi R^2$. Therefore the total amount of energy radiated to space—i.e., the luminosity—of a star with surface temperature T and radius R is

$$L = \text{constant} \times T^4 \times 4\pi R^2$$

   Using this formula, calculate the radius of a white dwarf with a temperature of 7000°K and a luminosity 1/1000th that of the sun; calculate the radius of a red giant with a temperature of 4000°K and a luminosity 1000 times that of the sun.
5. What kinds of stars become supernovas? Explain the circumstances that can lead to the formation of supernovas.

6. After a supernova explosion, what is left behind at the site of the explosion? What is seen surrounding the site?
7. Explain why astronomers first referred to pulsars as LGM's—Little Green Men.
8. Why are elements heavier than iron so much rarer in the Universe than those lighter than iron?
9. Why is it impossible for the pulsar in the Crab Nebula to be a white dwarf? Why must it be a smaller object?
10. Calculate the density of a neutron star whose radius is 10 kilometers and whose mass is 2 solar masses. Assume the neutron star is a sphere. Compare it to the density of a white dwarf. If you had the shape of a sphere and were as dense as a neutron star, how large would you be?
11. Why is a "black hole" black?
12. If you were observing a star in the process of becoming a black hole, what would you see? If you fell into a black hole, what would happen to you? Could you tell us on the earth about your experience?

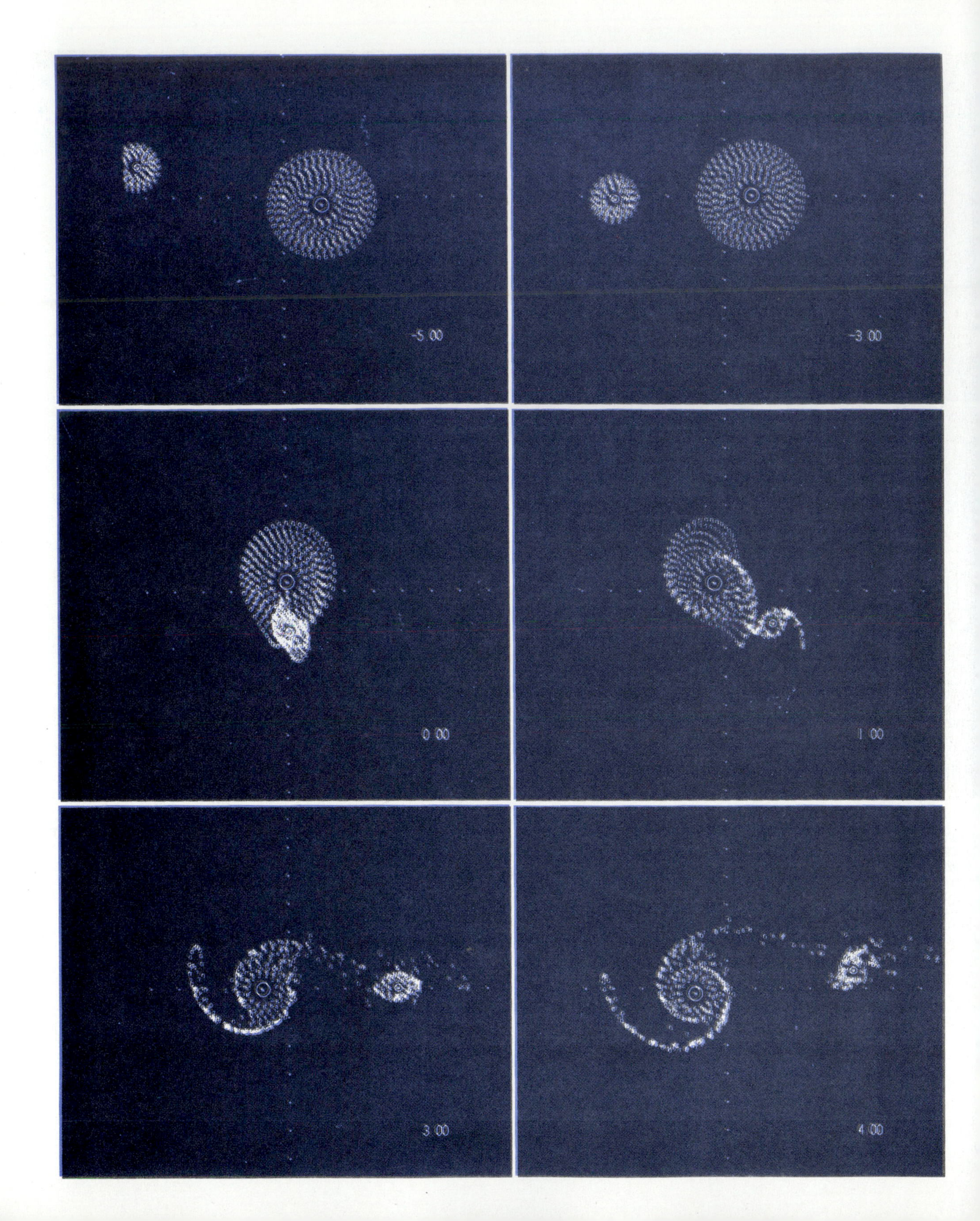

−5.00
−3.00
0.00
1.00
3.00
4.00

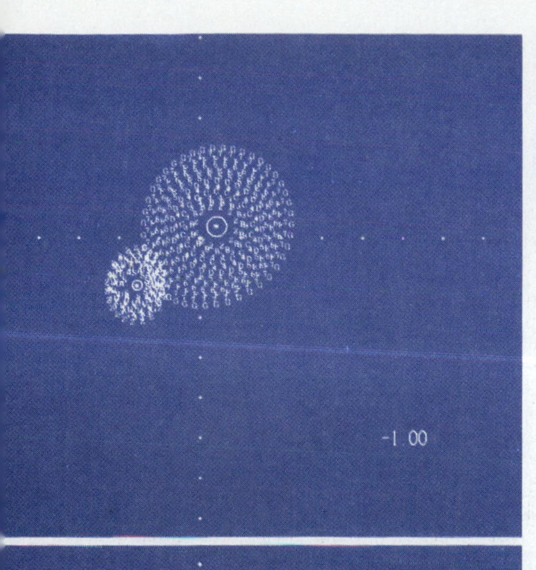

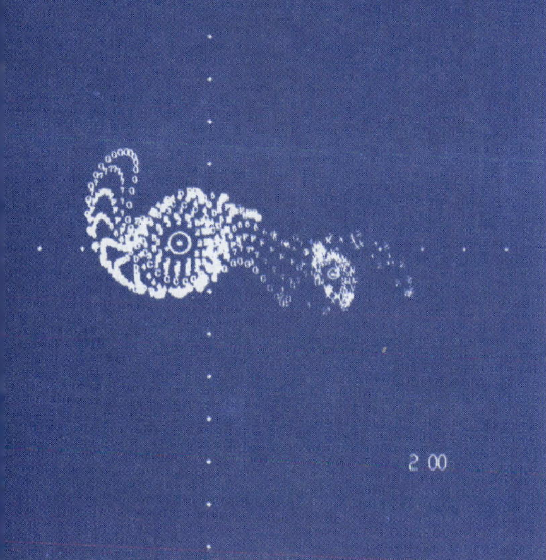

GALAXIES

PART THREE

GALAXIES

10

Stars are clustered in groups called galaxies, just as individuals are clustered together in nations. Galaxies are the nation-states of the heavens. Like stars, they come in many sizes and varieties. Most of the different varieties of stars are simply stages in the lifetime of a typical "normal" star. Is this also true for galaxies? Are elliptical, spiral, and barred spiral galaxies simply stages in the lifetime of a typical "normal" galaxy? Or are they completely distinct species of galaxies, as unrelated as elephants and crocodiles?

In the case of stars, astronomical discoveries of the last 30 years have revealed the answers to these questions. In the case of galaxies, far less progress has been made. Astronomers are not even certain of the answers to basic questions, such as how a galaxy changes its shape and luminosity during the course of time.

Rapid advances may be made in the next decade because we are obtaining new information about galaxies by studying them in the infrared and X-ray regions of the spectrum. More sensitive telescopes will see farther out into space, and farther back in time. With the aid of the 94-inch Space Telescope, to be launched later in the 1980s, and with mammoth, ground-based telescopes still on the drawing boards, we hope to be able to look back many billions of years to a time when galaxies were first forming. Then it will be possible to compare the very young galaxies we see at these great distances with the middle-aged galaxies and old galaxies we see nearby. This information may at last lead astronomers to an understanding of the life cycle of the galaxies.

The Sombrero, NGC 4594.

FIGURE 10.1
Elliptical galaxies of types E1 (NGC 4278), E3 (NGC 4406) and E6 (NGC 3115).

TYPES OF GALAXIES

Galaxies come in a variety of shapes. Edwin Hubble, an American astronomer who spent decades studying galaxies with the 60-inch and 100-inch telescopes at Mount Wilson, found that all could be classified into four major types: elliptical, spiral, barred spiral, and irregular.

Elliptical

An elliptical galaxy looks like a partly flattened, luminous sphere. Some of these galaxies are nearly perfect spheres; others are moderately flattened.[1] Examples of the elliptical galaxies are shown in Figure 10.1.

Hubble labeled all elliptical galaxies according to their degree of flattening. He designated the spherical galaxies "E0," the slightly flattened galaxies as "E1," and so on down to "E7," for the flattest, most elliptical ones. The classification is illustrated in Figure 10.2. The three elliptical galaxies shown in Figure 10.1 are types E1, E3, and E6, respectively.

Elliptical galaxies were 20 percent of Hubble's sample. However, elliptical galaxies were underrepresented in Hubble's count, probably because many are small and relatively dim objects. They may, in fact, constitute the majority of all galaxies. The main property of elliptical galaxies, apart from their shape, is the fact that they contain only old stars. No young stars, such as blue giants, are found in them. Some elliptical galaxies are relatively large and massive, but surprisingly faint in proportion to their mass. Their relative faintness results from the fact that the old stars that make up their population are not very luminous.

Elliptical galaxies also appear to have relatively little interstellar gas and dust compared to spiral galaxies. Although there is a component of matter called the hidden mass (page 256), there are no star-forming clouds of gas and dust. One explanation is that most of the atoms in an elliptical galaxy already have been gathered together to make stars. Another explanation is that such galaxies have been swept relatively free of gas and dust by their rapid passage through an intergalactic gaseous medium. Within the galaxy, this intergalactic medium has the effect of a strong wind that cleans out the interstellar gas (page 262).

[1] An elliptical galaxy may be less regular in shape, resembling a flattened sphere with one side pushed in.

Spiral

Hubble classified 50 percent of the galaxies he observed as spiral. They were his largest single class of galaxies. Spiral galaxies have the shape of a flattened disk of matter, with a small, spherical nucleus bulging out of the center. The stars in a spiral galaxy are concentrated in the central nucleus and in the spiral arms that radiate out from the center. These arms give the spiral galaxy its name.

Because of the way the spiral arms are curved, the spiral galaxy gives the impression to the observer that it is rotating like a pinwheel in a Fourth of July fireworks display. Of course, because of their great size, they rotate too slowly for us to see the wheeling movement during a single night or even a year. In our own galaxy, a typical example of a spiral galaxy, the pinwheeling motion carries the sun around the center of the Galaxy once every 250 million years. To a cosmic observer watching the heavens for several hundred million years, the spiral galaxies would look like a fireworks display.

Hubble classified spiral galaxies into three basic types—Sa, Sb, and Sc—according to the degree of openness of the spiral arms. Figure 10.3 illustrates the sequence of shapes. The Sa type has a number of spiral arms very close together and almost overlapping, so that the spiral pattern can hardly be seen. Figure 10.4*a* shows an Sa galaxy (NGC 2811). In Sb spirals, the arms are well defined, although still relatively close together (Figure 10.4*b*). In Sc type spirals there are only a few arms, each being clearly separated from the others (Figure 10.4*c*) (p. 248).

The mass of the galaxy is most highly concentrated in the disk in the case of the Sc type, and least concentrated in the disk in the Sa type. The trend towards concentration of matter in the disk of the galaxy is shown in the sequence of three edge-on galaxies shown in Figures 10.5*a*, *b*, and *c* (p. 248).

Spiral galaxies are relatively large and massive. The average spiral galaxy contains about 10^{11} stars, including young stars, middle-aged stars, and old stars.

FIGURE 10.2
The shapes of elliptical galaxies E0 and E2–E7.

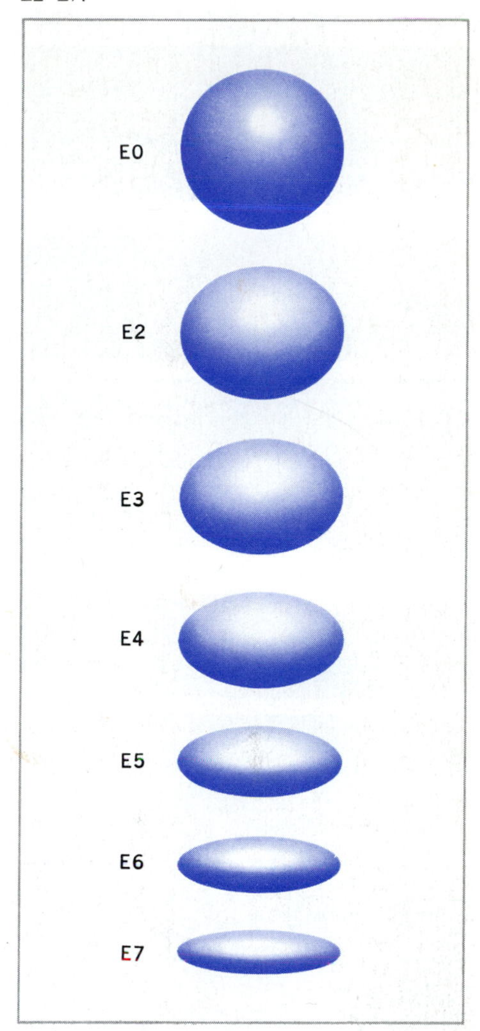

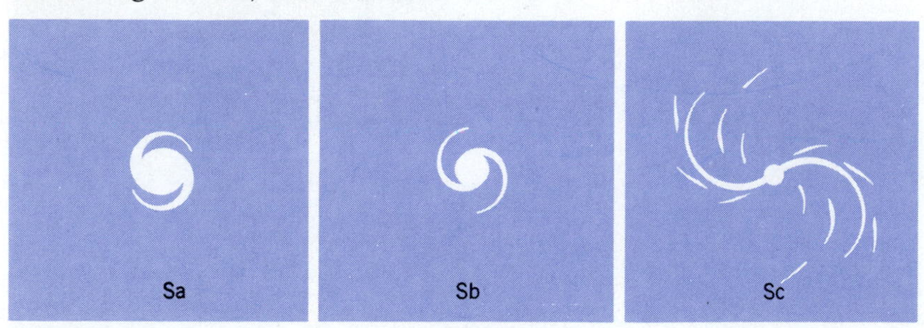

FIGURE 10.3
The shapes of spiral galaxies.

A substantial fraction of the stars in a spiral galaxy are located quite far out from the central disk. A small fraction of these stars are concentrated in clusters. An average spiral galaxy contains several hundred such clusters, randomly scattered in the space around the disk, as shown in Figure 10.6. This spherical volume, containing scattered clusters and isolated stars, is called the *halo* of the galaxy. Recent discoveries indicate that it also contains a substantial amount of matter—probably more than the mass of the visible galaxy—in an invisible form. The nature of the invisible matter in the halo is not yet known.

FIGURE 10.4
Face-on photographs of spiral galaxies of types: (a) Sa (NGC 2811); (b) Sb (NGC 3031); (c) Sc (NGC 628).

(a)

(b)

(c)

(a)

(b)

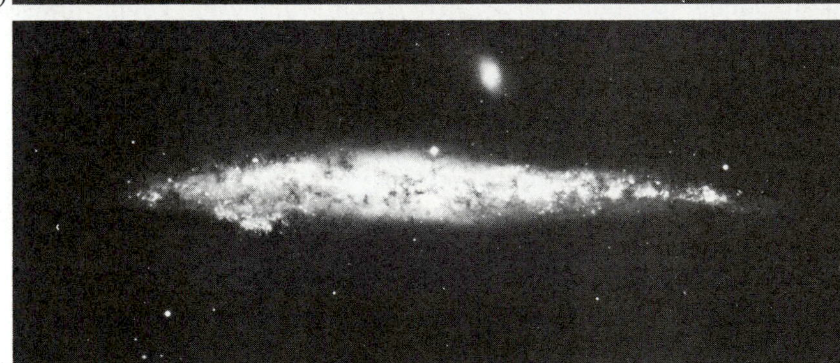
(c)

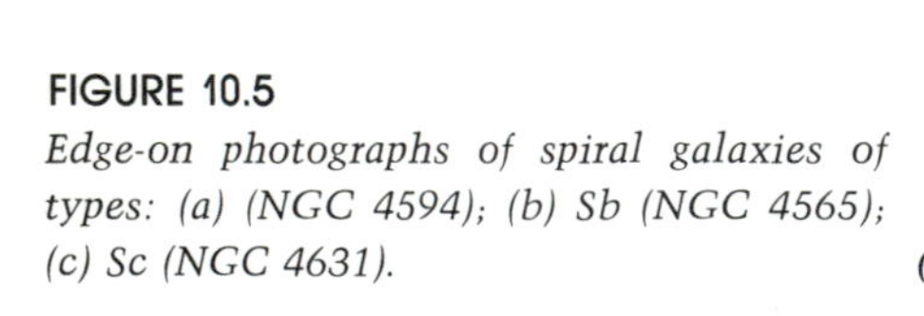

FIGURE 10.5
Edge-on photographs of spiral galaxies of types: (a) (NGC 4594); (b) Sb (NGC 4565); (c) Sc (NGC 4631).

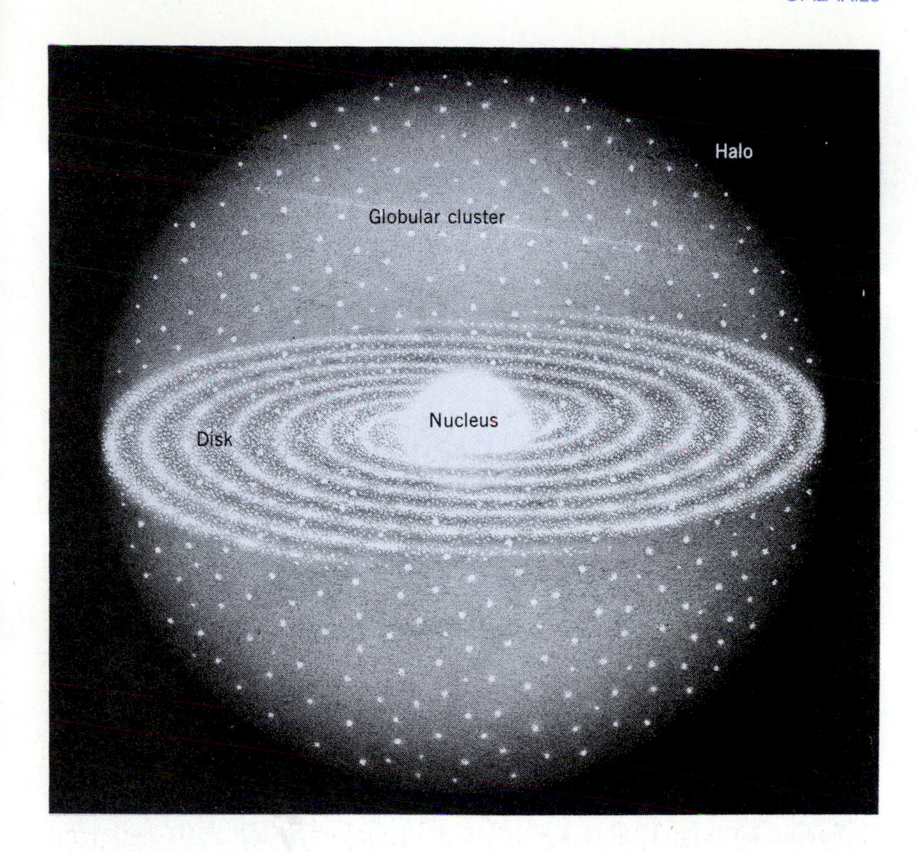

FIGURE 10.6
The structure of a spiral galaxy.

The clusters in the halo of a galaxy are called *globular clusters* because of their even, spherical shape. A typical globular cluster contains about one million stars and is approximately 100 light-years in diameter. A photograph of M13, a globular cluster in the halo of our galaxy about 20,000 light-years from the sun, is shown in Figure 10.7 (p. 250). Many globular clusters may also be seen in the photograph of the Sa/Sb spiral in Figure 10.5*a* (NGC 4594). Most of the points of light immediately surrounding the nucleus of this galaxy are globular clusters.

Globular clusters are believed to be formed at an early stage in the life of a galaxy, before the gas of the galaxy has collapsed to a thin disk as the result of rotation. One reason for this belief is the fact that these clusters only contain old stars.

As noted, globular clusters contain no new stars. Apparently, most of their gas and dust either was used up a long time ago in forming stars or was swept out of them as they passed through the galactic disk. As we will see in the next chapter, a plot of the stars in the globular clusters on an H-R diagram provides a powerful method for determining the age of the galaxy.

FIGURE 10.7
The globular cluster M13.

Barred Spiral

About 30 percent of all spiral galaxies were classified by Hubble as barred spirals. They resemble Sc spirals, with two arms radiating outward. However, instead of coming out of a small central nucleus, the arms project from the ends of a bar-shaped concentration of matter. This bar gives the galaxy its name.

Hubble classified barred spirals into three types—SBa, SBb, and SBc—depending on the degree of openness of the spiral arms, following the same scheme for classifying ordinary spirals. Barred spirals are similar to ordinary spirals in every respect, such as the types of stars they contain, their mass and luminosity, their halos, and their distribution of gas and dust. Figure 10.8 illustrates the sequence of shapes. Figure 10.9 shows barred spirals of the types SBa, SBb, and SBc.

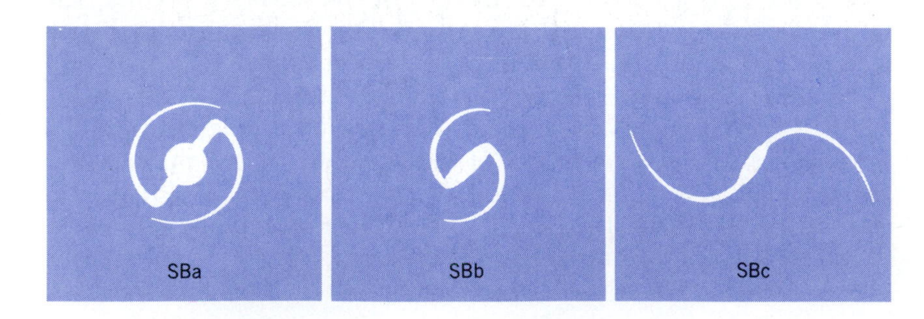

FIGURE 10.8
Types of barred spirals.

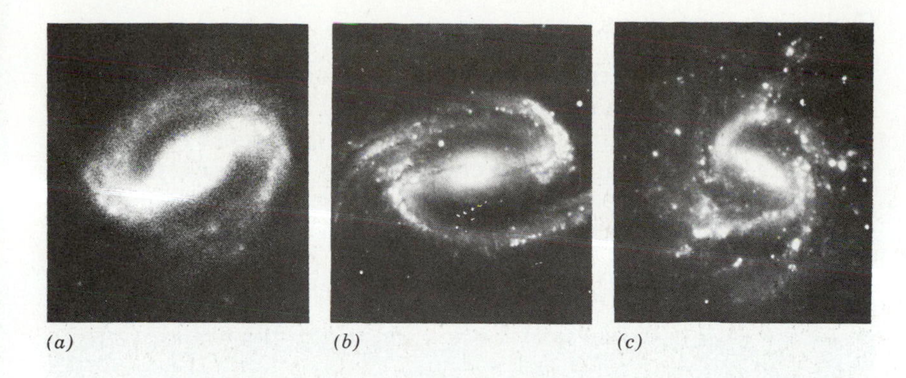

FIGURE 10.9
Barred spiral galaxies: (a) SBa (NGC 175); (b) SBb (NGC 1300); (c) SBc (NGC 1073).

Dwarf Galaxies

A small fraction of all galaxies do not fit into one of these categories. They do not have a clearly defined geometric form, probably because they are smaller. As a general rule, they have about a thousandth the mass of a typical spiral or elliptical galaxy, and only 10^{10} stars. Frequently, a dwarf galaxy is attached to a spiral galaxy, held captive by the gravitational force of the massive spiral. The Magellanic Clouds, which are bound to our galaxy, are examples of captive dwarf galaxies. Figure 10.10 shows the large Magellanic Cloud.

FIGURE 10.10
An example of a dwarf galaxy: the large Magellanic Cloud.

Colliding Galaxies

Several hundred apparent collisions or near collisions between galaxies have been photographed (Figure 10.11). Sometimes the region around the colliding pair is the source of an intense emission of radio waves which can be detected by radio telescopes. Colliding galaxies are one cause of the radio galaxies—galaxies emitting large amounts of energy in the form of radio waves—which are described in Chapter 12.

FIGURE 10.11
NGC 2535/36 is a peculiar object which appears to be two galaxies in collision according to a computer study. See pages 242 and 243.

Collisions between galaxies seem to be much more possible than collisions between stars; in fact, not a single pair of colliding stars has ever been observed or photographed since the telescope was invented. Stellar collisions are rare because the size of the average star is small in comparison to the average distance between stars. In our galaxy the distance between stars is about 30 trillion miles, while the size of a typical star is only a million miles; thus, the ratio of size to average distance between stars is less than a millionth. The size of a typical galaxy, on the other hand, is about 100,000 light-years, and the average distance between galaxies is a few million light-years, so that the ratio of size to distance is about a tenth. This is roughly the same as the ratio of size to distance for molecules in a gas at standard conditions, and collisions are correspondingly frequent. It is no surprise that when we look into the sky at any given time we see no colliding stars, but we do observe a number of galaxies either colliding with their neighbors or on collision courses.

Galactic Cannibalization. Recent evidence seems to confirm a suspicion that some of the largest galaxies in the Universe have reached their giant size by colliding with and consuming other galaxies nearby. These are the group known as the cD galaxies, which belong to the supergiant elliptical galaxies. The cD galaxies are unusual in having several nuclei inside them, which appear to be the nuclei of other galaxies that came too close in a collision and were swallowed up. The cD galaxies generally lie at the centers of clusters of galaxies, where the population of galaxies is most numerous and collisions between galaxies would be most common. In one cluster of galaxies, Abell 407,[2] the central galaxy possesses nine separate nuclei, each as luminous as the other galaxies in the cluster. A study of the properties of the other galaxies in the cluster suggests that the nine nuclei are in fact former elliptical galaxies that were cannibalized.

Anomalously Luminous Galaxies

A small fraction of the galaxies in the sky have very unusual properties. These galaxies belong to the group called by Hubble *peculiar galaxies,* which make up one or two percent of the galactic population of the heavens. Some peculiar galaxies look as though they have been distorted by an unusual event, such as the collision between galaxies as described above. In other cases, the distortion seems to be connected with an anomalous release of energy at the center of the galaxy. M82 – one of the strangest shaped galaxies – is an example. The unusual appearance of M82 is shown in Figure 10.12 (p. 254). Although astronomers had studied M82 for years, they did not know of its unusually high energy output until recently, because they had observed M82 only in the visible band of wavelengths. In 1969 the infrared radiation coming from M82 was measured for the first time, and was found to be 2×10^{45} ergs/sec. The total luminosity of our galaxy, summed over all wavelengths, is about 2×10^{44} ergs/sec. This means that M82 is emitting at least 10 times more energy in the infrared alone than our galaxy emits at all wavelengths.

[2] George Abell surveyed and mapped clusters of galaxies as a part of the Palomar Sky Survey.

FIGURE 10.12
M82 photographed in the red light of the hydrogen line at 6563 Å.

THE MASSIVE GALACTIC HALO

Until recently, astronomers believed that a galaxy consists of a large number of stars plus some gas and dust in the space between. Evidence has recently come to light that this relatively familiar part of the galaxy is only a small part of the galaxy's true extent. In many, and perhaps all, galaxies, the cluster of stars that makes up the visible part of the galaxy is now known to be a relatively small, luminous kernel embedded in a much larger and more massive cloud of invisible matter.

The cloud of invisible matter is sometimes called the *massive galactic halo.* In the case of a typical spiral galaxy such as the Milky Way or Andromeda, the massive halo extends out about 300,000 light-years from the center; that is, it is five to ten times the size of the visible galaxy. Figure 10.13 shows the relationship between the visible part of a typical galaxy and the surrounding massive halo.

Evidence for a Massive Halo

The evidence for the presence of massive halos came from measurements on the motions of stars in our galaxy and in neighboring spiral galaxies. Each star in a galaxy orbits around the center of the galaxy like a planet orbiting around the sun. The speed with which the star moves in its orbit depends on the force of gravity

FIGURE 10.13 (opposite)
A diagram of the massive galactic halo surrounding the visible part of a typical spiral galaxy (M74).

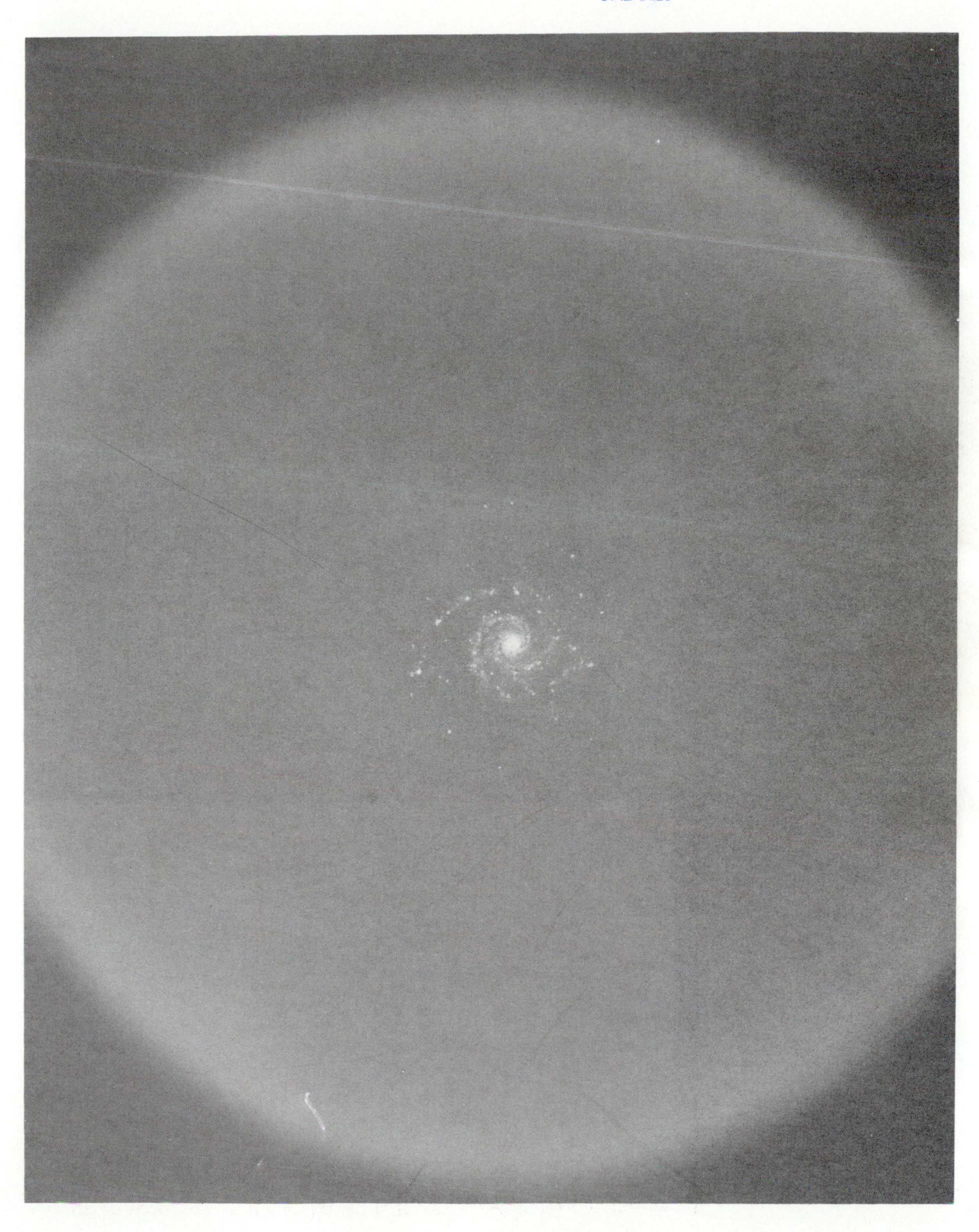

pulling it toward the center. This connection between the speed of the star and the force of gravity pulling on it can be seen in the following way. Suppose the star is moving very rapidly; then it will break the grip of gravity and fly off into space, leaving the galaxy. On the other hand, suppose the star is not moving; then it will be pulled straight in toward the center of the galaxy. If the star is in a stable orbit—that is, neither flying off to space nor falling in toward the center—it must be moving at just the right speed to counter the inward pull of gravity. In other words, as we set out to show, the speed of a star in its orbit is related to the force of gravity acting on it. The gravitational pull acting on a star is related, in turn, to the total amount of matter—stars, dust, and everything else—contained within the star's orbit. The larger the mass within the orbit, the stronger the pull on the star.

Thus, if we measure the speeds of outlying stars—stars orbiting at the outer boundary of the galaxy—we can determine the galaxy's mass.

The results of the measurements reveal a surprising fact: in all galaxies studied, the motions of the outlying stars are too fast to be explained by the amount of mass observed in these galaxies in a visible form—that is, as stars. At the speed at which these outlying stars are traveling, if only the visible mass were present, the stars would fly off into space at a tangent. Some form of matter is holding these stars in their orbits, even though it cannot be seen.

In the case of our galaxy and certain others, the same reasoning can be extended well beyond the "visible" edge of the galaxy, by measurements of the motions of nearby dwarf galaxies. These dwarf galaxies appear to be captives of the main galaxy, held in orbits by its gravitational attraction. In effect, they are satellites of the galaxy.

For our galaxy, the captive or satellite galaxies are the Magellanic Clouds and four other captive dwarfs. The rapid motions of the captive galaxies indicate that the mass associated with the Milky Way Galaxy continues to increase beyond the Galaxy's nominal boundary, all the way out to about 300,000 light-years from the center. This result is the basis for the conclusion that our galaxy possesses a massive halo of invisible matter extending out at least 300,000 light-years. The measurements indicate that the mass in this extended galactic halo is probably at least three times as great as the mass in the visible part of the Galaxy, and may be as much as ten times greater.

Presumably, other galaxies have comparable amounts of mass in their extended halos. This mass concealed in the extended halos of the galaxies is a part of the "hidden mass" of the Universe, whose presence has important cosmological implications discussed in Chapter 13.

THE EVOLUTION OF GALAXIES

Imagine an early stage in the life of the Universe, when space was occupied by dispersed matter and there were no stars or galaxies present. In the swirling motions of the materials within this cosmic cloud, now and then a number of atoms would come together by accident, forming a temporary condensation. If the condensation were massive enough—that is, if it included a sufficient number of atoms—the gravitational attraction of the atoms for each other would be strong enough to hold them together and prevent them from separating. Thus, the condensation, instead of dispersing to space again, would remain bound together as an isolated cloud, distinct from the rest of the gaseous matter that filled the Universe. Such a condensation, if it were large enough, would be the beginning of a galaxy.[3]

Formation of Protogalaxies

A galaxy in the process of condensing, but before any stars have formed in it, is called a *protogalaxy.* Once the protogalaxy has formed, it begins to contract, and its density rises as a result of the continuing inward force of its own gravity pulling matter toward the center. Throughout this period, pockets of gas continually form and dissolve in the swirling and eddying motion within the cloud. Whenever the density of one of these pockets of gas is high enough, it turns into a protostar. The protostars contract and heat up until nuclear reactions begin at their centers, at which point they have reached the Main Sequence line in the H-R diagram and have become full-fledged stars.

From Protogalaxy to Galaxy

When numerous stars have been formed in a protogalaxy, it becomes a galaxy. During the course of time, more and more of the gas of the original protogalaxy is swept up into stars, and after some time, the galaxy contains many stars and very little gas.

[3] This is the same process that was described in Chapter 8 as leading to the formation of stars. Apparently, gravitational condensation proceeds in two stages. First, most of the matter in the Universe condenses into galaxy-sized clouds, as described here, and then, as described in the next section, within each galaxy-cloud, smaller clouds condense into stars. (These stages may themselves be preceded by condensations into clouds with the masses of clusters or superclusters of galaxies (p. 262).

The smaller stars burn slowly and live a long time, but the most massive stars burn out quickly and are destroyed in supernova explosions. Eventually, the galaxy consists only of relatively small, slow-burning red stars and little else. The massive hot stars have disappeared, as has most of the interstellar gas and dust. As the last of the small and faint stars fade away, the galaxy becomes a graveyard of black dwarfs, neutron stars, and black holes in space. But many aeons are required to reach this stage; a small, red star like Barnard's Star, for example, lasts for 10 trillion years, which is a thousand times longer than the time the Universe has existed thus far. Thus, no completely dead galaxies exist yet in the Universe; the majority of the galaxies we see are still in the prime of their star-forming and star-burning vigor.

Explanation of Galactic Shapes

At one time astronomers thought that the galactic shapes catalogued by Hubble were stages in the evolution of a single "typical" galaxy, as red giants and white dwarfs are stages in the evolution of a typical star. However, the current view is that all galaxies were formed at approximately the same time, when the Universe was a billion years old or somewhat less. In this case, why do galaxies have different shapes? What basic property makes one different from another?

This question brings us to the frontiers of research on galaxies. No one is certain of the answer, although at the present time one explanation is more widely accepted than any other. According to this explanation, spherical galaxies are galaxies that were spinning very slowly when they were first formed, so that there was no tendency for them to flatten out as they condensed. Ellipticals, on the other hand, are galaxies that were spinning at a moderate rate when they formed, and therefore flattened out to some degree when they condensed. A spiral galaxy, according to the same theory, is a galaxy with a still greater amount of initial spin, which caused some of the matter within it to flatten out into the shape of a disk. A disk is, after all, essentially the same as a very flat ellipsoid. The nucleus of the spiral is, according to this view, composed of the matter that was near the center of the original protogalaxy, and was, therefore, not appreciably affected by the rapid rotation of the outer regions. This part would condense, shrink down, and collapse inward on itself without much flattening, and would be similar in its properties to a small spherical galaxy. In support of this picture, the nucleus of a spiral galaxy does have the same types of stars as a spherical galaxy.

What is the explanation for the arms in spiral galaxies? A theory with many adherents assumes that the gaseous matter of the central disk is spread out more or less uniformly throughout the disk. According to the calculations based on this theory, if matter is spread out in the disk in that way, density waves, or ripples of density, will appear and will travel through the disk. The theory predicts that as the density waves move around the disk, they give the illusion of a rotating spiral. The differences in density from crest to trough in this spiral pattern could be as little as 50 percent. However, owing to the very strong tendency of stars to form in regions of higher density, the amount of star formation that takes place in regions with slightly higher density than average can be vastly greater than that which takes place in the intervening troughs of low density. Where star formation is rapid, and many hot young stars are being formed continuously, the luminosity will be very great. Thus, to the observer recording the visible light from this disk a clear pattern of luminous spiral arms will be apparent.

CLUSTERS OF GALAXIES

Enormous though a single galaxy is, it does not constitute the largest collection of matter known in the Universe. Galaxies themselves occur in clusters, held together, once again, by the force of gravitational attraction that each galaxy in the cluster exerts on the others.

Some clusters contain only a few galaxies. An example is the cluster of galaxies in the direction of the constellation Pegasus, shown in the photograph in Figure 10.14 (p. 260). (Astronomers have long used constellations, or groups of stars that seem to form figures in the sky, to impose some order on the huge number of stars in the night sky. Much like signposts, they tell us what part of the heavens we are gazing at.) In Figure 10.14 the small circular spots are individual stars situated in our own galaxy along the line of sight to the Pegasus cluster.

The Local Group, of which our galaxy is a member, is an aggregate of galaxies about 5 or 6 million light-years across that includes the Milky Way, the Andromeda galaxy, and around 20 other objects of galactic size.

Some clusters of galaxies contain many hundreds or thousands of individual galaxies. An example of a large cluster is the giant group of galaxies in the direction of the constellation Hercules. A part of the Hercules cluster is shown in the photograph opposite

[4] The "spike" is caused by the struts in the tube of the telescope.

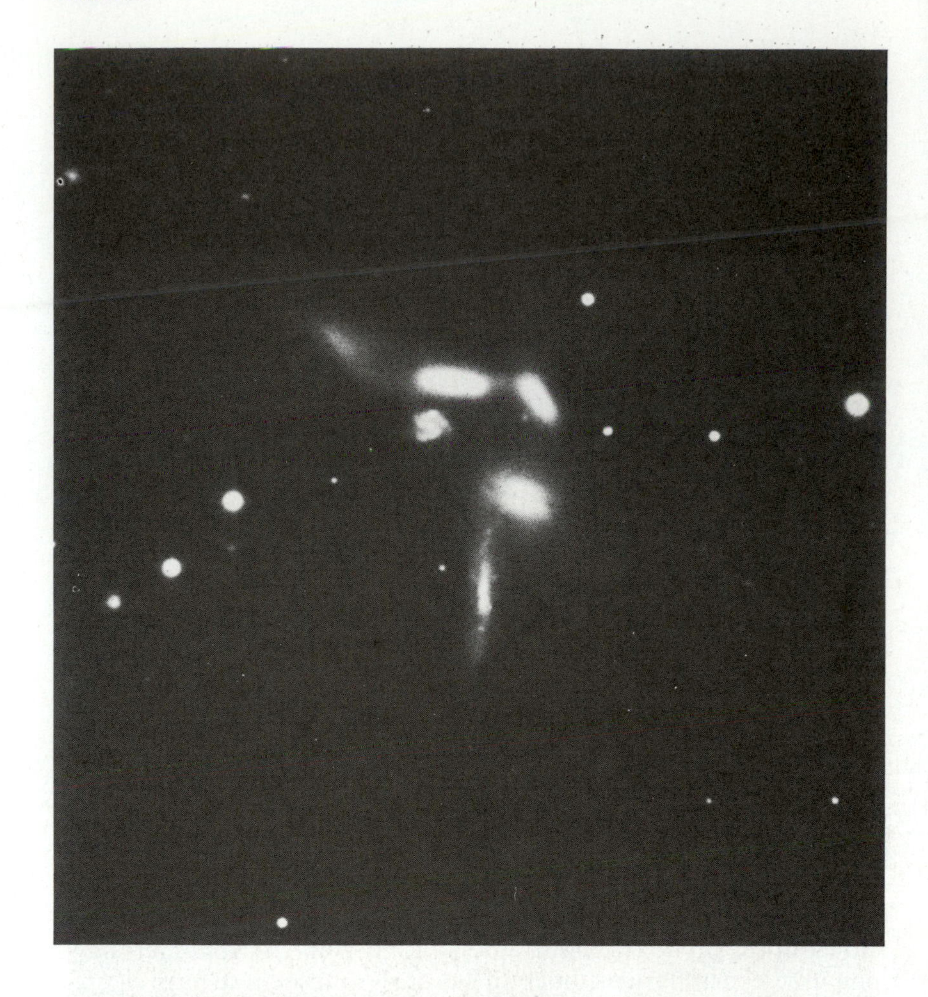

FIGURE 10.14
A cluster of five galaxies in Serpens.

page 69. The central region of the Coma cluster, another large cluster of galaxies in the direction of the constellation Coma Berenices, is shown in Figure 10.15.

The Intergalactic Medium

The galaxies in a cluster are embedded in a gas called the *intergalactic medium.* The amount of matter contained in the intergalactic medium, added to the dark matter in the massive halo surrounding each galaxy, makes up the hidden mass of the Universe (pp. 254 and 256).

Since the gas in the intergalactic medium presumably has never belonged to any galaxy, and therefore has not been processed through the chain of nuclear reactions occurring in stars, it should have no heavy elements and should be composed almost entirely of primordial hydrogen and helium from the Big Bang (Chapter 13).

FIGURE 10.15
The core of the Coma Cluster, a part of a supercluster about 300 million light-years from us. The object with the cross of light is a star in our own galaxy, but every other object is a galaxy in the Coma Cluster. About 300 galaxies can be seen in the photograph; each contains tens to hundreds of billions of stars.

However, a study of X-rays from the Perseus cluster of galaxies, carried out with X-ray telescopes on British and American satellites, has produced the surprising result that the intergalactic medium contains iron atoms, and therefore is not primordial. Iron can only be explained readily as a consequence of nuclear reactions in stars, leading to supernova explosions and thus to the presence of heavy elements in the interstellar medium. How did the iron get from the interstellar medium within a galaxy to the intergalactic medium in the space between galaxies?

The answer probably has to do with an effect called *ram pressure stripping.* All the galaxies in a cluster move through the intergalactic medium at high speeds, ranging up to several million miles per hour. The intergalactic medium resists this motion, and the resistance strips away much of the gas in the galaxy. The stripped-away gas, which includes atoms of iron and other heavy elements made previously in the stars of the galaxy, mixes with the intergalactic medium and becomes a part of it.

Evidence for the phenomenon of ram pressure stripping is found in X-ray images of M86, a galaxy in the Virgo cluster. M86 happens, at this time, to be moving rapidly through the center of the Virgo cluster at a speed of three million miles per hour. The X-ray images of M86, made with the Einstein Satellite, show a plume of matter trailing to the rear of the galaxy, apparently composed of gas from M86 that has been pushed out of it by the resistance of the intergalactic medium through which it is passing.

Superclusters

A recent study of approximately one million galaxies revealed that clusters of galaxies are not the largest aggregates of matter in the Universe. The clusters themselves tend to be grouped into *superclusters.* A supercluster can be as large as several hundred million light-years in diameter, and may include many thousands of separate galaxies. The superclusters have rather irregular forms. In general, instead of being roughly spherical, they are two-dimensional—that is, like huge sheets and filaments of condensed matter.

Our galaxy, and the other members of the Local Group, are among a large number of clusters of galaxies—perhaps as many as 50—that make up the so-called Local Supercluster. The Local Group is one of the outlying clusters near the edge of the Local Supercluster. The Virgo cluster, a very populous group of galaxies very roughly 60 million light-years away from us, forms the center of the Local Supercluster. At the center of the Local Supercluster, the population of galaxies is about 10 to 100 times denser than it is in, for example, the Local Group.

An indication of the appearance of the Local Supercluster is given in Figure 10.16, which shows the positions of 2175 galaxies within a distance of about 250 million light-years from our galaxy. The very dense concentration of galaxies to the right in the diagram is the Virgo cluster, lying at the center of the Local Supercluster. The empty shaded area represents the part of the sky in which our view of other galaxies is blocked by the Milky Way itself.

Voids

The same observations that reveal the existence of superclusters also reveal another interesting fact: the spaces separating super-

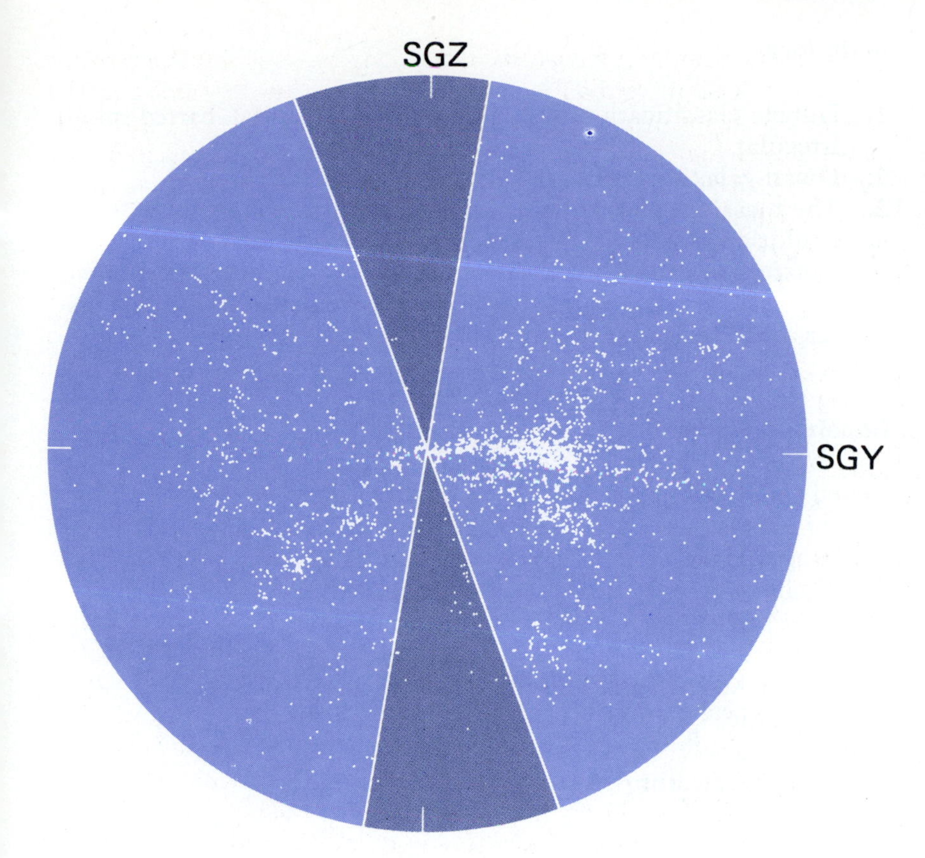

FIGURE 10.16
A plot of positions of several thousand galaxies, showing their tendency to clump into superclusters separated by voids.

clusters from their neighbors are surprisingly empty of matter; they are voids in space. These huge spaces that separate superclusters are hundreds of millions of light-years across, and yet contain essentially no galaxies.

Presumably, both the voids and superclusters grew from small irregularities in the distribution of matter in the Universe when it was young, and galaxies and stars had not yet formed (see Chapter 13). If, by chance, one region had a slightly greater density than average at that time, the force of gravity acting on the particles in this region would also be greater than average, and would tend to pull the particles within it still closer together. That is, if a slight condensation formed by accident in the gases of the early Universe, the condensation would tend to become more and more compact, and its density would become greater and greater. The process of continuing contraction in a condensed region is the same as the gravitational condensation of stars described in Chapter 8.

According to this picture, the condensed regions in the early Universe developed into superclusters, and the space between them developed into voids.

Main Ideas

1. Hubble classification of galaxies: elliptical, spiral, barred spiral, irregular.
2. Dwarf galaxies, colliding galaxies, peculiar galaxies.
3. The massive galactic halo; evidence for its existence.
4. Evolution of galaxies; explanation of galactic shapes.
5. Clusters of galaxies; the Local Group; the Hercules cluster.
6. Superclusters and voids; the Local Supercluster; the Virgo cluster; large-scale structure of the Universe.

Important Terms

barred spiral galaxy
cluster of galaxies
colliding galaxies
dwarf galaxy
elliptical galaxy
galactic disk
globular cluster
halo (galactic)
hidden mass
intergalactic medium
irregular galaxy
massive galactic halo
peculiar galaxy
protogalaxy
spherical galaxy
spiral arms
spiral galaxy
supercluster
void

Questions

1. What is a galaxy? Briefly describe the geometry of the major types of galaxies. Describe the distributions of stars, gas and dust in each type. Draw a diagram of a spiral galaxy, labeling the important regions.
2. Give a brief account of the formation of a galaxy.
3. What changes would you expect in the color of a galaxy during its lifetime—i.e., from the time of its formation early in the history of the Universe, to the present time? Base your answer on information in Chapters 8 and 9.
4. How is the shape of a galaxy explained by the rate of spin of the protogalaxy that formed it? Explain how this theory of galactic shapes accounts for the presence of the galactic nucleus in a spiral galaxy.
5. Why would the rate of star formation increase in colliding galaxies?
6. Discuss the main reason why astronomers were lead to the belief that galaxies may have massive halos.
7. Explain the evidence for the presence of a halo of dark matter extending beyond the visible edge of our galaxy.

8. Suppose you calculated the average density of the matter in a cube 10 million light-years across, by smearing out the galaxies in the cube so that their ingredients filled the space uniformly. Would you get the same answer regardless of where in the Universe the cube was located? Why? What would your answer to this question be if the cube were 100 million light-years across? One billion light-years across? Use the information in this chapter in constructing your reply. What does your answer tell you about the large-scale structure of the Universe?

THE MILKY WAY GALAXY: A TYPICAL SPIRAL

11

Only one galaxy is close enough to us to permit a detailed study of its contents. This is our own galaxy, the Milky Way—a typical spiral of Sb type. A great deal of important information—the composition of the gaseous matter in galaxies, the way in which this matter condenses to form stars, the evolution of stars, and the determination of their ages—all comes from the detailed study of the Galaxy of which our sun is a member.

On a clear night in the country, away from artificial light, we may see our galaxy from the inside, in an edge-on view, as a luminous band of light stretching from horizon to horizon. This band of light was called the Milky Way by the ancients. For that reason our galaxy is usually called the Milky Way Galaxy. Galileo was the first to discover that the Milky Way is composed of myriads of individual stars. He made this discovery in 1610 with the famous one-inch telescope with which he also found the mountains on the moon, the phases of Venus, and the satellites of Jupiter.

PROPERTIES OF THE MILKY WAY

The Milky Way Galaxy is a normal Sb type spiral. It is an aggregate of approximately 200 billion stars, held together by mutual forces of gravitational attraction. This vast number of stars is arranged in the shape of a flattened disk, about 100 thousand light-years in diameter and 5000 light-years in thickness. The galactic nucleus of

The Horsehead Nebula, a dense cloud of dust at a distance of 1100 light-years.

our spiral is hidden from our view by intervening clouds of dust in the space between the stars. We are certain of its existence partly because it emits infrared radiation that has been detected, and partly because we can see galactic nuclei in other galaxies that seem to resemble the Milky Way Galaxy in other respects.

The disk of the Galaxy has relatively dense, irregularly shaped clouds of gas and dust. If these clouds are very dense they obscure the stars behind them, and are observed as dark areas in the Milky Way. An obscuring cloud known as the Horsehead Nebula is shown facing page 267 and in Figure 11.1. Sometimes the atoms in a cloud

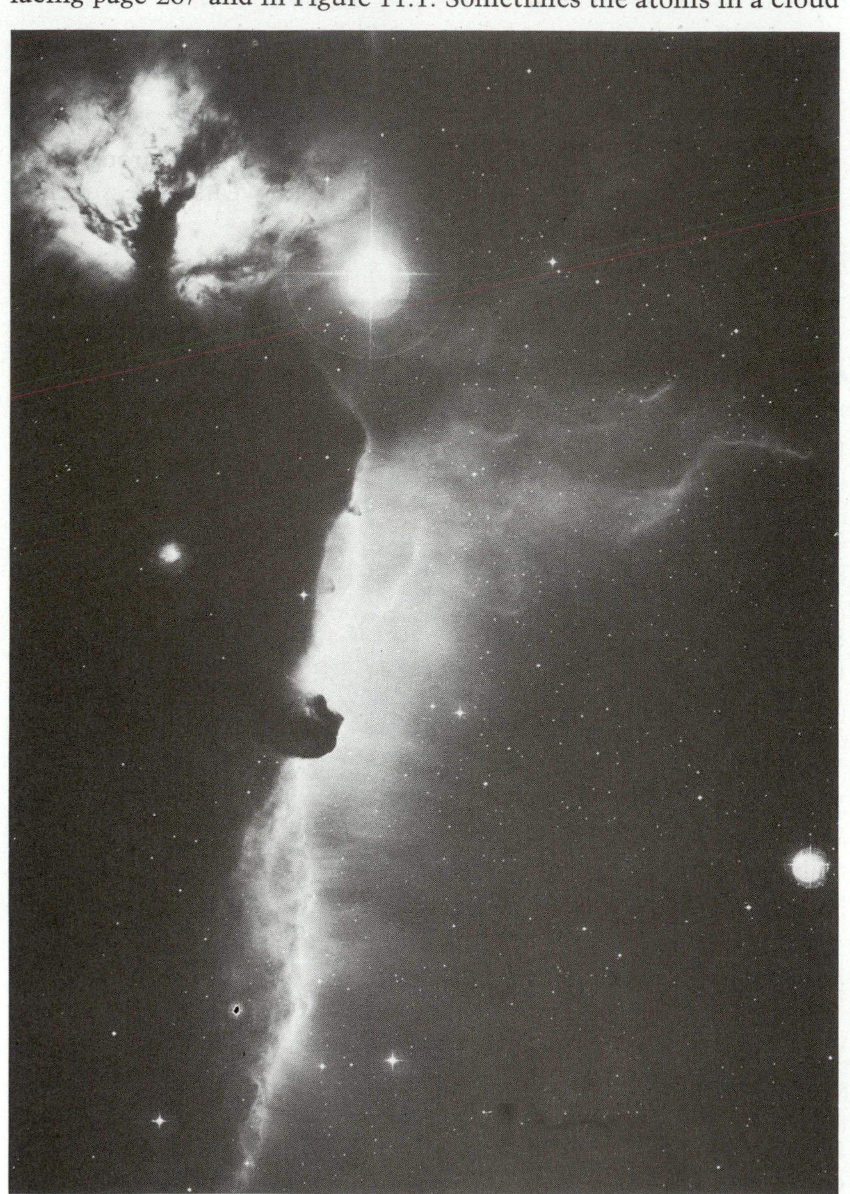

FIGURE 11.1
A region of the Milky Way containing the Horsehead Nebula and part of the Orion Constellation.

of gas and dust are excited by the radiation from young, hot stars embedded in them. The excited atoms reemit their energy in the form of a diffuse glow, forming a so-called emission nebula. Color Plate 16 shows the Trifid Nebula, an emission nebula located at a distance of 2300 light-years in the direction of the constellation Sagittarius. The dark lanes that separate the Trifid Nebula into three lobes are lanes of obscuring dust lying between us and the glowing regions.

Spiral arms radiating out from the center of the Galaxy contain most of the bright stars in the Galaxy, along with much of the gas and dust. The sun is located in one of these arms, approximately 30,000 light-years from the center. The entire Galaxy rotates on an axis through the center perpendicular to the plane of the disk, with the inner regions rotating at a faster angular rate than the outer ones. The sun and planets complete one trip around the center of the Galaxy in roughly 250 million years. The sun has gone around the Galaxy nearly 20 times during the 4.6 billion years of its existence. When it was last in its present position, the dinosaurs had not yet appeared on the earth.

THE SPIRAL STRUCTURE OF THE GALAXY

How do we know that the Milky Way Galaxy has spiral arms? It is a simple enough matter to determine the structure of other galaxies than our own by photographing them through a large telescope. But the solar system is immersed in the Milky Way Galaxy, and its view of the multitude of stars around us—except for relatively close neighbors—is blocked by intervening clouds of dust. Charting the shape of the Milky Way Galaxy from the vantage point of our solar system is as difficult as constructing a street map of a city from the central square on a foggy night.

The most complete answer has come from observations in radio astronomy. Radio waves, unlike waves of visible light, are not absorbed by the clouds of dust that exist between our solar system and the distant parts of the Milky Way Galaxy. It is this same property that makes radio waves valuable in communications on the earth; if they were not able to pass through clouds of particles, radio broadcasts would be blacked out on every cloudy day. Radio waves in space, reaching the earth with relatively little interference, have provided most of the information that we possess regarding the spiral structure of the Milky Way Galaxy.

Stars cannot be the source of these radio waves. The sun, for example, emits only a millionth of its energy in this form. The radio waves in the Galaxy are produced primarily by the interstellar medium. If our eyes were sensitive to radio, all the interstellar hy-

FIGURE 11.2
A radio map of the Galaxy produced by Leiden Observatory from observations of the 21-centimeter line.

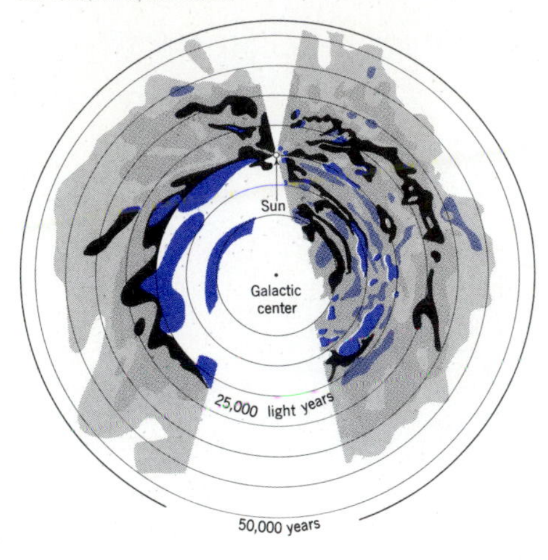

drogen in the Galaxy would seem to us to be emitting a soft glow. By charting the intensity of this "radio glow" in all directions around us, we could construct a picture of the distribution of hydrogen in the Milky Way Galaxy. Radio astronomers have done this using large antennas. The results indicate that the hydrogen in the Milky Way Galaxy is concentrated in distinct lanes separated by wide spaces containing relatively little hydrogen. The radio picture of the Milky Way Galaxy, shown in Figure 11.2, suggests that the Milky Way Galaxy has a spiral-arm structure, although it is not as clear as in photographs of other spiral galaxies taken in visible light.

The 21-Centimeter Line

Just as hydrogen atoms radiate energy at characteristic wavelengths in the visible region, they also radiate energy at characteristic wavelengths in the radio region. The most intense hydrogen line in the radio region has a wavelength of 21.1061 centimeters, or about eight inches.[1] Consequently the spiral arms of the Galaxy, where hydrogen is concentrated, emits a strong glow of 21-centimeter radiation.

In most directions in space we receive radio waves from several spiral arms simultaneously. This fact increases the difficulty of mapping the Galaxy by radio. Referring to Figure 11.3, imagine that a radio astronomer has oriented the radio antenna so that it is pointed along the dotted line. As Figure 11.3 shows, the astronomer will receive radiation from the hydrogen located in two spiral arms—labeled A and B—which lie in the line of sight when observing in this direction. How can the two signals be separated? The task is made feasible by the rotation of the Galaxy, which gives the spiral arms a relative motion with respect to our solar system. The relative motion produces a Doppler shift in the wavelength of the 21-centimeter line emitted by each arm. If an arm of the spiral is moving *toward* us, or the sun is moving *toward* that arm, the 21-centimeter line emitted by this arm will be shifted toward shorter wavelengths, and the astronomer will observe a blue shift. If the relative movement is *away* from the sun, the shift will be toward longer wavelengths; that is, the line will display a red shift.[2]

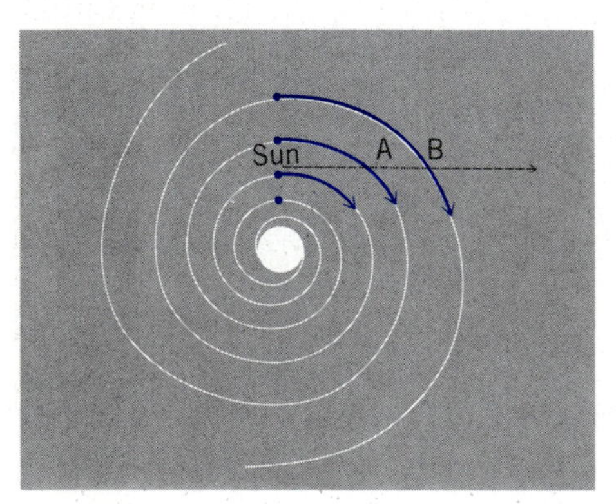

FIGURE 11.3
Velocities of the sun, and two gaseous clouds A and B in other spiral arms of the Galaxy.

[1] Although referred to as a radio wavelength, 21 centimeters is shorter than the wavelengths used in normal radio communications. Twenty-one-centimeter radiation lies in the region of the electromagnetic spectrum used by aircraft and ship radars for navigation.

[2] If this radiation were in the visible region, the shifts would be toward the blue end or the red end of the spectrum, respectively. By analogy with the visible region, astronomers refer to these shifts as blue shifts and red shifts, regardless of the region of the spectrum that is actually involved.

The amount of relative motion, and therefore the amount of the Doppler shift, is usually different for every arm. Thus, the astronomer will receive radiation of two different wavelengths—both in the neighborhood of 21 centimeters—from the two spiral arms A and B that lie in his line of sight.

In the case shown in Figure 11.3, representing our own galaxy, the rotation of the entire Galaxy is clockwise. Therefore the sun is moving toward the right. The velocities of the sun and points A and B are approximately the same, but because these velocities are not in the same direction as the sun's velocity, the velocities of A and B have a component along the line of sight to the sun. This component produces the Doppler shift.

Figure 11.3 shows the velocities of the sun and gaseous clouds A and B in two other arms of the Galaxy. According to the diagram, the sun has a relative velocity toward both A and B at this particular moment. Therefore, the 21-centimeter radiation from A and B will be Doppler-shifted toward shorter wavelengths; that is, it will be blue-shifted. However, the velocity of the sun relative to the hydrogen gas in Arm A will not be as great as it is relative to the hydrogen gas in Arm B because of the difference in the directions of the velocities of the two arms, and thus the blue shift in the spectrum of the gas in Arm A will be less than in Arm B.

Figure 11.4 shows the spectrum of the 21-centimeter radiation that astronomers detected with a radio telescope, in a case similar to that illustrated in Figure 11.3. The most striking feature of this

FIGURE 11.4
Twenty-one-centimeter signal received in a case corresponding to Figure 11.3.

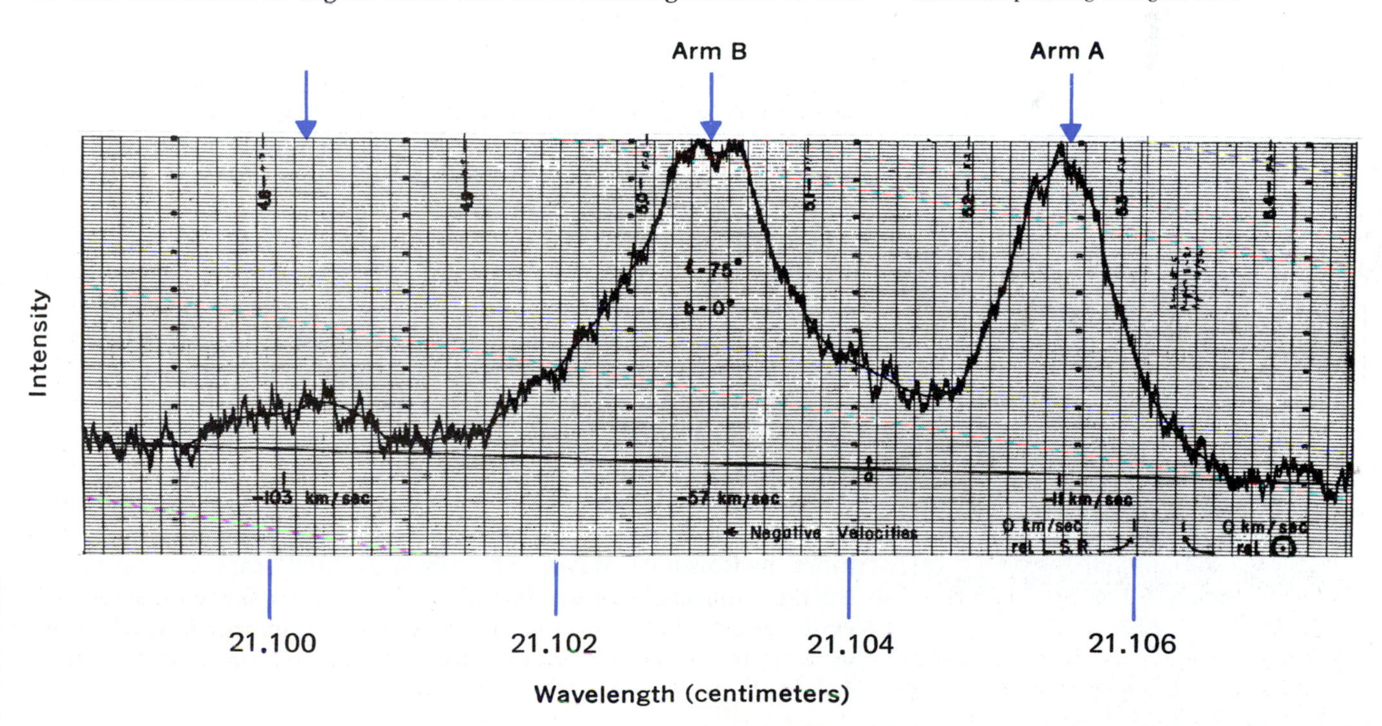

spectrum is that it has distinct peaks at two wavelengths and a somewhat distinct peak at a third. The presence of these peaks provides the evidence that the matter of the Galaxy is concentrated into spiral arms. If it were spread out in the continuum, there would be no peaks in the spectrum because there would be a continuous variation of Doppler shifts across the face of the Galaxy.

A second interesting feature of the spectrum is that one peak is at a wavelength slightly shorter than 21.106 centimeters. This peak is the radiation from Arm A, shifted a small amount toward the blue, as expected. A second peak appears at a still shorter wavelength. This peak comes from Arm B. Its shift toward the blue is several times greater than for Arm A, indicating that Arm B is farther out in the Galaxy than A. A third and exceedingly faint peak is also visible at a still shorter wavelength (arrow). This third peak of radiation must emanate from an arm very far out in the Galaxy. The peak is weak partly because the arm is far away and partly because the density of the matter in this arm is low.

THE INTERSTELLAR MEDIUM

The stars of the Galaxy move through a tenuous sea of gaseous matter called the interstellar medium. The density of this gaseous material is so low that it constitutes a vacuum one million times better than has ever been achieved in any laboratory on the earth.

Most of the interstellar medium consists of atoms and molecules of hydrogen, which make up 94 percent by number of the atoms in the Universe. If the proponents of the Big-Bang cosmology (Chapter 13) are correct, the Universe had its beginning in an explosive event that occurred about 15 billion years ago. Nearly all the hydrogen now in the interstellar medium dates back to that early period. Helium makes up most of the remaining 6 percent of the atoms in the interstellar medium. The 90 other elements in the periodic table make up about 1 percent. This final small fraction consists mainly of atoms of the elements carbon, nitrogen, oxygen, neon, magnesium, silicon, and iron. These substances, as well as still heavier and scarcer elements ranging up to uranium, have been manufactured in the centers of stars by nuclear reactions, and then sprayed into the interstellar medium by supernova explosions.

Presumably the amounts of carbon and the heavier elements, and helium as well, must be steadily increasing with time as more and more hydrogen is drawn into newly formed stars and processed through the cycle of nuclear reactions that accompanies a star's evolution. By the same token, the amount of hydrogen in the Universe must be steadily diminishing as it is used up in nuclear reactions, unless a fresh source of that basic substance exists. At the present time, however, there is no evidence for fresh sources of hydrogen in the Universe.

Molecules in the Interstellar Medium

In addition to atoms of hydrogen and helium, the interstellar medium also contains a large variety of molecules, some of which are quite complex. These molecules are formed when atoms stick together in collisions. They occur mainly in dense clouds of gas in the Galaxy, where collisions are most frequent. Two atoms of hydrogen can form a molecule of hydrogen, or they can combine with an atom of oxygen to form a molecule of water. Carbon and oxygen atoms can form a molecule of carbon monoxide (CO) or carbon dioxide (CO_2).

Carbon monoxide molecules have turned out to be a very important tool of observation for astronomers studying the structure of our galaxy. CO is found mainly in the spiral arms of the Galaxy, where the density of the gas is greatest, and, therefore, the likelihood of formation of molecules is also greatest.

Carbon monoxide has been detected throughout a large part of the Galaxy. It serves as a tracer for star formation, because wherever it is abundant, its presence signals the existence of a relatively dense cloud of matter. Since a high density is conducive to the formation of stars, stars tend to be born in such relatively dense clouds. About 1000 such clouds, ranging up to 150 light-years in size, are believed to exist in the Milky Way. The Orion Nebula contains a particularly dense concentration of carbon monoxide (Figure 8.3, p. 196). The Orion Nebula and similar clouds are the seedbeds of star formation in our galaxy.

The observations also reveal the presence of alcohol (C_2H_5OH), formaldehyde (H_2CO), and other organic molecules. More than 50 kinds of molecules—many of them organic, that is, containing a backbone of carbon atoms—have been discovered in space. Some of these molecules have never been seen on the earth. One of the largest and most complex interstellar molecules discovered thus far is cyanooctatetrayne (HC_9N), which has the following structure.

$$N{\equiv}C{-}C{\equiv}C{-}C{\equiv}C{-}C{\equiv}C{-}C{\equiv}C{-}H$$

The rich variety of molecules existing in the Galaxy is one of the great surprises of modern astronomy, and has even generated speculations that life could originate in space.

Hydrogen cyanide is of particular interest in this connection. Molecules such as this have been known for some time to be possible precursors of living matter. From a chemical point of view, they are suitable building blocks for amino acids and other biologically important molecules that provide the foundation of all known terrestrial life (Chapter 20).

Could the chemical evolution of life have commenced, not on the earth as many biochemists believe, but in outer space before the solar system existed? Perhaps the evolution of life has been going on in space throughout the history of the Universe, and is still going on there today. Unfortunately for these interesting possibilities, the interstellar evolution of life seems unlikely because of the low density of the molecules in space. In these conditions, the collisions between molecules which give rise to the chemistry of life are a rare event, and chemical evolution is likely to be too slow to produce a living organism, even when a billion years or more are available. The absence of liquids in space, especially liquid water, is another obstacle to the initiation of chemical evolution.

Finally, in addition to atoms and molecules, the interstellar medium also contains numerous small particles of solid matter known as *interstellar dust.* Interstellar dust makes up approximately one percent by weight of the matter in the interstellar medium. It is concentrated almost exclusively in the central plane of the galactic disk. The size of the dust particles is about one ten-thousandth of a centimeter. This is the same as the size of smoke particles. The composition of the dust is uncertain, but it is known to include bits of carbon and rocky materials. Crystals of ice are also present in the densest regions.

CLUSTERS OF STARS

Most of the stars in the Galaxy move about freely among their neighbors. However, some are bound together into groups called clusters, containing as few as five or six stars and as many as one million. Each star in a cluster is bound to the others by the gravitational force of the entire cluster. The cluster as a whole moves about in the Galaxy, with its individual member stars orbiting around the center of the cluster like a swarm of bees. A star cluster is formed when many stars condense simultaneously out of one large cloud of matter. All the stars in the cluster condense out of the cloud in a very short interval of time. They may not all appear at precisely the same moment; there may be a spread of as much as several million years between the first star and the last to form in a cluster; but a million years is still a very short time compared to the lifetime of most stars. Thus, all the stars in a given cluster can be regarded as having been born simultaneously.

Clusters are very valuable because they are groups of stars with a common birthday. If a random sample of the stars in the sky is selected, and their H-R diagram is plotted, we obtain a very blurred picture of a star's life owing to the circumstance that the sample of the family of stars usually contains stars of all different ages. A sharper picture would result if separate H-R diagrams could be plotted for each stellar age group. Clusters provide the opportunity to plot such diagrams. As we will learn in the following sections, there is evidence that our galaxy contains clusters of stars whose birth dates span the full period of years from the formation of the Milky Way Galaxy to the present era. The comparison between the H-R diagrams of Milky Way clusters of various ages and the H-R diagrams calculated from the theory of stellar evolution described in Chapters 8 and 9 provides the most important single observational check on that basic theory, and also provides a means for measuring the age of the Galaxy and the age of some of its stars with a remarkable degree of precision.

Galactic Clusters. Thousands of star clusters are scattered throughout the galactic disk. Because they are located within the galactic disk, these star clusters are known as *galactic clusters.* They are also called *open clusters,* but that term will not be used in this book.

An example of a galactic cluster is shown in Figure 11.5, taken through the Palomar Mountain 200-inch telescope. This cluster, designated M67, is located in the direction of the constellation Cancer at a distance of 2500 light-years.

FIGURE 11.5
The galactic cluster M67.

Galactic Clusters and the H-R Diagram. Significant information can be obtained by plotting the H-R diagram for the stars of a single galactic cluster. The age of the cluster can be measured with surprising accuracy in this way, and at the same time observational confirmation can be obtained for the broad features of the theory of stellar evolution developed in Chapters 8 and 9.

How can a young galactic cluster be distinguished from an old one? The answer can be found in its H-R diagram. You will remember that the stars are arranged along the Main-Sequence line of the H-R diagram according to their masses. The most massive stars—the ones that are extremely bright and very blue—are at the upper end of the Main Sequence and the least massive stars—the ones that are dim and red—are at the lower end (see Figure 7.13). The key to determining the age of a galactic cluster lies in the fact that very massive stars in the cluster—the ones that would be plotted at the top of the Main Sequence—have very short lifetimes. For example, a star with 10 solar masses lives on the Main Sequence for only 10 million years before its fuel is burned up. This is one one-thousandth of the life of the sun. If the H-R diagram for a cluster shows stars at the upper end of a Main Sequence, where stars of 10 solar masses are to be found, it follows immediately that this cluster cannot be more than 10 million years old. It is a very young and newly formed cluster, in comparison to the roughly 10-billion-year age of our galaxy.

Let us carry this reasoning further. Suppose that the H-R diagram for a cluster shows no extremely bright, blue stars but does contain a number of moderately bright stars of bluish-white color, lying fairly far up on the Main Sequence but not at the very top. For example, suppose that the brightest stars in the diagram are approximately 100 times more luminous than the sun, and have temperatures around 10,000°K. Reference to the H-R diagram in Figure 7.13 indicates that these stars would have about three solar masses. A star with three solar masses lives about 300 million years. If this cluster were considerably older than 300 million years, those stars would not be present on its H-R diagram. On the other hand, if it had lived for substantially less than 300 million years, stars that are still brighter and bluer would be present on its diagram, but they are not. Therefore, the cluster must be approximately 300 million years old.

Of course, the reasoning would be invalid if this cluster had no star greater than three solar masses when it was first formed. That is a possible occurrence for a small cluster with, say, only five or six stars. However, if the cluster is relatively large, with 100 stars or more, we can assume that the full range of stellar masses will be represented in it, from the least massive to the most massive stars.

FIGURE 11.6
H-R diagrams for galactic clusters:
(a) Pleiades.
(b) Praesepe.

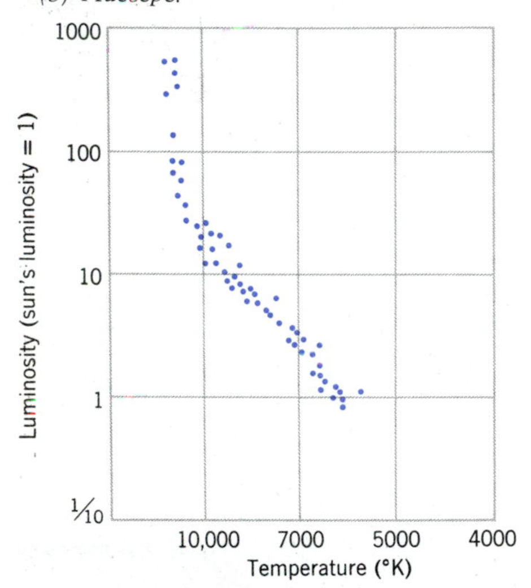

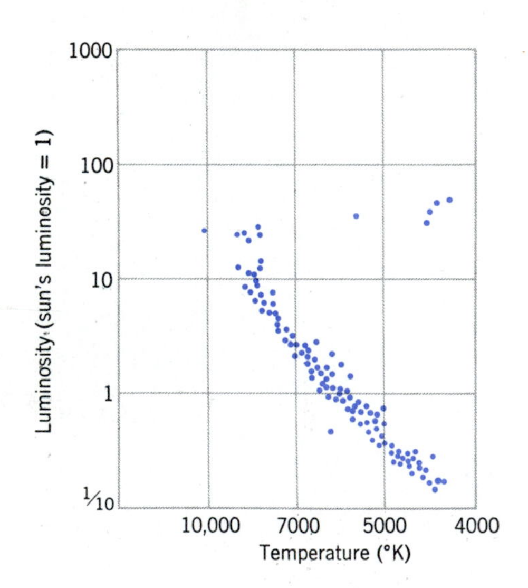

Three Examples. As an example of a very young cluster we take the Pleiades, a very familiar cluster containing six stars clearly visible to the naked eye (see Color Plate 10)[3] This cluster is 400 light-years away. Examination through a telescope reveals that the cluster contains more than 100 stars in all. The H-R diagram for most of the stars in the Pleiades cluster is shown in Figure 11.6*a*. Notice the six extremely luminous and blue stars at the top of this diagram. They are the same six stars that are visible to the naked eye. A comparison with Figure 7.13 indicates that the brightest of these stars is approximately 6 solar masses. A star of this mass lives for 60 million years. This makes the Pleiades cluster roughly 60 million years old.

Now consider the H-R diagram for the Praesepe cluster, located in the direction of Cancer at a distance of 500 light-years. This cluster also contains approximately 100 fairly bright stars. Its H-R diagram is shown in Figure 11.6*b*. We see that the extremely

[3] In the color insert each of the six stars is surrounded by an intensely blue sphere. This sphere does not belong to the star itself but is dust which reflects the light from the star at its center. The six stars themselves emit most of their energy as blue and ultraviolet light because they are extremely hot.

luminous and hot stars that marked the youth of the Pleiades cluster are missing from this diagram. The brightest stars it contains are 70 times as luminous as the sun, corresponding to a mass of 3 solar masses. As noted above, a star of this mass lives for 300 million years, which must, therefore, be the approximate age of the Praesepe cluster.

As a final example, consider the H-R diagram for the cluster M67 located in the constellation Cancer at a distance of 2500 light-years. This cluster has 80 fairly bright stars. A glance at its H-R diagram, Figure 11.7 indicates that M67 is far older than the Pleiades or the Praesepe clusters, for nearly all the bright blue stars in the cluster have disappeared[4] and the brightest stars still present in abundance

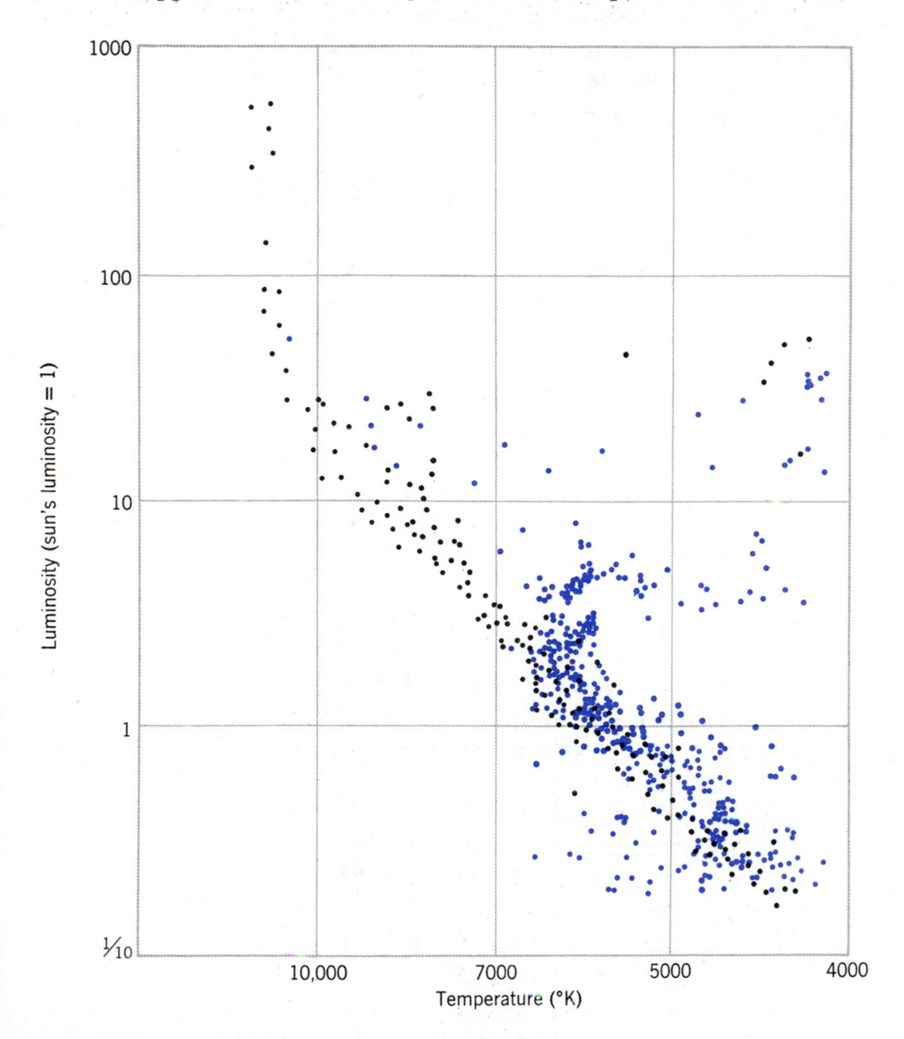

FIGURE 11.7
M67, with the Pleiades and Praesepe clusters added. Luminosity and temperature values are approximate.

[4] A few hot, blue stars, called blue stragglers, still remain on the Main Sequence. No satisfactory explanation has been given for these stars, whose properties indicate that they long since should have come to the ends of their lives.

FIGURE 11.8
The globular cluster M3.

have clearly peeled off to the red giant region, away from the Main Sequence. When the calculated positions of the stars in the cluster are compared with their observed positions on the H-R diagram, the best fit for the cluster as a whole is obtained for an age of about 10 billion years, indicating that M67 appeared when the Galaxy itself was relatively young.

Globular Clusters

Although the galactic clusters are the most numerous star clusters in the Galaxy, they are not the largest. Clusters containing up to one million stars, known as *globular clusters,* are found outside the plane of the galactic disk in the surrounding region of space known as the galactic halo. Our galaxy contains more than 100 globular clusters, scattered at random throughout the spherical volume of the halo. Although most are located outside the galactic disk, the globular clusters are still regarded as members of the Galaxy because they are bound to it by the gravitational force of the matter in the galactic disk. Figure 11.8 shows the typical globular cluster M3.

The Ages of the Globular Clusters. All globular clusters are believed to be very old, nearly as old as the Galaxy in some cases, because of their location far outside the galactic disk. This belief is based on our ideas regarding the history of the Galaxy. According to one theory, when the Galaxy was young, it is thought to have had the form of a large spherical cloud of gaseous matter. Clusters of stars could have begun to condense out of the cloud at that time.

As time passed, the Galaxy contracted to its present disk-shaped form. Some groups of stars, which were the first to appear in the Galaxy, were left behind as the cloud contracted. These were the globular clusters. After perhaps a billion years the Galaxy would have completed its contraction and would have the shape it has today, with its visible matter concentrated in a very flat disk. However, the globular clusters, which had formed during the first billion years, would have remained outside the disk in the galactic halo, where they were born. From that time to the present, the appearance of the Galaxy has not changed appreciably.

This line of reasoning has led astronomers to the conclusion that the stars in the globular clusters are very old, and some were probably among the first stars to form in the Milky Way Galaxy. Therefore the ages of the globular clusters will provide a clue to the age of the Galaxy.

The H-R Diagrams of Globular Clusters. As before, we look to the H-R diagrams of the globular clusters for information regarding their ages. What will these H-R diagrams look like? If the clusters are really very old, nearly all their stars should have peeled off the Main-Sequence line. Inspection of the H-R diagram for M3, a typical globular cluster in our galaxy, shows this to be the situation (Figure 11.9). All stars in M3 have left the upper part of the Main Sequence and are scattered about on the diagram in various stages of intermediate or advanced evolution. Some stars presumably have disappeared from the diagram in the aftermath of a supernova explosion. The only exceptions are the very small stars at the extreme lower end of the Main Sequence, whose lifetimes range up to a trillion years. These stars will anchor the bottom of the Main Sequence as long as we and our descendants exist in this Galaxy.

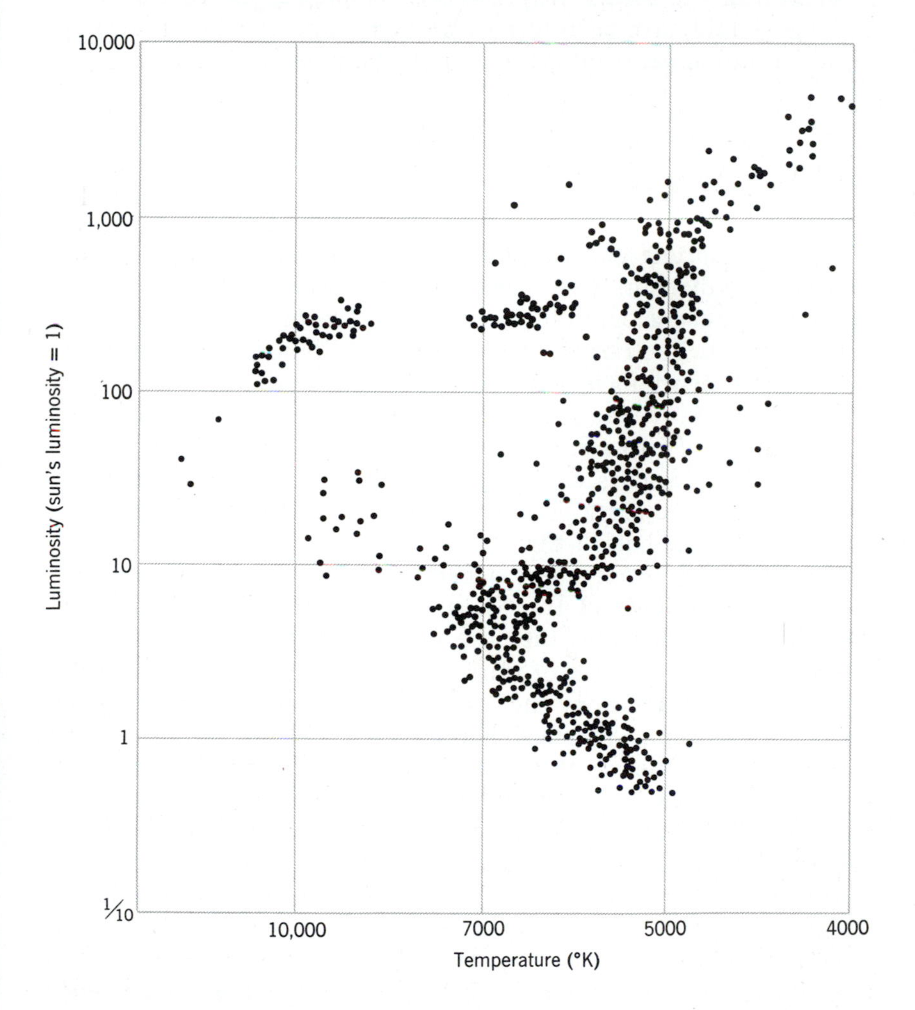

FIGURE 11.9
H-R diagram for the M3 globular cluster.

M3 is a representative example of globular clusters. We have selected it because it has been more carefully studied than any other globular cluster. All globular clusters in the Milky Way Galaxy have H-R diagrams which resemble that of M3; all have ages, deduced from the position of the turn-off point on their H-R diagrams, of billions of years. In some globular clusters the stars are relatively metal-poor, indicating that these were formed very early in the history of the Galaxy, before many supernova explosions had occurred. These metal-poor and very old globular clusters turn out to have ages of about 14 billion years. Presumably this is also the age of the Galaxy.

Other globular clusters, with stars that are relatively metal-rich, are somewhat younger, with ages ranging down to 10 or 11 billion years. It is not clear at this time how the rather late formation of these metal-poor globular clusters fits into current theories on the formation of the Milky Way Galaxy, which suggest that all such clusters should have been formed at approximately the same time, within a billion years or so after the Galaxy itself formed.

STELLAR POPULATIONS

When Hertzsprung and Russell plotted the first versions of their now-famous diagram, they used stars that were relatively close to the sun because accurate distances could be most easily obtained for these stars. Their original diagrams looked more or less the same as the one that we have shown on page 170, with many stars lying on the Main Sequence and a few stars in the red giant and white dwarf regions. However, the H-R diagram for the globular cluster M3, which we discussed in the previous section, presents a different appearance, mainly because this diagram has hardly any stars in the upper one-half of the Main Sequence above the position of the sun. Second, it has a very densely populated track of stars leading from the middle of the Main Sequence along a curved path upward and to the right into the red giant region. Third, it has a moderately large number of stars running more or less horizontally across the diagram from the red giant region back to the upper region of the Main Sequence. These differences indicate or suggest that the stars of the M3 cluster and other globular clusters represent a type of star population different from stars such as the sun and its neighbors. The globular cluster stars are not the only stars that are different, as a group, from stars like the sun and its neighbors. Very old clusters also have this type of H-R diagram. Individual isolated stars that are moving around in the halo have, when their data are collected and put on an H-R diagram, the same kind of diagram as stars in a globular cluster. Finally, there is evidence that many stars in the nucleus of the Galaxy would fit on the same special diagram.

These differences are very striking. It is just as though you were to make up a diagram on which you plotted height against weight for the entire population of a country, and were to find that there were two completely distinct distributions showing up in your plot. You would probably conclude that there were two distinct populations or types of people living in that country. Astronomers have come to the same conclusion about the stars. They call the stars whose H-R diagrams resemble that of the sun's neighbors the *population I* (one) stars. The stars whose H-R diagrams resemble those of the globular clusters, galactic halo stars, and many galactic nuclei stars, are called the *population II* (two) stars.

Other differences between the two populations have been observed, in addition to the basic difference in their H-R diagrams. The most important of them is the difference in their chemical composition or ingredients, as determined from the study of their spectral lines using the methods explained in the chapter on atomic spectra. According to their spectra, population II stars have relatively little of the metals and heavy elements. Stars of population I, on the other hand, have substantial amounts of heavy elements, ranging up to as much as several percent in extreme cases. What is the meaning of the difference between the two star populations? What types of stars are contained in each population? A clue to the answer has already been provided in the previous section, where we found that globular clusters are relatively old groups of stars.

Age is the essential difference between the population I and population II stars. All stars in population I are relatively young; all stars in population II are relatively old. By relatively old, we mean dating back to the early years of the Galaxy.

If this is so, population II stars were formed out of pristine materials of the original Universe, which were, according to the most widely accepted cosmological theories, a mixture of hydrogen and helium atoms (Chapter 13). Therefore, these stars should contain nothing but hydrogen and helium.[5] As we have seen in Chapters 8 and 9, heavy elements, extending from helium, carbon, nitrogen, and oxygen up through iron to the heaviest substances such as gold, lead, and uranium, all are manufactured in the bodies of stars by the nuclear reactions that take place during their lifetimes. These elements are then spewed out to space in final explosions that mark the deaths of these stars, there to mix with the primitive hydrogen and helium gases of the original Universe. Later in the history of the Universe and the history of this Galaxy, other stars

[5] Since the stars of population II possess small but measurable amounts of carbon, nitrogen and other heavier elements, these population II stars cannot have been the *very* first stars to appear in our galaxy. Another group of stars, made of primordial hydrogen and helium, must have existed before them. These stars are called *population* III by some astronomers. Presumably, the deaths of the population III stars would contribute small amounts of heavy elements to the interstellar matter, later to be condensed into the bodies of population II stars.

are formed out of this mixture, and the contents of these stars will contain some amounts of the heavier elements. The later a star is formed, that is, the closer to the present time, the more of the heavier elements it will have among its ingredients, because the abundance of these elements is rising steadily in the course of time as more and more stars come to the ends of their lives, explode, and contribute their contents to the space around them. Another way of looking at the difference between the two populations is to say that population II stars are the first generation of stars to be formed in a galaxy, and population I stars are a mixture of all later generations.

One other point should be mentioned. Population II stars never are associated with concentrations of gas and dust. For example, the globular clusters are almost free of gas and dust, in contrast to the stars located within the galactic plane, which are surrounded by, and often embedded in, dense concentrations of gaseous material. Also, population II stars rarely contain blue giants, but population I stars usually include an appreciable number of blue giants. The explanation of this set of properties is also clear when we realize that population II stars are the first to form in a galaxy and are very old. Since they are old, they cannot include blue giants, for blue giants are very hot, massive stars which burn up their resources quickly and disappear in a time that is typically some millions of years or so. By the same token, population II stars would not be embedded in or surrounded by gas and dust, for they are a group of stars that have been around for a long time, and all the gas and dust that was originally present in their neighborhood has been used up in forming stars. Population I stars, on the other hand, exist in regions that are richly abundant in gas and dust, and in which star formation is still going on at the present time.

Main Ideas

1. The structure and contents of the Milky Way Galaxy; nucleus, disk, halo.
2. Techniques for using radio waves to map the Galaxy.
3. The interstellar medium; elemental and molecular composition.
4. Properties of galactic clusters and globular clusters.
5. Determination of the age of a cluster from its H-R diagram.
6. Populations I and II and their interpretation.

Important Terms

cluster of stars
galactic cluster
galactic nucleus
globular cluster
interstellar dust
interstellar medium
nebula
populations I and II (stars)

Questions

1. Draw the Galaxy to scale showing the location of the sun. Indicate the shape and size of the nucleus, the disk and the halo and the distribution of the globular clusters.
2. Explain how radio astronomers use 21-centimeter radiation to map the disk of the Galaxy.
3. What molecules have been found in the interstellar medium? What is their significance for life?
4. What are the major differences between the two types of star clusters found in the Galaxy?
5. From your study of stellar evolution in Chapters 8 and 9 and the discussion of the ageing of clusters in this chapter, plot an H-R diagram showing successive positions of the stars in a cluster, as the cluster ages from birth to an age of 10 billion years.
6. Suppose globular clusters contained the elements heavier than helium in the same proportion in which they are found in the Cosmos as a whole. What would this imply regarding the importance of stellar evolution in the synthesis of the elements, relative to the importance of the nuclear reactions occurring shortly after the Big Bang? Explain.
7. What are the differences between population I and population II stars? Interpret these differences in terms of the history of the Galaxy.
8. What are the principal differences between the H-R diagram for the globular cluster M3 and the H-R diagram in Figure 7.12?

RADIO GALAXIES AND QUASARS

12

Out of the description of galaxies in the previous chapters emerges a concept of what may be called a normal galaxy. A normal galaxy may be regular or irregular in shape, but it has, in general, the properties that one would expect for a cluster of many millions of individual stars. The energy emitted from a normal galaxy is what we would expect to find if we added up the radiation emitted from many separate stars with a range of masses. The distribution of wavelengths in the radiation from galaxies also fits this picture. Furthermore, we occasionally see a supernova flare up in one of the normal galaxies, which suggests that stellar life flows on in its familiar way in these galaxies, each star passing from the Main Sequence to the red giant region and then expiring as a supernova or a white dwarf.

But recent discoveries suggest that an important element is missing from this picture. Remarkable events take place in the centers of many galaxies, in which enormous amounts of energies are created in extremely small volumes of space. The amount of energy produced, and the very small size of the region in which it is produced, are such that one cannot explain these events readily as the result of nuclear reactions taking place in multitudes of separate stars. Apparently galaxies are not simply clusters of stars. In many galaxies a powerhouse exists in the nucleus, whose energy output dwarfs the emission from all the billions of stars surrounding it; and yet this powerhouse appears to be no larger in size than our solar system. The resolution of these seemingly paradoxical properties is one of the frontier problems in contemporary astronomy.

Centaurus A: a radio galaxy (NGC 5128).

FIGURE 12.1
Cygnus A, a strong radio source.

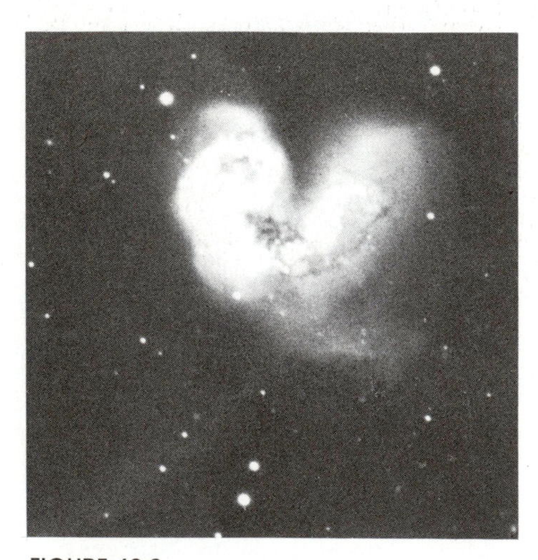

FIGURE 12.2
NGC 4038/9. A radio galaxy located in the constellation Corvus.

RADIO GALAXIES

The discovery of *radio galaxies* was the first clue to the fact that some galaxies may have unusual sources of energy not explainable directly by nuclear reactions in stars. Radio galaxies are objects that look like galaxies but emit intense radio signals. The radio emission from the most powerful radio galaxies is equal to, and in some cases greater than, the entire output of energy from our galaxy at all wavelengths.

Grote Reber, one of the two pioneers of radio astronomy, discovered the first radio galaxies in 1940, using a 31-foot antenna that he built with his own hands and at his own expense, in his backyard. Reber observed three separate sources, located in the directions of the constellations Cassiopeia, Sagittarius, and Cygnus. The source of the radio signals from Cassiopeia turned out later to be a supernova remnant—a cloud of agitated gas, blown out into space in the aftermath of a supernova explosion in the Milky Way some time ago. The Sagittarius source turned out to be the center of our galaxy. Subsequently it became known that all normal galaxies, including ours, emit a small amount of radio waves. The radio emission from normal galaxies was not mentioned previously because it is millions of times weaker than the emission in the visible region from a normal galaxy.

The third radio source observed by Reber, located in the direction of the constellation Cygnus, was another galaxy one billion light-years from us. This source was far more interesting, for when allowance was made for its great distance, the relatively faint radio signals detected by Reber turned out to be equivalent to an outpouring of 2.3×10^{45} ergs/sec of energy at radio wavelengths. That emission is six times greater than the total amount of energy radiated into space from the stars in our galaxy at all wavelengths, and several million times greater than the radio emission from our galaxy.

Photographs of the sky taken with the Palomar Mountain telescope reveal that at the center of the radio source discovered by Reber, called Cygnus A, is a strange-looking object whose shape has not been explained (Figure 12.1), although there is evidence that its bifurcation is due to a bisecting ring of dust, as in the case of NGC 5128 (page 285).

Thousands of radio sources have been discovered since Reber's time. Some look like two galaxies in collision, as in the case of NGC 4038/9 (Figure 12.2), while others are distinguished by enormous jets of matter that shoot into space from the center of the galaxy. An example is M87, (Figure 12.3), in which one jet is clearly visible. Some astronomers believe that a second jet of matter is pointed in the opposite direction. M87 is located in the direction of the constellation Virgo at a distance of 30 million light-years. The main jet, clearly visible in the photograph, is 3000 light-years in length. Radio observations of very high resolution, made with the VLBI and VLA instruments described in Chapter 5, also indicate the presence of narrow jets of energetic particles and radiation emerging from the centers of several radio galaxies (Color Plate 14).

A third example of a strong radio galaxy is NGC 5128, also called Centaurus A, which is shown opposite the opening page of this chapter. This object presents one of the strangest appearances in the sky. A luminous body, looking like a normal elliptical galaxy, is bisected by a broad lane of dark matter. The dark lane was once thought to be a spiral galaxy, viewed edge-on, in the process of colliding with an elliptical galaxy behind. However, recent measurements of the velocity of the gases in NGC 5128 suggest that this object is not the result of a collision, but that the dissecting dark lane is the central disk of the galaxy, formed rather late in its evolution.

FIGURE 12.3
The radio galaxy M-87 in a short exposure showing the massive jet of ejected material.

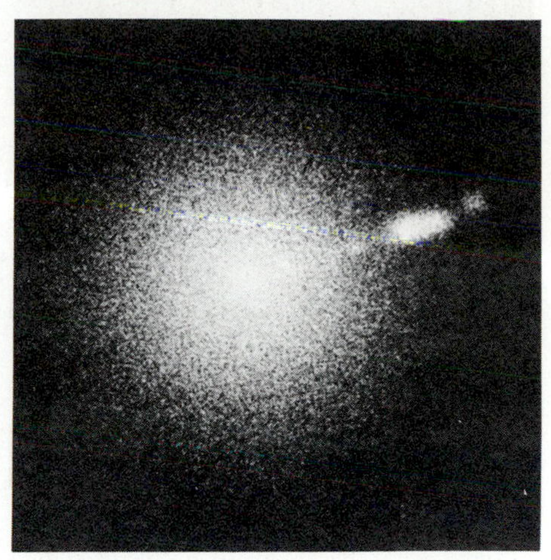

The Twin-Lobed Shape of a Radio Galaxy

The radio image of a galaxy can be formed by plotting the strength of the signals received from each part of the galaxy on a map of the sky. The radio image usually exceeds the diameter of the visible galaxy. In most cases the radio emission comes from two large regions, one on each side of the visible galaxy, and each situated at a considerable distance from the visible object.

The size of each region and the separation of the two regions from each other vary from one radio galaxy to another. On the average, the two radio sources are situated about a million light-years apart, and each source is about 300,000 light-years in diameter. The visible galaxy always lies on or near the line between the centers of the two radio sources.

The Cygnus radio galaxy is an example. A photograph of the visible object associated with this galaxy was shown in Figure 12.1. Figure 12.4 shows the double pattern of radio signals emitted from the Cygnus galaxy, with the contours shaded in intensity in proportion to the strength of the radio emissions. The visible galaxy, or pair of galaxies, is midway between the centers of the two radio sources (arrow).

The radio source associated with NGC 5128 or Centaurus A is also a double source, as the diagram of radio contours in Figure 12.5 demonstrates. A photograph of NGC 5128 has been superimposed on the diagram with the correct location and size to properly relate the optical object to the radio sources on either side of it.

What is the explanation for the twin-lobed appearance presented by the typical radio galaxy? The hypothesis tentatively accepted

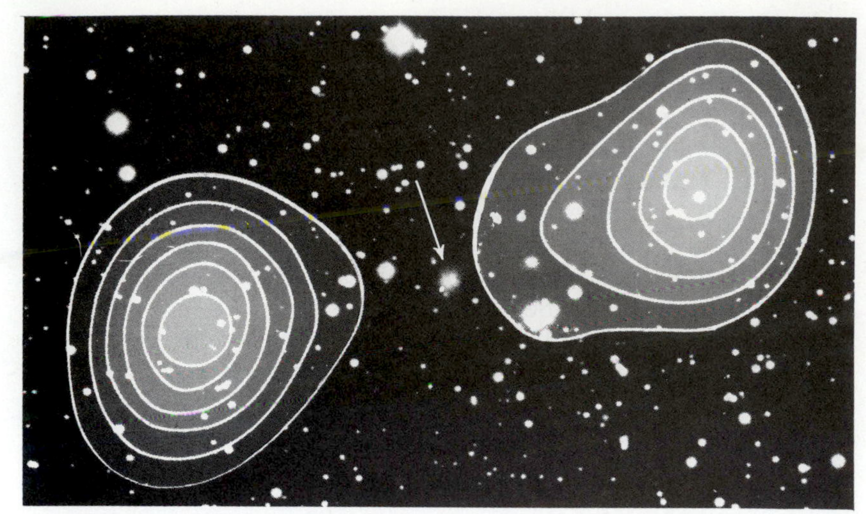

FIGURE 12.4
The radio image of Cygnus A superimposed on a photograph of the galaxy (arrow).

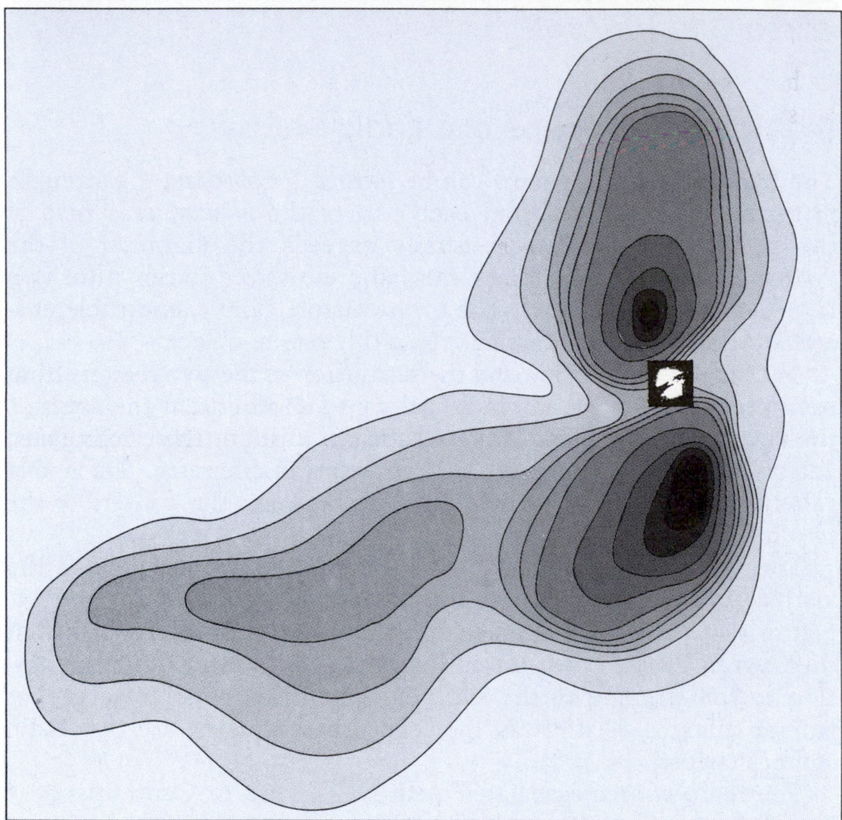

FIGURE 12.5
The radio image of NGC 5128 or Centaurus A.

by astronomers is that violent events are occurring in the centers of the visible galaxy, ejecting two vast clouds of matter that expand rapidly outward in opposite directions from the scene of the disturbance. These two outward-moving clouds of matter betray their presence by the powerful radio signals they emit. The twin-lobed radio image of a galaxy such as Cygnus A or NGC 5128 is, according to this theory, the boundary of the double cloud of matter ejected from the visible galaxy.

In the photograph of the radio image of NGC 5128, the two clouds of matter, marked by their radio emissions, might be expected to be symmetrical with respect to the position of the galaxy. However, the lower cloud of matter appears to be streaming out far to the left, and to a lesser degree the upper cloud of matter also appears to be distorted to the left. The explanation of this asymmetry is thought to be that NGC 5128 is moving to the right through the intergalactic medium at a high speed. The resistance of the intergalactic medium tends to sweep the clouds of matter to the rear—that is, to the left.

Explanation of Radio Emission

Why do expanding clouds of matter generate radio waves? The answer is believed to be that conditions within each cloud are chaotic and disturbed. As a consequence, many atoms are stripped of their electrons, so that matter within each cloud contains a large number of free electrons. It can be assumed that a magnetic field is present in the cloud, since magnetic fields are commonly present in galaxies.[1] One of the laws of electromagnetism states that an electrically charged particle moving in a magnetic field must travel in a circle. If the circular path of the charged particle is viewed edge-on, it appears to be vibrating up and down. In Chapter 5 we saw that a vibrating electric charge generates a train of electromagnetic waves. This must be the case for the circling electrons in the two radio-source regions around a radio galaxy (Figure 12.6).

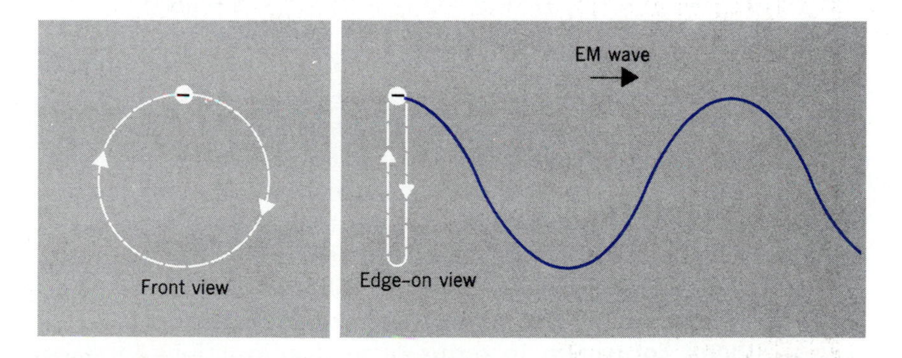

FIGURE 12.6
The emission of an electromagnetic wave by a circling electron.

[1] Although these magnetic fields are relatively weak—no more than one ten-thousandth of a gauss or less than a thousandth of the earth's field—they play a critical role in the explanation of emission from radio galaxies.

The stripped atoms carry a positive charge and also generate electromagnetic waves, but because of their greater mass and inertia they vibrate at much lower frequencies that cannot be detected. Nearly all the radio-wave emission is produced by the electrons.

FIGURE 12.7
A Seyfert galaxy (NGC 4151), photographed with successively longer exposures from (a) to (c).

(a)

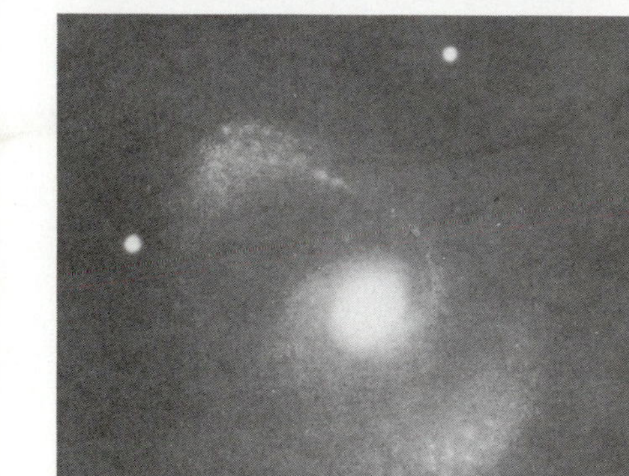

(b)

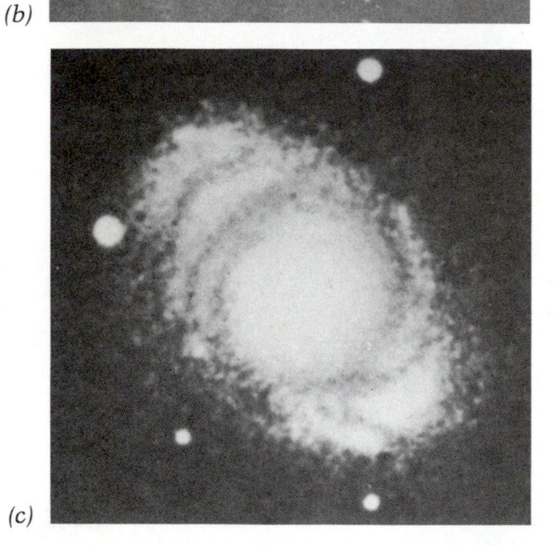

(c)

SEYFERT GALAXIES

Seyfert galaxies were named after their discoverer, Carl Seyfert, who first identified them and pointed out some of their unusual properties in 1944. Seyfert galaxies, like radio galaxies, have properties that cannot be explained readily in terms of the emission of radiation from a multitude of stars. They are distinguished by the fact that their total emission of energy at all wavelengths is extremely large—many times greater than the total emission of energy from an ordinary galaxy like ours. Moreover, this energy is emitted from an extremely small and brilliant central nucleus.

It was the property of an exceptionally small and brilliant nucleus that originally attracted the attention of Seyfert to these objects. He noted that, although this group of galaxies has a spiral structure resembling the Milky Way Galaxy, their nucleus appears to be far smaller than the nucleus of a typical spiral galaxy. As noted, the nucleus is also unusually brilliant; its total output of visible energy being greater than the output from our entire galaxy.

Because of the brilliance of the central nucleus in a Seyfert galaxy, when the Seyfert galaxy is photographed with a short exposure, the nucleus is the only object to appear on the photograph. It looks like a small, fuzzy, starlike object (Figure 12.7*a*).

When the exposure is lengthened, the spiral arms in the disk begin to appear (Figure 12.7*b*).

With a very long exposure, the spiral-arm structure becomes fully visible, and the fuzzy, starlike object is revealed as a complete spiral galaxy (Figure 12.7*c*).

Seyfert galaxies emit up to 100 times more energy than our galaxy. That fact presents the basic problem: what is the source of the enormous energy output from a Seyfert galaxy?

You might think at first that Seyfert galaxies emit more energy simply because they are larger than ordinary galaxies, and contain more stars. But this is not true, for the typical Seyfert galaxy—or rather the nucleus of the galaxy, which emits most of its energy—is less than a light-year in diameter. This is far less than the diameter of our galaxy. The energy emitted from Seyfert galaxies seems difficult to explain in terms of nuclear reactions in stars. They are an example of the objects mentioned at the beginning of this chapter, that contain mysterious sources of energy confined to extremely small volumes of space.

The emission from a Seyfert galaxy has other peculiar properties. In particular, the energy output sometimes varies by a factor of two or more over times as short as hours and even minutes. Individual stars vary rapidly in luminosity in this way, but not large ensembles of stars such as one would expect to find in the nucleus of a normal galaxy.

QUASARS

Quasars are another group of extraordinary objects that release enormous amounts of energy, even exceeding the energy released by Seyfert galaxies. As in the case of Seyfert galaxies, the powerhouse for this energy is confined to a tiny volume, in some cases as small as a few light-hours. Quasars are the most difficult objects in the sky to explain in terms of the properties of large numbers of stars clustered in galaxies.

How Quasars Were Discovered

The discovery of quasars is a fascinating vignette in modern astronomy. The objects that later came to be known as quasars had been showing up on photographic plates for many years as ordinary, relatively faint stars which, everyone assumed, were probably located in our galaxy. Around 1960 it was first noticed that these otherwise ordinary stars appeared to be the sources of radio waves. While radio waves are frequently observed to come from galaxies and nebulas, no star had been observed to emit strong radio signals. In fact, it seemed impossible to explain how radio waves could be generated in large amounts in any ordinary star.

The strange new objects were dubbed "radio stars." Following the normal procedure in investigating a new phenomenon in the sky, astronomers set about getting the spectra of some of the "radio stars." By 1963, observers at Palomar Mountain had succeeded in obtaining the spectra of two of the "stars." Astronomers expected that the spectra would be somewhat unusual, but their expectations were surpassed, for the spectra of the "stars" were different from the spectrum of any star previously observed. The wavelengths of the lines in the spectra did not agree with the wavelengths of the lines for any known element ever observed on the earth or in the heavens.

Did the "radio star" contain an exotic substance, never before seen? Physicists knew enough about atoms and nuclei to be sure that this was almost impossible. But what was the explanation?

The Quasar Red Shift

Maarten Schmidt, an astronomer at Palomar Mountain, found the answer. He looked at the pattern of the lines in the spectrum of 3C273—the brightest of the "radio stars"—and noticed that the spacing of these lines resembled the spacing of the lines in the Balmer spectrum of the hydrogen atom, but their wavelengths were different. All the lines in the 3C273 spectrum were shifted toward the red end of the spectrum relative to the Balmer lines. Schmidt wondered whether the mysterious spectrum of 3C273 might be the familiar Balmer spectrum of hydrogen, but with the wavelength of each line displaced toward the red by the same proportionate amount. This would keep the pattern of the lines unchanged, but move each individual line to a new location at a longer wavelength.

Schmidt checked his idea by comparing the 3C273 spectrum with the Balmer spectrum. He found by trial and error that he could multiply the wavelengths of the Balmer lines by 1.158 and obtain the wavelengths of the lines in the 3C273 spectrum. For example, the second line in the Balmer series has a wavelength of 4861 Å in the blue region of the spectrum. Schmidt multiplied 4861 by 1.158 and obtained a new wavelength of 5632 Å, which would be in the yellow-green part of the spectrum. There is a yellow-green line in the spectrum of 3C273 at just this wavelength. The third line in the Balmer series has a wavelength of 4340 Å, placing it in the violet. Multiplying 4340 by 1.158, Schmidt obtained a new wavelength of 5026 Å, which would be in the blue-green part of the spectrum. There is a blue-green line in the spectrum of 3C273 at precisely 5026 Å. In the same way, Schmidt was able to match the fourth Balmer line to the fourth line in the 3C273 spectrum. Figure 12.8 shows three of the shifted Balmer lines in the spectrum of 3C273, compared to the same lines in an unshifted hydrogen spectrum.

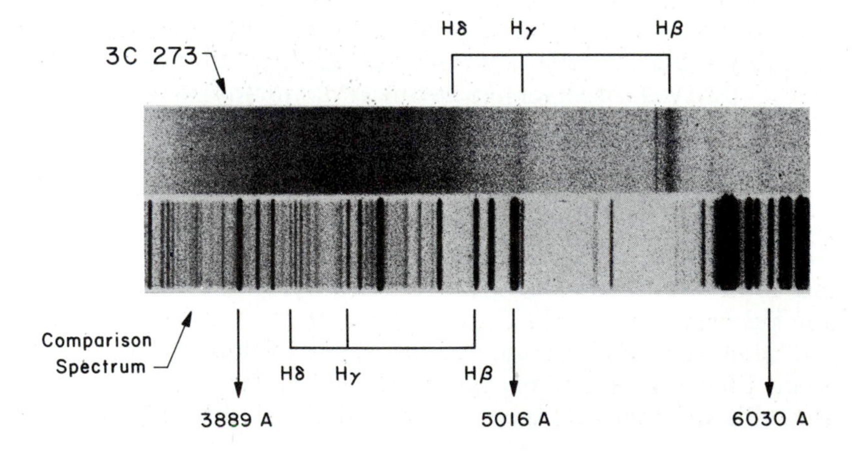

FIGURE 12.8
The spectrum of the Quasar 3C273, compared with an unshifted spectrum.

The precise agreement of four distinct lines in the two spectra was not likely to be a coincidence. Clearly, the unusual lines in the 3C273 spectrum were nothing more than the Balmer spectrum shifted to the red by 15.8 percent.

Quasar Distances

What was the cause of the shift to the red in the spectrum of 3C273? Astronomers had observed red shifts like this in other spectra, and had a ready explanation. Whenever a star or a galaxy is receding from the observer at a high speed, its light is shifted in wavelength toward the red end of the spectrum by an amount proportional to its speed of recession. This shift in wavelength is called the Doppler effect. It was explained in Chapter 5. Applying the theory of the Doppler shift to 3C273, Schmidt concluded that this "star" was moving away from the earth at the enormous speed of 28,400 miles per second. If Schmidt's interpretation was correct, the other "radio stars," which had larger red shifts than 3C273, must be moving still faster. Some of them, in fact, had red shifts that indicated that they must be moving at speeds greater than 90 percent of the speed of light.

If the "radio stars" were moving this fast, they must be relatively distant objects. They could not be in our own galaxy, since if they were, their rapid motion across our line of sight would cause their positions to change drastically from year to year against the background of the fixed stars. Yet some of these "stars" had been observed for years without displaying any noticeable change in position.

Perhaps the radio stars were outside our galaxy but located in other galaxies quite close to us, so that they could be picked out as individual stars in these galaxies, but were far enough away so that the apparent motion of these stars would be small. Unfortunately for this idea, none of the "radio stars" seemed to be connected with any of the galaxies in our neighborhood. Thus Schmidt's brilliant explanation of the spectrum of 3C273 eliminated the original mystery, only to replace it with another: Why did these "stars" have such a large red shift?

Quasar Luminosities

As we will see in Chapter 13, astronomers have discovered a relationship between the red shift or speed of recession of an object and its distance from us. According to this relationship, an object receding from the earth at a speed of 28,400 miles per second must be 2 billion light-years away. If the relationship also holds true for the "radio stars," 3C273 must be 2 billion light-years from us. Knowing the distance of 3C273, we can calculate its absolute magnitude or luminosity from its apparent magnitude, which is +13. An object of the 13th magnitude, located at a distance of 2 billion light-years, turns out to have a luminosity of 10^{46} ergs/sec or roughly 50 times the luminosity of our galaxy.

All the "radio stars" for which spectra could be obtained turned out to have very large red shifts; hence, they were apparently billions of light-years distant. Therefore, when the apparent magnitudes of these objects were translated into absolute magnitudes, all turned out to be extremely luminous.

But an object with an energy output of 10^{46} ergs/sec over an extended period cannot be a single star. Stars have energy outputs that are typically many billions of times smaller. The only known objects with luminosities comparable to this over long times are galaxies. In fact, some quasars have luminosities considerably greater than typical spiral galaxies like the Milky Way or Andromeda. It looked as though these "radio stars" were not stars at all, but galaxies located very far away, so that the details of their galactic structures were not visible in photographs, and therefore they looked like stars.

Since the strange objects looked somewhat like stars, but were not stars, they became known as "Quasi-Stellar Sources," or QSS's, soon shortened to quasars.

After the discovery of the first quasars, astronomers found numerous other starlike objects that had similar optical properties—that is, very large red shifts, and unusual emission spectra—but were not strong radio sources. These objects received the separate name of "Quasi-Stellar Objects," or QSO's. Today the two objects are usually grouped together under the common label of quasars.

These arguments, and others such as that on page 300, have convinced many astronomers that quasars are exceedingly distant and, therefore, exceedingly luminous objects.

FIGURE 12.9
Quasar 3C196.

The Visible Image of a Quasar

As we have seen above, although 3C273 and other quasars are intensely luminous objects, emitting far more energy than our galaxy, they appear as very faint objects in the sky because of their great distance from us. None of the quasars is visible to the naked eye, nor can they be photographed except with the aid of large telescopes. The photographs reveal them to be somewhat fuzzy starlike objects. Figure 12.9 shows quasar 3C196, located approximately 9 billion light-years from us.

An interesting feature appears in photographs of the closest and brightest quasar, 3C273. In the photograph of 3C273 in Figure 12.10, obtained with the 200-inch telescope, a jet of matter may be seen extending to the lower right. This jet, whose presence has not been explained, is 250,000 light-years long.

FIGURE 12.10
Quasar 3C273. The image is larger than the image of 3C196 because the plate has been exposed to reveal the faint jet.

The Radio Image of a Quasar

When quasars were thought to be nearby "radio stars," they were regarded as relatively weak radio sources. As soon as their great distances were established by the measurement of their red shifts, it was realized that they must be extremely powerful emitters of radio energy.[2]

When the strength of their radio emissions is corrected for their distance from us, some quasars outrank the most powerful radio galaxies known. The question immediately arises: Do quasars, as radio sources, have the striking twin-lobed appearance of a typical radio galaxy? The answer is that many of them do. An example is Quasar 3C47, located 4.5 billion light-years from our galaxy. The plot of contours of equal radio brightness from this quasar, shown in Figure 12.11, is its radio image. The "X" placed on the radio image marks the position of the optical object. As in the case of radio galaxies, the radio image is a double source, with the optical object located close to the line running between the centers of the two sources. The centers of the two radio sources in the radio image are one million light-years apart.

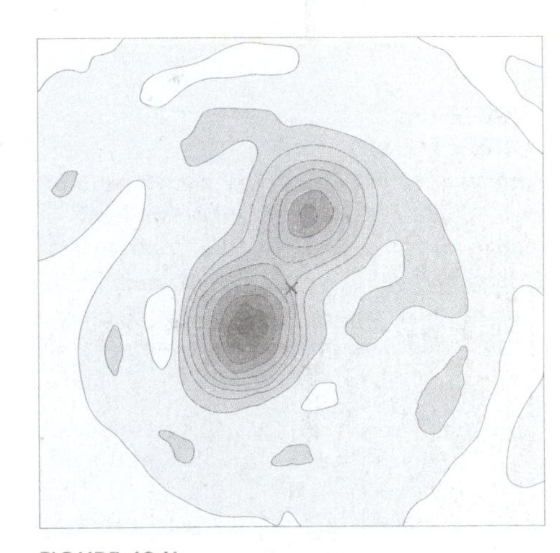

FIGURE 12.11
The radio image of Quasar 3C47. As with radio galaxies, the radio image of the quasar consists of two sources lying on either side of the optical object.

The Small Sizes of Quasars

Quasars, like Seyfert galaxies, combine their exceptional luminosity with a very small size. In fact, the energy emitting region in typical quasars is even smaller than in Seyfert galaxies. The light-emitting region in Quasar 3C273 is less than a light-year in diameter, and in some quasars it appears to be as small as a few light-hours.

The evidence for the small size of quasars comes from the fact that large changes in their brightness have been observed over short periods of time. Two examples of the variations in the luminosity of quasars are shown in Figure 12.12. Figure 12.12*a* shows the measured intensity of Quasar 3C454.3. At one point, marked by the arrows, a change of a full magnitude occurs over an interval of less than one week. Figure 12.12*b* shows larger changes in the intensity of X-rays from Quasar 3C273 in periods of a few hours.

[2] The remarks in this section refer only to quasars that are strong radio sources. Many quasars do not emit a detectable radio signal.

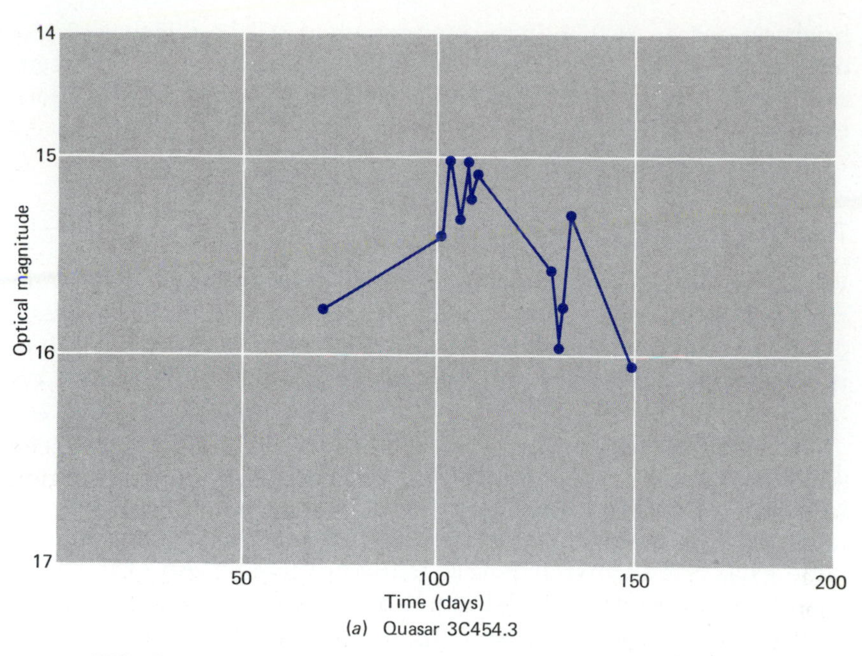

FIGURE 12.12(a)
Rapid time variation in the energy output from Quasar 3C454.3.

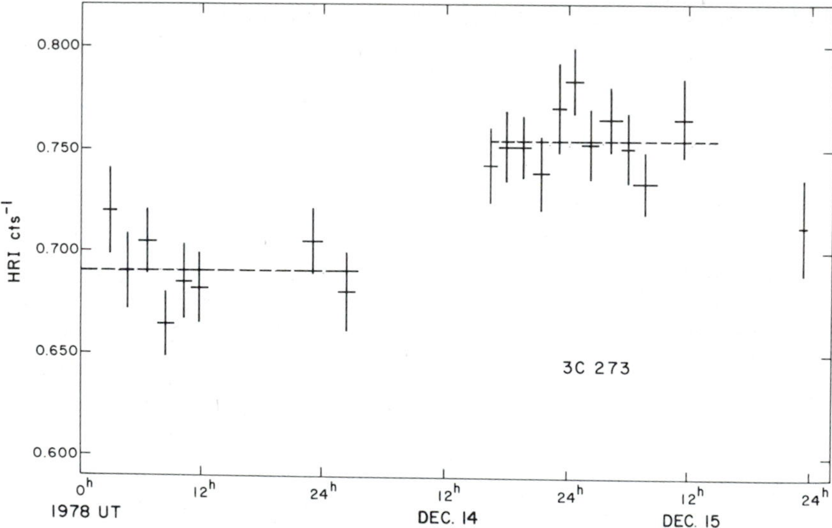

FIGURE 12.12(b)
Variation in the intensity of X-rays emitted from Quasar 3C273 in 1978, showing a change of approximately 10 percent in the 12-hour period from midnight of December 14 to noon of December 15. (P. Henry, Harvard-Smithsonian Astrophysical Observatory.)

To see why these rapid variations mean that quasars are very small, imagine an orchestra so large that the back row is a mile behind the front row. At the speed of sound, it takes five seconds for a note to travel from the back to the front of this orchestra. When the conductor drops his baton to signal the opening, members of the orchestra begin to play in unison. However the notes from the different sections of the orchestra reach the audience as much as five seconds apart because of the limited speed of sound. As a consequence, the sounds that reach the audience are blurred. In other words, a very large orchestra cannot play a crisp note.

Now replace the orchestra with a quasar that emits light as the orchestra emits sound. Suppose, for example, the quasar is 100 light-years across. This means that the packets of light emitted from different parts of the quasar can reach an observer on the earth as much as 100 years apart. If the entire quasar changes its brightness suddenly, in a matter of minutes or hours, the observer on the earth will not see the sudden change; he will see this change develop slowly over the course of 100 years. That is, the quasar cannot emit a "crisp note" because of the limited speed of travel of light.[3]

Thus, if the visible light output from a quasar is seen to vary in a few weeks, that indicates that the light-emitting part of this quasar must be no more than a few light-weeks in size. The fact that the X-rays from a quasar have been found to vary in hours, as was seen in the case of Quasar 3C273, means that the X-ray emitting region of this quasar cannot be more than some light-hours in size.

The size of our solar system is only a few light-hours. It is remarkable that Quasar 3C273 should be emitting the energy of billions of suns from a region not much larger than the solar system.

This extraordinary combination of properties—an enormous output of energy, combined with an exceedingly small size—makes the quasar one of the most puzzling objects ever discovered by astronomers.

ARE QUASARS GALAXIES?

Quasars emit energy comparable to the energy emitted by galaxies. Is it possible that they *are* galaxies? At first, it seems obvious that they cannot be, because of their very small size. But in spite of this fact, many astronomers think quasars may belong to the family of galaxies after all.

Connection Between Quasars and Seyfert Galaxies

The first line of evidence leading to that conclusion is the fact that the properties of quasars differ from those of Seyfert galaxies only in degree, and Seyfert galaxies, in turn, seem to differ only in degree from ordinary galaxies. In other words, if ordinary galaxies, Seyfert galaxies, and quasars are arranged in a sequence, the Seyfert galaxies bridge the gap between ordinary galaxies and quasars in all important respects. With regard to *brightness,* for example, the brightest Seyfert galaxies are as bright as some quasars while the least bright

[3] We are indebted to Professor John Thorstensen for this analogy.

FIGURE 12.13
This sequence of photographs illustrates the transition in appearance from the photograph of a quasar (a) to the photograph of a Seyfert galaxy (b) and a normal galaxy (c).

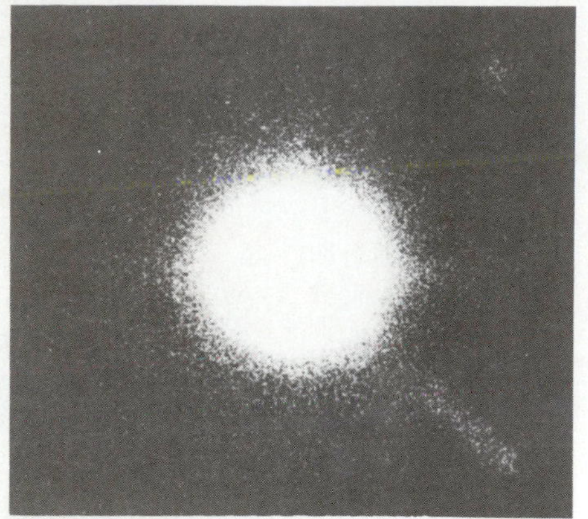

(a) Quasar 3C273

(b) Seyfert galaxy NGC4151

(c) Spiral galaxy NGC488

Seyfert galaxies have about the same brightness as ordinary galaxies. With regard to visual appearance, Seyfert galaxies also tend to bridge the gap between quasars and ordinary galaxies, for, as we noted above, if a Seyfert galaxy is photographed with a short time exposure, the only part of it that shows up on the photographic plate is its brilliant nucleus, which appears as a fuzzy, starlike object, very much similar to a quasar; but as the exposure time is increased, we see that the Seyfert galaxy comes to resemble a normal spiral galaxy. The sequence in Figure 12.13 shows the smooth transition in visible image from quasars to Seyfert galaxies and then to spirals.

Finally, the similarity between quasars and radio galaxies with respect to their twin-lobed radio images also suggests that quasars belong to the family of galaxies (see page 295).

As a consequence of this evidence, a consensus is emerging among astronomers that quasars are galaxies that happen to have unusually bright nuclei.

Direct Evidence for Galaxies Surrounding Quasars

The notion that quasars are unusually bright nuclei of galaxies was confirmed in 1982 by a team of astronomers at the California Institute of Technology. Using a sensitive electronic imaging device (a charge-coupled device; see Chapter 4) in place of photographic plates, they showed that a faint glow of starlight surrounds Quasar 3C48 and extends out a substantial distance from the quasar itself. The starlight comes from two wisps of luminosity on either side of the quasar.

Because of the sensitivity of the new instrumentation, the astronomers were able to obtain the spectrum of the light from the faintly luminous regions near the quasar. The spectrum revealed characteristic lines that are commonly found in the spectra of hot, young stars. Furthermore, the spectral lines in the luminous wisps had very closely the same red shift as the quasar itself, proving that the luminous wisps are at the same distance as the quasar, and not just close to the same line of sight. These results indicate that Quasar 3C48 lies at the center of a galaxy.

The first quasar ever discovered, 3C273, was also discovered in 1982 to have a faint glow of nebulosity around it (Figure 12.14). The intensity of the glow decreased from the center outward in the manner expected for the light from an elliptical galaxy. As in the case of Quasar 3C48, the light from the quasar in 3C273 had very closely the same red shift as the light from the surrounding nebulosity, proving that the quasar and the nebulosity belonged to the same object.

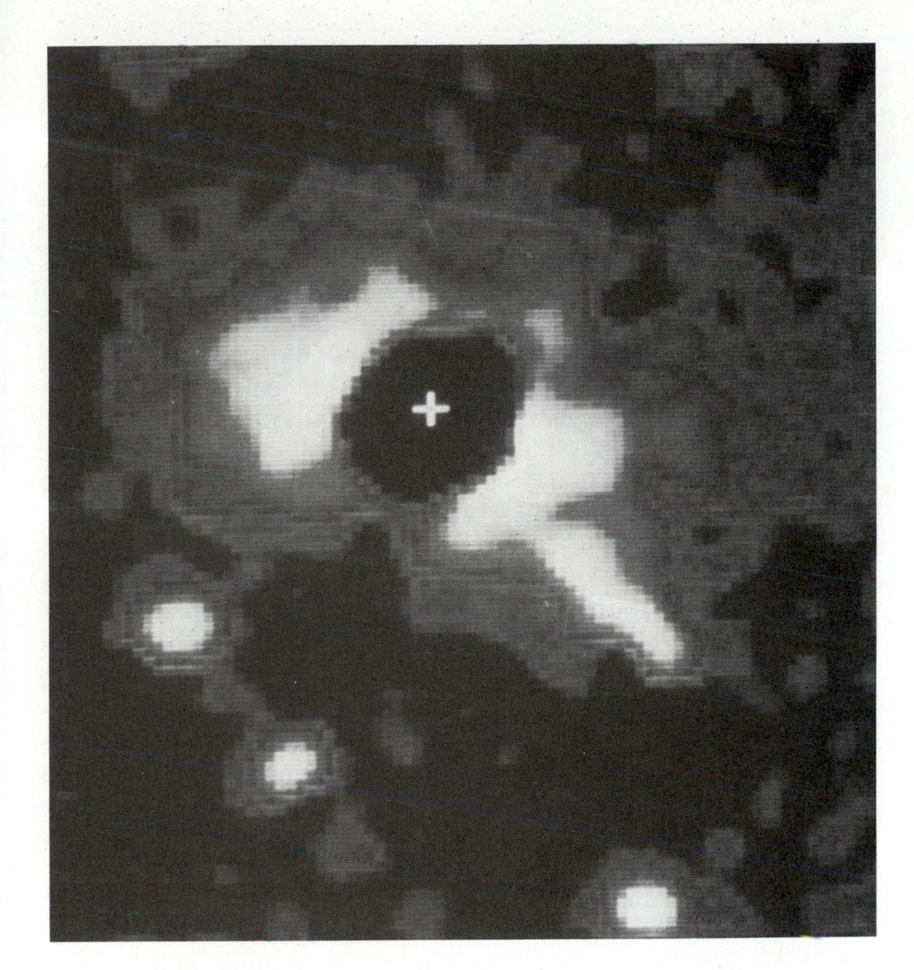

FIGURE 12.14
The nebulosity surrounding Quasar 3C273. Light from the quasar at the center, marked by the cross and the black oval, was blocked off, allowing the faint surrounding region to emerge.

THE NATURE OF THE QUASAR ENERGY SOURCE

Now that the connection between quasars and galaxies has been fairly well established, one interesting question still remains. What is the nature of the quasar powerhouse? How can a very small region in space produce such an enormous output of energy?

One possible answer is a very dense galactic nucleus in which stars collide frequently. When two stars collide they may coalesce into one large star, and the coalescence of small stars into larger ones, in turn, may lead to a high frequency of supernovas. One supernova per day, exploding in a dense galactic nucleus, would account for all but the brightest quasars. The energy released by the supernovas would be bolstered by the energy emitted from the pulsars or neutron stars found at the centers of some of these exploding stars.

Massive Black Holes as Energy Sources

Another possibility is that quasars are galaxies with massive black holes at their centers. A black hole could explain the enormous output of energy from a quasar in the following way. The black hole's gravity would draw in gaseous matter, and entire stars as well, from the space around it in the galaxy. As this material moved inward toward the black hole, it would pick up speed under the pull of gravity. Some of the material, moving on a collision course with the center of the black hole, would disappear into its interior; other material, not on a direct collision course, would be attracted to the vicinity of the black hole and circle it at a distance. This circling material would form a halo of hot gas surrounding the black hole. The gas would be very hot because it was accelerated to high speeds as it drew close to the black hole. Being so hot, the gas would radiate an enormous amount of energy. In this way the black hole would become the powerhouse of the quasar.

Why would a black hole form at the center of a galaxy? One answer is that it could form directly, by the gravitational collapse of a dense cloud of gas at the center of the galaxy. Since the centers of galaxies are apt to be especially dense concentrations of matter, this is a plausible assumption. Another possibility is that the stars at the centers of galaxies are closely packed together, and collisions between them are therefore relatively frequent. These collisions would cause the stars in the galactic nucleus to coalesce into a smaller number of massive stars. Finally, each massive star would come to the end of its life and form a black hole. In the course of additional time, the black holes would collide with each other. In each collision, the larger black hole would swallow the smaller one, and itself become larger in the process. As a black hole became larger in this way, its chance of swallowing other black holes would become still greater. Eventually, all the black holes in the center of the galaxy would coalesce into a single massive black hole sitting at the center of the galaxy, containing the mass of millions of separate stars, and pouring out enormous amounts of energy as it drew in surrounding gas and stars. A third possibility is that a stellar-sized black hole would form at the galactic center by the death of a massive star, and then grow into a massive black hole over the next few billion years by swallowing up gaseous matter.

Evidence for the Black-Hole Theory of Quasars

Because a black hole, by its nature, is extremely compact, the black-hole theory satisfies one of the two basic requirements for an explanation of quasars. Another reason for favoring massive black holes as an explanation of quasars is the fact that a black hole can produce enormous amounts of energy. Calculations show that the gravitational pull of a black hole will accelerate nearby gaseous matter to speeds close to the speed of light, as the gas spirals in toward the black hole. As a result, as much as 20 percent of the rest mass of the particles in the stream of gas can be converted to energy. In nuclear reactions, such as those that produce the energy in stars, less than one percent of the rest mass of the reacting nuclei is converted to energy. In other words, black holes as energy-producing machines are at least 20 times more efficient than stars. Short of the annihilation of matter and antimatter, which converts 100 percent of the rest mass into energy, nothing known to science compares with the effectiveness of the black hole as a cosmic powerhouse.

The Mass of the Black Hole. How massive must a black hole be to produce the energy output observed for quasars? Several lines of argument suggest that the mass must be in the neighborhood of 100 million to a billion solar masses. This conclusion is in agreement with the evidence on the size of the energy-emitting region in quasars, because a black hole with a mass of a billion suns will have a radius of 1.6 billion miles, according to the formula on page 235. In other words, it will be about 5 light-hours in diameter. The measurements on the rapid variation of X-rays from quasars indicate that the size of the quasar energy source may be in just this range.

Thus, massive black holes with masses ranging up to about a billion solar masses meet two important requirements for an explanation of quasars: they are very compact, and they are very powerful energy sources.

Further Observational Support. The black-hole theory of quasars leads to another prediction that agrees with observation. If a quasar is a massive black hole, the inner regions of the quasar, close to the boundary of the black hole, will be at the highest temperature. Calculations show that the gas in this inner region should have a temperature of millions of degrees or more. Therefore, the radiation it emits should have predominantly short wavelengths, in the X-ray region of the spectrum.

Farther from the center of the quasar, and therefore farther from the black hole, the surrounding gas should be somewhat cooler, and the wavelength of the radiation it emits should be somewhat longer. Thus an observer would detect radiation at ultraviolet and visible wavelengths from these regions. At still greater distances, the temperature should be still lower, and the emitted radiation should have wavelengths in the infrared region. Finally, still farther out, the flow of energy should disturb the charged particles in the surrounding medium, producing radio waves as described on pages 291–293.

These points of agreement between the observed properties of quasars and the predictions of the massive-black-hole theory have led to growing support among astronomers for the view that quasars may be massive black holes located at the centers of otherwise "normal" galaxies.

THE SIGNIFICANCE OF QUASAR DISTANCES

Quasars have another special property in addition to their small size and enormous energy output: *All quasars are relatively far away.* Most quasars are billions of light-years from us, and even the very closest quasar observed thus far is at the considerable distance of 800 million light-years. What is the significance of this fact?

One possible explanation is that the earth is in a special region in the Universe that happens to contain no nearby quasars; that is, it is in the center of a hole in the quasar distribution, so to speak. However, science since the time of Copernicus has rejected the idea that the earth occupies a special place in the Universe: in fact, most astronomers believe that no place in the Universe is different, in a large sense, from any other place.

Another possible explanation for the great distance of quasars is connected with an important fact in astronomy: *When you look out into space, you look back in time.* This is so because light travels at a finite speed and takes a finite time to travel across space. Thus, when we see an object that is a billion light-years away, we are observing light that was emitted by that object one billion years ago. During the past billion years, this light has been traveling steadily across space towards the earth at the rate of approximately 6 trillion miles a year. When the light from the object finally reaches the earth and is collected by telescopes, we see the object not as it is today, but as it was a billion years ago.

Thus, the fact that the most powerful quasars are very far away may mean simply that they existed mainly in the remote past, so the only way we can still see them is by looking far back into the past—that is, by looking far out into space.

But why should quasars have existed mainly in the past? The black-hole theory of quasars provides a possible answer. According to that theory, a quasar produces a large amount of energy only as long as matter exists around it, close enough to be drawn in toward the central black hole and heated to very high temperatures. After a time, the black hole will presumably clean out its galaxy, stripping it of all available matter. Now, without a halo of hot, radiating matter around it, the black hole becomes quiescent. It still sits in the center of the galaxy, but it is no longer an intense energy source. It is no longer an active quasar.

In other words, a quasar may have a limited lifetime as a highly luminous object. It is very bright for a while, and then it fades away.

If we cannot see any nearby quasars at the present time, the reason may be that they all flared up early in the history of the Universe, consumed the stars and gas around them, and then became quiescent, so that now they are practically invisible. The only quasars we can still observe are those far enough away so that we see them as they once were in the distant past.

If this is so, it follows that many nearby galaxies may have relatively inactive black holes at their centers, that may have been fully luminous quasars at one time. Seyfert galaxies provide some support for this idea. Seyferts look like ordinary galaxies in some respects, and in other respects—mainly their very luminous central regions—they also resemble quasars. Perhaps a Seyfert is a galaxy whose central black hole is in a period of transition toward an inactive state, and is therefore intermediate in luminosity between a quasar and an ordinary galactic center.

The idea that galaxies with massive central black holes are commonplace has been given some further support by recent observations of the centers of several radio galaxies. The observations, which use the extraordinarily high resolution provided by the VLBI and VLA instruments (see Chapter 5), reveal intense streams of particles and radiation coming from the centers of these galaxies. The particles are concentrated in two very narrow jets of hot matter that appear to be squirting out of a compact object located in the galactic center. Color Plates 14 and the cover show some of these spectacular high-resolution images. Calculations on the properties of spinning black holes predict just this phenomenon—two narrow jets of fast-moving particles coming from a rotating black hole, ejected in opposite directions from the poles of rotation.

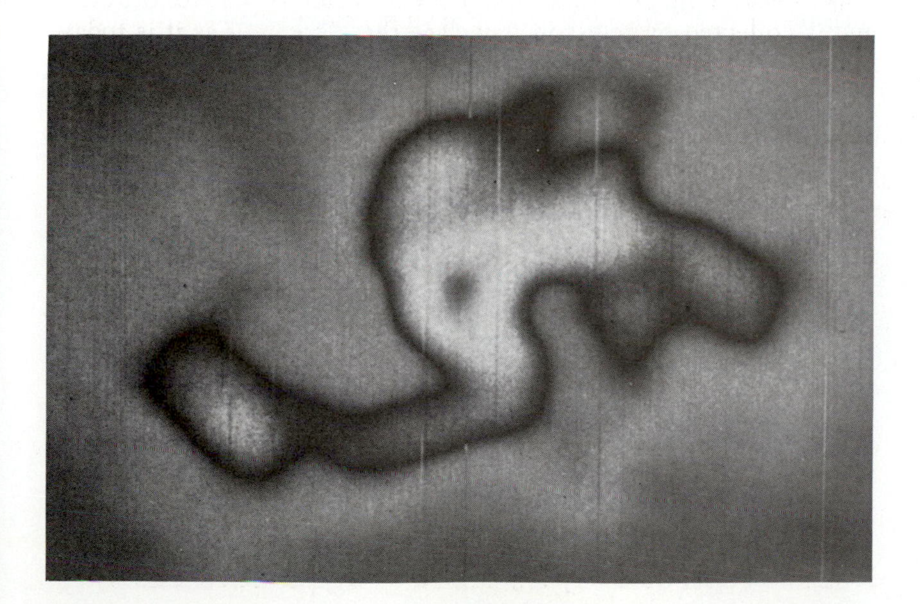

FIGURE 12.15
A radio image of the center of the Milky Way Galaxy, obtained with the VLA instrument in New Mexico. Two S-shaped jets of gas are seen to emerge from an extremely compact core in the Galactic center.

The Milky Way Galaxy also shows signs of the presence of an unusual source of energy located in a compact object at the Galactic center. Observations of the Galactic center at radio wavelengths reveal rapid variations in the intensity of the source over short periods of time, indicating that the unusual energy source may be as small as a few million miles in diameter. Studies of the motions of gas clouds near the Galactic center provide additional evidence for a compact energy source. Is there a black hole in the Galactic center? The evidence is suggestive, but further observations are needed to decide the matter.

Main Ideas

1. Energetic events in the centers of galaxies.
2. Radio galaxies; their twin-lobed structure.
3. The mechanism of radio emission.
4. Properties of Seyfert galaxies.
5. Quasars; their red shifts and distances.
6. Unusual properties of quasars; intense luminosity and small size.
7. Evidence that quasars are located at the centers of galaxies.
8. The origin of quasar energies; massive black holes as quasar energy sources.
9. The meaning of quasar distances.

Important Terms

black hole
light-week
quasar
radio galaxy
radio image
redshift
Seyfert galaxy
synchrotron emission

Questions

1. What is the main characteristic of a normal galaxy?
2. Describe the radio image of a typical radio galaxy.
3. What is synchrotron radiation? How is it produced? In what kinds of objects is it important?
4. How many years have elapsed since the jet in M87 was formed, assuming that it has always been moving outward at a speed of 15,000 miles per second?
5. What properties distinguish a Seyfert galaxy from a normal galaxy?
6. What are the principal reasons for believing that quasars belong to the family of galaxies? What is the evidence that quasars are located at the centers of galaxies?

7. Explain the reasoning leading to the conclusion that quasars are extremely distant objects.
8. If the luminosity of Quasar 3C273 were produced by stars like the sun, how many stars would 3C273 have to contain to account for its luminosity? What would the average distance between the stars in 3C273 be in this case, assuming this quasar has a diameter of one light-month? Compare your answer to the diameter of a solar-mass star. Would collisions between stars be frequent or rare in this quasar? What do you think would be likely to be the product of a collision between two stars? Can you propose an alternative theory for quasar energies on the basis of your answer, making use of the information on the evolution of massive stars presented in Chapter 9?
9. Explain how a black hole can become an energy source. Briefly describe three items of observational support for the black-hole theory of quasars.
10. As we look out in space, we look back in time. For example, if Quasar 3C273 is 3 billion light-years away we see this quasar as it was 3 billion years ago. How would our galaxy look if viewed from a distance of 5 billion light-years? Fourteen billion light-years?
11. Suppose all quasars in the Universe formed about one billion years ago. What would be the average distance of quasars from us in that case?

COSMOLOGY 13

In this book we have followed astronomers as they trace the history of the basic ingredient, hydrogen, through a series of events in which, first, galaxies are formed out of the parent cloud of gaseous hydrogen, then stars are formed out of the hydrogen within each galaxy, and last, the heavier elements are formed out of hydrogen within each star. This history answers many questions about the origin of the world. It explains how stars come into being, how they obtain their energy, and how the elements of the Universe are created. But the very success of the astronomer in reconstructing the life story of the stars increases our desire to know the answers to even more fundamental questions: How did the Universe start? Who or what created the hydrogen in the Universe at the beginning? And how will the Universe end? What will happen when the supply of hydrogen is exhausted, and the old stars go out, one by one?

These matters are the domain of the field of scientific investigation known as *cosmology*. Cosmology is concerned with the nature and origin of the entire Universe—its structure today, its past, and its future. To study cosmology, you must stretch your concepts of space and time even more than you have done earlier in this book. You must adopt a point of view so broad that the tremendous life span of a galaxy seems a mere detail and the passage of a billion years is like a day.

To cosmologists, the birth of each star is a minor incident in the life of the Universe. They reflect on the innumerable births and deaths of all the stars that have existed since the Universe began, and ask themselves the meaning of this pattern of details. Does the

E. P. Hubble observing at the prime focus of the 200-inch telescope.

life story of a single star have a significance for the Universe as a whole?

When we consider that our galaxy contains more than 100 billion stars, and billions of other similar galaxies exist around us, it seems at first thought that a single star cannot possibly tell us anything significant about the entire Universe. But this conclusion is incorrect. Every star that has lived has unalterably changed one aspect of the entire Universe. This aspect is the amount of hydrogen that exists in the Universe. Hydrogen is the essential cosmic ingredient. It is the primary source of the energy by which stars shine, and it is also the source of all the elements in the Universe. As soon as a star is born it begins to consume some of the hydrogen in the Universe, and continues to use up hydrogen until its death. Once hydrogen has been burned within that star and converted to heavier elements, it can never be restored to its original state. With the passage of time, and the appearance of successive generations of stars, the supply of hydrogen in the Universe must grow smaller.

Globular clusters provide evidence for the slow change of hydrogen into heavier elements in stars. As we noted in Chapter 11, the H-R diagram for these clusters indicates that they were formed early in the history of the Universe. At that time, only a relatively few supernova explosions could have occurred and the abundance of heavy elements must have been very low. The spectra of the globular clusters confirm this idea. They show that the stars in these clusters have far less of the heavy elements than stars in the spiral arms of the Galaxy, which were formed later.

The implications in these observations are clear: Hydrogen is disappearing. As the old stars burn out, fewer and fewer new stars can be formed to replace them. Stars are the source of energy by which all beings live. When the light of the last star is extinguished, life must end throughout the Universe.

The depletion of the cosmic supply of hydrogen is the feature in the life story of stars that is most interesting to cosmologists. Minute by minute and year by year, the supply of hydrogen continually decreases as a result of nuclear reactions that occur in stars. As a consequence, the Universe is running down and changing irreversibly.

Reflecting further on the situation, the cosmologist turns the clock back in his imagination and asks himself what the Universe must have been like billions of years ago. Clearly, there must have been more hydrogen in the Universe at that time than there is today, and less of the heavier elements. At the present time approximately 75 percent of the mass in the Universe consists of hydrogen. A billion years ago this number would have been slightly different; there would have been more hydrogen and less of the heavier elements, because some of the stars that have contributed to today's abundance of those elements had not yet been born. Four and one half billion years ago, around the time when the sun and earth were formed, there would have been still more hydrogen and still less of the heavier elements. Turning the clock back still further, we would eventually come to a time when the Universe contained nothing but hydrogen—no helium, no carbon, no oxygen, and none of the other elements out of which, for example, the earth and the creatures on it are composed. This point in time must have marked the beginning of the Universe.

In other words, projecting the present conditions in the Universe backward in time, we are forced to conclude that it had a beginning, and projecting the present conditions forward in time, we can see that the Universe must eventually come to an end. That is the cosmological significance to the life story of the stars.

THE EXPANDING UNIVERSE

Other evidence suggests that the Universe has been changing in an irreversible way. Between 1912 and 1914, V. M. Slipher undertook a study of the spectrum of the light emitted by 15 nearby spiral galaxies. Familiar lines, such as the lines of hydrogen, appeared in these spectra, but in most cases the wavelengths of the lines were shifted toward the red end of the spectrum by large amounts. Slipher interpreted this red shift as a Doppler effect (pages 152–153). His measurements of the red shift indicated that most of these galaxies were moving away from the earth at high speeds, in some cases as much as several million kilometers per hour.

According to astronomical knowledge, our planet and its parent star, the sun, are indistinguishable from countless other planets and stars in the heavens. Why should all the galaxies move away from us? We would expect them to move randomly, so that at any moment half the galaxies in the Universe would be moving toward us and half would be moving away. But according to Slipher's measurements, this was not so; the entire Universe was moving away from one special point in space, and the earth was located at that point.

In the 1920s and 1930s, astronomers Edwin Hubble and Milton Humason, using the 100-inch Mount Wilson telescope, then the largest telescope in the world, succeeded in measuring the speeds of many other spiral galaxies. These galaxies were too faint to have been seen by Slipher with his smaller instrument. The observations confirmed Slipher's discovery: without exception, all the distant galaxies in the heavens were moving away from us at high speeds.

After World War II, the power of the 200-inch telescope on Palomar Mountain was brought to bear on the problem of the receding galaxies and again Slipher's discovery was confirmed; nearly every distant galaxy within the range of this instrument was retreating from the earth at an enormous speed.

Many more measurements have been made on Palomar Mountain and elsewhere down to the present day, and no exception has been found to the pattern discovered by Slipher. Regardless of the direction in which we look out into space, all the distant objects in the heavens are moving away from us and from one another. The entire Universe appears to be exploding.

Is the Earth at the Center of the Expanding Universe?

The discoveries of Slipher, Humason, and Hubble contain a puzzling implication; if all the galaxies in the heavens are moving away from the earth, our planet must be at the center of the Universe. That notion, commonly held in previous times, was challenged by Copernicus 500 years ago, and very few people accept it today. Why does modern astronomy lead to a picture of the world that was abandoned by men of science many centuries ago?

The answer is that if you were sitting on a planet in one of the other galaxies in the Universe, you would see the other galaxies around you receding in exactly the same way an observer in our galaxy sees our neighbors moving away. Your galaxy would seem to be at the center of the expansion, and so would every other galaxy; but, in fact, there would be no center.

To understand the meaning of this statement more clearly in the real world of three dimensions, imagine a very large, unbaked loaf of raisin bread. Each raisin in the bread is a galaxy. Now place the unbaked loaf in the oven; as the dough rises, the interior of the loaf expands uniformly, and all the raisins move apart from one another. The loaf of bread is like our expanding Universe. Every raisin in the interior sees its neighbors receding from it; every raisin seems, from its point of view, to be at the center of the expansion; but there is no center.

To make the analogy more precise, we would have to imagine a loaf of raisin bread so large that you could not see the edge from the interior, no matter where you were located; that is, the loaf of bread, like the Universe, would be infinite.

THE BIG-BANG COSMOLOGY

The picture of the expanding Universe has an important implication. If the galaxies are moving apart, at an earlier time they must have been closer together than they are today. At a still earlier time, they must have been still closer together. Continue to move backward in time in your imagination. The outward motions of the galaxies, reversed in time, bring them closer and closer; eventually, they come into contact; then their materials mix; finally, the matter of the Universe is packed together into one dense mass under enormous pressure, and with temperatures ranging up to trillions of degrees. The dazzling brilliance of the radiation in this dense, hot Universe must have been beyond description. The picture suggests the explosion of a cosmic bomb. The instant in which the cosmic bomb exploded marked the birth of the Universe. The corresponding theory of the evolution of the Universe, which traces its history from the moment of the explosion through subsequent aeons, is called the *Big-Bang* cosmology.

THE HUBBLE LAW

When Hubble and Humason started to follow up on Slipher's initial discovery of the moving galaxies, they divided the task between them. While Humason used the 100-inch telescope to measure the red shifts of the galaxies and estimate their velocities, Hubble concentrated on measuring their distances. His objective was to find out the pattern in their movement. Were the most distant galaxies moving away faster than the nearby ones, or slower? This connection between the distance to a galaxy and its velocity was a necessary first step toward understanding the meaning of Slipher's strange result. But the measurement of the distances to other galaxies was exceedingly difficult.

Measuring the Distances to Galaxies

In the 1920s, when Hubble and Humason began their work on the spiral galaxies, astronomers did not know what these luminous spirals were. Were they mammoth clusters of stars located at enormous distances, or relatively small clouds of gaseous matter? Until astronomers decided between these possibilities for the luminous spirals, they had no hope of deciphering the meaning in their rapid motions.

A few astronomers held the first view; they argued that the spirals[1] were island universes or true galaxies, enormously large and enormously distant, each containing many billions of stars. In their opinion, the galaxy to which the sun belonged was only one island universe of stars among many that dotted the vastness of space. But other astronomers preferred the second theory, which held that the luminous spirals were small, nearby objects—little pinwheels of gas, swirling in the space between the stars of the Milky Way.

Some proponents of this view even argued that each spiral was a

[1] At the time these objects were called "spiral nebulas" because no one knew whether or not they were true galaxies. The term "spiral galaxy" came into use later, largely as a result of Hubble's work.

newborn solar system, with a star forming in the center of the spiral and a family of planets condensing out of the streamers of gas around it.

Hubble settled the controversy. Using the 100-inch telescope, he photographed several nearby spirals with great care, and showed that each one contained enormous numbers of star-like points of light. Hubble found that among these points of light, some varied rhythmically in their light output, in the same way as a certain type of star in our galaxy called a Cepheid variable. He concluded that the points of light were indeed true stars. Hubble's observations proved that the spirals were true island universes or galaxies—great clusters of stars, very much like our galaxy.

Furthermore, since the spiral galaxies contained so many stars they must be very large; yet their apparent size, as seen in the telescope, was quite small. The implication was that they were extremely far away—far beyond the boundaries of our galaxy.

Exactly how far away were the spirals? Hubble thought that if he knew the answer to that question, he could solve the mystery of Slipher's retreating galaxies. Hubble used a simple method for judging distances called the method of the Standard Candle. This method is used by every person who drives along a narrow road on a dark, moonless night. If a car approaches traveling in the opposite direction, the driver judges how far away it is by the brightness of its headlights. If the lights are bright, the car is close; if they are dim, the car is far away.

Following the same reasoning, Hubble judged the distance to other galaxies by the brightness of the stars they contained. He used the automobile driver's rule of thumb: the fainter the stars in the galaxy, the more distant it is.

An accurate measurement of galactic distances by this method is complicated by the fact that some stars in a galaxy are much brighter than others. Hubble used a certain kind of star known as a cepheid variable, whose true brightness was known from the properties of similar stars in our own galaxy. This method worked out to distances of about 10 million light-years. Beyond that point, the Cepheid variables in other galaxies were too faint to be seen. For still greater distances, Hubble developed other methods, such as using the brightness of the entire galaxy as an indication of its distance. Appendix A describes these and other methods of distance determination.

In this way, Hubble arrived at values for the distances to about a dozen nearby galaxies. The majority were more than a million light-years away, and the distance to the farthest one was seven million light-years.[2]

These distances were staggering; they were far greater than the

[2] A light-year is approximately 6 trillion miles.

size of our visible galaxy, which is 100,000 light-years. Until Hubble made those measurements, no one knew how big the Universe is.

Next, armed with his list of distance measurements, Hubble turned back to Slipher's values for the speeds of these same galaxies, augmented by Humason's more recent observations. He plotted speed against distance on a sheet of graph paper, and arrived at the amazing relationship known as Hubble's law: *the farther away a galaxy is, the faster it moves.* Hubble's measurements indicated that the relationship was a simple proportion. That is, if one galaxy is twice as far away as another, it will be moving away twice as fast; if it is three times as far, it will be moving away three times as fast; and so on.

Hubble's Law can be stated mathematically in the following form; let v be the velocity of recession of a galaxy, and let x be its distance from us. Then

$$v = Hx$$

H is a constant of proportionality called the *Hubble constant.* It has units of velocity over distance and is often expressed in kilometers per second per million light-years.

The Hubble law is illustrated by Figure 13.1 (p. 314), showing a number of galaxies together with their spectra. In each case the spectrum of the galaxy is the tapering band of light in the middle, with laboratory spectral lines of known wavelength above and below for purposes of comparison. The clearest feature in the spectrum of each galaxy is a pair of calcium absorption lines that appear at the left in the top spectrum. The normal unshifted positions of these lines in the laboratory are marked by the two black lines at the top of the diagram. In the spectrum of a fairly close galaxy, 55 million light-years away in the direction of the constellation Virgo, these lines are shifted toward the red—which means to the right on this diagram—by a small and barely perceptible amount. A careful study of the spectrum shows that this small shift to the red corresponds to a speed of recession of 1200 kilometers per second.

The next galaxy is very roughly 700 million light-years from us in the direction of the constellation Ursa Major. The calcium lines in the spectrum of this galaxy are seen to be shifted to the red by a considerably greater amount. Accurate measurements of the red shift of this galaxy indicate that it is receding from us at a velocity of 15,000 kilometers per second.

The three remaining galaxies in the diagram are still more distant. The last one, barely visible in the photograph, is about 2.8 billion light-years away, close to the limit of the range of visibility with the 200-inch telescope. The pronounced red shift of the calcium lines in its spectrum corresponds to a velocity of recession of 61,000 kilometers per second, or more than 20 percent of the speed of light.

FIGURE 13.1
Galaxies at various distances and their spectra, showing the corresponding red shifts.

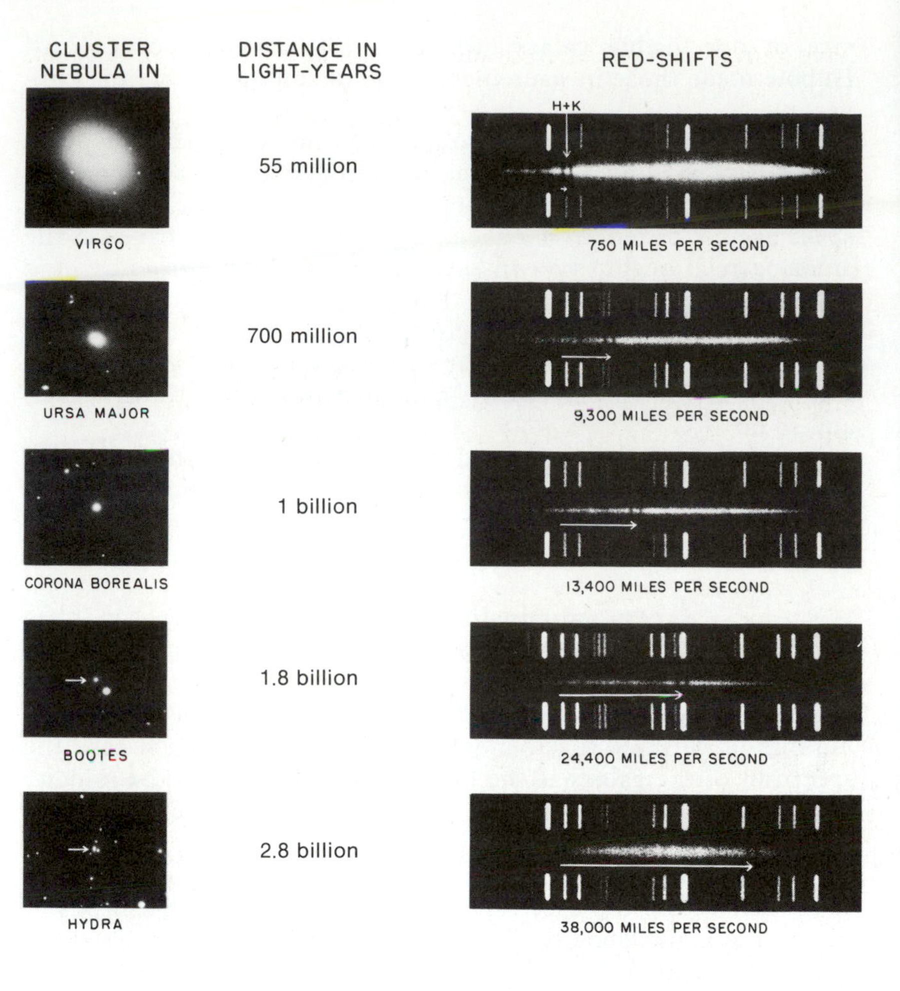

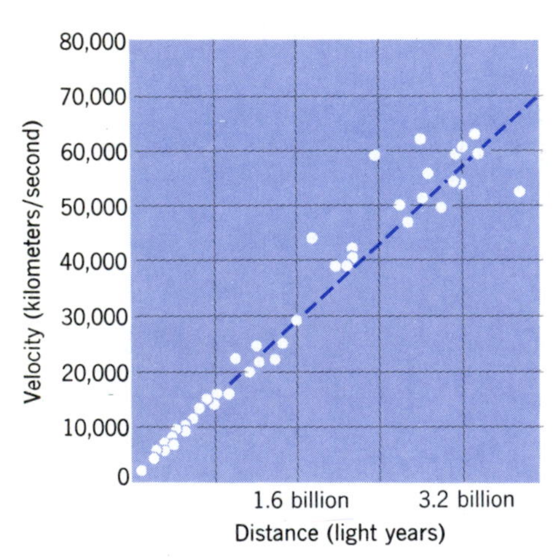

FIGURE 13.2
Velocity versus distance for 46 galaxies.

A plot of velocity versus distance for a number of galaxies is shown in Figure 13.2. The straight line gives the best average fit to the points plotted on this graph. The line corresponds to a Hubble constant of approximately 22/km/sec/10^6 light-years.

An Analogy to the Hubble Law

Hubble's law seems mysterious; why should a galaxy recede from us at a higher speed simply because it is farther away? An analogy will help to make the meaning of the law clear. Consider a lecture hall whose seats are spaced uniformly, so that everyone is separated from his neighbors in front, in back, and to either side by a distance of, say, three feet. Now suppose the entire hall expands rapidly, so that the walls move outward and all the seats move away from one another. Imagine that the hall doubles its size in a short time. If you are seated in the middle of the hall, you will find that your neighbors have moved away from you and are now at a distance of six feet. However, a person on the other side of the hall, who was originally at a distance from you of, say, 300 feet, is now 600 feet away. In the interval of time in which your close neighbors moved three feet farther away, the person on the other side of the hall increased his distance from you by 300 feet. Clearly, he is moving away from you at a faster speed than your close neighbors.

This is the Hubble law. It applies to the Cosmos and every other uniformly expanding object, including an inflating balloon and a loaf of bread rising in the oven. All uniformly expanding objects are governed by this law; if the seats in the lecture hall moved apart in any other way, they would pile up in one part of the hall or another; similarly, if galaxies moved outward in accordance with any law other than Hubble's law, they would pile up in one part of the Universe or another.

THE AGE OF THE UNIVERSE

Knowing the Hubble constant, we can calculate when the cosmic explosion—the Big Bang—occurred. In other words, we can calculate the age of the Universe. Suppose, for example, the Hubble constant is very large. That means that the galaxies are receding from one another very rapidly. Then they must have been packed together a short time ago; that is, the cosmic explosion must have occurred very recently, and the Universe is relatively young. If the galaxies are receding very slowly, a great deal of time must have elapsed since they were close together; in other words, the explosion occurred a long time ago, and the Universe is relatively old.

According to the latest measurements of the speeds and distances of the receding galaxies, the Hubble constant is roughly 22/km/sec/10^6 light-years.[3] That is, a galaxy at a distance of 1 million light-years will be receding at a velocity of 22 km/sec, and a galaxy 2 million light-years distant will be receding at 44 km/sec. This value of the Hubble constant corresponds to a value of about 15 billion years for the age of the Universe. If allowance is made for the fact that when the Universe was young it probably was expanding more rapidly than it is today, the age of the Universe drops to only 10 billion years.

The age of the Universe can be determined in a completely independent way by estimating the ages of the globular clusters in our galaxy. Since globular clusters were presumably among the first

[3] Astronomers making independent examinations of the available observations have arrived at values of H ranging from approximately 15 to 30 km/sec/10^6 light-years. The value quoted here lies in the middle of this range.

stars to form in the Galaxy, their ages should give a good indication of the age of the Galaxy. Chapter 11 explained how the turn-off point to the red giant region in the H-R diagram can be used to determine the ages of star clusters. The results for globular clusters in the Milky Way indicate ages ranging up to 14 billion years. Assuming that the Milky Way Galaxy (and other galaxies) formed when the Universe was about one billion years old, this method yields an age of 15 billion years for the Universe.

A third method for estimating the age of the Universe depends on the fact that supernova explosions create certain rare kinds of radioactive elements, in addition to the more abundant and familiar substances. One of these radioactive elements, rhenium, decays into the element osmium with a half-life of about 60 billion years. That means that if an object contains a large number of rhenium atoms, about half of them will change to osmium atoms in 60 billion years. This fact leads to the method for determining the age of the Universe. Suppose the amount of osmium in the Universe at the present time is much smaller than the amount of rhenium; that implies that the Universe is young compared to 60 billion years, since otherwise a large part of the rhenium in the Universe would have changed to osmium by now. As the Universe grows older, the amount of osmium steadily increases, and the amount of rhenium steadily decreases. Thus, a measurement of the relative amounts of the two substances at any given moment tells how old the Universe is.

The actual calculation is complicated by the fact that all the rhenium was not made at one time; some has been made continuously throughout the history of the Universe, at a rate that depends on the rate at which stars are formed. Different assumptions on the rate of star formation lead to somewhat different results for the age of the Universe.

Allowing for this and other uncertainties, the age of the Universe, as determined by the rhenium-osmium method, lies between 11 and 17 billion years.

Although the three measurements of the age of the Universe yield different results, they overlap when allowance is made for the large uncertainties in each measurement. With allowance for these uncertainties, it seems reasonable to take 15 billion years as a round-number approximation to the age of the Universe.

THE PRIMORDIAL FIREBALL

The reasoning that leads to the Big-Bang theory is indirect; that is, in the astronomer's imagination, the outward motions of the galaxies are projected backward in time until a point is reached at which all galaxies are packed together, and the density and temperature of the Universe are very high. It is as if we saw a cloud of fragments spreading through the air, and inferred from their paths that an explosion had occurred earlier. The inference is reasonable, but less convincing than direct evidence, such as an eyewitness account or a recording of the event would be.

Discovery of the Cosmic Fireball Radiation

In 1948 theorists investigating the possibility of a Big Bang calculated that if a cosmic explosion really did occur, the Universe must have been filled with an intense radiation in the first moments of the explosion. In fact, this radiation would resemble the fireball that forms when a hydrogen bomb explodes. The intensity of the fireball would diminish as the Universe expanded, but a small remnant of the primordial fireball radiation should still be present in the Universe today. Furthermore, it should be detectable with sensitive radio telescopes.

In 1965, Arno Penzias and Robert Wilson of the Bell Laboratories discovered that the earth is bathed in a faint glow of radiation coming from every direction in the heavens. Their measurements revealed that the puzzling radiation arrived at the earth with equal intensity from all parts of the earth. It did not seem to come from the direction of the sun, a galaxy, or any other particular object in the sky; the entire Universe seemed to be the source. This was just the characteristic expected for the cosmic fireball radiation.

Penzias and Wilson were not looking for proof of the Big-Bang cosmology when they made this observation; they were measuring the intensity of radio waves in a large antenna which had been set up for a completely different purpose (Figure 13.3) (p. 318). They were unable to explain the source of the uniform radiation until a friend told them about a lecture he had heard on the possibility of finding radiation left over from the fireball that filled the Universe at the beginning of its existence. Then they realized they had detected the cosmic fireball. The rest is scientific history.

The Spectrum of the Cosmic Fireball

Radio astronomers have made many measurements on the fireball radiation since it was first discovered. The measurements show that the spectrum of the radiation—that is, the change in its intensity with wavelength—very closely matches the spectrum of the radiation emitted by a heated object (Figure 13.4) (p. 319). This characteristic pattern of wavelengths, called a black-body spectrum, was described in Chapter 7. The same spectrum is also observed for the radiation produced by the hot gases resulting from an explosion. The

FIGURE 13.3

Arno Penzias (right) and Robert Wilson in front of the antenna that detected the primordial fireball radiation.

pattern of wavelengths measured by the radio astronomers is the fingerprint of the Big Bang. The agreement between the spectrum of the Penzias-Wilson radiation and a black-body spectrum has convinced nearly all astronomers that the Big Bang really did occur.

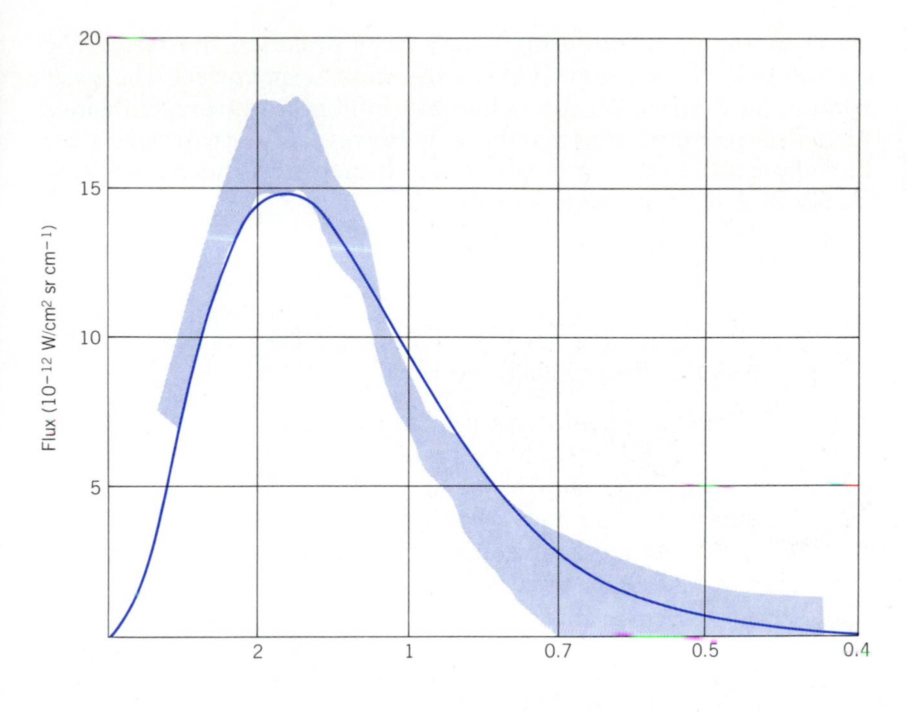

FIGURE 13.4
Comparison between the measured spectrum of the primordial fireball radiation (shaded area) and the theoretical spectrum of radiation from a heated gas (solid line). The wavelength scale increases to the left. The shading indicates the range of experimental uncertainty. The theoretical curve represents radiation from a gas at a temperature of 3 degrees above absolute zero. Shortly after the Big Bang the temperature of the fireball radiation was much higher, but it decreased with time as the Universe expanded.

Demise of the Steady-State Theory

An alternative explanation for the expanding motion of the galaxies, called the Steady-State cosmology, was favored by many astronomers for a time because it accounts for the measured red shifts of distant galaxies without introducing the need for a cosmic explosion. According to the Steady-State cosmology, matter is created continuously at a modest rate throughout the Universe. This freshly created matter, which appears in the Universe in the form of atoms of hydrogen, fills up the empty spaces left by the retreating motion of the galaxies. In the course of time, the fresh hydrogen atoms condense under the force of gravity to form new galaxies and, within these, new stars and planets. Thus as the galaxies move away from one another, newly formed galaxies always are present to take their places, so that the Universe, while expanding forever, is nonetheless always the same, steady and unchanging.

Adherents of the Steady-State theory have suggested that the radiation discovered by Penzias and Wilson also can be explained without assuming a Big Bang. They suggest that this radiation may be a superposition of waves from many separate quasars scattered throughout the Universe. The numerous distinct beams would be blurred into a uniform background by the limited angular resolution of radio telescopes, so that the quasar radiation would appear to arrive at the earth uniformly from every direction, imitating the cosmic fireball radiation. If this suggestion were correct, the spectrum of the Penzias-Wilson radiation would bear some resemblance to the spectrum of quasar radiation. However, the two spectra are distinctly different. As a result of that discrepancy, the Steady-State theory no longer has wide support.

Additional Evidence for the Big Bang; the Abundance of Primordial Helium

The fireball radiation and its spectrum provide the strongest evidence in favor of the Big-Bang theory. However, additional evidence comes from measurements on the abundance of helium in the Universe. According to Chapter 9, helium is formed from hydrogen in the interiors of stars during their lifetimes. As noted on page 321, the Big-Bang cosmology predicts that helium was also formed from hydrogen in the early stages of the Big Bang. These circumstances lead to a criterion for judging whether the Big Bang occurred. If nearly all the helium in the Universe turned out to be primordial, that would be evidence in favor of the Big Bang. If it turned out that all the helium that exists today has been manufactured in the interiors of stars, and none is primordial, confidence in the Big-Bang theory would be weakened.

The line of attack on the problem consists in measuring the helium content of young and old stars. The young stars have been formed from an interstellar medium containing the primordial helium, if any, plus all the helium that was added to the Universe subsequently in many generations of stellar evolution. The old stars were formed when the Galaxy was young, before the interstellar medium had been enriched by helium formed in stellar interiors. Their helium content is primordial only. Therefore the comparison of the helium content in the two groups of stars tells how much helium is primordial, and how much has been added as the product of reactions in stellar interiors.

The helium content of young stars can be determined most directly from the intensities of the helium absorption lines in their spectra. These lines are formed in the atmosphere of the star, and their intensity only gives the amount of helium in the star's outermost layer; however, in a young, unevolved star, the helium is dispersed uniformly, and the amount in the atmosphere is an accurate indicator of the amount in the entire star. Of course, only the hot stars—O and B type—can be used for the purpose, because helium lines only appear with significant intensity in the spectra of these stars. The spectroscopic studies show that about 30 percent of the mass of young stars consists of helium.

When we come to old stars, the helium absorption lines cannot be used in the same way to determine helium content, because a population of old stars does not include O and B types, which are massive and live only a short time. However, the ages of old stars located in globular clusters can be determined in another way, from the H-R diagrams for globular clusters. Computations on stellar structure show that the position of a star on the H-R diagram depends to some degree on its chemical composition. In particular, the position on the diagram is strongly dependent on the helium content of the star. By matching the observed and theoretical H-R diagrams for a globular cluster with care, it is possible to determine not only the age of the stars in the cluster, but also their helium content. The result is a helium content between 22 and 26 percent. In other words, the helium content of old stars is a little less than the helium content of young stars, but close to it. This agrees with the prediction of the Big-Bang theory that most of the helium in the Universe was made shortly after the Big Bang, and only a small amount was contributed subsequently by nuclear reactions in stars. The quantitative agreement between the predicted and observed amounts of primordial helium is impressive. These findings significantly strengthen the case for the Big-Bang cosmology.

CONDITIONS IN THE EVOLVING UNIVERSE

Why did the Universe explode 15 billion years ago? The Big-Bang cosmology does not attempt to answer this question, nor does it provide information on conditions in the Universe prior to the cosmic explosion. However, accepting the Big Bang as a starting point, cosmologists can predict the major events in the subsequent evolution of the Universe with some confidence.

The Early Stages: Formation of Primordial Helium

In the first moments after the Big Bang, the density of matter and energy in the Universe were very great, and the temperature ranged up to trillions of degrees. After a few minutes, the temperature decreased to about a billion degrees, and protons and neutrons began to stick together in groups of four to form helium nuclei. The amount of helium formed at that early time is not accurately known, but calculations suggest that about 25 percent of the matter in the Universe was transformed into helium in the first few minutes after the Big Bang. This is the amount of *primordial helium* in the Universe. The calculated abundance of primordial helium is in good agreement with the measured amount deduced from the H-R diagrams of globular clusters, as described above. The agreement provides additional support for the Big-Bang cosmology.

It might be expected that once helium formed, other heavier elements would then be built up until the whole periodic table of the elements existed. However, calculations on the early history of the Universe indicate that this did not occur. The principal reason is that a nucleus with five neutrons and protons, which would be the natural successor to helium in a chain of nuclear reactions, does not exist. The next stable nucleus after helium is lithium, which has a total of six neutrons and protons. The gap between a four-particle nucleus and a six-particle nucleus turns out to be an impossible one to cross under the conditions of rapidly falling temperature and density that existed in the first minutes after the Big Bang. Thus, the synthesis of heavier elements beyond helium ceased when the Universe was about 30 minutes old, and was not resumed until a billion years later, when star formation began.

Subsequent Evolution of the Universe

After helium was formed, the expansion of the Universe continued and its temperature continued to drop. The matter of the Universe existed at that time mainly in the form of hydrogen nuclei, helium nuclei, and electrons. Atoms did not yet exist, because whenever an electron was captured into an orbit around a nucleus to form an atom, it was knocked out of its orbit almost immediately, under the smashing impact of the violent collisions that occur at such high temperatures.

However, when the Universe was about a few hundred thousand years old, the temperature had dropped to 3000 degrees Kelvin and collisions were no longer violent enough to dislodge electrons from their orbits. From that time on, the hydrogen and helium in the Universe were present mainly in the form of neutral atoms.

With the passage of time, the gaseous clouds of primordial hydrogen and helium cooled and condensed into galaxies and, within the galaxies, into stars. The formation of stars probably began shortly after the formation of the first galaxies, when the Universe was between a few hundred million and a billion years old. At this time, the synthesis of the heavier elements resumed. After about 15 billion years of continuing expansion, star formation, and heavy-element formation, the Universe reached the state in which it exists today.

What will happen in the future? One possibility is that the expansion of the Universe will continue indefinitely. As the galaxies fly apart and the distances between them increase, space will grow emptier, and the density of matter will dwindle. At the same time, as the hydrogen within each galaxy is used up, fewer new stars can be formed. The old stars will go out, one by one, the galaxies will grow dim, and the Universe will fade into darkness (Figure 13.5).

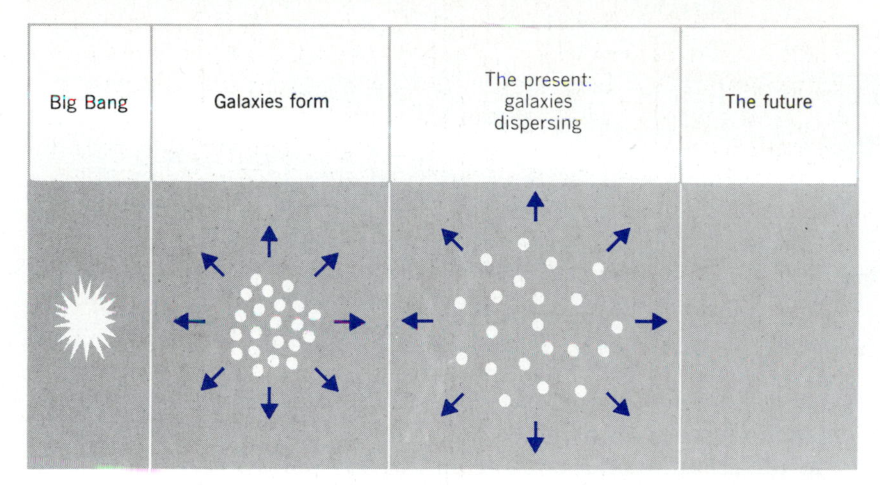

FIGURE 13.5
The future according to the Big-Bang cosmology.

THE OSCILLATING COSMOLOGY

The Big-Bang theory can also lead to a very different prediction regarding the future. Consider the picture of a Universe expanding in the aftermath of the great explosion. In the Big-Bang theory it was assumed that the expansion would continue forever, but the contrary may be true because of the effect of gravity. This force, acting throughout the Universe, pulls back on the outward moving galaxies and slows their retreat. If the pull of gravity is sufficiently strong, it may be adequate to bring the expansion of the Universe to a halt at some point in the future.

What will happen then? The answer is the crux of this theory. The elements of the Universe, held in a balance between the outward momentum of the primordial explosion and the inward force of gravity, stand momentarily at rest; but after the briefest instant, always drawn together by gravity, they commence to move toward one another. Slowly at first, and then with increasing momentum, the Universe collapses under the relentless pull of gravity. Soon the galaxies of the Cosmos rush toward one another with an inward movement as violent as the outward movement of their expansion when the Universe exploded earlier. After a sufficient time, they come into contact; their gases mix; their atoms are heated by compression; and the Universe returns to the heat and chaos from which it emerged many billions of years ago (Figure 13.6) (p. 324).

And after that? No one knows. Some astronomers say that the world will never come out of this collapsed state. Others speculate that the Universe will rebound from the collapse in another explo-

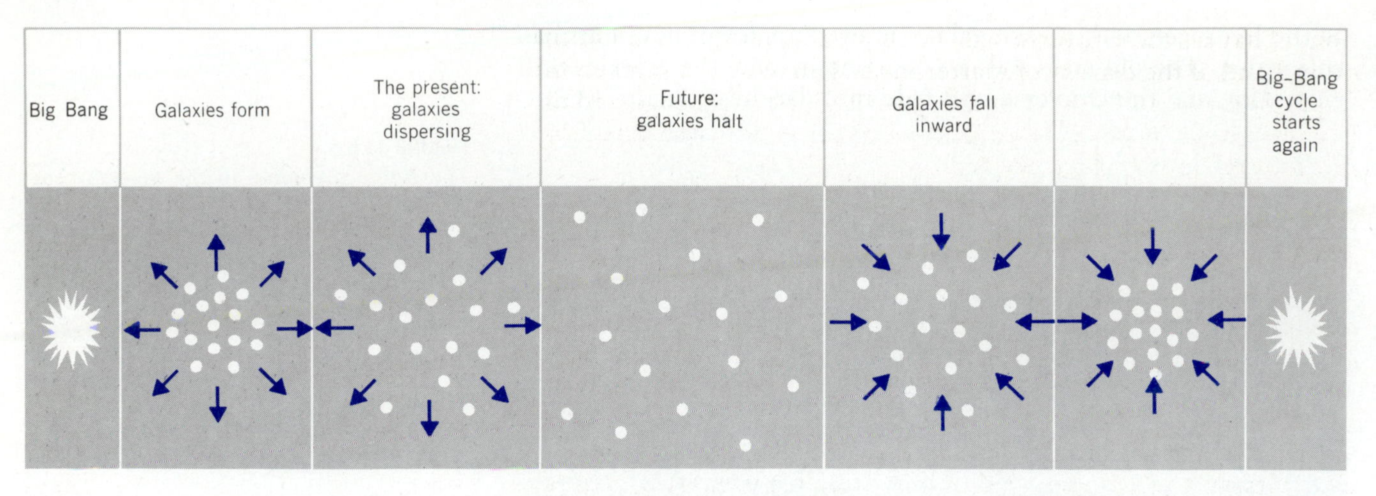

FIGURE 13.6
The Oscillating cosmology.

sion, to become an entirely new world in which no trace of the existing Universe remains.

In the reborn world, once again the hot, dense materials will expand rapidly outward in a cosmic fireball. Later, when the primordial gases have cooled sufficiently, galaxies and stars will form out of them. Gradually the expansion will slow down under the pull of gravity; eventually a new collapse will occur, followed by still another creation; and after that, another period of expansion, and another collapse. . .

This theory envisages a Cosmos that oscillates forever, passing through an infinite number of Big Bangs in a never-ending cycle of birth, death and rebirth. It unites the astronomical evidence for an explosive beginning of the world with the concept of an eternal Universe.

The Oscillating theory has the advantage over the Big-Bang theory of being able to answer the question: What preceded the Big Bang? The answer offered by the Oscillating theory is that prior to the Big Bang the Universe was in a state of increasing density and temperature. As the Universe approached the state of maximum density and temperature, all the complex elements that had been made within stars during the preceding cycle were melted down, so to speak, into the basic hydrogen out of which they had originally been manufactured. At the moment of maximum compression, another Big Bang occurred and the Universe was born anew.

Density of Matter in the Universe

How can the theory of an Oscillating Universe be tested? The answer is straightforward. If the density of matter in the Universe is sufficiently great, the gravitational attraction of the different parts of the Universe on one another will be strong enough to bring the expansion to a halt, and reverse it to commence a renewed contraction.[4] That is, the Universe will be in an oscillating state. On the other hand, if the density of matter in the Universe is not great, the force of gravity will not be sufficient to halt the expansion, and the Universe will continue to expand indefinitely into the future, as predicted by the Big-Bang theory.

In other words, the density of matter in the Universe is a critical factor in deciding between the two cosmologies. What is the critical density of matter required to slow down and reverse the expansion? A calculation shows that the present expansion of the Universe will be halted if the density of matter is approximately 10^{-29} g/cm^3 or greater. This density corresponds to one proton or hydrogen atom in a volume of 4 cubic feet.

How does the critical value of the density compare with the observed density of matter in the Universe? The matter whose density can be most readily estimated is that which is present in the galaxies in a visible form, as luminous stars and dense concentrations of gas. If we were to smear out the visible matter in the galaxies into a uniform distribution filling the entire Universe, the density of this smeared-out distribution of matter would be roughly 5×10^{-31} g/cm^3. That is, the visible matter present in all the galaxies of the Universe is too small by a factor of 30 to halt the expansion.

Since energy and radiation are equivalent to matter by Einstein's formula, $E = mc^2$, we must add to the above figure the contribution from various types of radiant energy in the Universe, such as starlight and the primordial fireball radiation. These forms of energy turn out to increase the average density of matter by one or two percent, which is not enough to affect the outcome.

Cosmological Significance of the Hidden Mass

What about matter that is unobservable because it is not luminous? For example, this matter could exist in the galaxies in the form of nonluminous gas in the space between the stars, or as dead stars, or as stars of very low mass and negligible luminosity. It could also be present in the form of gas in the space between the galaxies.

In Chapter 10 evidence was described for the presence of a large amount of nonluminous mass in the Universe—the so-called hidden mass. This hidden mass, contained partly in the space in and around galaxies and partly in the intergalactic medium, may include as much as 10 times more matter than the amount, represented by the density of 3×10^{-31} g/cm^3, that is present in the Universe in a visible form. In other words, the total density of matter in the Universe, in both visible and invisible forms, is about 5×10^{-30} g/cm^3,

[4] A high density means that on the average, particles in the Universe are relatively close to one another, and, therefore, their mutual gravitational attraction is strong.

or one-half of what is needed to bring the expansion of the Universe to a halt.

This result seems to indicate that the Oscillating cosmology is not correct, and the Universe will expand forever. However, the measurement of the hidden mass may be in error by a factor of two or more. In addition, the theoretical estimate of the critical mass required to halt the expansion of the Universe may also be inaccurate by a factor of two or more, because of uncertainties in the rate at which the Universe is currently expanding—that is, in the measured value of the Hubble constant.[5] In view of these uncertainties, it can only be said that there appears to be enough hidden mass in the Universe to have a strongly decelerating effect on the motions of the galaxies, but whether or not the amount of hidden mass is sufficient to bring the expansion to a halt is still an open question.

Cosmic Abundance of Deuterium

An independent line of reasoning leads to still another way of measuring the density of matter in the Universe and testing the Oscillating cosmology. Theoretical studies of the Big Bang indicate that in the first few moments after the explosion neutrons and protons collided in great numbers in the Universe, and fused to form nuclei of deuterium or heavy hydrogen. Most of the deuterium was lost quickly, because it collided with other particles and fused into nuclei of helium. However, if the density of matter in the Universe was low, fewer collisions would occur, and more deuterium would remain. This deuterium would survive to the present day as a part of the interstellar gas. In other words, the amount of deuterium in the interstellar gas today is an indicator of the density of matter in the Universe at an early time. Theoretical studies can then provide a reliable connection between the density of matter in the Universe at that early time and the density of matter today.

The deuterium measurements, which were carried out with ultraviolet observations made from the Copernicus satellite (see page 201), indicate a relatively low density of matter. The results suggest that the density of matter in the Universe at the present time is 10^{-30} g/cm^3. This is about one-tenth of the critical value required to halt the expansion of the galaxies. It is also about 20 percent of the value deduced from the hidden-mass measurements.

The deuterium result is tentative, because the deuterium detected in the interstellar gas may not all be primordial; that is, it may not all have been produced in the first moments after the Big Bang. Some deuterium could instead have been manufactured in the

[5] The faster the Universe expands—that is, the larger the Hubble constant—the larger the mass that is necessary to halt the expansion.

interiors of stars during the subsequent history of the Universe. Although the deuterium nucleus is too fragile to survive in the deep interior of a star, some deuterium could be manufactured in the outer layers of a star during a supernova outburst. However, other light and fragile nuclei, such as lithium, would also be manufactured in supernovas at the same time, and a search for lithium reveals that it is very scarce in the interstellar gas. The scarcity of lithium suggests that little deuterium has been made in stars, and most is primordial. This suggests in turn that the deuterium present in the Universe today is mostly primordial, and is therefore a reliable indicator of the density of matter in the Universe.

Neutrinos and the Oscillating Universe

The deuterium method for estimating the density of matter in the Universe yields a result that is about one-fifth the value yielded by the hidden-mass method. Is there any significance to this discrepancy? It would seem that there is none, because of the large uncertainties inherent in both methods. However, a reason exists for believing that the difference may in fact be meaningful. The estimates based on the deuterium abundance only include matter present in the Universe in the form of "ordinary" particles—that is, neutrons, protons, and electrons. However, there is another possible candidate for the hidden mass that is not a form of "ordinary" matter and would not be counted in the deuterium measurements. This is the elusive particle known as the neutrino (page 201).

Recent evidence suggests indeed that neutrinos may constitute the bulk of the mass in the Universe. According to calculations on the early conditions in the expanding Universe, neutrinos were created in great numbers when the Universe was young and its temperature was very high. These primordial neutrinos—relics of the Universe's first moments—are believed to still be present in great abundance in the Universe; in fact, it is estimated that there are approximately one billion neutrinos for every atom of hydrogen in space. But in spite of the theoretical abundance of neutrinos, until recently these particles were not thought to contribute significantly to the density of matter in the Universe, because their rest mass—that is, the mass a neutrino has when motionless—was thought to be zero.

However, according to experiments carried out in the USSR, the rest mass of the neutrino, while small, is not zero. The experiments suggest that the neutrino rest mass is only one hundred-thousandth of the mass of the electron, but even that very small mass, when multiplied by the enormous number of neutrinos in the Universe, would make these elusive particles the major component of the hidden mass, and the dominant form of matter in the Cosmos. The implication in this result is that if the mass density of the Universe does eventually turn out to be sufficient to halt the outward movement of the galaxies, the form of matter responsible for that effect is likely to be the neutrino.

OPEN AND CLOSED UNIVERSES

The Big-Bang Universe, in which the expansion of the galaxies continues without limit, is called an *open universe.* The Oscillating Universe, in which galaxies only recede to a limited distance from one another before coming together again, is called a *closed universe.* It is also possible that there can be a halt to the expansion, followed by a collapse, but the Universe never rebounds from the collapse. That is, the cycle is not repeated, and the Universe does not oscillate. This case would also be an example of a closed universe.

Open universes and closed universes are very different. For example, an open universe has an infinite volume; if you set out from the earth on a straight line course through space and travel forever, you will never return to your starting point. A closed universe, on the other hand, has a finite volume and a finite mass. If you set out from the earth on a straight-line course, through space, and continue to travel for a very long period of time, you will eventually return to your starting point from the opposite direction.

This last property of a closed universe seems peculiar, but an analogy with motion on the surface of a sphere helps to make it plausible. Suppose you are a two-dimensional person, living on the surface of a sphere. The surface of the sphere is your universe; it is a two-dimensional world. In this two-dimensional universe there are no boundaries; that is, you can travel through your universe forever without meeting a barrier or an edge. But if you set out on a "straight-line" course and travel for a sufficiently long time, you will travel around your "universe" and return to your starting point. The surface of this sphere is a closed universe in two dimensions.

On the other hand, if you are a two-dimensional person living in a two-dimensional world that consists of a flat plane of infinite extent, like an infinitely large sheet of paper, you can travel forever on a straight-line course and never return to your starting point. The infinitely large sheet of paper is an open universe in two dimensions.

How can we extend these ideas to the three-dimensional world of real life?

It is easy to do this mathematically, but impossible to do it in a way that can be visualized clearly without mathematics. The reason is that the surface of a sphere is a two-dimensional world carved out of a three-dimensional space. Since we live in a three-dimensional world, we have no trouble in visualizing the surface of a sphere, as well as the space out of which it is, so to speak, carved. Now let us consider the three-dimensional world. In this world, the analogue to the surface of the sphere is a three-dimensional *volume* carved out of a four-dimensional space. But what is a four-dimensional space? How can we visualize it? We cannot, any more than we can visualize the *curvature* of a three-dimensional volume. These concepts are among the few ideas in science that cannot be explained in an entirely nonmathematical way. They can be made plausible with the help of the two-dimensional analogies, but a deeply rooted, intuitive understanding, based on everyday experience, is not within our grasp.

THE BOUNDARIES OF THE KNOWABLE

It seems likely that the ultimate cosmological theory will resemble one of the two forms: Big-Bang or Oscillating; but which one? Until that choice is settled, we will be unable to answer the questions with which this chapter opened. Was there a beginning? Will there be an end? Is the physical Universe eternal?

There the matter rests for the moment. Astronomers have exposed very interesting details in the history of the Universe—the birth of stars, the assemblage of the elements within the stars out of the three basic particles of matter, and dispersal of these elements to space in supernova explosions—but science has been unable to solve the fundamental problems of beginning and end.

Main Ideas

1. The definition of cosmology.
2. Evidence for the expanding Universe; the red shift.
3. The Big-Bang cosmology.
4. The Hubble law and the meaning of the Hubble constant.
5. Methods for determining the age of the Universe.
6. Discovery of the cosmic fireball radiation; significance for the Big-Bang cosmology.
7. Other evidence for the Big Bang; primordial helium.
8. The history of the Universe after the Big Bang.
9. The oscillating cosmology.
10. The fate of the Universe; cosmic density of matter.
11. Significance of cosmic deuterium abundance; the hidden mass; primordial neutrinos.
12. Open and closed universes.

Important Terms

Big-Bang cosmology	Hubble constant
closed universe	Hubble law
cosmic abundance of deuterium	neutrinos
cosmic density of matter	open universe
cosmic fireball radiation	oscillating cosmology
cosmic fireball spectrum	primordial helium
cosmology	red shift
hidden mass	steady-state cosmology

Questions

1. What is the cosmological significance in the life story of the stars?
2. Describe the evidence indicating that the Universe is expanding. How does the Steady-State cosmology account for the same facts without the assumption of an expanding Universe?
3. The observed red shift for Galaxy 3C295 is 36 percent. What is its velocity relative to us? What is its distance? Describe the condition of the materials of the solar system when the light we now receive from this galaxy set out on its journey.
4. Explain why we are not necessarily at the center of the expanding Universe, in spite of the fact that nearly all galaxies seem to be moving away from us.
5. As we look out in space, we look back in time. Suppose that the light from a quasar at a distance of 8 billion light-years showed a red shift, while the light from a quasar at a distance of 5 billion light-years showed a blue shift. Could this pattern of Doppler shifts be explained by any of the three cosmologies? If so, which one? Explain your reasoning.
6. Explain how the age of the Universe can be determined from the Hubble Law.
7. Plot a graph of velocity vs. distance based on the hypothetical observations of galaxies listed below. Draw a smooth curve through the set of points.

Distance (light-years)	*Velocity (km/sec)*
500 million	12,900
1 billion	26,500
2 billion	56,000
4 billion	124,000
8 billion	296,000

Is the graph a curved line? In which direction is it curved—up or down? What does this graph tell us about the rate at which the Universe was expanding approximately 8 billion years ago, compared to the rate at which it is expanding today? Does this information support or contradict the Big-Bang cosmology? Explain.

Construct a straight-line approximation to the beginning part of the graph by connecting the origin to the 2-billion-year point. Calculate the Hubble constant from this straight-line approximation. Determine the age of the Universe from the value of the Hubble constant you have calculated. In making the calculation, assume that H has had the same value since the Big Bang; i.e., the Universe has been expanding at a constant rate. If you calculated the age of the Universe again, allowing for the fact that the graph is curved and H is not constant, would your value for the age of the Universe be increased or decreased? Explain.

8. If the radius of the observable Universe is 15 billion light-years and the average distance between galaxies is 2 million light-years, how many galaxies are contained in the observable Universe? If each galaxy contains 100 billion stars and the average mass of a star is 10^{33} grams, what is the total mass of the observable Universe? What is the average density of matter in the Universe? Compare your result with the density required to halt the expansion of the Universe. What is your conclusion?

9. If the Steady-State cosmology did not appear to be excluded by scientific evidence, what would be your personal preference among the three theories? What are the reasons for your preference?

10. If you were located in a galaxy at a distance of 10 billion light-years from the Milky Way Galaxy, what would be the appearance of the sky in all directions?

LEVERRIER
JUNO
VESTA
PALLAS
CERES
MARS
EARTH
ASTRÆA
MERCURY
VENUS
S
JUNE
JULY
AUG
SEP
OCT
NOV

THE SOLAR SYSTEM

PART FOUR

THE SUN

14

The sun was formed 4.6 billion years ago, is nearly halfway through its life, and will not change its properties appreciably until it moves off the Main Sequence in five or six billion years to become a red giant. It is an ordinary body in the cosmic hierarchy, similar to countless other G2 stars on the Main Sequence in its general characteristics; but it has one unique feature: it is 300,000 times closer to us than any other star.

The closeness of the sun gives it a considerable astrophysical interest. The sun may be one point among many on the H-R diagram from the viewpoint of the stellar evolutionist, but the solar astronomer, who studies its properties in detail, finds it to be a complicated and interesting object. More is known about the solar interior than about the interior of any other star. The sun's surface is a tempestuous region, marked by violent outbursts known as flares whose origin remains largely unexplained. These solar eruptions produce effects in the earth's atmosphere that have major consequences for the inhabitants of this planet, including radio blackouts and possibly changes in climate. The sun also provides us with our only opportunity to take a close look at a stellar atmosphere. The precise measurement of the intensities and detailed shapes of the lines in the solar spectrum, made possible by the sun's proximity to the earth, provides an observational checkpoint for calculations of absorption lines in stellar atmospheres. The information yielded by these calculations is the basis for the interpretation of all stellar spectra. Thus, the sun is the indirect source of a large body of astrophysical knowledge. A summary of its properties is given in Table 14.1 (p. 336).

A solar prominence.

TABLE 14.1
Properties of the Sun

Quantity	Value	Method of Measurement
1. Average sun-earth distance	92,956,000 miles 149,598,000 kilometers	Radar reflection from Venus
2. Angular diameter	32′	
3. Radius	432,000 miles 696,000 kilometers	Angular size and distance to the earth
4. Mass	1.99×10^{33} g	Orbits of the planets
5. Average density	1.41 g/cm³	$\rho = \frac{\text{Mass}}{\text{Volume}}$
6. Solar constant	1.947 cal/min/cm² 1.358×10^6 ergs/sec/cm²	High-altitude aircraft measurements
7. Luminosity	3.90×10^{33} ergs/sec	Solar constant and sun-earth distance
8. Surface temperature	5800°K	Luminosity and radius ($L = 4\pi R^4 \sigma T^4$; σ (Stefan-Boltzmann constant) $= 5.7 \times 10^5{}_{\text{cgs}}$)
9. Spectral type	G2	
10. Apparent magnitude	−26.8	Photometer
11. Absolute visual magnitude (M_v)	4.79	Apparent magnitude and sun-earth distance
12. Bolometric correction (B.C.)	0.07	Surface temperature
13. Bolometric magnitude	4.72	$M_v -$ B.C.
14. Rotation Period	Sunspots / Photosphere	Motion of sunspots. Doppler shift in photosphere spectrum
Equator	25.0 (days) / 26.0	
30°	26.4 / 27.3	
60°	— / 32.5	
80°	— / ~35	
15. Magnetic field	~ 1 gauss averaged over surface, fluctuating and irregular. Hundreds of gauss over disturbed areas. Thousands of gauss in sunspots.	Zeeman effect
16. Composition: most abundant elements listed by percent of total solar mass (approximate)	H ~75; Si 0.1 He ~25; N 0.1 O 1.0; Mg 0.09 C 0.4; Ne 0.07 Fe 0.16	Solar absorption spectrum

THE SUN'S INTERIOR

Theoretical studies of stars of one solar mass have been carried out by many theoretical astrophysicists under a variety of assumptions, and agreement has been reached regarding the general conditions that exist in the interior of the sun. Typical results of these calculations are shown in Figures 14.1 and 14.2, which represent the temperature and density of the sun at various points between the center and the surface. Figure 14.1 shows that the temperature decreases from a central value of approximately 15 million degrees to a value that appears to be zero at the surface. In reality, the surface temperature is about 6000°K, but this value would be less than the thickness of a pencil line if represented on the million-degree scale of the graph in Figure 14.1.

Figure 14.2 indicates that the density within the sun falls off very sharply with increasing distance from the center. The central density is about 150 g/cm^3, or 13 times the density of lead. Halfway from the center to the surface, the density has decreased to 1 g/cm^3, which is the density of water. At the surface, the density is 10^{-7} g/cm^3, or approximately one ten-thousandth of the density of air at the earth's surface.

As a result of the rapid falloff in the density of the sun, most of its mass is concentrated in a relatively small volume, approximately 90 percent of the sun's mass being contained in the inner half of its radius. The average density of the sun is 1.4 g/cm^3, or somewhat greater than the density of water.

FIGURE 14.1
Temperature at various depths in the sun's interior. The arrow indicates the surface.

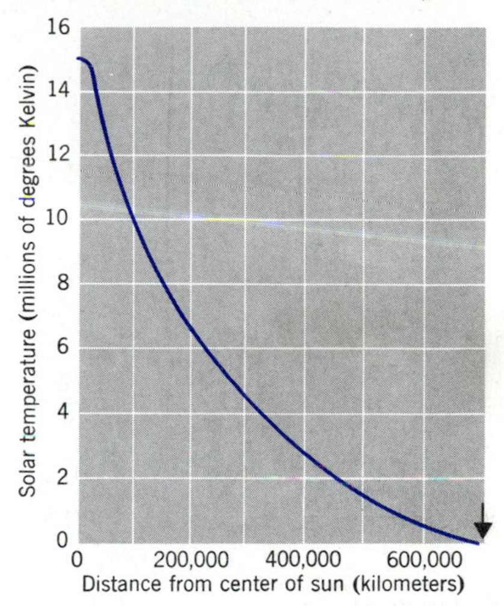

Composition of the Sun

It is likely that the sun was uniform in composition initially, with the same mixture of elements and the same relative abundances throughout its interior. Its main ingredients were primordial hydrogen and helium plus a small amount of heavier elements.[1] Hydrogen made up about 75 percent of the mass of the primitive sun, helium made up most of the remaining 25 percent,[2, 3] and elements heavier than helium constituted roughly one percent.

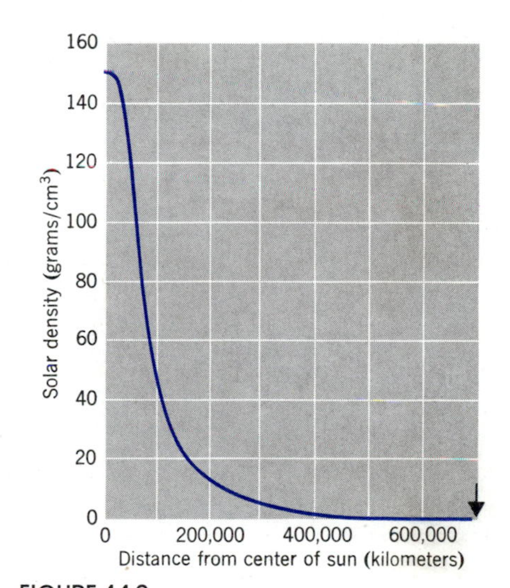

FIGURE 14.2
Density at various depths in the sun's interior.

[1] In addition to primordial helium, dating back to the Big Bang, the sun must also contain helium that was manufactured out of hydrogen in other stars and added to the interstellar medium later. The available evidence appears to favor the view that most of the helium in the Cosmos, and presumably in the sun, is primordial.

[2] This value is an average between an estimate of 30 percent deduced from the intensities of helium lines in the chromosphere (page 345) and 20 percent deduced from the abundance of helium in solar cosmic rays.

[3] Expressed in numbers of atoms, hydrogen and helium constitute roughly 94 and 6 percent, respectively.

Once nuclear burning began, helium started to build up in the center of the sun and the concentration of hydrogen began to diminish, as a consequence of the steady conversion of protons to helium nuclei in fusion reactions. Gradually, a zone of helium-rich material spread outward from the center. Today, after four and a half million years of steady burning, the concentration of hydrogen at the center has been depleted by half, from 75 percent to approximately 35 percent. At the same time, the concentration of helium at the center has risen from 25 to 65 percent.

In the lifetime of the sun thus far, only five percent of the sun's total mass has been converted from hydrogen to helium. The reason is that although the change in composition near the center is drastic, it has occurred in a space that makes up only a small fraction—about one-fiftieth—of the sun's total volume.

The Solar Neutrino Experiment

Although these temperature and density values are considered to be close to the true conditions within a star of solar mass, confidence in their detailed accuracy has been diminished somewhat by the results of a recent experiment, in which an attempt was made to measure the number of neutrinos emitted from the sun's interior. These massless, chargeless particles are created in the core of the sun in nuclear reactions such as those on page 202, and escape to space directly because of their negligible interaction with matter. In the experiment, a tank-car-sized neutrino detector consisting of 600 tons of liquid perchloroethylene (C_2Cl_4), was buried in a mine in order to screen out interfering cosmic rays. This substance, a widely used cleaning fluid, is relatively effective in recording the presence of neutrinos, which are captured by the chlorine atoms in the fluid.

After running the experiment 15 times over a period of several years, physicists reported that the number of neutrinos emitted from the sun was substantially smaller than the number predicted (Figure 14.3). The predictions were based on laboratory measurements of nuclear reaction rates, combined with theoretical studies of the temperature and density at the center of the sun.

In order to bring the calculations on the number of solar neutrinos into agreement with the results of the experiment, it would be necessary to reduce the computed temperature at the center of the sun by about 1.5 million degrees. This reduction represents a change of only 10 percent in the temperature values shown in Figure 14.1; however, a 10 percent correction is far greater than the uncertainty that the astronomers had previously attached to their calculations of conditions in the interior of stars of solar mass.

The discrepancy between theory and observation indicated by the solar neutrino experiment poses a serious problem in astronomy.

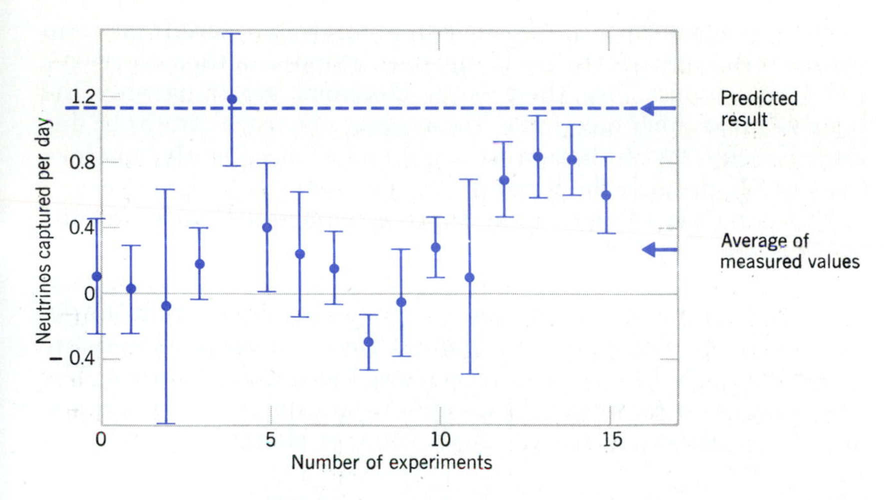

FIGURE 14.3
Results of the solar neutrino experiment. The average of the observed values is one-quarter of the theoretical prediction. Each circle is a measurement; the vertical lines are estimates of the possible error in the results.

Although the experiment is difficult, most astronomers and physicists consider that it has been done carefully and its result must be taken at face value. A possible explanation of the experiment is that the neutrino has unsuspected properties that have not been revealed previously. Another possibility is that the internal structure of the sun is more complex than previously assumed. One theoretical study suggests that occasionally, perhaps every few hundred million years, the core of the sun becomes unstable and expands, the temperature at the center drops, and the nuclear reaction rate and neutrino production decrease accordingly. The sun's surface temperature and luminosity diminish also. This means that for a time the central temperature of the sun, the neutrino production rate, and the solar luminosity all are below their "normal" values. That could be the condition that prevails today, and would explain the result of the neutrino experiment.

Later, according to the studies, conditions in the sun's interior and on the surface return to normal. After perhaps 200 million years the sun's core becomes unstable again, and the entire cycle repeats. It has been suggested that the periodic changes in solar energy output that take place during this cycle might be the cause of the major ice ages on the earth.

Although a generally accepted explanation of the solar neutrino experiment has not yet been provided, the experiment and its possible theoretical interpretations are of great interest, partly because they have forced astronomers to reopen a nearly closed chapter in stellar evolution, and partly because they may have interesting implications for the history of the earth.

The Zone of Convection

In the deep interior of the sun the temperatures range up to many millions of degrees. In this range of temperatures, collisions between atoms are sufficiently violent to eject many electrons from their orbits. Light atoms are completely stripped of their electrons, and heavy atoms lose their outer electrons, retaining only the tightly bound inner electrons. These inner electrons cannot be dislodged easily by absorption of a photon. Consequently, photons pass readily through the inner part of the sun.

Nearer to the surface of the sun the temperature falls, and the heavier atoms, such as iron, begin to recapture their outer electrons. The outer electrons in an atom are bound to the nucleus by relatively small forces and, therefore, can be easily separated from the nucleus by the absorption of a photon. For this reason, photons are strongly absorbed by atoms that possess their outer electrons. The appearance of these absorbing atoms in appreciable numbers below the sun's surface tends to block the flow of photons coming from the interior.

If photons are the only means of carrying energy up to the surface of the sun, the blocking of these photons will cause the temperature to drop sharply at some depth below the surface. In this situation, the outer region of the sun now consists of a layer of relatively cool gas resting on a hotter interior.

This layer of cool gas reacts in the same way as a pot of water placed on a hot stove. The water at the bottom of the pot, heated by contact with the stove, expands, and rises to the surface. At the surface, the water loses some of its heat to space, cools, and descends to the bottom of the pot. There it is reheated and rises again. The result is a circulating current of water, which carries heat from the bottom of the pot to the surface (Figure 14.4).

In the same way, the gas at the bottom of the cool outer layer of the sun is heated by its contact with the hot gas in the interior, expands, and rises toward the surface. At the surface, the hot gas loses its heat to space, cools, and descends again into the interior. As a consequence the entire outer layer of the sun breaks up into ascending columns of heated gas and descending columns of cooler gas.[4] As with the pot of water on the stove, these circulating currents of gas carry heat or energy upward to the surface from the interior of the sun.

The transport of energy by circulating currents of gas or fluid is called *convection*. The currents that carry the heat upward are called *convection currents*, and the region of the sun in which this large-scale upward and downward movement of gases occurs is called the *zone of convection*. It extends from a depth of about 150,000 kilometers upward to the surface of the sun (Figure 14.5).

FIGURE 14.4
Rising currents in a heated pot of water carry heat to the surface.

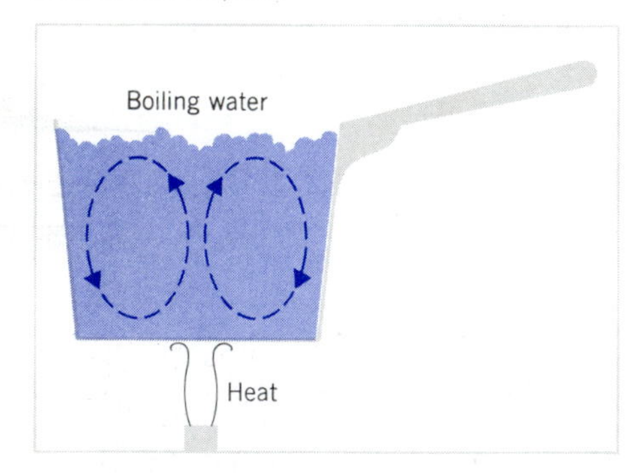

[4] In the case of a pot of water on a hot stove, if the stove is very hot, bubbles of water vapor—that is, steam—form at the bottom of the pot. These bubbles rise rapidly to the surface because of their buoyancy. That is, the water begins to boil. In the case of the sun, the materials are already in the gaseous state, and boiling does not occur.

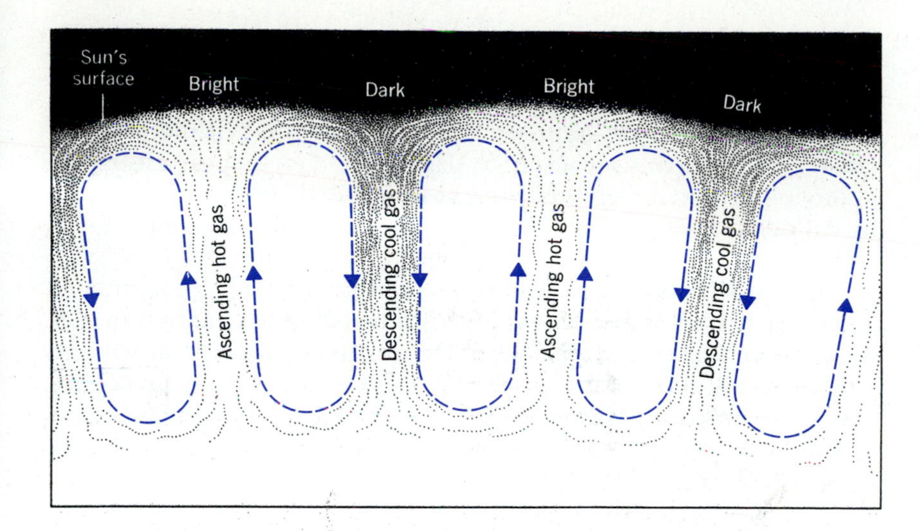

FIGURE 14.5
Upward- and downward-moving columns of material in the zone of convection beneath the sun's surface. Gas heated at the lower boundary ascends, loses a part of its heat to space, and descends.

At depths greater than 150,000 kilometers, energy is transported within the sun by radiation, that is, by the flow of photons. At 150,000 kilometers, the outward flow of radiant energy is blocked to a substantial degree by the absorption of photons, and convection sets in. From that depth out to the surface, energy is transported partly by convection and partly by radiation. Above the surface, radiation again becomes the sole means of energy transport.

Tiers of Convection. The currents in the zone of convection are shown in Figure 14.5. The diagram simplifies the structure of the zone by showing single columns of gas rising without a break from the lower boundary of the zone to the surface. The actual situation is likely to be more complex. It is believed that several tiers of convection currents—perhaps three in number—carry heat successively upward. The lowest tier contains massive convective currents, perhaps 200 to 300 thousand kilometers in diameter. At an intermediate depth, these large convection currents give way to a second tier of currents. The convection currents in the second tier are about 30,000 kilometers in diameter and 15,000 kilometers deep. Above the second tier lies a third tier. The currents in the third tier are roughly 1000 kilometers across and 1000 to 2000 kilometers deep. The tops of the upward-moving columns of gas in the third tier make up the visible surface of the sun.

The Photosphere

The visible surface of the sun is called the *photosphere*. The photosphere is the sun's disk as observed visually or with a telescope. It has a uniform appearance when viewed with the eye or a small telescope, but inspection with a large telescope under good visual conditions reveals that it has a granulated texture. The granules are relatively small—up to 1500 kilometers in diameter—and are difficult to observe under ordinary conditions because of the blurring effect of the earth's atmosphere. However, they appear clearly in photographs taken from instruments carried above the atmosphere in balloons (Figure 14.6a), or from ground-based telescopes under good seeing conditions.

Doppler-shift measurements indicate that in the bright center of a granule the gas moves upward, and at the dark boundary it moves downward. Apparently the granules are the tops of the ascending columns of hot gas in the uppermost tier of the zone of convection. The Doppler-shift measurements, combined with photographs such as that in Figure 14.6a are proof that the zone of convection actually exists in the sun.

Solar Granules and Solar Convection

The formation of rising columns of gas above a hot surface is a familiar phenomenon in the earth's atmosphere. Frequently the sun's radiation heats the surface of the earth to a higher temperature than the air immediately above. The air, warmed by contact with the ground, expands, becomes buoyant, and rises in columns of heated gas, forming a zone of convection in the atmosphere. The top of the zone of convection is usually at an altitude of about

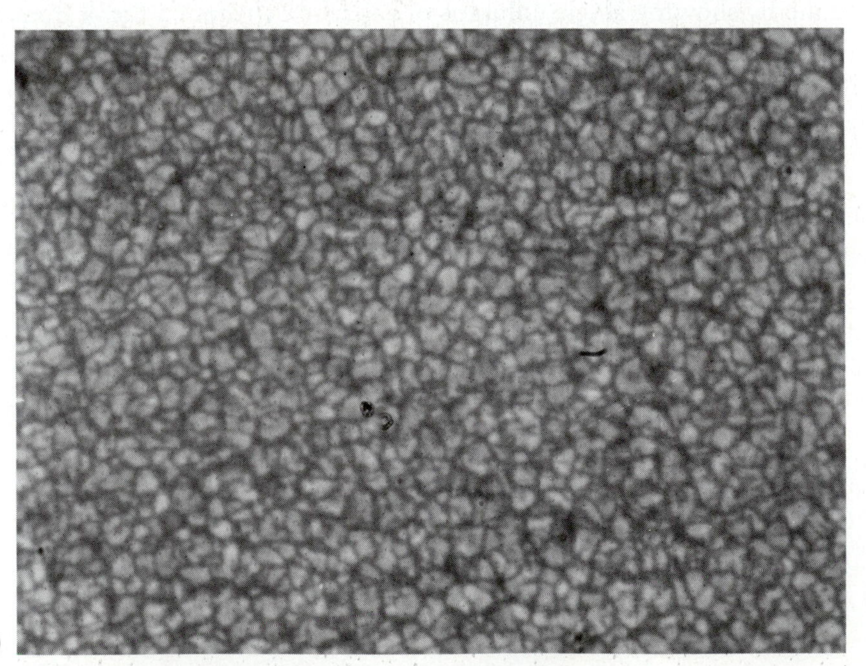

(a)

FIGURE 14.6a
Solar granules photographed from a balloon at 80,000 feet.

30,000 feet. As the columns of warm air ascend their temperature drops, and the moisture they have carried with them condenses into droplets of water, forming clouds and rain.

Sometimes, because of conditions in the atmosphere, condensation into clouds occurs only at the top of each upward-moving column of air. When this happens, the top of a column can be seen clearly as an isolated puff of cloud. Figure 14.6*b* shows a photograph of a field of cloud puffs formed in this way. Each cloud is analogous to a brightly glowing granule in the photosphere. Figures 14.6*c* and *d* compare the terrestrial clouds with solar granules and with convection cells produced in the laboratory by placing a pan of fluid over a uniformly heated surface.

Supergranules. Measurements of the Doppler shift in the solar spectrum reveal large-scale movements in the gas at the surface of the sun, similar to the movements of the gas in solar granules, but extending over much greater distances and persisting for longer times. The entire surface of the sun is broken up into a pattern of cells by these movements (Figure 14.6*e*).

The large cells are called *supergranules*. Each supergranule is about 30,000 kilometers in diameter, includes roughly 300 granules within its boundaries, and lasts for about one day. The supergranules are the second tier of convection cells in the sun.

(*b*)

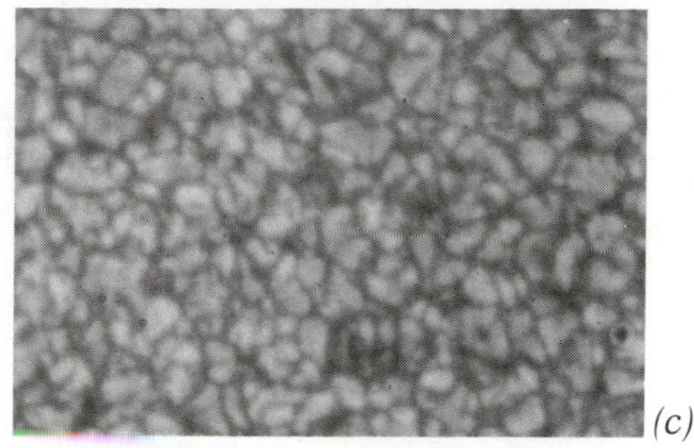
(*c*)

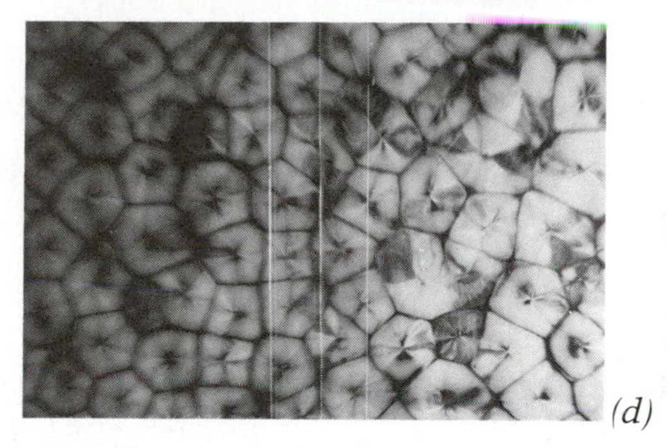
(*d*)

FIGURE 14.6b, c, and d (above)
Comparison of convection cells in (b) the earth's atmosphere showing an overhead view of altocumulus clouds, (c) the sun and (d) a laboratory experiment with a pan of fluid heated at the bottom.

(*e*)

FIGURE 14.6e (left)
A photograph utilizing the Doppler shift to reveal large-scale movements of material at the surface of the sun. Light regions are material moving toward the observer and dark regions are material moving away. The alternating pattern of light and dark indicates that the sun's surface is broken up into cells of moving material about 30,000 km in diameter.

The Solar Atmosphere

The region of tenuous and essentially transparent solar gas lying above the photosphere is called the *solar atmosphere*. The outer boundary of the solar atmosphere is not clearly defined. The atmosphere extends out to a distance of about 5 million kilometers from the sun, if its limit is considered to be the point at which the density of the solar gas has decreased to the density of the gas in the space between the planets (Figure 14.7).

FIGURE 14.7
A cross section of the sun showing the structure of the solar interior and atmosphere. The spacing of the dots represents the density of helium nuclei. Thicknesses of the photosphere and chromosphere are exaggerated by a factor of ten. Other regions are drawn to scale.

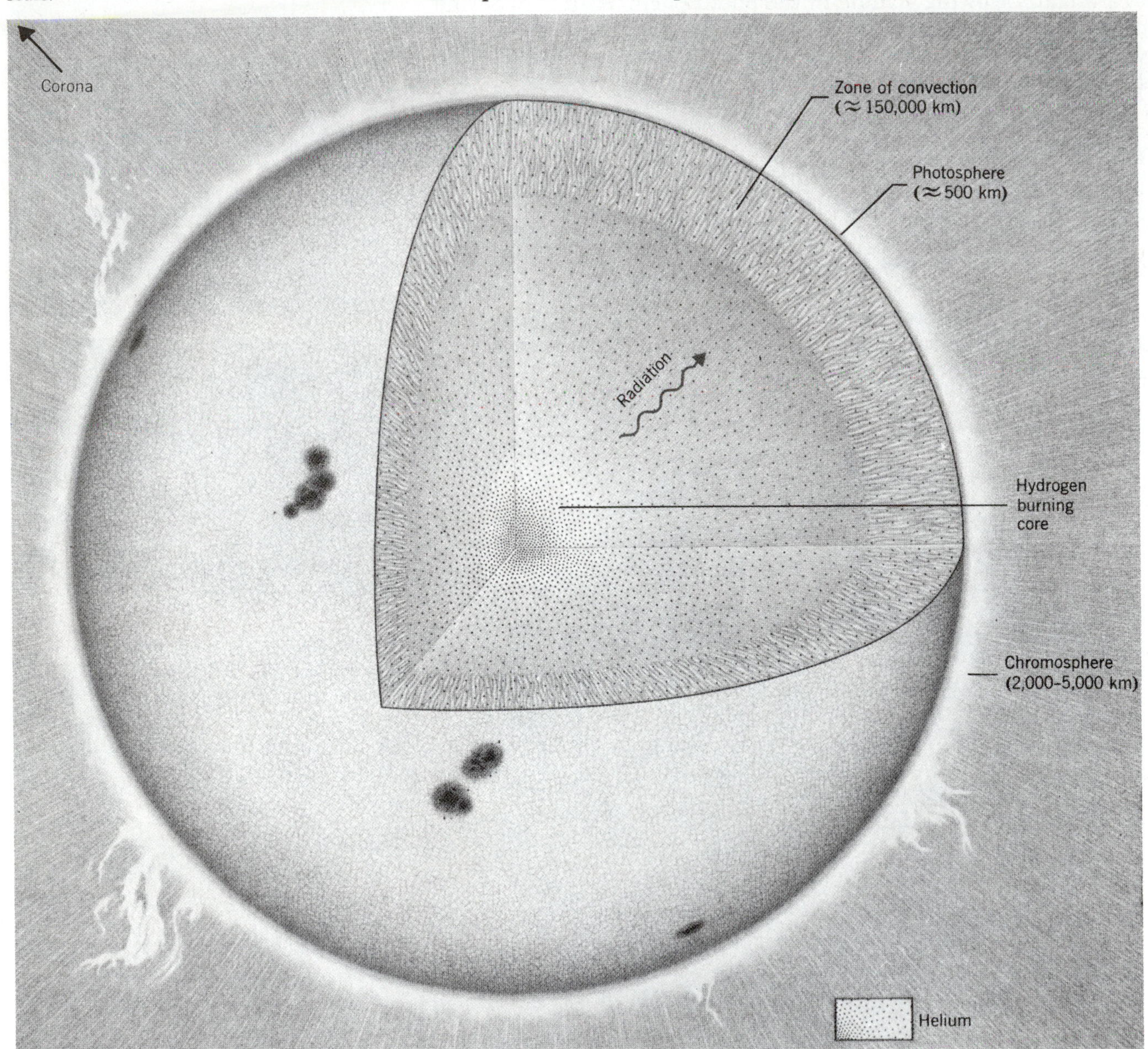

The solar atmosphere is divided into two regions called the *chromosphere* and the *corona*. Both regions are invisible under ordinary conditions because their faint luminosity is masked by sunlight that has been scattered in the earth's atmosphere or in the telescope itself.

These scattered photons create an apparent halo of light around the sun that is enormously brighter than the true solar atmosphere. A street lamp, viewed on a foggy night, possesses a similar halo of light scattered by the water droplets making up the fog.

During the brief moments in a total eclipse when the face of the sun is completely covered by the moon, the halo of scattered light disappears, and the solar atmosphere becomes visible as a luminous aureole surrounding the moon's black disk and extending out into the space around the sun to a distance of as much as 10 solar diameters or 14 million kilometers. The sudden appearance of this pearl-white luminescence, radiating as much light as the full moon and covering 100 times the moon's area, has a greater impact on the observer than any other visual display created by the movement of the celestial bodies (Figure 14.8).

The Chromosphere. Moving upward from the photosphere, the temperature falls from 6000°K to approximately 4000°K at a height of 500 kilometers above the surface, and remains at this relatively low temperature for a few thousand kilometers. This layer of relatively cool gas, the chromosphere, lying over the hotter gas beneath absorbs radiation at wavelengths characteristic of the atoms in the sun's composition. The chromosphere is the region in which the solar absorption spectrum is formed (page 132).

Photographs of the Chromosphere. The radiation from the chromosphere, summed over all wavelengths, is roughly 1000 times fainter than the radiation from the photosphere. For this reason the chromosphere is not visible in ordinary photographs of the sun taken in white light. However, if the light from the sun is passed through a filter that transmits light only in a limited band of wavelengths, a different situation may prevail. Suppose that the filter transmits light at the wavelength of the first Balmer line of hydrogen, at 6563Å, and blocks light at all other wavelengths. The 6563 Å line of hydrogen is one of the strongest absorption lines in the solar spectrum. This means that most of the 6563 Å photons coming from the lower levels in the photosphere are absorbed, and very few escape to space.

Each atom that absorbs a 6563 Å photon subsequently collapses to its initial state, emitting another 6563 Å photon, but this photon also is absorbed if the atom is at a depth where the density is substantial.

Only if the atom emitting the 6563 Å radiation is located in the chromosphere, where the density is lower and the gas is relatively

FIGURE 14.8
Visibility of the chromosphere during a total eclipse. (a) At the beginning of totality, the entire chromosphere is visible on the eastern limb of the sun. (b) A little later the moon's disk cuts off the light from the lower chromosphere. (c) Still later the entire chromosphere is concealed. The full sequence lasts 15 to 20 seconds.

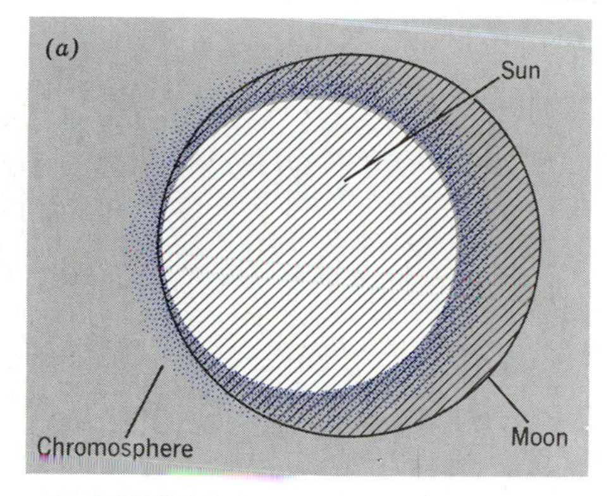

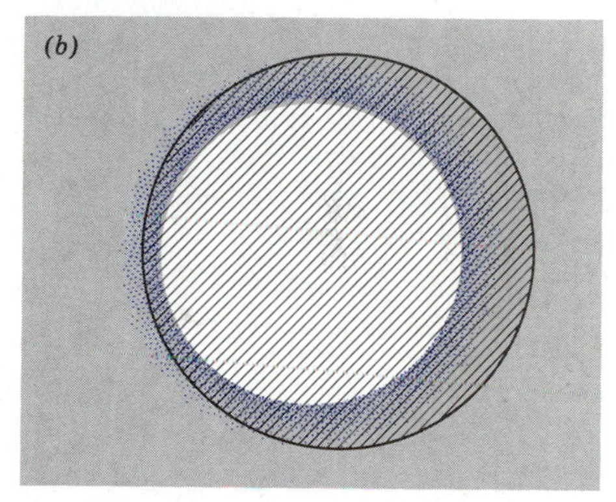

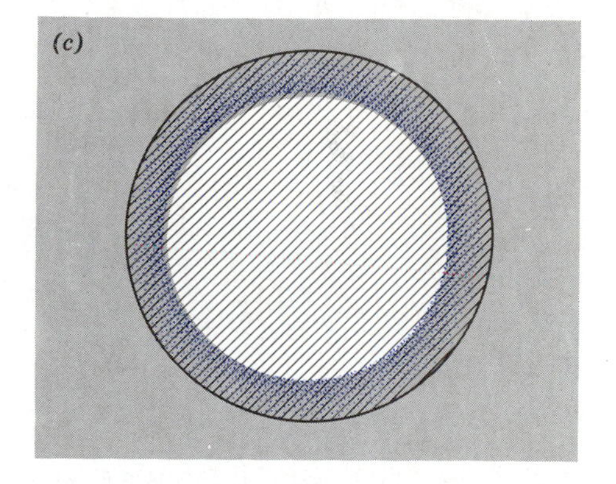

transparent, will a 6563 Å photon be likely to escape from the sun. Therefore, an observer, viewing the sun through a filter transmitting only 6563 Å radiation, does not see the photosphere; instead he sees the *chromosphere* (Figure 14.9).

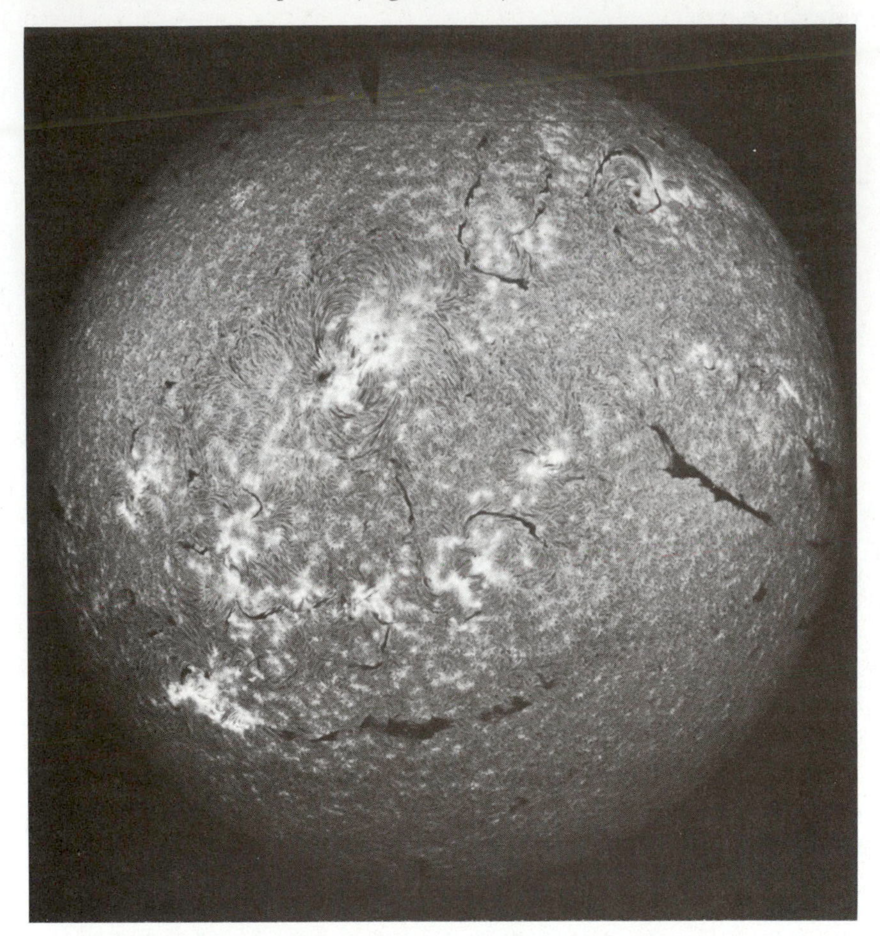

FIGURE 14.9
This photograph, taken at 6563 Å, shows the structure of the sun's chromosphere. The bright areas in the 6563 Å photograph are the disturbed hot regions in the chromosphere called plages (page 356).

The large plage area at upper left consists of two distinct regions separated by a distance of about 100,000 kilometers. Curved lines suggest that lines of magnetic force emerge from one group of sunspots and return to the other group, in a pattern resembling the magnetic field around a bar magnet. The photograph also shows dark wisps and streamers called filaments. These are relatively cool masses of gas far above the sun's surface, which absorb 6563 Å radiation strongly because of their lower temperature, and therefore appear black against the solar disk when the sun is photographed at this wavelength. If viewed at the limb of the sun against the dark of space, they appear highly luminous, and are seen to be identical with solar prominences (page 358).

It is possible to explore the structure of the photosphere and chromosphere at several levels by using slightly different wavelengths, all in the neighborhood of the 6563 Å line. Figure 14.10 shows the variation of absorption with wavelength in the vicinity of this line. If the filter transmits light at the precise center of the line, where the absorption is strongest, the photographs will show details of chromospheric structure several thousand kilometers above the surface. If the filter is modified to transmit light half or three-quarters of an angstrom away from the center of the line, the absorption will be somewhat less than at the center, and the photograph will show the structure of the chromosphere at a somewhat lower altitude. If the transmitted light is, say, two angstroms from the line center, where the absorption is relatively small, the photograph will show some of the structure of the underlying photosphere.

In this way, by varying the wavelength in the vicinity of a strong absorption line, the observer can see into the chromosphere to different depths and obtain a picture of the way in which the properties of the chromosphere change with height. Figure 14.11 illustrates the effect on a photographic image of variations in wavelength in the vicinity of the 6563 Å line.

FIGURE 14.10
Measurement of light transmitted through the chromosphere in the vicinity of the absorption line of hydrogen at 6563 Å. The shallow depression 2.5 Å to the short-wavelength side of the center of the 6563 Å line is the combined result of absorption by silicon atoms in the sun and water molecules in the earth's atmosphere.

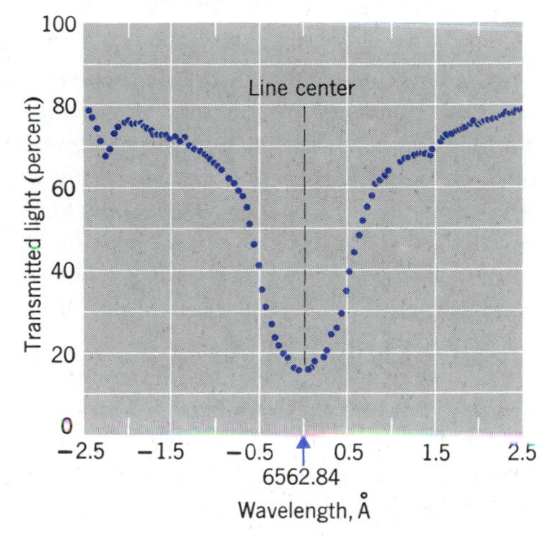

The Corona. Above an altitude of about 5000 kilometers in the chromosphere lies the region of the solar atmosphere called the *corona*. As seen during an eclipse, the visible corona extends out from the edge of the solar disk many millions of kilometers (Figure 14.12). When viewed from the ground, the luminosity of the corona fades into the background of scattered light from the sky at a distance of roughly 10 million kilometers from the sun, but photographs taken from a balloon at high altitudes, where the sky is darker, show a visible corona out to 30 solar radii or 20 million kilometers, and measurements of the influence of the corona on radio waves show a detectable effect halfway out from the sun to the earth. Other measurements made from satellites and space probes suggest that the corona has no outer boundary. A stream of gas called the *solar wind* flows out of the corona and into the solar system at all times, continuously immersing the earth and its sister planets in the tenuous gases of the solar atmosphere.

The Coronagraph. The chromosphere and corona can be studied at leisure, without limiting the period of observation to the few minutes of totality of a solar eclipse, by blocking out the interfering light from the disk of the sun artificially. A telescope can be modified for this purpose by placing an opaque disk, whose diameter is precisely equal to the apparent diameter of the sun, at the focus of the objective lens where the first image of the sun is formed. Telescopes constructed in this way are called *coronagraphs*.

The coronagraph has greatly increased the amount of information available to the solar astronomer. However, under the best conditions coronagraphs still are unable to detect the weak radiation from the outer corona, which is too faint to be seen except during a total solar eclipse. Because a total eclipse provides unique conditions for observing the outer corona, solar astronomers travel to remote and nearly inaccessible places, if necessary, to set up their instruments in a region in which the period of totality is greatest. Calculations based on the orbits of the earth and moon reveal that the longest period of totality in an eclipse is seven minutes and 40 seconds. Eclipses with periods of totality exceeding seven minutes are exceedingly rare, although three have occurred in this century, in 1937, 1955, and 1973. The next seven-minute eclipse will occur in 2150.

The Temperature in the Corona. The spectrum of the corona shows the lines of atoms that have lost many of their electrons. For

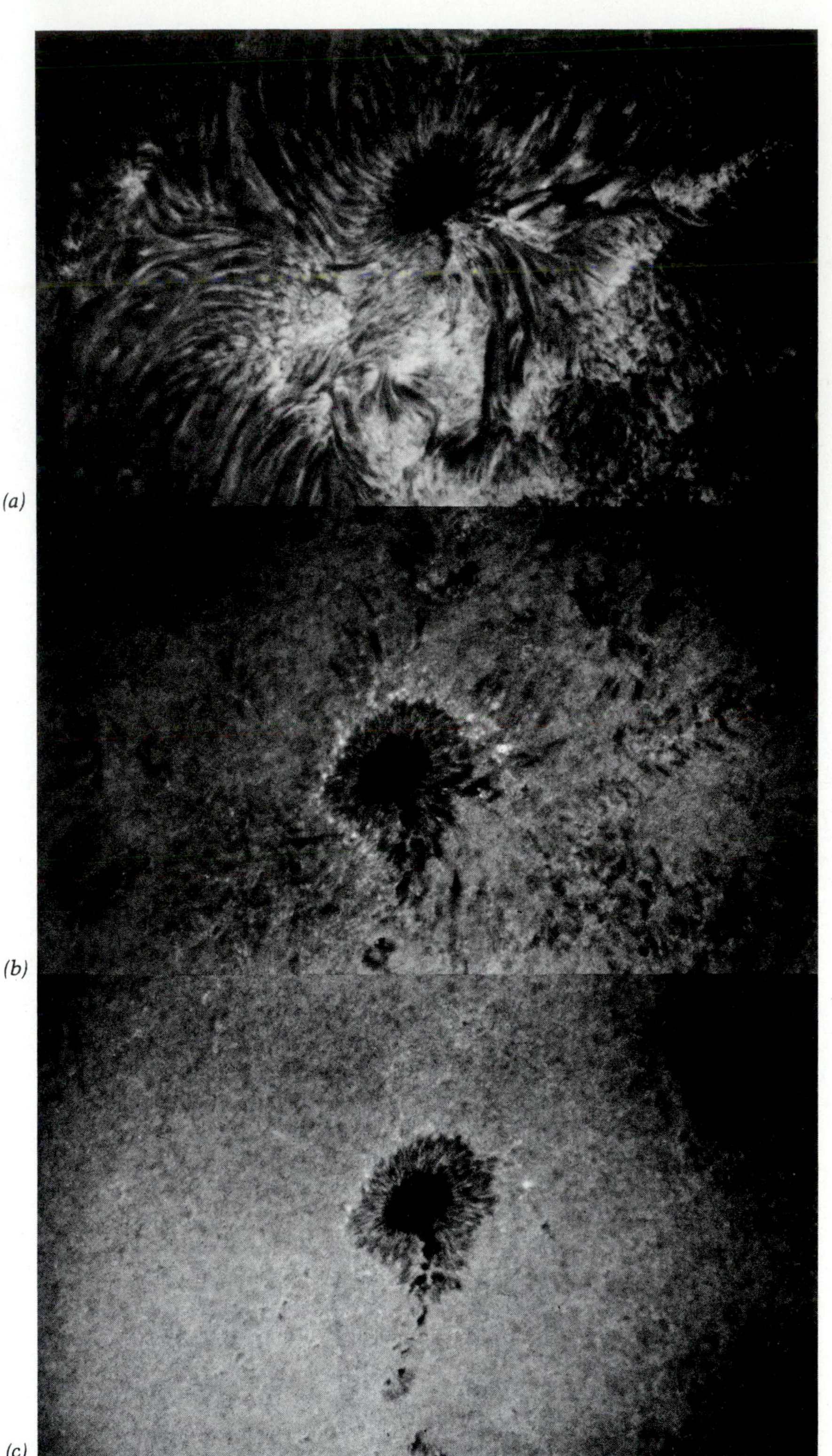

FIGURE 14.11

A sunspot seen at three wavelengths. Photographs (a), (b), and (c) show a sunspot observed through filters transmitting light at three wavelengths near the first Balmer line of hydrogen. The wavelengths were (a) 6563 Å, in the center of the absorption line, (b) ⅞ Å off center and (c) 2 Å off center. In each case the width of the transmitted wavelength was ¼ Å.

The photographs illustrate that we see into the chromosphere to a depth that depends on the wavelength of the light used in forming the image. In (a), because of strong absorption in the center of the line, the light from the photosphere is screened out, and only the structure of the chromosphere appears. Dark streaks reveal the direction and sometimes changes in polarity of magnetic fields in the chromosphere.

In (b) the absorption is less pronounced, and the structure of the photosphere emerges, together with features of the chromosphere, such as the dark jets of matter identified with spicules.

In (c) the wavelength is sufficiently far from the center of the line so that absorption is substantially reduced, spicules disappear, and further details of the photosphere appear, including granulation.

example, there are lines produced by transition in iron atoms that have lost 13 electrons. An enormous amount of energy is required to remove 13 electrons from an iron atom. The outermost electrons can be dislodged with relative ease, but the inner electrons are bound very tightly to the nucleus, and collisions of exceedingly great violence are required to eject them from their orbits. Collisions with the necessary degree of violence occur only at a temperature of 500,000°K or more. Other lines in the spectrum of the corona, such as the lines produced by transitions in calcium atoms that have lost 14 electrons, require a temperature of at least one million degrees.

The presence of these lines indicates that the temperature in large regions of the corona is close to one million degrees. This discovery led to one of the greatest problems in the study of the sun. What is the source of the energy that heats the corona to a temperature of a million degrees?

No completely satisfying resolution of the difficulty has been obtained thus far, but one factor that is believed to be important is the continual agitation of the gases of the corona by shock waves rising out of the turbulent, boiling surface of the sun. These shock waves are sharply defined sound waves, like a thunderclap or the boom produced by a supersonic aircraft. Each shock wave is a zone in the gas that has been compressed and heated. The compressed, heated wave travels through the atmosphere at approximately the speed of sound. The degree of heating produced by a single shock wave is very small, but if innumerable shock waves follow one another in rapid succession, the heating effect can accumulate and raise the temperature of the corona considerably. Intense magnetic fields on the surface of the sun may also contribute to the heating of the corona.

FIGURE 14.12
The solar corona 30 seconds after the start of totality during the eclipse of March 1970. Features are visible at a distance of 4.5 solar radii or 3 million kilometers. The sun's axis of rotation is about 40° counterclockwise from the vertical in the photograph.

SOLAR SURFACE ACTIVITY

Up to this point the sun has been treated as a typical Main-Sequence star, whose internal structure and evolution were discussed in detail because the sun's position on the Main Sequence, midway between the largest and smallest stars, made it particularly well suited to serve as an example of an average star. Throughout the discussion the questions of primary interest were the release of nuclear energy within the sun by the fusion of hydrogen, the transport of this energy to the surface, and its radiation into space. Theoretical studies of these processes, based on experiments in the nuclear physics laboratory combined with an enormous body of knowledge derived from stellar spectra, lead to predictions of the evolutionary tracks of stars of various masses that constitute one of the great syntheses of theory and observation in science.

These predictions seem to include all the basic properties of the sun as a star. They predict that a star with one solar mass, and roughly 5 billion years old, will have a luminosity of 4×10^{33} ergs/sec and a surface temperature of 6000°K. They also predict the details of the internal structure of a sun-sized star, including the amounts of hydrogen and helium at various distances from the center, the values of density and temperature at all depths from the surface to the center, and the nature of the zone of convection.

But they completely fail to predict the occurrence of solar "weather." They fail to predict violent storms that rage across the face of the sun in an 11-year cycle, geysers of hot gas that rise hundreds of thousands of miles into the solar atmosphere, and explosive outbursts of gamma rays, X-rays, ultraviolet radiation, and energetic protons that erupt sporadically from the surface.

We see the tempestuous solar "weather" only because of the closeness of the sun. If our star were thousands of light-years away, most of the activity on its surface would be undetectable. The reason is that the surface storms, violent as they are, contain only a millionth part of the sun's total output of energy. Consequently, they play a limited role in the broad scheme of evolution of Main-Sequence stars. However, they produce a great variety of displays that are scientifically interesting, strikingly beautiful, and of great practical importance to the inhabitants of the earth. A major branch of solar physics is devoted to the study of the pyrotechnics that play across the face of the sun, waxing and waning with the 11-year sunspot cycle.

Sunspots

Sunspots are the only sign of solar surface activity that can be detected with the naked eye under ordinary circumstances. When a large spot or group of spots is present on the face of the sun, it can be seen easily at sunset, or through a thin haze of clouds or a filter.

References to sunspots go back to a Greek observer in the fourth century B.C. There was little further mention of them during the long period in which western astronomy languished, the commonly held belief being that the sun was a perfect, unblemished sphere. The invention of the telescope in 1609 destroyed this illusion immediately. By 1611, four observers, among them Galileo, had independently studied the spots with the aid of the newly invented instrument.

When sunspots are photographed with high resolution under good seeing conditions, they look like irregularly shaped holes or craters in the sun's surface. Figure 14.13 is an example. It shows a photograph of a sunspot about 30,000 kilometers in diameter, taken

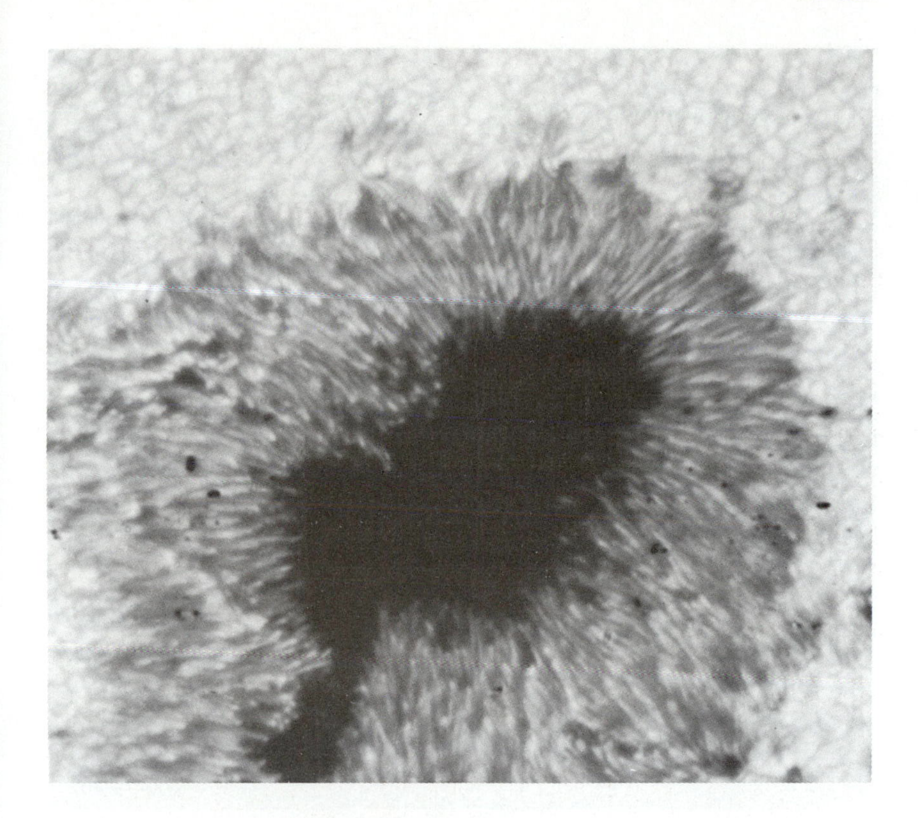

FIGURE 14.13
A sunspot photographed with a balloon-borne telescope. The dark umbra and bordering penumbra of the sunspot and the granules in the surrounding photosphere are visible.

through a telescope suspended from a balloon at an altitude of 80,000 feet. The black inner region of the spot is called the *umbra* and the surrounding fringe is called the *penumbra*. The granulation in the photosphere is also visible in this photograph.

The average size of sunspots is about 10,000 kilometers, but on rare occasions spots appear that extend across more than 150,000 kilometers of the sun's surface. The spots usually occur in pairs or in complex groups. Figure 14.14 (p. 352) shows a large sunspot group photographed with the 100-inch telescope on Mount Wilson that includes one enormous spot about 100,000 kilometers across.

Small sunspots persist for several days or a week, and the largest spots may last for many weeks, long enough to be carried across the entire face of the sun during the course of its rotation, and reappear about a month later on the opposite limb.

What are sunspots? Early observers of the spots through telescopes offered widely varying explanations. Some thought they were small planets circling the sun within the orbit of Mercury; others said they were mountains projecting above luminous clouds that covered the sun's surface; Galileo thought they were clouds drifting in the sun's atmosphere.

The answer turned out to be simpler than any of the theories proposed in early times. Sunspots are regions of the sun that are a

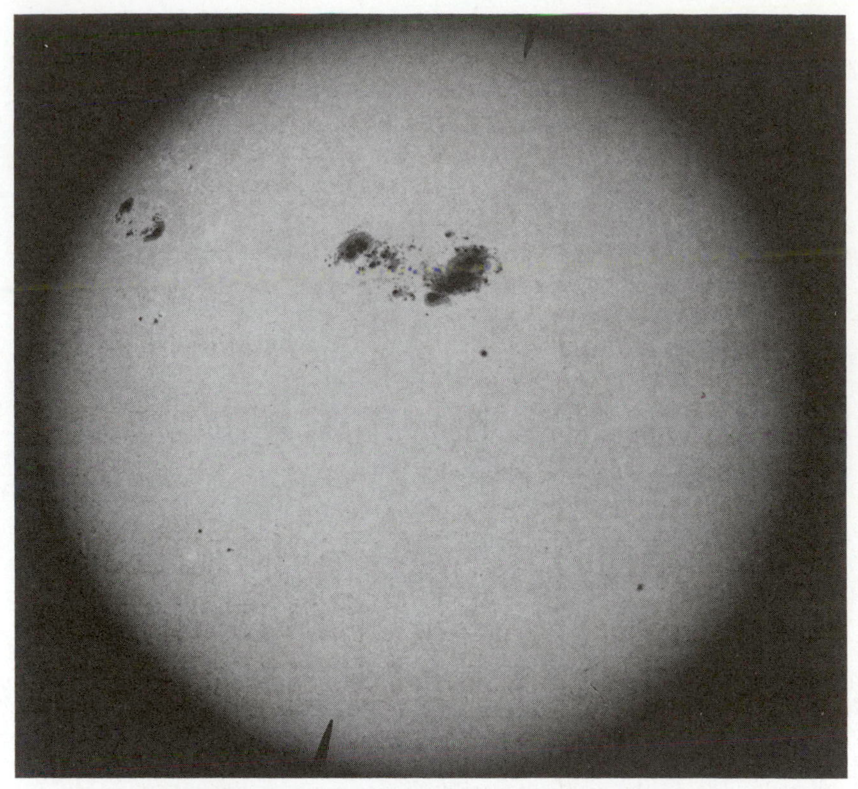

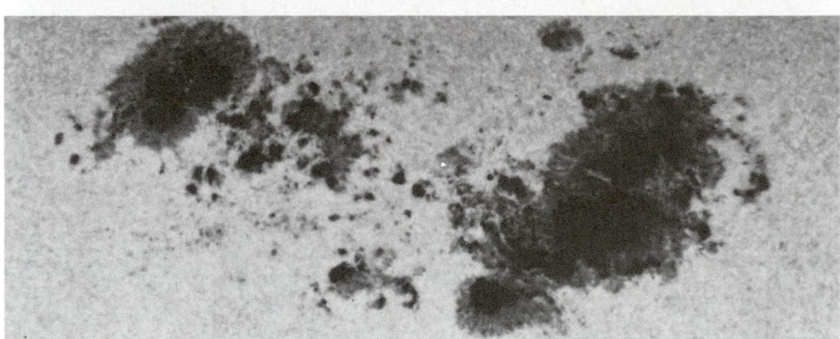

FIGURE 14.14
A large cluster of sunspots photographed with the 100-inch telescope in 1947 at sunspot maximum. The lower photograph is an enlarged view.

FIGURE 14.15
The sunspot cycle since 1760.

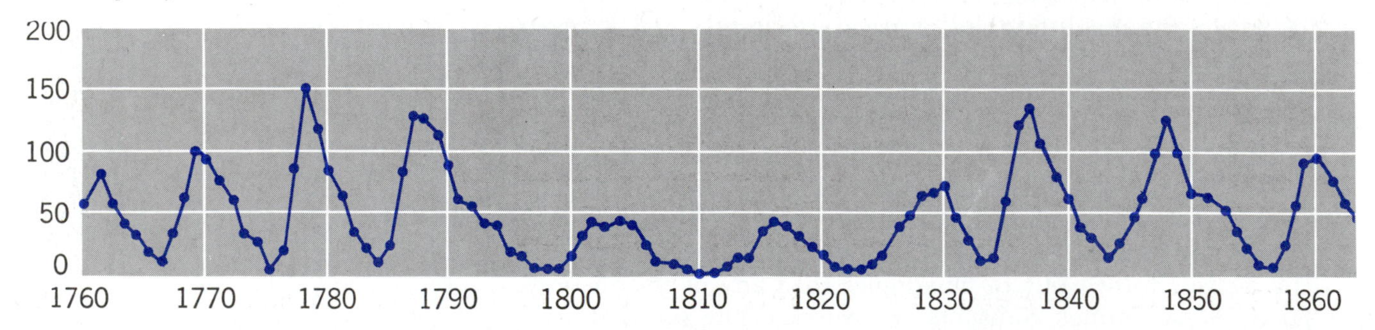

few thousand degrees cooler than the gas surrounding them. Consequently, they radiate less energy to space, and appear darker. The average temperature in a sunspot is about 4000°K, compared to 6000°K for the surface of the sun as a whole. Although the temperature difference is great enough to make sunspots appear black, the spots are intrinsically bright, the luminosity of a typical sunspot being hundreds of times greater than the light of the full moon. If a sunspot could be separated from the sun and viewed by itself in the sky, it would appear as an object of dazzling brightness.

The Sunspot Cycle. For about 200 years, astronomers have recorded the number of spots appearing on the face of the sun. Figure 14.15 shows the variation in the number of sunspots from 1760 to 1969. The graph clearly displays a cyclic rise and fall with maxima and minima recurring approximately every 11 years. The periodic change in sunspot number is called the *sunspot cycle*.

Although the average length of the sunspot cycle is about 11 years, the interval from one maximum in the cycle to the next has been as short as 7 or as long as 17 years. All the other manifestations of solar activity discussed later in this chapter, including flares, plages, and prominences, are keyed to the pattern of rise and fall in the sunspot numbers. In the years of sunspot maximum the surface of the sun is violently disturbed, and outbursts of particles and radiation of all wavelengths are a common occurrence. In the years of sunspot minimum, these outbursts are far less frequent.

The records of the positions of the spots also reveal the important fact that sunspots are almost entirely confined to the zone of latitudes between 40° and the sun's equator, and never appear near the poles. The spots are found at their highest latitudes at the start of a new sunspot cycle, immediately following the last minimum. Later in the cycle they tend to appear at lower latitudes, and the last spots of a given cycle usually lie close to the equator. No fully satisfactory explanation has been given either for the 11-year cycle or for the drift of the sunspots toward the equator during each cycle.

Very few sunspots were recorded by European astronomers during the 70 years from 1645 to 1715, although substantial numbers

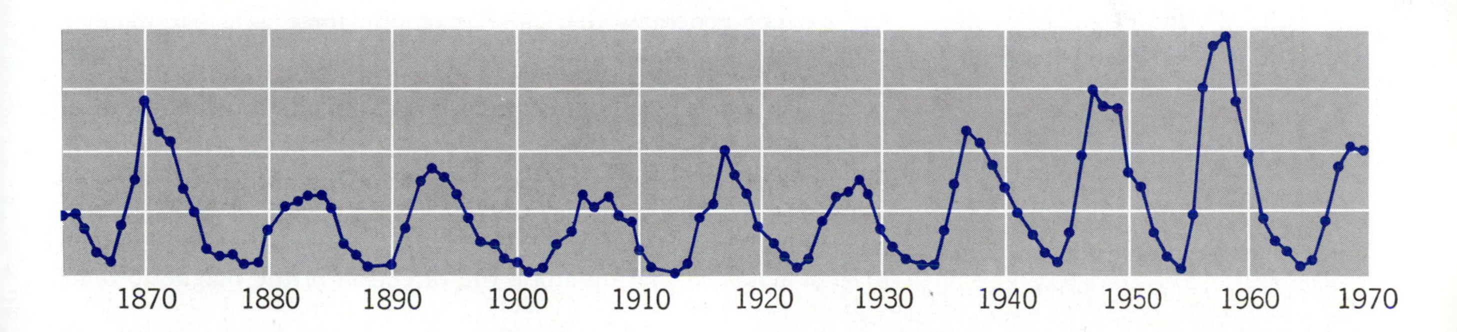

FIGURE 14.16
The splitting of a line into three separate lines (left) by the magnetic field in a sunspot (right). The black line on the right shows the position of the spectrograph slit with respect to the spot.

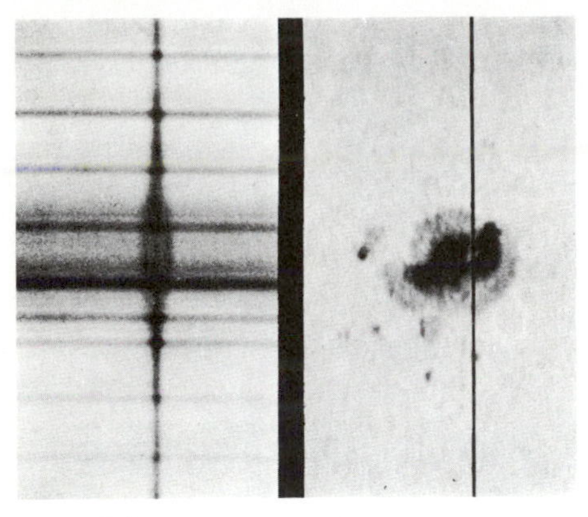

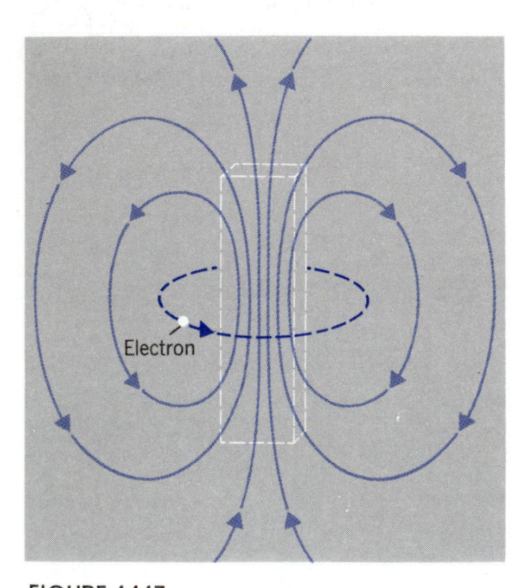

FIGURE 14.17
The magnetic force produced by an orbiting electron.

of spots had been reported by Galileo and other astronomers before 1645. Most nineteenth-century astronomers considered this early period of low sunspot activity to be the result of poor observations and incomplete records, but in the 1890s an English astronomer named E. W. Maunder reexamined the historical record and came to the conclusion that sunspots had, in fact almost entirely disappeared between 1645 and 1715.

For many years, the sunspot cycle was regarded as a nonvarying property of the sun, and Maunder's interpretation of the early sunspot records was considered questionable. However, evidence presented in 1976 suggests that his basic conclusion was correct: the average level of solar activity seems to have waxed and waned several times over intervals of hundreds of years. Furthermore, the changes in the level of solar activity appear to be correlated with changes in the climate of the earth. This correlation is difficult to understand. A direct causal link between solar activity and climate seems unlikely, since the energy contained in solar surface outbursts is a million times smaller, on the average, than the solar energy that heats the surface of the earth. The explanation may be that the sun's total output of energy—the solar luminosity—varies somewhat over periods of one hundred years. This variation, which could easily produce changes in the earth's climate, might be accompanied by simultaneous changes in the level of solar activity.

Magnetic Fields in Sunspots. If the light from a sunspot is passed through a spectrograph, some of the lines in the sunspot spectrum are found to be similar to the lines in the normal absorption spectrum of the sun, but others are strikingly different. The special lines are distinguished by the fact that each one is split into two or more closely spaced components. Figure 14.16 shows how a single line divides into three components in the interior of a sunspot.

The splitting of a spectral line into several separate lines, called the Zeeman effect, is a well-known phenomenon in the laboratory. It is produced by a magnetic field acting on the atoms of gas that radiate the spectrum, in the following way. Each atom consists of electrons orbiting around a central nucleus. The orbiting electrons are moving charges of electricity, equivalent to rings of electric current. According to the laws of electromagnetism, a ring of electric current generates the same magnetic force as a bar magnet (Figure 14.17). The magnetic field in the sunspot, acting on the innumerable atom-sized bar magnets, distorts their structure and changes the energy levels of each atom. The shift in energy changes the wavelengths of the lines in the spectrum. The change in energy is small and, therefore, the change in wavelength is also small.

Why is each line split into several separate lines, instead of being just shifted in wavelength? The reason is that every tiny atom-magnet tends to line up along the direction of the magnetic field but, according to the laws of the atom, it can only line up in certain allowed directions, just as electrons in the atom can only occupy certain allowed orbits (see pages 135–137). Each allowed direction corresponds to a different energy for the atom, and therefore a different wavelength for the photon emitted in a transition. In the case of the particular element producing the line shown in Figure 14.16, the atom-magnets are allowed to line up with the magnetic field, *opposite* to the field, or at *right angles* to it. The three directions give rise to three energy levels and three separate lines.

A theoretical study yields a formula connecting the degree of splitting in the Zeeman effect, that is, the separation between the lines, and the strength of the magnetic field. In this way, the measured separation between the lines yields the information that magnetic fields in sunspots are as high as several thousand gauss.

A magnetic field of several thousand gauss is a very strong field. By comparison, the strength of the earth's magnetic field is less than one gauss.

The Origin of Sunspots. Why is a sunspot cooler than its surroundings? Heat normally flows from hotter to cooler regions; why doesn't it flow into the sunspot from the surrounding high-temperature gas, and eliminate the difference in temperature?

The answer is connected with the very strong magnetic fields that exist in the interiors of sunspots. The critical point in the explanation of sunspots is that these strong magnetic fields bend the paths of electrically charged particles. If the particle is inside the region of the magnetic field, its path is bent into a circle (page 289). If the particle is outside the field and its motion carries it toward the field, it is deflected at the boundary and prevented from entering (Figure 14.18).

FIGURE 14.18
Deflection of electrically charged particles at the boundary of a magnetic field.

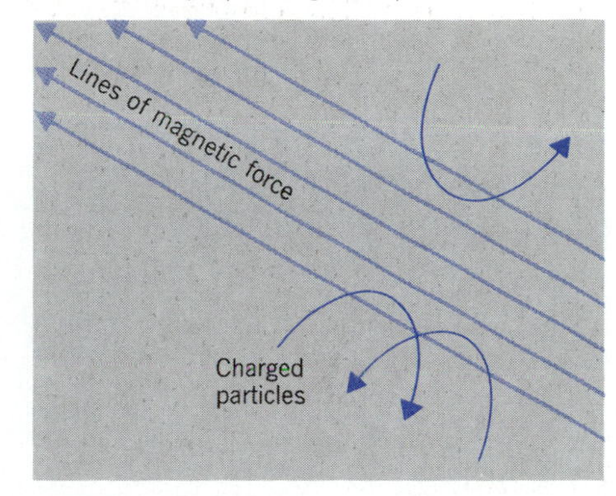

All the particles in the sun's interior are electrically charged. These particles carry heat from the interior of the sun to the surface in the form of convection currents. However, charged particles cannot readily enter regions where strong magnetic fields are present. Consequently the convection currents are partly suppressed, and heat is prevented from reaching the surface.

This explanation accounts for the fact that a sunspot is colder than the surrounding surface of the sun when it is first formed. However, it does not explain why the spot *remains* cold. Heat should leak into the region under the spot in the form of photons, which are not electrically charged and, therefore, cannot be affected by the magnetic field. Theoretical studies suggest that a sunspot should warm up to a normal temperature in a few days as a result of the heat carried into its interior by the flow of photons. But large sunspots live for many weeks. What keeps them cold? The answer is not clear. A complete theory of sunspots is one of the major challenges in solar astronomy.

FIGURE 14.19
The magnetic field over a pair of sunspots.

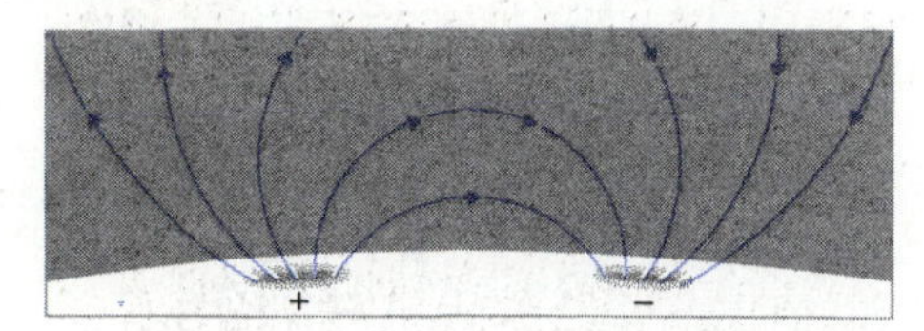

Sunspot Pairs. Sunspots often appear on the surface of the sun in pairs, aligned in approximately an east-west direction. Magnetic field measurements using the Zeeman effect show that the two sunspots often have opposite magnetic polarities, and are connected by lines of magnetic force that emerge from the surface at the position of one spot and reenter at the position of the other, as in Figure 14.19.

For a given sunspot cycle, and a given hemisphere on the sun, the polarity of the leading, or westernmost, sunspot is always the same.

Plages, Flares, Prominences, and Filaments

Plages. Figure 14.20*a* shows a sunspot group photographed near the limb of the sun on October 8, 1964. Figure 14.20*b* shows the same area photographed in 6563 Å radiation, revealing the structure of the chromosphere. The bright glow in the 6563 Å photograph indicates a highly disturbed condition in the gas above the sunspots. These bright patches are called *plages* (French: beaches). They are nearly always found above regions in the photosphere in which a strong magnetic field exists, regardless of the presence of sunspots.

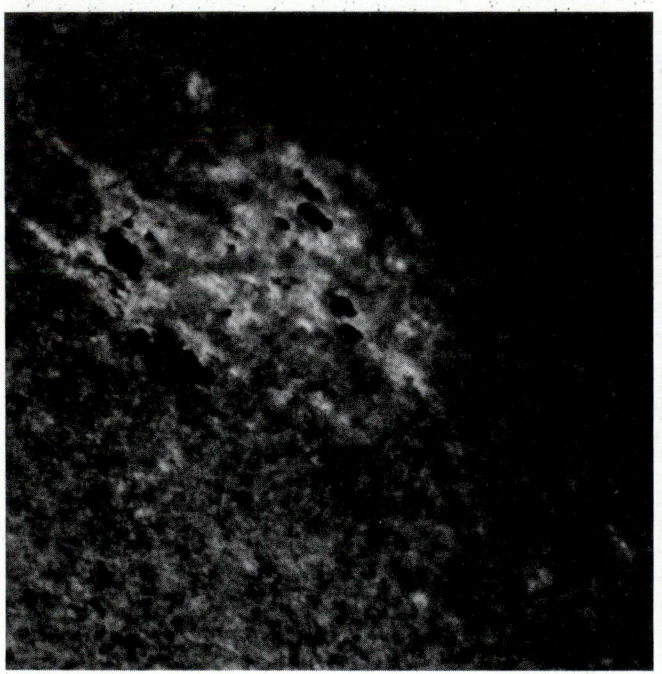

(a)

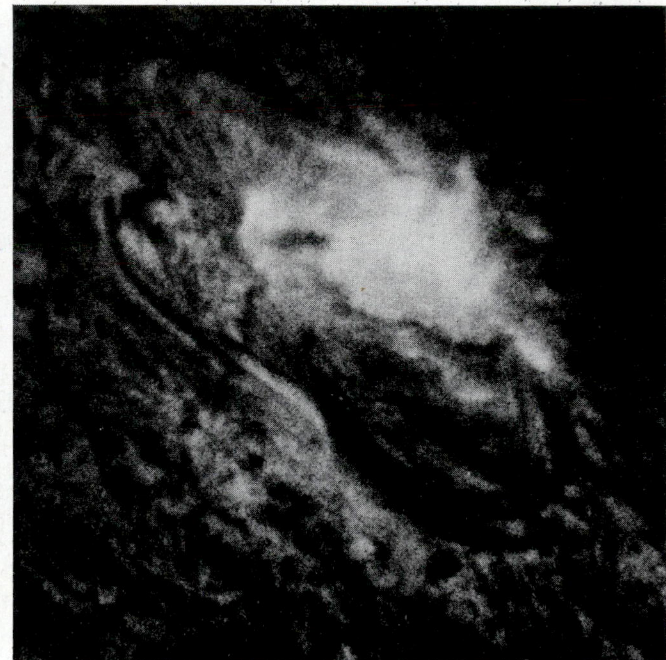

(b)

FIGURE 14.20
(a) A sunspot group photographed in white light near the limb of the sun on October 8, 1964. (b) The same region photographed in 6563 Å light, showing plage in the chromosphere above the sunspot group. See also Figure 15.9.

A plage usually precedes the appearance of sunspots in a high-magnetic-field region on the sun. It remains while the sunspots are visible, and persists for several weeks or more after the spots disappear.

A plage is a region in the chromosphere that apparently has been heated to incandescence by a local concentration of magnetic fields.

A satisfactory explanation of the heating action of the fields has not been provided. It is possible that the lines of magnetic force guide energetic charged particles to the region, forming small volumes of hot, dense gas; or the field lines may act as guides for traveling vibrations, similar to the sound waves caused by a thunderclap or a supersonic boom, conducting the vibrations upward into the chromosphere and corona (page 345).

Flares. Frequently during peak periods of the sunspot cycle, and less often at other times, the surface of the sun is marred by explosive outbursts of energy that hurl particles and radiation into the solar system. These outbursts, called *solar flares,* usually are observed in the vicinity of large, complex groups of sunspots such as the group in Figure 14.14.

Flares build up to peak intensity in a few minutes, and disappear in a period varying from 10 to 15 minutes to several hours, depending on the size of the flare. The burst of radiation and energetic particles produced by a large flare may play havoc with radio communications and cause substantial changes in the normal magnetic field of the earth. Large flares also create the luminous draperies and filaments of the aurora at latitudes as low as the southern United States and Mexico, which normally do not witness these spectacular atmospheric displays.

An exceptional series of solar flares occurred in August 1972. The flares originated in the disturbed region of the sun's surface that had first appeared a month earlier as a small and innocuous cluster of sunspots. A photograph of the sun on August 4 of that year is shown in Figure 14.21. The flare region is in the dead center of the disk. Figure 14.22 is a detailed view of a part of the disturbed region, showing the very rapid buildup of the large flare that occurred on that day. The visible luminosity of the flare grew to full intensity in 18 minutes, and disappeared an hour later.

FIGURE 14.21
A strongly disturbed region in the center of the sun's disk where the August 4, 1972, flare occurred.

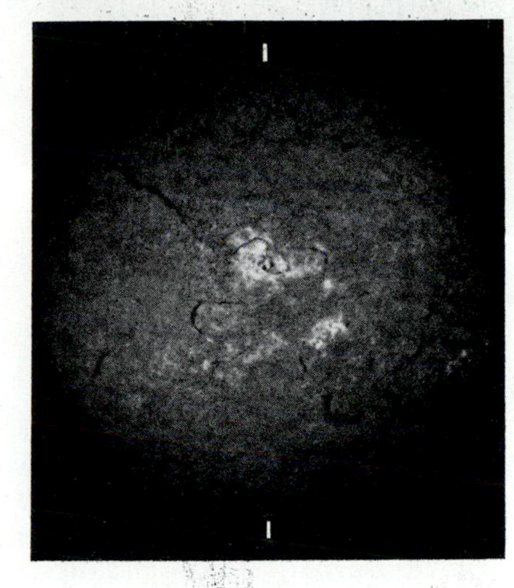

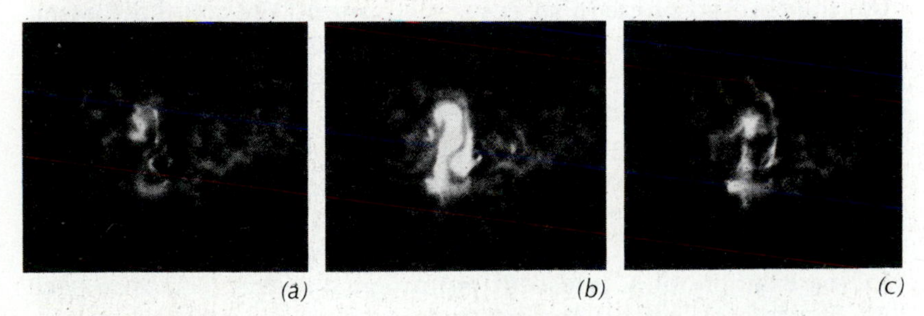

FIGURE 14.22
Buildup of an intense flare on August 4, photographed in 6563 Å light. (a) 0620; (b) 0638; (c) 0738 (Greenwich Mean Time).

A spectacular eruption of luminous gas, shown in Figure 14.23 (p. 358), occurred on August 11, when the disturbed region had rotated beyond the edge. The arrow indicates an ejected mass of material moving rapidly away from the sun. This material rose to a height of 250,000 kilometers above the surface in approximately 20 minutes.

High-energy protons ejected in the flare of August 4 ionized

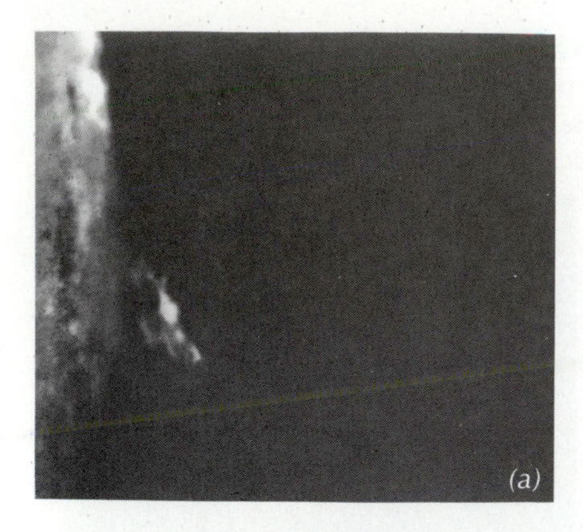

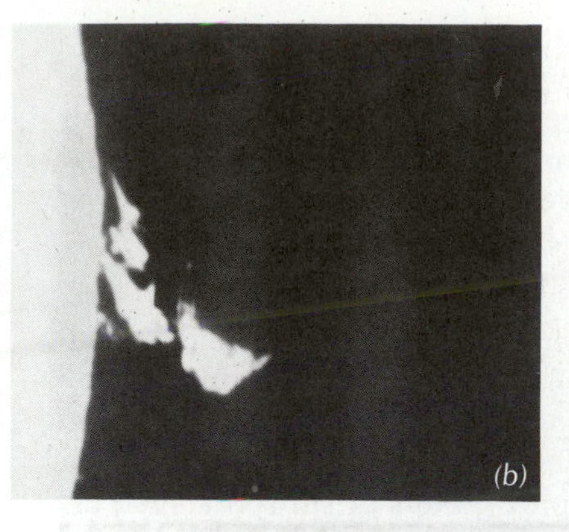

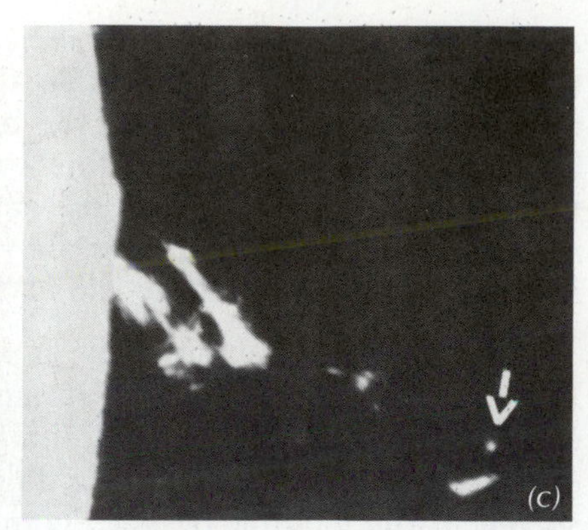

FIGURE 14.23
Ejection of a huge mass of material into space from the disturbed region at the limb of the sun on August 11. The arrow points to ejected matter 250,000 kilometers above the sun's surface. The times are (a) 2029; (b) 2040; (c) 2049 (Greenwich Mean time).

atoms and molecules in the earth's upper atmosphere and disrupted radio communications at high latitudes for several days. Slower-moving particles traveling outward from the sun strengthened the normal force of the solar wind (page 347) on arriving at the earth two days later. Some of the charged particles in the solar wind penetrated the earth's magnetic field, while others were turned aside by it. Both effects combined to create a major disturbance in the geomagnetic field, triggering strong currents of electricity in the ground and tripping circuit breakers in power lines in several places in the United States and Canada.

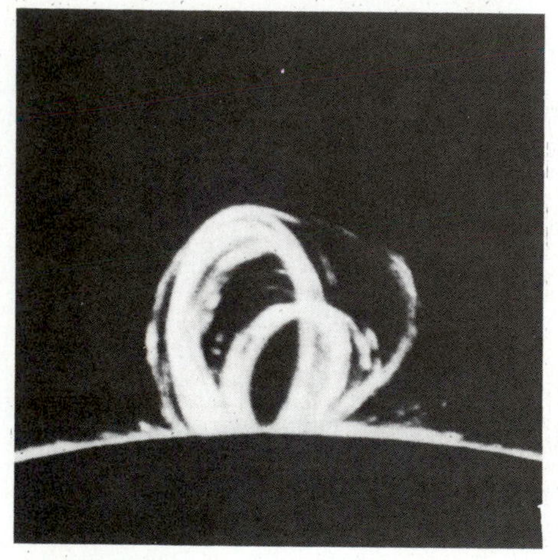

FIGURE 14.24
A giant loop prominence photographed in 6563 Å light. Although the appearance of the loop suggests that the material rises out of the surface along one branch of the loop and returns along the other, motion pictures show, surprisingly, that luminous regions condense out of the corona at the top of the loop and move downward along both branches.

Prominences. Prominences are masses of luminous gas that appear in the corona far above the sun's surface. Prominences consist of gas that is cooler and denser than the surrounding corona. They are luminous because at the relatively low temperatures and high densities that prevail in a prominence, its ions recapture electrons and emit photons.

Sometimes the material in a prominence seems to rise upward from the chromosphere in surges and eruptions; the cloud of ejected matter in Figure 14.23 is an example of an eruptive prominence. At other times, the material in the prominence streams downward from great heights, like luminescent rain. Often the glowing streams of gas form graceful curves that appear to be shaped by lines of magnetic force looping upward out of the chromosphere (Figure 14.24). Many prominences are relatively stable and quiescent, and seem to float for hours or days above the solar surface. A striking example of a quiescent prominence appears opposite page 335. Figure 14.25 shows another extraordinary prominence.

Filaments. Photographs of the chromosphere above a region of high magnetic field often show long dark streaks called *filaments.* These are regions of relatively cool and dense gas that appear dark against the solar disk in 6563 Å photographs, because they are

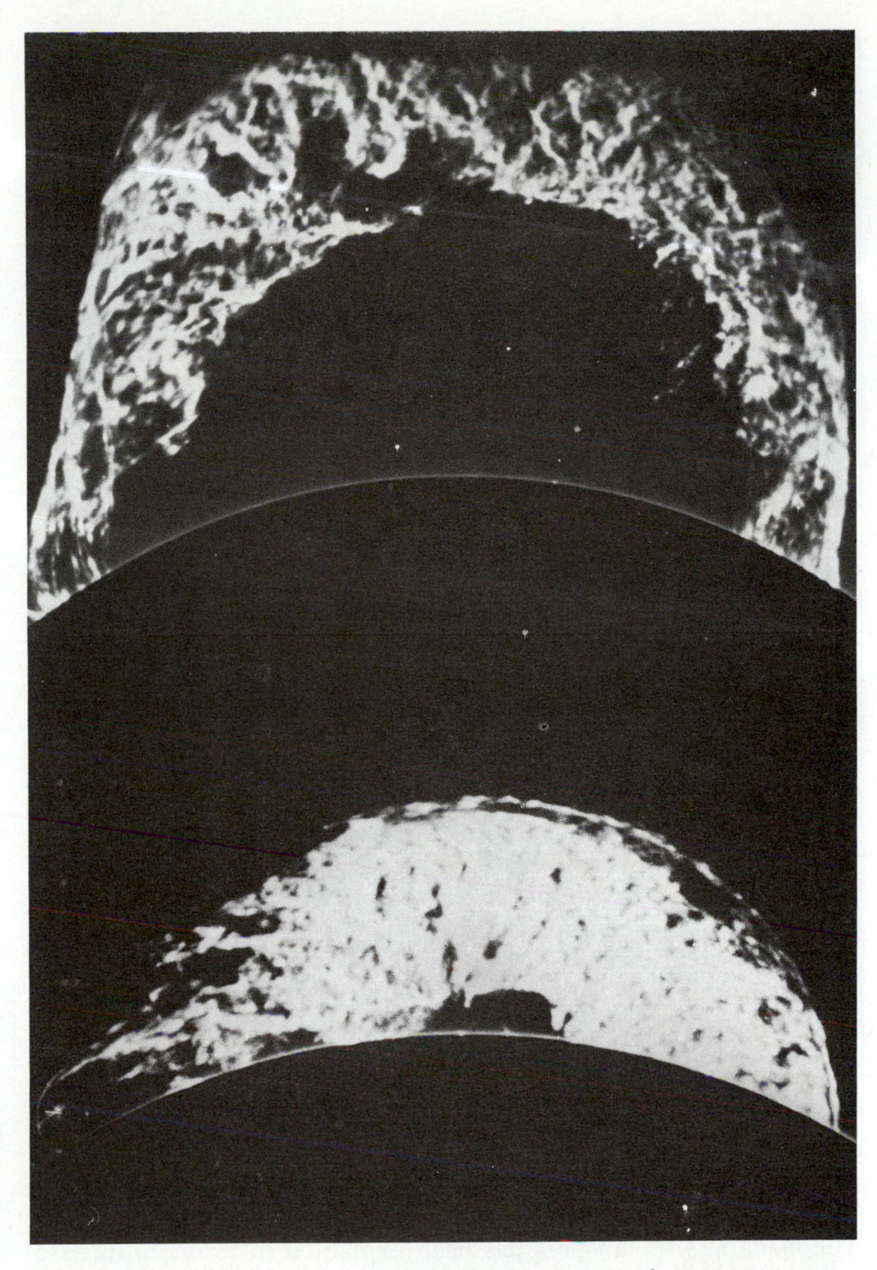

FIGURE 14.25
Two stages in the eruption of one of the greatest solar prominences ever photographed, on June 4, 1946. The eruption had commenced prior to sunrise. Below, shortly after sunrise; the prominence was already 200,000 kilometers above the sun's surface. Above, 40 minutes after, it had risen to a height of 500,000 kilometers, expanding at a speed of about 400,000 kilometers (250,000 miles) per hour.

cooler than their surroundings and absorb 6563 Å radiation more effectively. Several filaments are visible on the sun's face in Figure 14.11.

Filaments and prominences are the same objects. If a filament is followed to the edge of the solar disk during the sun's rotation and photographed silhouetted against space, it appears as a luminous prominence (Figure 14.26) (p. 360).

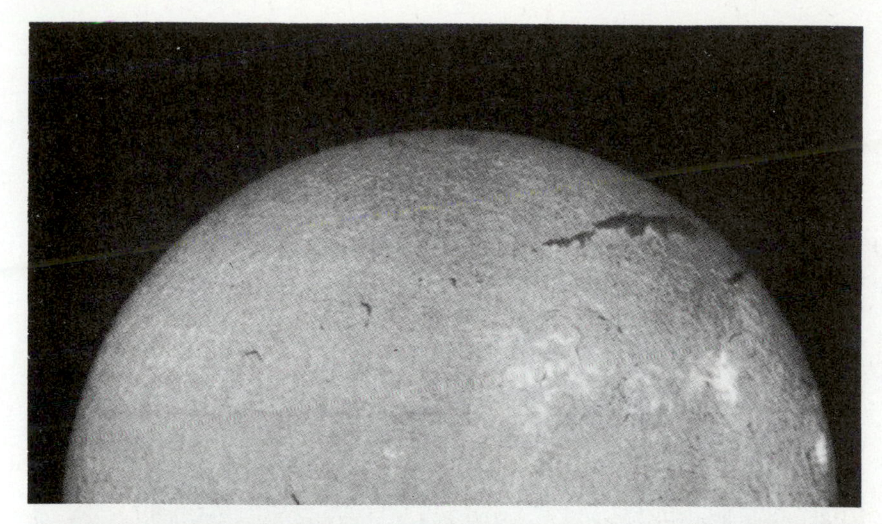

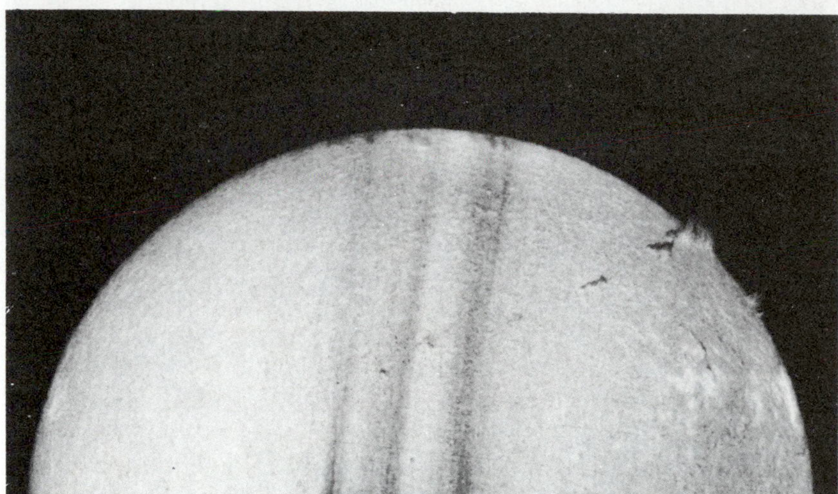

FIGURE 14.26
A filament on the face of the sun rotates to the limb to become a prominence.

The Sun's Rotation and The Solar Magnetic Field

The blackness of sunspots has been explained in terms of intense magnetic fields of thousands of gauss that exist in the interior of a typical sunspot. Less intense but still relatively large magnetic fields are also observed outside the sunspots, both in the photosphere and in the chromosphere above. In fact, sunspots, flares, prominences, and all other disturbances on the surface of the sun seem to be connected with the presence of magnetic fields that are concentrated in relatively small areas of the photosphere and chromosphere (Figure 14.27). The sun has a weak, irregular surface field

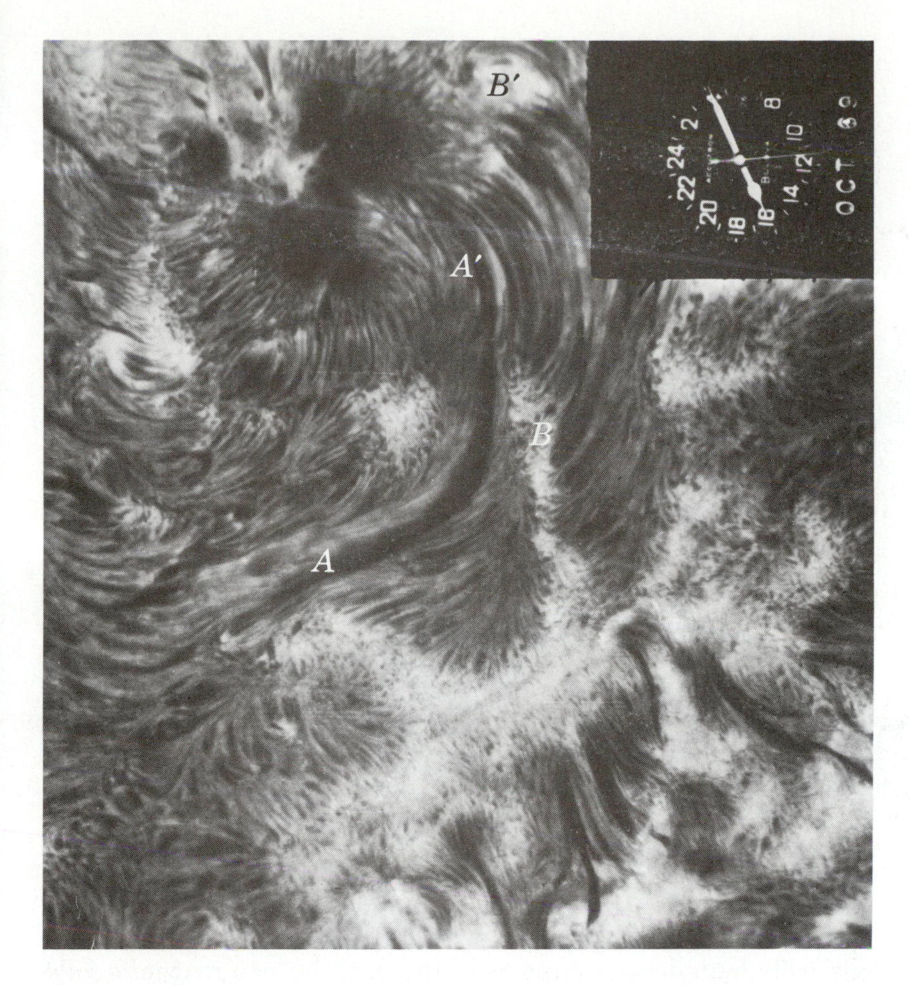

FIGURE 14.27
Chromospheric structure near a sunspot group. This 6563 Å photograph shows two sunspot groups (lower right and upper left). The bright areas are plages. The dark streak AA′ is a filament (page 358 and Figure 14.26). Numerous streaks and threads of material, dark because they absorb 6563 Å radiation strongly, originate in the plages. The threads indicate the magnetic field pattern in the chromosphere. An example is seen to the right of the larger sunspot group, where dark threads emerge from one plage (b) and re-enter the surface at another plage (B′). The regions B and B′ clearly have opposite polarities.

with a strength of roughly one gauss, which is too little to provide a simple explanation of the strong fields in sunspots and disturbed areas.

One explanation of the strong fields in sunspots that has won adherents among astronomers connects these intense magnetic fields to a peculiar property of the sun's rotation. The sun rotates on its axis approximately once a month, but the rotation is more rapid at the equator than at other latitudes. The period of rotation is roughly 26 days at the equator, 28 days at a latitude of 45°, and still longer at higher latitudes. A rigid body like the earth or the moon could not rotate in this way, but the sun, being a gaseous sphere, is capable of doing so. Important consequences for the solar magnetic field can be deduced from the fact that this variation in the sun's rate of rotation occurs.

The starting point in the discussion is the assumption that the sun's general field resembles a bar-magnet field at high latitudes (Figure 14.28). If the resemblance to a bar-magnet field were accu-

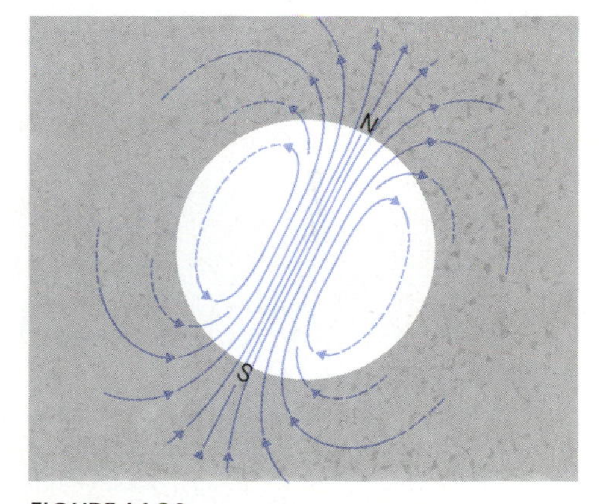

FIGURE 14.28
The lines of force of a bar magnet.

rate, the lines of force would emerge from the north pole, curve around in space and reenter at the south pole. At low latitudes, the lines of force would run largely in a north-south direction, as in the figure. We know that the magnetic force at the surface is highly irregular and variable at low latitudes, and does not bear any resemblance to the uniform north-south field shown in the figure, but it is possible that the irregularities in the field only exist near the surface, and that the magnetic field in the sun's interior has a more uniform appearance.

FIGURE 14.29
A line of magnetic force running from north to south is stretched into a long loop by the more rapid rotation of the sun at the equator.

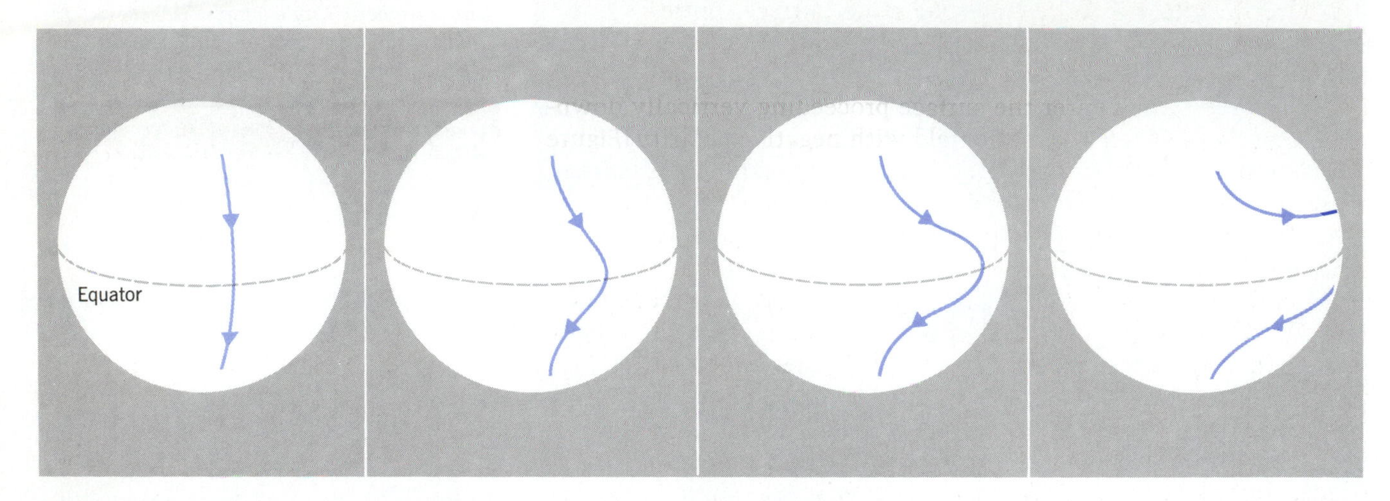

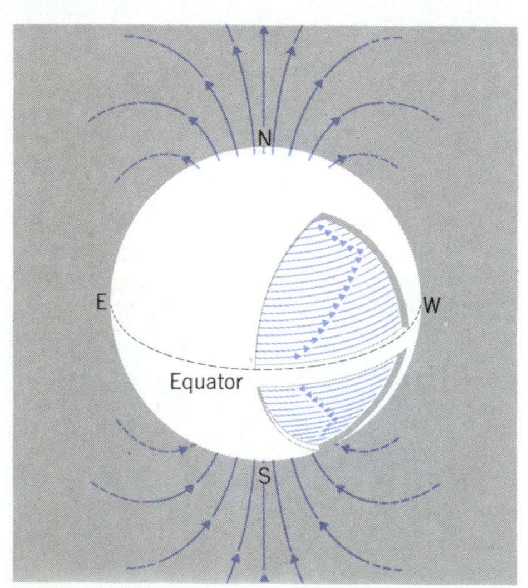

FIGURE 14.30
The magnetic field in the interior of the sun after many rotations. The field lines are stretched into an intense east-west field under the surface as a result of the sun's differential rotation.

Now consider the effect of the sun's differential rotation on a bar-magnet field in the interior. The interior of the sun contains electrically charged particles—mainly electrons and protons—at a relatively high density. Studies of the behavior of a magnetic field in a dense, gaseous mixture of electrons and protons show that the lines of magnetic forces are carried along with the flow of the particles in the gas. Thus, lines of force that would normally run from north to south in the sun are stretched out into long loops by the more rapid rotation of the gas at the equator, as in Figure 14.29. After the sun has rotated a number of times, the direction of these lines of force is changed from north-south to east-west. When several years have elapsed and many rotations have occurred, the lines of force become very tightly wrapped around the sun at low latitudes, creating an intense east-west magnetic field just below the sun's surface, as in Figure 14.30.[5]

Theoretical studies of magnetism also show that two adjacent lines of magnetic force tend to move apart, as if they had a physical reality and were acted on by a force of mutual repulsion. The repul-

[5] There is also evidence that the rate of rotation changes with depth, enhancing the effect of the differential rotation.

sion increases in strength when the field is strong. If the field is many thousands of gauss, this repulsive force can become larger than the force of the sun's gravity. The studies indicate that when this happens, a loop of magnetic force may burst out of the sun's surface (Figure 14.31).

In addition, because a magnetic field deflects charged particles the gas in a strong-field region has a lower density than the surrounding gas and becomes buoyant. The buoyancy adds to the upward force propelling the material to the surface.

Where the lines of force leave the sun's surface, there is an intense magnetic field directed vertically upward, that is, with positive polarity. Where the lines of force return to the sun at the other end of the loop, they enter the surface proceeding vertically downward, creating a strong magnetic field with negative polarity (Figure 14.32). This description corresponds to the appearance of pairs of sunspots with magnetic fields of opposite polarity, that are commonly observed in disturbed regions on the sun.

Figure 14.33 (p. 364) shows the sequence of events in a disturbed region. Prior to the commencement of the disturbance, the lines of magnetic force run evenly beneath the surface in an east-west direction (Figure 14.33*a*). In the first phase of the disturbance, lines of force rise to the surface and break through, forming a magnetic arch connecting two regions of moderately strong magnetic field and opposite polarity (Figure 14.33*b*). Charged particles are constrained to follow the direction of the magnetic arch, forming a pattern of moving gas in the chromosphere that reveals the direction of the lines of force when the region is photographed from above in 6563 Å light, as in Figure 14.27. Particles and sound waves are guided upward along the lines of force, heating the chromospheric gas to incandescence and forming plages.

In the second phase, the lines of force, continuing their upward movement, leave the surface of the sun at a nearly vertical angle and loop far out into the corona, occasionally reaching heights of 100,000 kilometers or more (Figure 14.33*c*). The regions in which the lines of force go through the surface vertically, and where the field strength is greatest, form the interiors of sunspots. The lines of force looping upward high into the corona sometimes lead to the luminous veils, streamers and loops classified as prominences.

In the next phase of the storm, the magnetic field diminishes in strength in the disturbed region, the sunspots disappear, and the region regains the appearance of Figure 14.33*b*, with magnetic arches and overlying plage.

Normally, an interval of 10 to 15 days elapses between the initial appearance of plages and the peak level of activity, with a gradual buildup of sunspots, prominences, and flares. Most of the sunspots are gone after about four weeks or one solar rotation. Filaments often form and erupt during this period. After two rotations the

FIGURE 14.31
A loop of magnetic force bursts out of the sun's surface.

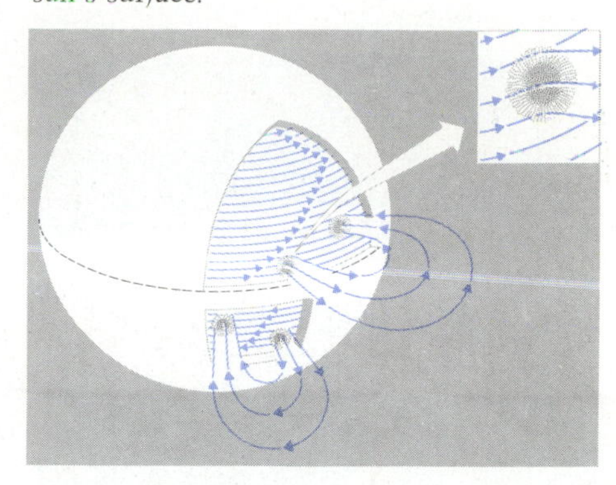

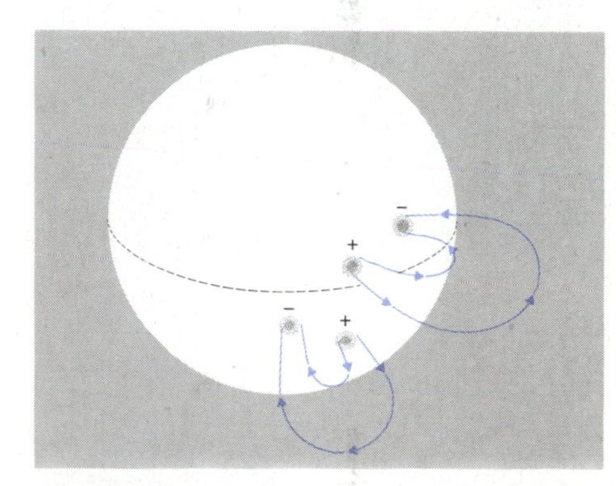

FIGURE 14.32
A pair of sunspots with opposite magnetic polarity remains where the loop of force broke through the surface. In the present sunspot cycle, the polarity of the leading (westernmost) spot is negative in the northern hemisphere and positive in the southern hemisphere. The polarities will reverse in the next sunspot cycle when the general magnetic field of the sun reverses.

FIGURE 14.33
Sequence of events in a disturbed region on the sun's surface. The lines indicate the local magnetic field.

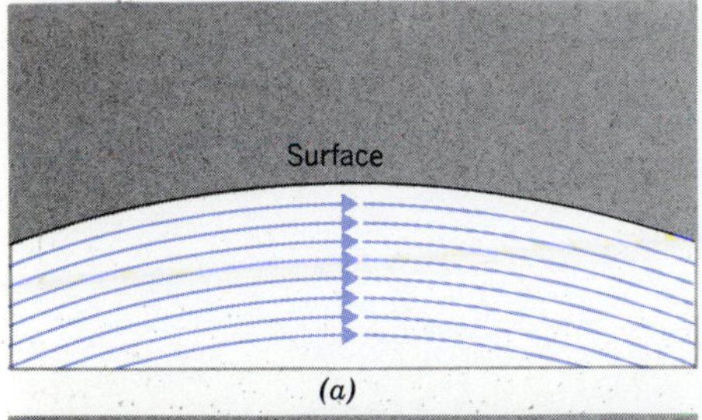

(a)

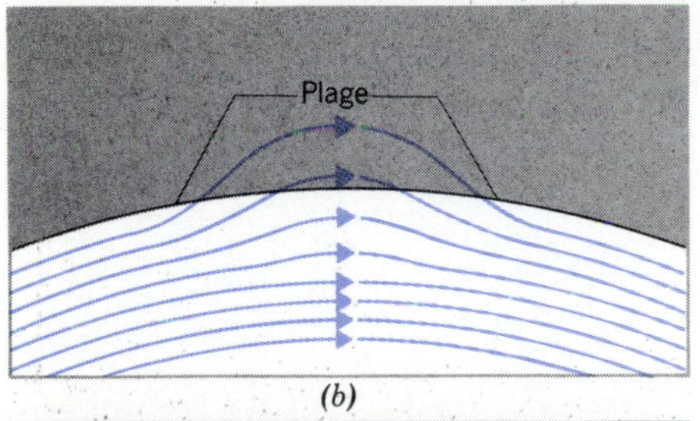

(b)

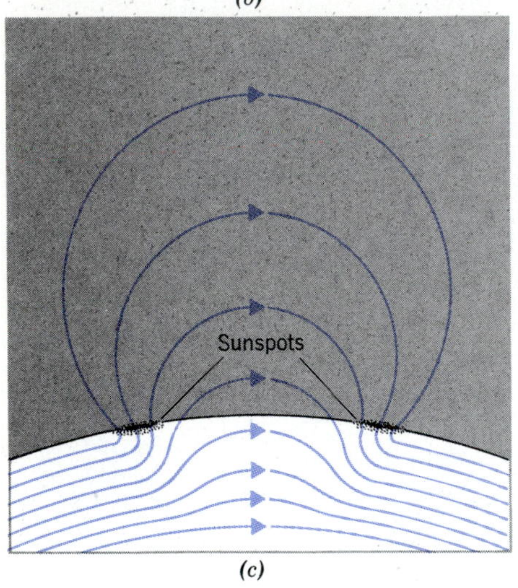

(c)

brightness of the plage is considerably diminished, and after four rotations it usually disappears.

In the final phase the magnetic field grows still weaker and spreads in space, sometimes persisting for more than a year. The superposition of many such old disturbed regions seems to be the explanation for the weak irregular field on the sun's surface.

Main Ideas

1. Conditions in the interior of the sun; composition, density, and temperature.
2. The meaning of convection; the zone of convection in the sun.
3. The photosphere; evidence of convection; granules.
4. The solar atmosphere; the chromosphere and the corona.
5. Sunspots; their properties and origin; the sunspot cycle.
6. Plages, flares, prominences and filaments.
7. The sun's rotation and magnetic field; an explanation of the intense magnetic fields on the sun's surface; sequence of events in a disturbed region.

Important Terms

chromosphere
convection currents
convection zone
corona
coronagraph
differential rotation
filament
flares
granules
magnetic polarity
penumbra
photosphere
plage
prominence
shock wave
solar atmosphere
solar magnetic field
solar neutrino experiment
solar wind
sunspot
sunspot cycle
umbra
Zeeman effect

Questions

1. With the aid of a diagram describe the solar interior. Include specifications of temperatures and densities at various levels. Indicate the regions in which energy is transferred outward by radiation and convection. Show where nuclear reactions play an important role.
2. How has the sun changed during its history? How will it evolve during the remainder of its life? What is the principal

reason for these changes? Mention changes in composition, luminosity, radius and surface temperature. How will the sun's life probably end?

3. Describe the process of convection. Give examples of convection on the earth. Where does convection occur in the sun? Why? How is convection in the interior of the sun manifested on the surface?
4. Why do photographs of the sun taken at a wavelength of 6563Å reveal details of chromospheric structure not visible in white-light photographs?
5. How does the spectrum of the corona differ from the spectrum of the chromosphere? Why is the temperature of the corona believed to be in the neighborhood of one million degrees?
6. How can granules in the photosphere be observed spectroscopically?
7. Describe a sunspot. Why are sunspots darker than the surrounding surface of the sun? Summarize the life history of a sunspot.
8. Describe the following: plage; flare; filament; prominence. Do these phenomena have a common cause? Explain.
9. Describe the sun's corona. What produces the radiation from the corona? Why is the corona normally invisible?
10. According to current theories of solar surface activity, two key properties of the sun give rise to sunspots, flares, and other disturbances. Describe these properties and the role they play in generating surface disturbances.

GENERAL PROPERTIES OF THE SOLAR SYSTEM

15

According to the available evidence, the Universe began its existence approximately 15 billion years ago as a very dense, hot cloud. The cloud cooled as it expanded, and after a time, stars began to form. Within each star the manufacture of heavier elements out of hydrogen and helium began through a series of nuclear reactions. In this way, the elements of the Periodic Table were assembled out of the basic building blocks of matter.

In the smaller stars, these elements were locked within the star permanently by gravity, but the larger stars ended their lives in the cataclysmic explosion of the supernova, radiating the heat of billions of suns, and spraying out to space the elements that had been manufactured within the star during its lifetime. There these elements mingled with the primordial hydrogen and helium to form an enriched mixture of gas. Later, new stars were born out of the mixture.

Four and a half billion years ago, after innumerable supernova explosions, the concentration of the heavier elements amounted to about one percent. Around that time, the sun and its family of planets condensed out of a cloud of gaseous matter located in one of the spiral arms of the Milky Way Galaxy. How did the planets form? The first task in our study of the planets is the investigation of their formation in this astrophysical context.

Halley's Comet, photographed in 1910.

THE ORIGIN OF THE PLANETS

The origin of the planets is less of a mystery than the origin of the Universe, but it is still not a clearly understood event. Fifty years ago, a commonly taught theory held that the planets came into being as by-products of a catastrophic event in which the sun collided with a passing star. The force of gravity tore huge streamers of flaming gas out of the bodies of the two stars during this encounter. As the intruding star receded into the distance, some of these streamers of gaseous material were attracted by the sun's gravity and captured into orbits circling around it. The earth condensed out of one of these streamers of hot gas to form a molten mass, on whose surface a crust formed and gradually hardened with the passage of time.

It is easy to calculate the probability that the earth and other planets originated in a collision between two stars. The likelihood of a collision between the sun and another star depends on the size of the sun, and on the distance between it and its neighbors. Stars, large though they are, are minute in comparison with the average distances that separate them. The sun, for example, is one million miles in diameter, but 24 *trillion* miles from its nearest neighbor. A calculation shows that because the stars are so far apart, the possibility of a stellar collision is extremely small. In fact, the calculations indicate that throughout the billions of years in which our galaxy has existed, there have probably been no more than one or two stellar collisions in the spiral arms of our galaxy.

Thus, the collision theory of the origin of planets implies that we are alone in this corner of the Universe.

A very different prediction comes out of the modern theory of the birth of stars. This theory asserts that whenever a star is formed, planets are likely to form around it. If the theory is correct, untold millions upon millions of planets exist in the Universe. All speculation about extraterrestrial life rests on that conclusion.

According to the second theory, when our solar system was young, it consisted of a cloud of gas, mainly hydrogen and helium, in which small grains of solid matter were embedded. The grains included bits of iron, rock, and ice substances formed from elements made earlier in other stars and dispersed to space in supernova explosions.

Such clouds, condensing out of the mists of space under the pull of gravity, are the first stage in the birth of every star. With the passage of time, the cloud continued to contract under its own gravitational pull; as it contracted, it grew hotter, until finally the temperature of the center reached the critical level of 20 million degrees needed for the ignition of thermonuclear reactions. This point in the contraction marked the birth of the sun. In our solar system that occurred about 4.6 billion years ago.

While the sun was forming, smaller condensations appeared in the outer regions of the cloud. These condensations, appearing by accident, were also held together by gravity, as the sun had been. The smaller knots of matter became the planets.

This theory provides a good explanation for the birth of the so-called *giant planets* of the solar system—Jupiter, Saturn, Uranus, and Neptune. The giant clouds are composed mainly of hydrogen and helium, the same gases that make up the bulk of the sun, and there is little question that they were formed in the same way as the sun, by the condensation of clouds of gaseous matter under the force of gravity. That conclusion is supported by the fact that about half the stars in the sky are multiple—two, three, or even four or more stars relatively close together—and all formed at the same time. This means that whenever a cloud of gas and dust starts to condense in space, the condensing cloud tends to break up into a number of separate clouds of various sizes. The biggest clouds form stars, while the smaller clouds form stars or planets, depending on their size.

Thus, both theory and observation support the idea that planets probably are common in the Universe.

Origin of the Earthlike Planets

The explanation of planetary origins based on gravitational condensation, which works so well for the giant planets does not work at all for the earth and its sister planets—Mercury, Venus, and Mars. These bodies, called the *earthlike planets,* are very different from the giant planets. They are largely composed of rock and iron, with very little hydrogen and helium. They are also much smaller than the giant planets; the earth, for example, weighs only 1/300 as much as Jupiter. It is the small size of the earthlike planets that creates a problem; they are just not massive enough to have condensed under their own gravity, as the sun and the giant planets did.

It might seem possible that the earth was formed in two stages. First, like Jupiter, it condensed as a ball of hydrogen and helium, with small amounts of heavier elements. Then, it lost its hydrogen and helium as a result of a solar flare-up. The difficulty with this hypothesis is that once a body as massive as Jupiter has formed, its gravitational force is so strong that hydrogen and helium cannot escape; at least, their rate of escape is so small that very little would have been lost thus far in the lifetime of the solar system. If a Jovian-sized planet had formed in the earth's orbit at the beginning of the solar system's history, it would still be there today.

How did the earthlike planets form? Suppose we go back to the beginning, before any planets had formed. At that time the entire solar system consisted of a cloud of hydrogen and helium, with the new sun at the center of the cloud. The cloud is called the solar nebula. As time went on, the outer part of the solar nebula lost its heat to space and cooled sufficiently for knots of condensed matter to develop in it under the attraction of gravity. These became the giant planets. But no planets as large as Jupiter formed in the inner part of the solar system at that time, because the gas there was still too hot; the heat was such that the random motions of the atoms in the gas dispersed any pocket of condensed gas as soon as it formed.

Now, according to the theory, the sun flared up in a sudden outburst of energy. Young stars are observed to flare up in this way frequently. The flare-up blasted the hydrogen and helium out of the inner part of the solar system, leaving behind only the bits of solid matter—mostly iron and rocky substances—that were too heavy to be affected by the sun's outburst. Subsequently, the earthlike planets formed out of those grains of iron and rock.

How that happened is a mystery. Perhaps collisions occurred occasionally between neighboring bits of rock, in the course of their circling motion around the sun. Some collisions were gentle, and the particles stuck together. In this way, in the course of millions of years, small grains of rock gradually grew into large ones, until finally some pieces of rock became large enough to exert a gravitational attraction on their neighbors. Once that happened, these large fragments quickly swept up all the materials in the space around them and developed into full-sized planets in a short time.

The complete process of formation of the earthlike planets went on over a period of perhaps 10 million years, proceeding with extreme slowness at first and then with rapidly increasing momentum in the final stages. At the end, the remaining matter in the solar system was gathered into the existing planets, and only a few atoms of gas remained in the space between. This is the situation in the solar system as it exists today.

Other Solar Systems. The condensation theory implies that planets are formed frequently when a star is formed and must, therefore, be very common objects in the Universe. In fact, there must be billions of them in our galaxy alone. Unfortunately, this prediction cannot be tested readily by astronomers, because a planet circling a neighboring star, and shining only by the reflection of that star's light, is lost in the glare of the light from the star. However, the Infrared Astronomy Satellite (p. 129) has provided indirect support for the currently accepted picture of solar system formation. The satellite recorded an unusual pattern of infrared radiation from the star Vega, roughly ten times brighter at very long wavelengths than would be expected from a star of Vega's type, and apparently emitted from an extended shell or ring of particles surrounding the star. A similar ring of particles is believed to have formed in our solar system at an early stage, as a part of the sequence of events leading to the condensation of the planets.

Composition of the Planets

We believe that when the solar system came into being, it was only a cloud of gaseous hydrogen mixed with small amounts of other substances. According to modern astronomy, this cloud contained the materials that are now in the bodies of the sun, the planets, and the creatures that walk on the surface of the earth. It was the parent cloud of us all. At its center existed a dense, hot nucleus that later formed the sun. The outer regions—cooler and less dense—gave birth to the planets.

Out of what materials were the planets formed? The bulk of the parent cloud must have been composed of the light gases, hydrogen and helium, because they are the most abundant elements in the Universe. Other elements relatively abundant in the Universe, although less so than hydrogen and helium, are carbon, nitrogen, and oxygen, and metals such as iron, magnesium, aluminum, and silicon. These substances must also have been present in relatively great abundance in the parent cloud of the planets. No doubt the remaining 80-odd elements were also represented, but in smaller amounts.

All the familiar chemical compounds of these substances would be formed in the cloud. The heavier elements—silicon, aluminum, magnesium, and iron—combine with oxygen to form grains of rock-like materials and metallic oxides. Hydrogen combines with oxygen to form molecules of water vapor; hydrogen also combines with nitrogen to form molecules of ammonia gas, and it combines with carbon to form methane, also called marsh gas, which is used extensively today for cooking. Carbon and oxygen combine to form carbon dioxide. Considerable amounts of each of these compounds must have appeared in the parent cloud.

When the planets first condensed out of this mixture of gases and solid matter, the bulk of their mass should have consisted of hydrogen and helium which made up about 98 percent of the solar nebula. The giant planets—Jupiter, Saturn, Uranus, and Neptune—are in fact composed mostly of these very light elements, and are small-scale models of the sun and stars, although their internal temperatures are not high enough for nuclear reactions to occur. There is little question but that the giant planets were formed in precisely the same way that a star is formed, by condensation out of gaseous matter under the force of their own gravity. Our planet—along with

Mercury, Venus and Mars, called earthlike planets—is very different. It is a rocky ball of matter, almost entirely solid except for some molten material at the center, and does not resemble the sun at all in bulk composition.

A SURVEY OF THE PLANETS

The basic properties of the nine planets are listed in Table 15.1. Columns 1 and 2 of Table 15.1 indicate that the planets revolve around the sun in more or less regularly spaced orbits, the distance from one planet to the next increasing by approximately a factor of 1.5 to 2 in most cases (Figure 15.1) (pp. 374–375).

A conspicuous gap lies between the orbit of Mars and the orbit of Jupiter, which are located, respectively, at 1.5 and 5.2 times the distance from the earth to the sun. A planet should be located midway between Mars and Jupiter, at a distance of, perhaps, three times the earth's orbital radius, but none is found there. Instead, a swarm of rock and iron fragments known as the asteroid belt circles the sun in that neighborhood at an average distance of 2.9 earth-orbit radii.

The orbits of all the planets are within a few degrees of the plane

TABLE 15.1
Properties of the Planets.

Name	1 Distance from Sun 10^6 miles	2 Distance from Sun Relative to Earth's Distance (A.U.)[a]	3 Inclination to Ecliptic	4 Eccentricity of Orbit[b]	5 Revolution Period (earth–years)
Terrestrial Planets					
Mercury	36	0.39	7°00′	0.21	0.24
Venus	67	0.72	3°24′	0.01	0.62
Earth	93	1.00	0°00′	0.02	1.00
Mars	142	1.52	1°51′	0.09	1.88
Giant Planets					
Jupiter	483	5.20	1°18′	0.05	11.86
Saturn	886	9.54	2°30′	0.06	29.46
Uranus	1780	19.18	0°46′	0.05	84.01
Neptune	2790	30.07	1°07′	0.01	164.97
Pluto	3670	39.44	17°19′	0.25	248.4

[a] The earth's mean distance from the sun—92,600,000 miles—is known as the astronomical unit (A.U.).
[b] Difference between the closest and farthest distance to the sun, expressed as a fraction of the average radius of the orbit.

of the orbit of the earth called the *ecliptic.* The orbits of the innermost planet, Mercury, and the outermost planet, Pluto, are exceptions (column 3 in Table 15.1). All orbits are close to perfect circles, with the exception, again, of Mercury, Pluto, and to a lesser extent, Mars (column 4). The distance from Pluto to the sun varies by approximately one billion miles during the course of the planet's year. The farthest reach of Pluto's orbit marks the outer boundary of the solar system, except for the minor bodies known as comets. Some cometary orbits reach out one-fifth the distance to the next-nearest star.

In the space between the planets there is a tenuous cloud of gas, mostly hydrogen, with a density of 100 to 1000 atoms per cubic inch. Beyond the boundary of the solar system the density of matter drops to its interstellar value of 10 atoms per cubic inch.

Length of Year. The length of a planet's year—that is, the time required to complete one circle around the sun—increases with distance from the sun, varying from 88 days for Mercury to two and a half centuries for Pluto (column 5). The relationship between a planet's distance to the sun (R) and the length of its year (T) is given by:

$$\frac{T\text{ (planet)}}{T\text{ (earth)}} = \sqrt{\frac{R^3\text{ (planet)}}{R^3\text{ (earth)}}}$$

6 Rotation Period (hours or days)	7 Mass (earth's mass = 1)	8 Diameter (miles)	9 Density (water = 1)	10 Effective Temperature[c]	11 Incident Solar Energy (relative to earth)[d]	12 Number of Satellites
59d	0.05	3,000	5.4	450	6.7	0
243d	0.82	7,600	5.1	283	1.9	0
23h 56m 4.1s	1.00	7,930	5.52	240	1.00	1
24h 37m 22.6s	0.12	4,270	3.97	220	0.43	2
9h 50.0m	317.80	89,000	1.33	100	0.04	16[e]
10h 14m	95.2	75,000	0.68	75	0.01	23[e]
10h 49m	14.5	30,000	1.60	50	0.0031	5
15h	17.2	28,000	2.25	40	0.001	2
6.39d	0.002	4,000(?)	0.5	40	0.0006	1

[c] Temperature at the surface if there were no atmosphere.

[d] The earth receives on the average 2 calories per square centimeter per minute.

[e] These numbers may increase as spacecraft observations continue.

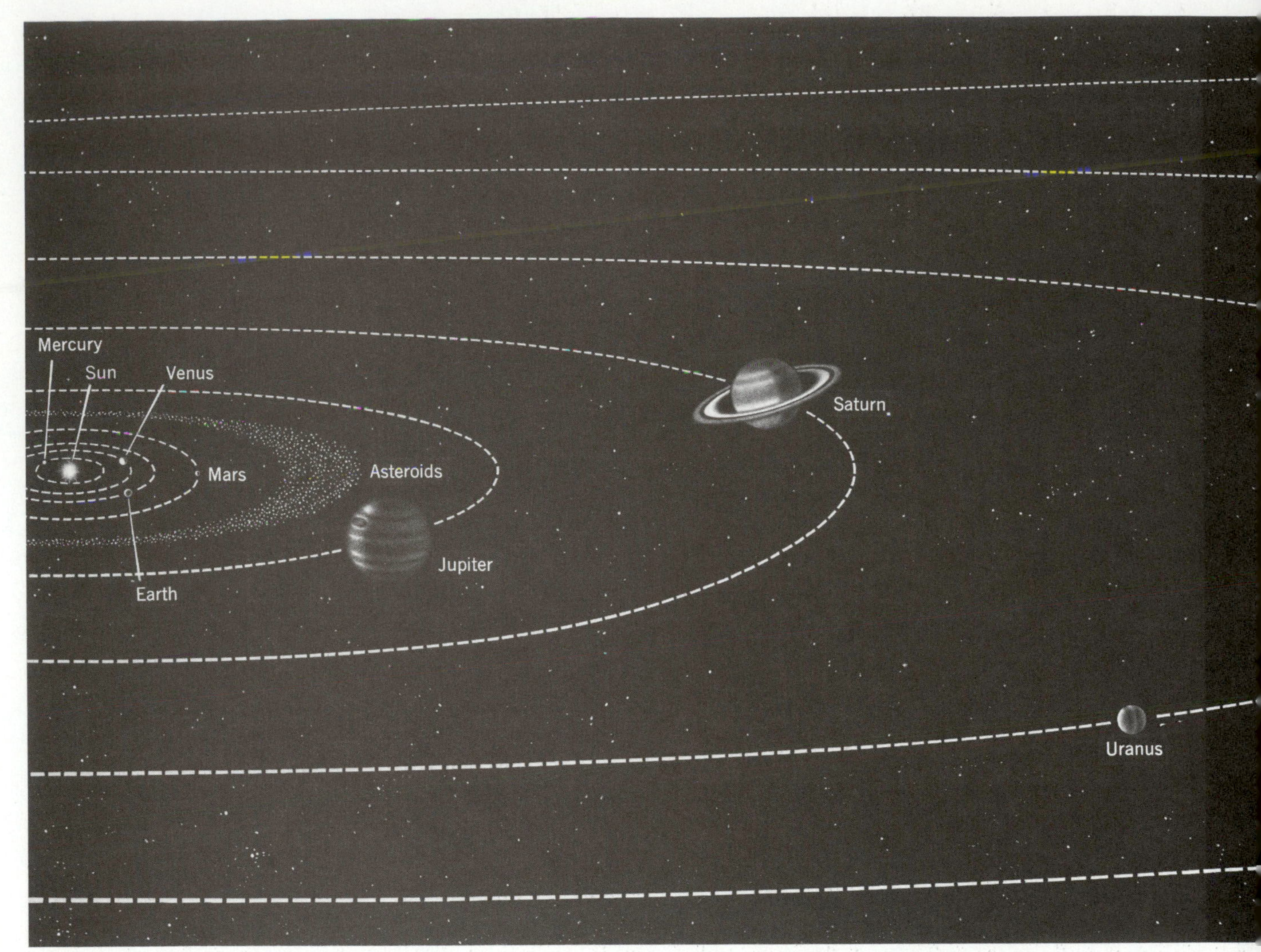

FIGURE 15.1
The spacing of the orbits of the planets.

Length of Day. The length of a planet's day—that is, the time required to complete one rotation about its axis—varies from 9 hours and 50 minutes for Jupiter to 243 days for Venus (column 6). The similarity in the lengths of the day for the earth and Mars probably is a coincidence. The significance of the long period of rotation of Venus is discussed below. All giant planets rotate very rapidly on their axes. The rotation produces a pronounced bulge at the equator which is particularly large for Saturn, amounting to 3000 miles or 9.5 percent of the radius of the planet. In contrast, the earth's equatorial bulge is only 14 miles, because the earth rotates more slowly on its axis and is also a smaller planet.

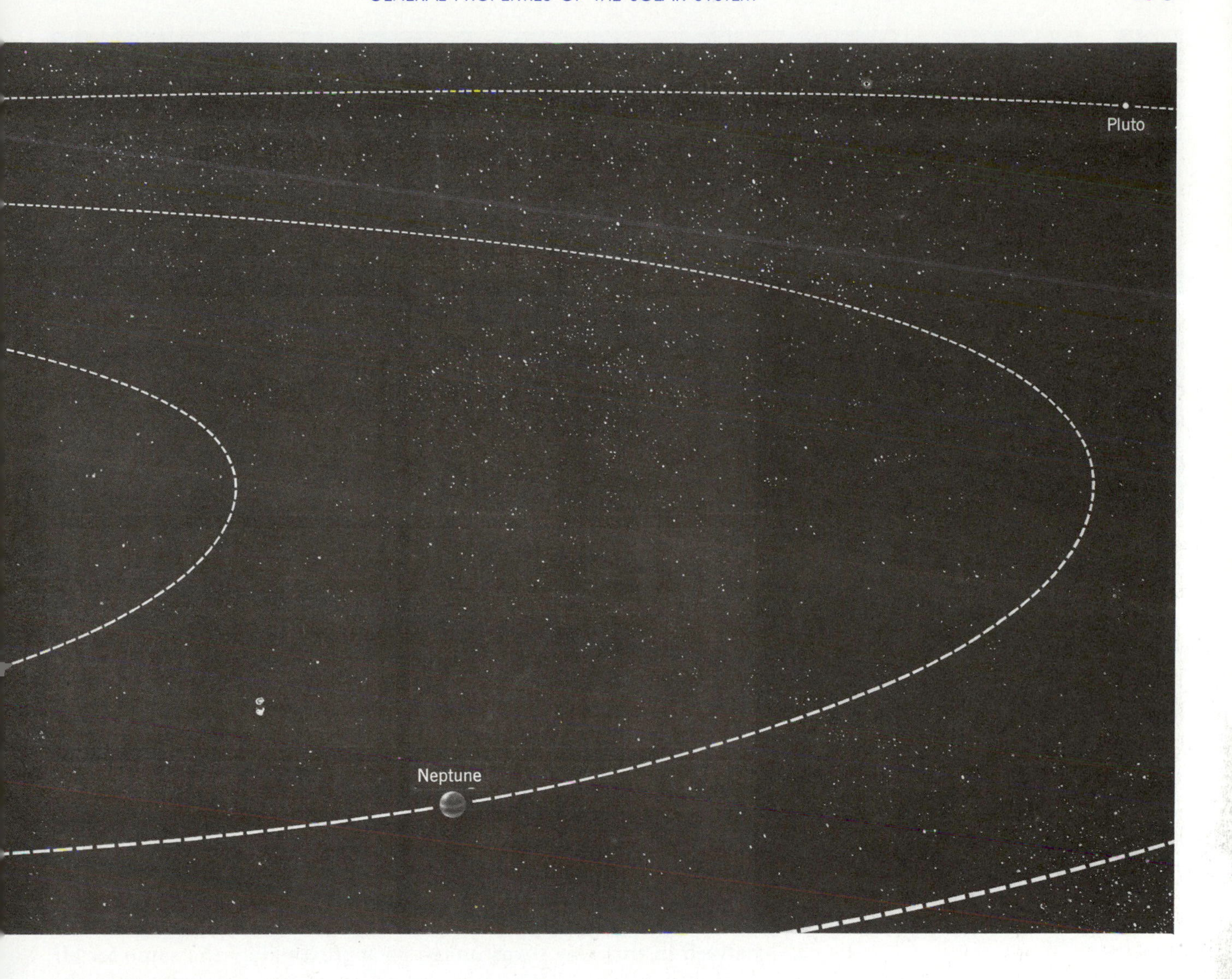

The Terrestrial Planets

The nine planets divide into two groups differing greatly in their size, mass, and composition (Figure 15.2) (p. 376). Mercury, Venus, and Mars resemble the earth in being composed almost entirely of rocky materials and iron, and are known, together with the earth, as the *terrestrial planets.* The moon is only slightly smaller than Mercury and is sometimes included with the terrestrial planets, being composed of similar materials.

Three terrestrial planets—the earth, Mars, and Venus—are discussed in detail in subsequent chapters. This section surveys the properties of the terrestrial planets as a whole.

FIGURE 15.2
The sun and the planets: relative sizes. The sun and planets, arranged in order of distance from the center of the solar system, are shown here in proportion to their actual sizes. The sun is 860,000 miles in diameter, or roughly 10 times the size of the largest planet, Jupiter.

Table 15.1 lists hypothetical surface temperatures for each planet, calculated from the intensity of the solar radiation reaching the planet at its average distance from the sun (column 10). In the case of the earth and Venus, the actual surface temperatures are substantially higher because the relatively dense atmospheres of these planets act as insulating blankets. Mars has a very thin atmosphere, and Mercury has substantially no atmosphere; hence the true surface temperatures for these bodies are close to the ones listed in the table.

The average densities of the terrestrial planets vary considerably according to column 9 of Table 15.1, suggesting a substantial difference in composition. However, when allowance is made for the increase in density in the interior of each planet caused by the pressure of the overlying layers, the so-called "uncompressed density" derived in this way turns out to be approximately the same for all terrestrial planets. For each of these planets, the uncompressed density is somewhat greater than the density of surface rocks on the earth, as would be expected for a planet composed of rocky materials plus a substantial admixture of iron.

Mercury. Mercury is one-third the size of the earth and has one-twentieth the earth's mass. It completes an orbit in 88 days. The length of its day is 59 earth-days.

A close view of Mercury was obtained in 1975 from the Mariner spacecraft (Figure 15.3). The Mariner photographs revealed a cratered moonlike surface. Mercury resembles the moon in being waterless, nearly airless, and almost certainly without life.

Venus. Venus is the earth's nearest neighbor, and also has nearly the same size and mass as the earth. The planet circles the sun inside the earth's orbit, completing one circuit in 225 earth-days. As Venus revolves around the sun, it goes through phases similar to those of the moon. The phases of Venus are visible in Figure 15.4, taken with the 36-inch telescope at Lowell Observatory. At the "full Venus" the planet is on the opposite side of the solar system from the earth, and its face is fully illuminated by the sun. At the "half-Venus" the planet has moved halfway around its orbit toward the earth. The "new Venus" is on the same side of the sun as the earth; because it is directly between the earth and the sun, it can hardly be seen although at this point in its orbit, it is at minimum distance and assumes its largest apparent size.

The planet's rotation on its axis is opposite to the direction of its movement around the sun. That is, looking down on the north pole of Venus from "above," the planet will be seen to revolve about the sun in a counterclockwise direction—as is the case for all planets in this solar system—while rotating on its axis in a clockwise direction. This type of motion is called *retrograde rotation.* Venus is the only planet in the solar system whose rotation is unmistakably retrograde.

Why is the rotation of Venus so peculiar? When the planet was born, it probably rotated on its axis at about the same speed as the earth. However, being closer to the sun, it felt the effect of the sun's gravity more strongly. The pull of solar gravity must have slowed down the rate of rotation until the planet was rotating on its axis at very closely the same rate at which it revolved around the sun. This means that Venus would present the same face to the sun at all times. The moon, which is controlled by the earth's gravity, rotates on its own axis in the same period of time in which it repeats one

FIGURE 15.3
The cratered moonlike surface of Mercury photographed by a spacecraft from an altitude of 11,800 miles.

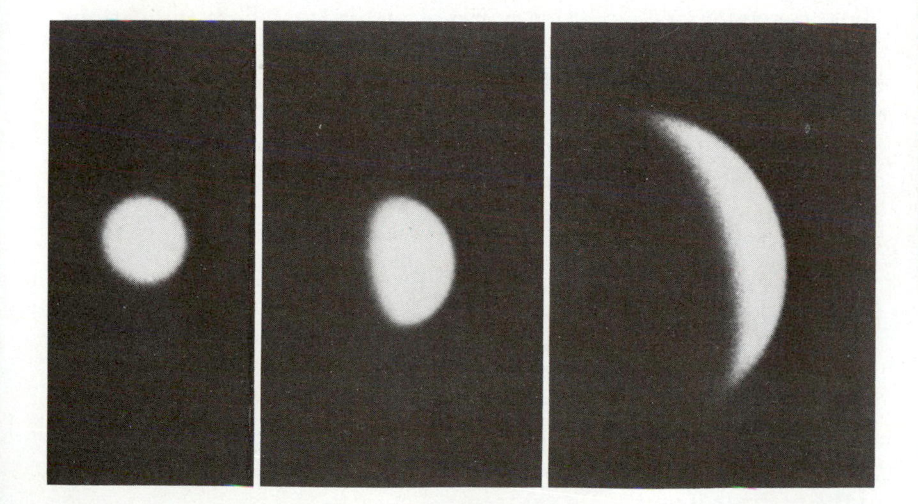

FIGURE 15.4
The phases of Venus.

FIGURE 15.5
The change of seasons on Mars.

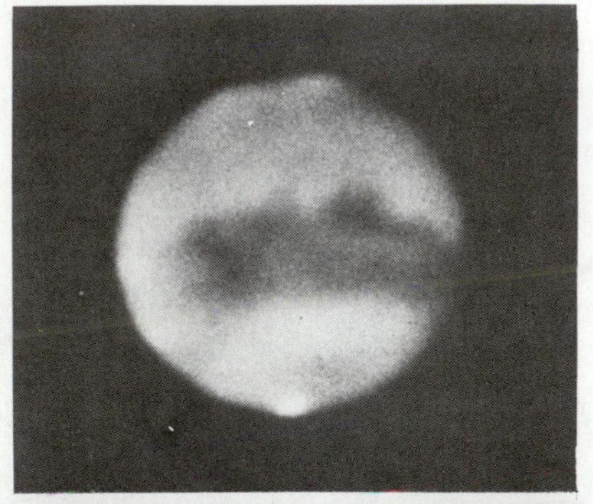

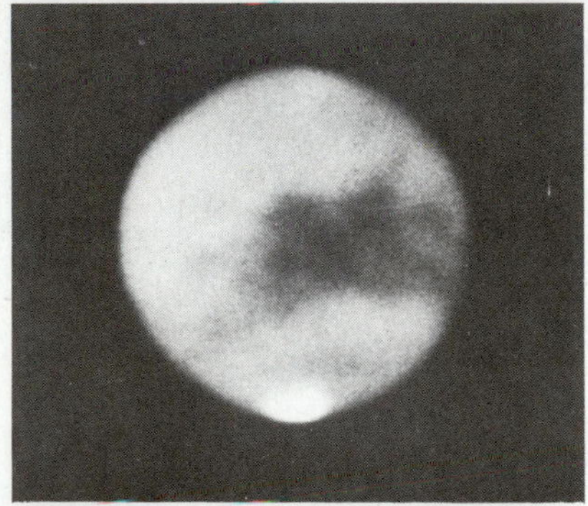

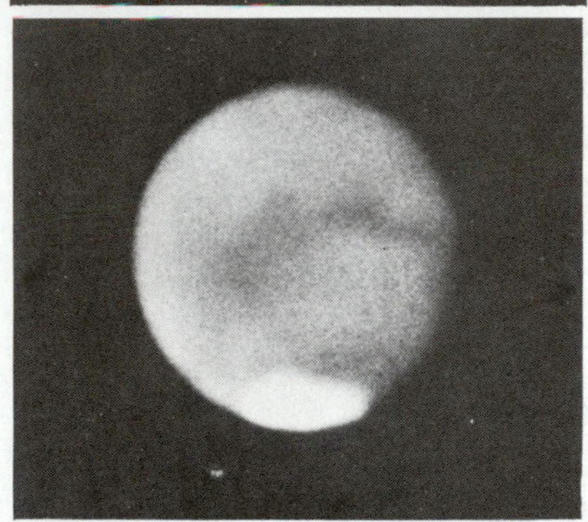

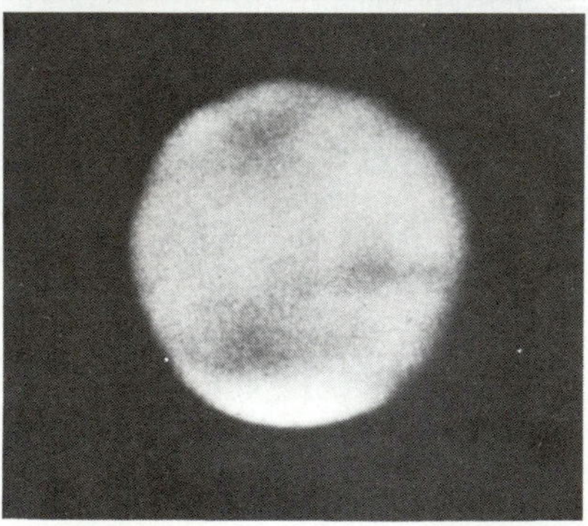

revolution around our planet, and presents one face to the earth at all times for this same reason.

This would explain why the length of the Venus day is so much longer than the length of our day. The explanation for the retrograde rotation is probably connected with the fact that the rotation rate observed for Venus is precisely such as to assure that the planet will present the same face *to the earth* every time it comes closest to the earth in its orbit. Apparently, once the sun had slowed the rotation rate of Venus down to 200 days or so, the weaker pull of the earth was then effective enough to swing the same face of Venus around to the earth whenever the two planets were at their minimum distance during their orbits around the sun.

Venus is a particularly interesting planet because of its similarity to the earth in size, mass and distance to the sun. Chapter 18 contains a more detailed description of Venus as a planetary body.

Mars. Mars revolves around the sun at a distance of 142 million miles, one and one-half times as distant as the earth. The length of the planet's year is 687 days or 1.9 earth-years. It rotates on its axis in 24 hours and 37 minutes. Its axis of rotation is inclined at an angle of 24° to the plane of its orbit. By a coincidence, the length of the Mars day and the inclination of the Mars axis to the plane of the planet's orbit are nearly the same as the corresponding quantities for the earth.

The inclination of the Mars axis means that the planet has seasons. The most conspicuous feature in earth-based photographs is the polar cap, resembling the cover of ice and snow at the poles of the earth (Figure 15.5). The polar cap grows in size from 200 miles in the upper photograph to a maximum of 2000 miles in the lower photograph. Spacecraft measurements have indicated that the caps are composed of a thin layer of frozen carbon dioxide (dry ice) covering a less extensive but thicker cap of water ice. These indications of the presence of water suggest that Mars is a possible abode of life. The planet is discussed in greater detail in Chapter 19.

Mars has two small moons named Phobos (Figure 15.6) and Demos. They are approximately 10 miles in diameter. Phobos circles the planet in a close orbit that carries this moon around Mars in $7\frac{1}{2}$ hours—shorter than a Martian day.

The Giant Planets

The giant planets—Jupiter, Saturn, Uranus, and Neptune—are 5 to 10 times larger than the earth, and far more massive, but considerably lower in density. In general, their density is about the same as that of water; Saturn, in fact, is less dense than water; it would float in the bathtub if you could get it in.

The giant planets are less dense than the earth and its neighbors because they contain large amounts of the lightest elements, hydrogen and helium. As a consequence, the structure of a giant planet may be entirely different from that of a terrestrial planet. Jupiter and Saturn are starlike in composition; they consist almost entirely of hydrogen and helium, and lack a well-defined surface. Uranus and Neptune have lost much of their light gases, and probably are composed of ices of water, ammonia, and methane, deposited in a deep mantle surrounding a rocky core.

FIGURE 15.6
Phobos, one of two moons of Mars, photographed in 1977 from a distance of 300 miles by the Viking orbiter. Phobos is 14 miles in diameter. It lacks the red color of its parent planet and may be a captured asteroid.

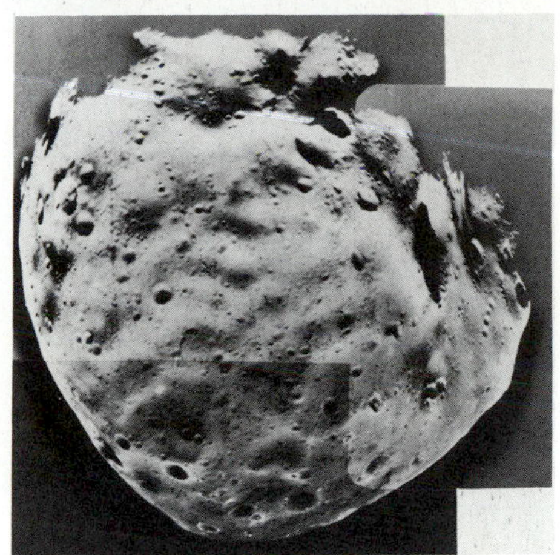

Jupiter. Jupiter is the largest of the giant planets, and the first to be observed from a spacecraft (Color Plate 20). Its average distance from the sun is 480 million miles. Jupiter completes a revolution around the sun in approximately 12 earth-years. The length of its day is 9 hours and 50 minutes. The rate of spin of Jupiter is very rapid for a planet of its size, and leads to a pronounced bulge of material at the equator. The other giant planets also have large equatorial bulges.

Theoretical estimates of the internal structure of Jupiter indicate that the pressure at the center of the planet is about 10 million pounds per square inch. The temperature at the center may be as high as 50,000°K. The high temperature is a result of the condensation of this very massive planet under the force of its own gravity. If Jupiter were roughly 50 times more massive, its temperature would have risen to a level sufficient to ignite nuclear reactions, converting it into a small star.

It might be expected that the interior of Jupiter would be in a gaseous state resembling the interior of a star, in view of the high temperature at its center. However, calculations on the properties of hydrogen, when subjected to a pressure of 10 million pounds per square inch and a temperature of 50,000°K, indicate that the extreme pressure forces the atoms into the state of a liquid metal, like lithium or sodium at high temperatures. This liquid metal core of hydrogen is believed to extend out to a distance of 40,000 miles from the planet's center (Figure 15.7).

A thick envelope of highly compressed, gaseous hydrogen and helium overlies the hydrogen core, extending upward with gradually diminishing density. The gaseous envelope is topped by two thick decks of clouds. The lower deck consists of water droplets and ice crystals. The upper layer of clouds consists of crystals of frozen ammonia compounds at a temperature of −300°F. These clouds of frozen ammonia present the visible face of the planet to the observer.

Photographs taken from spacecraft show turbulent cloud formations that appear to be violent hurricane-like storms extending over thousands of miles (Figure 15.8). The photographs also reveal details of the great Red Spot, about 40,000 kilometers long and 13,000

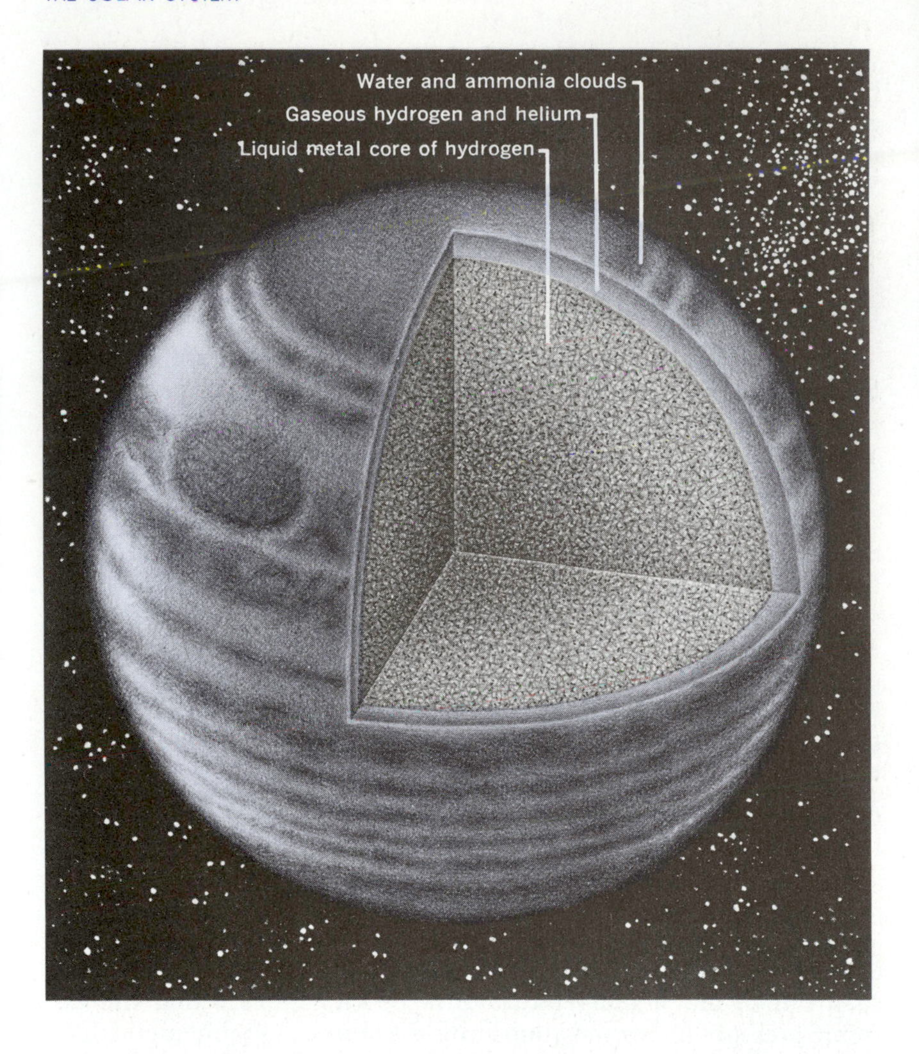

FIGURE 15.7
The structure of Jupiter.

kilometers wide (Color Plate 2). The Red Spot is probably a mammoth storm center or hurricane. However, the Red Spot has persisted for at least 300 years, whereas hurricanes and storms on our planet play themselves out and disappear after a few days at most because their energy is dissipated into turbulence in the surrounding air.

What mechanism feeds energy into the Red Spot to keep it intact for centuries? No one knows. There is another puzzle in the Red Spot: Hurricanes and storms tend to wander over a surface, whereas the Red Spot is anchored in one place. An object lying beneath the clouds on Jupiter must be the Red Spot's anchor. What is this object? That question is also unanswered.

Conditions beneath the clouds of Jupiter are concealed from our view, but below the level of the cloud tops the temperature must rise, just as it does on the earth or any other planet with a fairly

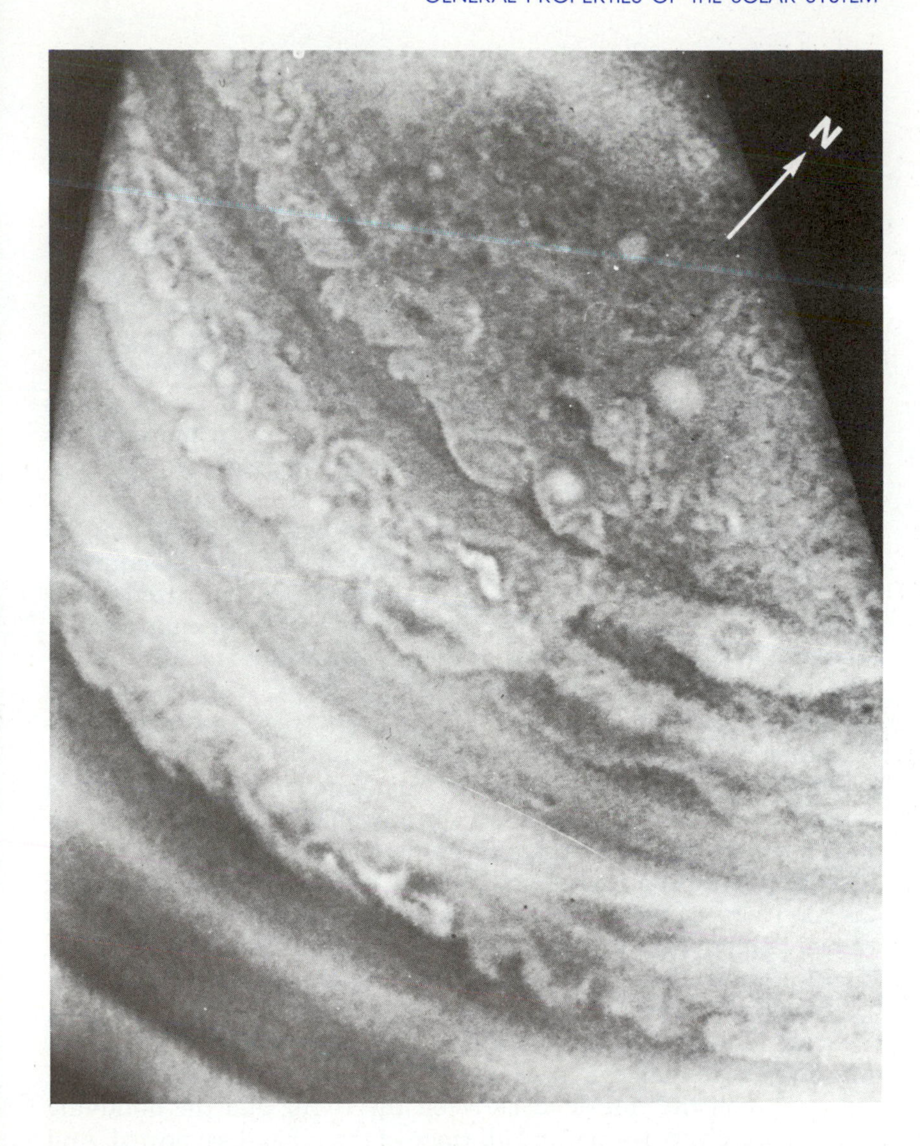

FIGURE 15.8
Cloud formation and storms of Jupiter.

dense atmosphere. The rise in temperature is the result of the greenhouse effect, in which the atmosphere acts as an insulating blanket, sealing in the planet's heat and increasing its temperature (see page 446). We suspect that Jupiter does not have a well-defined surface but, instead, an atmosphere that grows steadily denser with increasing penetration into the interior of the planet. With or without a surface, Jupiter probably has a region, at some depth below its clouds, in which the temperature passes through a comfortable range for the development and support of life.

Conditions for Life. Is it possible that life exists on Jupiter? The planet contains an abundance of hydrogen as well as the compounds of hydrogen with relatively common elements such as carbon, nitrogen, and oxygen. These compounds are methane (CH_4), ammonia (NH), and water (H_2O). All are believed to have been present in abundance in the primitive atmosphere of the earth, and to have played a critical role in the events that led to the development of life on our planet. Their importance in evolution on the earth has ended, and they have long since escaped, but their continued presence on Jupiter leads us to wonder whether the initial steps along the path to life have not also occurred on that planet. Laboratory experiments described in the next chapter show that these four gases—ammonia, methane, hydrogen, and water vapor—when mixed together and energized by an electric spark, produce copious amounts of amino acids, the molecular building blocks of living matter. This fact has given rise to a plausible theory of the origin of life by chemical evolution.

According to the theory, when the earth was a young planet, its atmosphere contained substantial amounts of methane and ammonia. Strokes of lightning in earth thunderstorms energized these gases in the atmosphere and produced a rich yield of amino acids. The amino acids accumulated in the oceans. Random collisions between molecules, occurring repeatedly in the oceans over millions of years, linked small molecules into large ones, and finally produced a molecule on the threshold of life.

The life-giving gases—ammonia, methane, and hydrogen—have largely vanished from the earth's atmosphere because ultraviolet rays from the sun break them up and release their hydrogen atoms. The light hydrogen atoms escape from the earth because the pull of its gravity is too weak to hold them. Jupiter's gravity, however, is so strong that not even hydrogen atoms can escape. This is the explanation for the abundance of methane and ammonia, as well as hydrogen itself, on Jupiter.

Scientists have never been able to check this theory of the origin of life by searching the fossil record, because the earliest forms of life, lying on the threshold between the living and the nonliving worlds, were very fragile, and all trace of them has disappeared. And the theory cannot be checked easily in the laboratory by putting together life's molecular building blocks, because such experiments necessarily lack a key ingredient. The missing ingredient is *time.* Nature required several millions of years of ceaseless, random experimentation to discover the chemical pathways to life on the earth. Scientific ingenuity has not yet been equal to the task of imitating these experiments.

But Jupiter, which still has its ammonia and methane, has been the scene of experiments on the origin of life for four billion years, and these experiments are still continuing. Even the lightning discharges are present: the spacecraft photographs showed a brightening in some of the stormiest regions of the Jupiter atmosphere that suggests lightning on a scale dwarfing terrestrial experience. With all conditions of the laboratory experiments duplicated in Jupiter's atmosphere, it seems possible that the building blocks of life may exist there in great numbers. Many interesting products of their repeated collisions may also be present. Jupiter may afford a unique opportunity to find out how life arose on the earth, and how it may arise elsewhere in the Cosmos.

The Moons of the Giant Planets

Forty-six moons have been found thus far in orbits around the giant planets. Many were discovered in recent years by spacecraft. The majority of these satellites are small, irregular bodies of rock resembling asteroids ranging in size from 70 miles down to 4 miles. Some of these smaller moons of the giant planets probably were captured from the asteroid belt. Others, however, are large bodies, in one case as large as the planet Mercury. These larger moons of the giant planets are fascinating worlds of rock and ice, which must be included among the most interesting objects in the solar system.

The Galilean Satellites

Four large moons orbit Jupiter, in addition to a dozen or so smaller ones. Their names, in order of increasing distance from Jupiter, are Io, Europa, Ganymede, and Callisto. These four moons, which were discovered by Galileo in 1610 when he first turned his telescope on Jupiter, are known as the Galilean Satellites. Their relatively rapid motions around their parent planet suggested to Galileo that Jupiter was a solar system in miniature. Galileo concluded that the earth was not the center of all motion. His conclusion was a powerful argument for the view of Copernicus that the earth was not the center of the Universe, but moved instead around the sun.

Io. Io is the innermost of the Galilean Satellites. Spacecraft examination of the surface of Io in 1979 revealed one of the greatest surprises in the history of planetary exploration: this satellite has active volcanoes. Several volcanoes were observed on Io by the Voyager spacecraft, and six of these are still active. There is evidence that volcanic eruptions take place continuously at one point or another on Io's surface.

The volcanoes on Io were a great surprise because Io, while fairly large for a moon, is quite small in comparison to volcanically active bodies such as the earth or Venus. Volcanoes on a planet like the earth are believed to be the result of heat released in the interior of a planet by the decay of uranium and other radioactive substances.

This heat raises the temperature of the rocks in the interior to the melting point. However, if the planet is small, the heat produced in its interior flows to the surface and is lost to space, and the planet becomes volcanically inactive in a relatively short time. The earth's moon became volcanically inactive 3 billion years ago, probably as a consequence of its small size and rapid rate of loss of internal heat. Io, being smaller than the moon, should have lost its internal heat at a faster rate and become volcanically inactive even earlier. Why is Io still volcanic? What is the source of its heat?

The answer is connected with the pull of Jupiter's gravity. The gravitational attraction of Jupiter, tugging at the near side of Io harder than the far side, raises tides on the surface of the satellite similar to the tides raised on the earth by the sun and the moon (Chapter 2). The height of the tides depends on Io's distance to Jupiter. Since the pull of the other Galilean satellites causes this distance to vary continually, the heights of the tides also vary. These continually changing tides bend and flex the body of the satellite, causing internal friction and heating. The frictional heat accumulated as this flexing motion, repeated in orbit after orbit, is sufficient to heat the rocks in the interior of Io to the melting point. Calculations show that in proportion to the size of Io, the heat produced within it by Jupiter's tides is considerably greater than the heat produced within the earth by radioactive elements.

Europa. Europa is slightly smaller than the earth's moon. It is an extraordinarily smooth body, resembling a billiard ball when photographed from a distance. The highest "mountain" on Europa is only a few hundred feet above the average level of the surface. The planet's polished, white appearance is produced by a surface coating of ice. However, ice and water cannot make up a large fraction of its composition, because the density of Europa is 3 g/cm^3, which is approximately the same as the density of rock. A comparison with the moon's density of 3.3 suggests that Europa is made 90 percent of rock and iron, and about 10 percent of water. The tidal forces produced by Jupiter must heat the interior of Europa in the same manner in which they heat Io, although not to as great a degree. Consequently the water content of Europa probably exists in liquid form. It may form a deep global sea, capped by a thin crust of ice.

Ganymede and Callisto. Ganymede and Callisto are about as large as the planet Mercury, and are the second and third largest moons in the solar system. Studies indicate that they consist of roughly equal parts of rock and water, the latter probably in the form of ice. Ganymede and Callisto are too far away from Jupiter to be heated appreciably by its tidal forces, but they contain enough rocky materials—and probably, therefore, enough radioactive elements—so that radioactive heating could have been appreciable at an earlier time. However, because of the relatively small size of these satellites compared to a planetary body like the earth, much of this heat should have escaped to space by now, leaving Ganymede and Callisto cold and geologically dead.

This expectation is confirmed in the case of Callisto, whose surface is heavily scarred with meteorite craters, indicating, as in the case of our moon, an ancient and well-preserved surface, unmarked by volcanism and lava flooding (Figure 15.9). Ganymede, however, seems to be geologically active, with a younger surface marked by many strange grooves and light-colored streaks (Figure 15.10). The cause of the geological activity and strange surface markings on Ganymede is uncertain. Ganymede is a challenging example of the difficulty of understanding the geology of a planetary body made largely of ice.

FIGURE 15.9
The ancient cratered surface of Callisto.

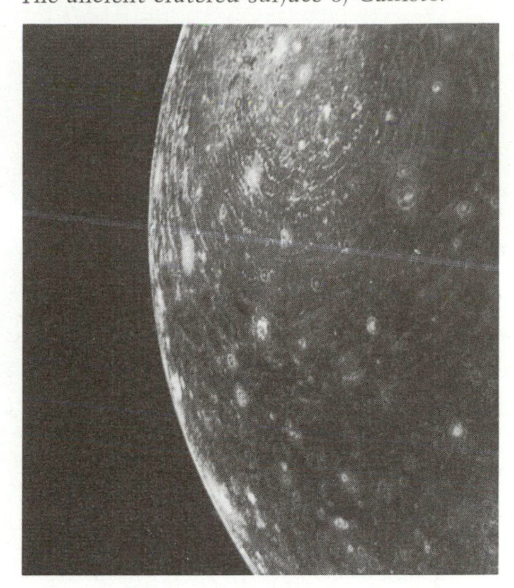

The Moons of Saturn. Saturn has at least 22 small moons and one large one. The small moons have densities only moderately greater than water, indicating that they are composed largely of ice plus smaller amounts of rock. The largest moon of Saturn, named Titan, has a diameter of 3180 miles, substantially larger than the planet Mercury. Its density – 1.9 grams per cubic centimeter – suggests a 50-50 mixture of rock and ice or water. Titan is unique among the moons of the solar system because it possesses its own atmosphere.

Conditions on Titan. Titan's atmosphere and heavy cloud cover prevent us from observing conditions on its surface, but spacecraft observations suggest a surface temperature of −300°F. The atmosphere consists mainly of nitrogen, but there is also a substantial amount of methane and probably some ammonia. Planetary scientists disagree over probable conditions on the surface. Some calculations suggest nitrogen and ammonia are only present as gases; others indicate that the gases condense to liquids and solids, with oceans of liquid methane lapping at the shores of ice continents and occasional showers of methane rain. It is interesting to note that this planet-sized moon has an abundance of the compounds – ammonia, methane, and water – that play a critical role in experiments on the origin of life on the earth. However, because of the low temperature on its surface, water could not be present in liquid form. Thus, the onset of chemical evolution seems unlikely.

Saturn. Saturn's average distance from the sun is 886 million miles – nearly twice that of Jupiter. It completes a revolution around the sun in approximately 30 years. The length of a day on Saturn is approximately 10 hours. Saturn is the second largest planet in the solar system, being five-sixths the size of Jupiter and about a third as massive.

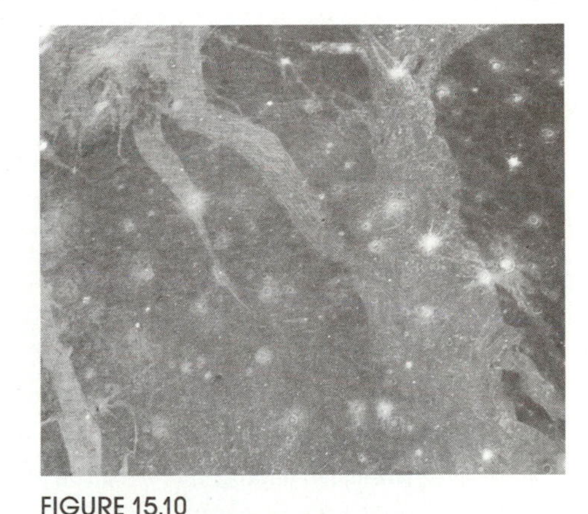

FIGURE 15.10
A region on the surface of Ganymede, showing light streaks and grooves. The light areas may be the result of eruption of ice or quickly freezing water on the surface.

The Rings of Saturn. The rings of Saturn are an extraordinary phenomenon familiar to every amateur astronomer. The four distinct rings visible from the earth occupy the region between 46,000 miles and 85,000 miles from the center of Saturn. The innermost ring of the four is 9500 miles above the planet's surface. The outermost ring ends 1600 miles inside the orbit of Janus, the closest of Saturn's moons. The rings are paper-thin relative to their diameter, with estimates of their thickness ranging from 4 inches to 2 miles. Consequently, when the rings are viewed edge-on to the earth, they disappear, although when viewed face-on they are dazzlingly bright (Figure 15.11).

FIGURE 15.11
Photographs showing various orientations of the rings of Saturn.

Spacecraft photographs have revealed that the ring system of Saturn consists of at least 95 and possibly hundreds, of closely spaced ringlets, the entire system resembling a phonograph record (Figure 15.12). Stars can be seen through the rings, indicating that they are not solid sheets. Spacecraft observations have confirmed that the rings are made of particles ranging up to the size of baseballs and are probably composed of water and other ices.

The origin of the rings lies in the gravitational force exerted by Saturn on its satellites. Since the force of gravity increases in strength with decreasing distance, the near side of a moon feels a somewhat stronger gravitational pull toward its parent planet than the far side. The stronger attraction on the near side tends to wrench the material on this side out of the body of the moon. The force of

FIGURE 15.12 (opposite)
A computer-enhanced photograph of the rings of Saturn, taken from a spacecraft at a distance of five million miles. The tiny dot at upper left is a small moon discovered by the spacecraft.

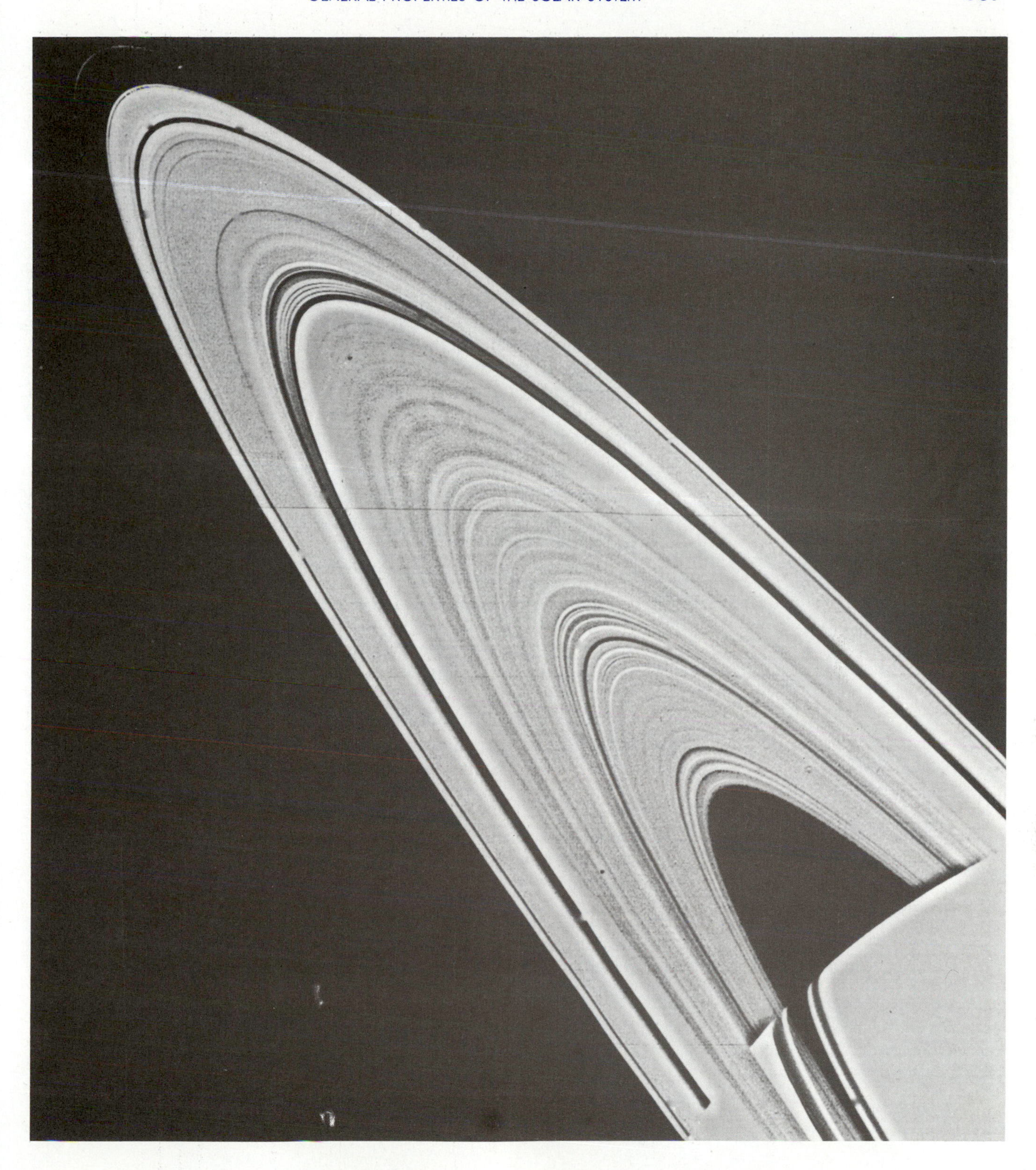

the moon's internal gravity, holding it together, resists this effect. If the moon is too close to its planet, the excess pull on the near side becomes too great to be counteracted and the moon is torn apart. Each of the resulting fragments then circles in orbit around the planet as a miniature moon.

A moon must keep a minimum distance from its parent planet in order to stay intact. The minimum distance is called the *Roche limit.* A calculation shows that the rings of Saturn are inside the Roche limit for that planet. Either the rings are fragments of a moon that spiraled in too close and was torn apart; or, more likely, they are grains of material that were within the Roche limit around Saturn originally and, therefore, were prevented from collecting into a single object when the moons of Saturn were first forming.[1]

All planets and stars possess Roche limits. The Roche limit for the sun is one million miles from its center, or approximately 500,000 miles above its surface. No planet could form within this distance. The Roche limit for the earth is 10,000 miles from its surface, well inside the orbit of the moon.

The Ring of Jupiter. Spacecraft measurements made in 1979 revealed that Jupiter, like Saturn, is surrounded by a huge ring of small particles, 160,000 miles in diameter. The existence of the ring had been unsuspected because, unlike Saturn's rings, it is too faint to be visible through earth telescopes.

Uranus and Neptune. These planets were probably formed by gravitational condensation, in the same manner as Jupiter and Saturn. In spite of their smaller size and lesser degree of gravitational compaction, they have higher densities than Jupiter and Saturn, indicating that they have lost a substantial fraction of their hydrogen and helium. Their structures resemble that of Jupiter but with nearly all the hydrogen-helium envelope removed. Uranus has the unusual property that its axis of rotation is tilted approximately 90° to the plane of its orbit, so that the planet rotates, so to speak, on its side. Because the highly elliptical nature of Pluto's orbit sometimes carries it inside Neptune's orbit, Neptune is currently the outermost planet in the solar system, and will remain so until 1999.

Pluto. Pluto was the ninth and last planet to be found in the solar system. It was discovered in 1930 when the observation of Neptune revealed that a gravitational force of unknown origin was perturbing its orbit.

The orbit of Pluto carries it farther from the sun than that of any other planet. Because Pluto is so far away, we have been able to

[1] Artificial satellites orbiting the earth within the Roche limit are preserved from distintegration by their rigid construction.

learn very little about it. New observing techniques that minimize the blurring effect of the earth's atmosphere have led to a more accurate measurement of the diameter of Pluto. The planet is approximately 2000 miles in diameter—close to the size of our moon—but has a considerably lower mass and density. Observations in 1978 indicate that Pluto itself has a moon, which has been named Charon, about 1000 miles in diameter. From the orbit of Charon it is possible to deduce the mass of Pluto, which turns out to be about one-seventh the mass of our moon. The mass and diameter measurements yield a density for Pluto of about 0.5 g/cm^3. This surprisingly low value, half the density of water, suggests that Pluto is composed mainly of a mixture of ice, water, methane, and ammonia. There is direct evidence that at least part of Pluto's surface consists of frozen methane. There may also be a methane atmosphere.

Before the discovery of Charon, Pluto's mass was assumed to be considerably greater. The newly determined values of its mass are too small to account for the gravitational force perturbing Neptune's orbit. The discrepancy has led some astronomers to assume that the solar system contains a tenth planet, circling the sun far outside Pluto's orbit, and completing its orbit every 1000 years.

MINOR BODIES

The solar system contains a large number of small objects that are held in orbit by the sun's gravity. These minor bodies of the solar system—asteroids, meteorites, and comets—are negligible in mass compared to the planets. However, they can create striking visual effects if they come close to the earth. They also have a special scientific interest because they provide information on the primordial state of condensed matter in the solar nebula before it aggregated into the full-sized planets.

The Asteroids

Between the orbits of Mars and Jupiter there is a gap in the distribution of the planets. We might expect to find a planetary body located outside the orbit of Mars, about three times the earth's distance from the sun; but instead we find only a large number of small bodies—planetesimals—circling in a ring. These are called *asteroids.* The largest of the known asteroids is Ceres with a diameter of 480 miles. Three other asteroids—Pallas, Vesta, and Hygiea—have diameters greater than 200 miles. The remaining asteroids—estimated to number tens of thousands—are far smaller.

Asteroids are discovered by taking a long-exposure photograph of stars and noting any objects which move relative to the stars (Figure 15.13).

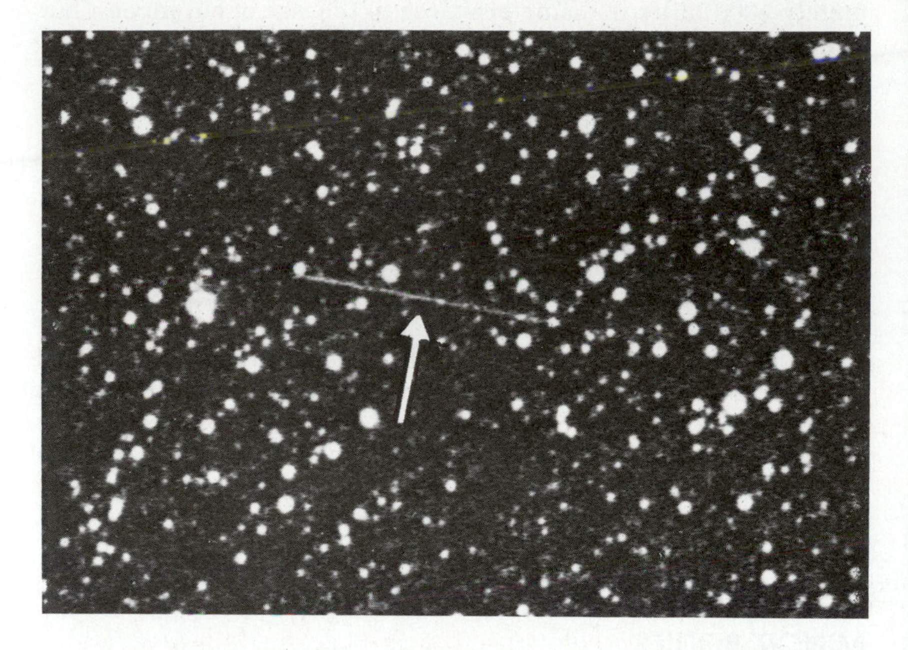

FIGURE 15.13
Discovery of the asteroid Icarus by its motion against the background of the fixed stars.

Asteroids whose orbits take them inside the orbit of the earth are called Apollo asteroids. Some Apollo asteroids come dangerously close to our planet. Hermes went by at a distance of 400,000 miles in 1937, and Icarus approached within 4 million miles of the earth in 1968. The visit of Hermes was a near miss for earth inhabitants; if it ever strikes the earth—as it may in the future—the force of its impact will liberate the energy of 10 million hydrogen bombs, and may destroy a substantial fraction of the population of the earth. We know from the geological record that the land areas of the earth have not been struck by Hermes-sized objects during the last few hundred million years. However, the early years of the earth's record have been wiped out, and during that period such collisions may have been frequent. The great craters and circular maria of the moon show the marks of collisions with asteroids and planetesimals the size of Hermes and larger.

The unusual properties of asteroid orbits provide a clue to the origin of these bodies. Many orbits are inclined at very large angles to the plane of the ecliptic (Figure 15.14). The orbits also tend to be highly elliptical in contrast to the near-circular orbits of the planets (Figure 15.15). The orbit of Icarus, for example, carries it to within a distance of 19 million miles from the sun and out to a distance of 180 million miles. These peculiarities can be explained by the powerful gravitational force of Jupiter. Jupiter's gravity affects all

FIGURE 15.14
Orbit of Icarus. Icarus is about one mile in diameter. Its orbit is inclined at 23 degrees to the ecliptic; the revolution period is 409 days; the orbital eccentricity is 0.83, and at perihelion the asteroid is only about 19 million miles from the sun—within the orbit of Mercury.

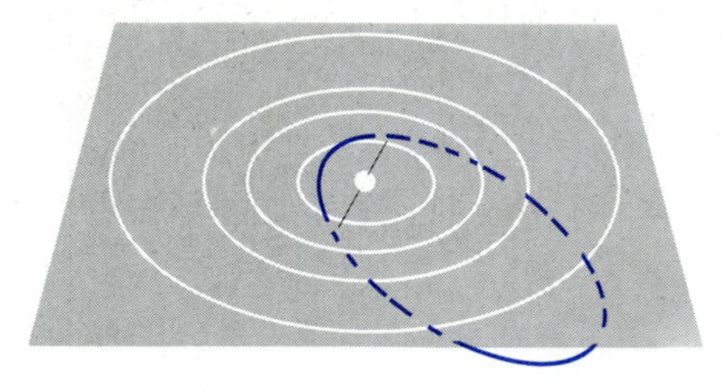

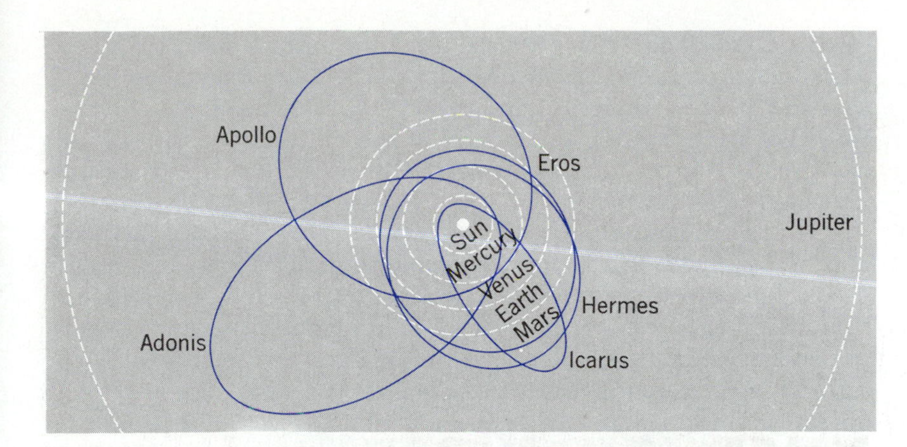

FIGURE 15.15
The orbits of various asteroids.

nearby objects, and occasionally pulls an asteroid out of its normal orbit, and may set it on a collision course with another asteroid. If a collision occurs, fragments of the two asteroids will leave the scene of the collision traveling through space in many different directions and with different speeds and will fall into new orbits around the sun, including orbits that make a large angle to the plane of the ecliptic. Some may be slowed down by the collision and fall in toward the sun, as presumably was the case with Icarus, while others are hurled farther out into the solar system. This collision theory would explain the great variety to be found in the asteroid orbits; it also explains the fact that they have never accumulated into a single large planetary body during the history of the solar system.

Meteorites

Among the fragments ejected from the asteroid belt in a collision, some fall by chance into orbits crossing the orbit of the earth. It is believed that most of the meteorites that hit the earth have this origin. The examination of meteorites that survive the searing passage through the earth's atmosphere reveals that most are pieces of rock and iron with a rather complex physical and chemical history, suggesting that they were broken off from larger bodies during repeated collisions.

Most meteorites consist of rocky materials primarily, but fragments of pure iron and nickel are not uncommon. The meteorite that blasted out the Arizona meteorite crater was a large block of iron weighing approximately one million tons (Figure 15.16).

FIGURE 15.16
An iron meteorite from the Arizona Crater.

Meteorites range in size from blocks of material weighing many tons to nearly invisible grains of rock and iron called micrometeorites. If the meteorite is the size of a grain of sand or larger, it leaves a fiery trail of incandescent matter behind as it passes through the earth's atmosphere. When we see this trail in the night sky, we call it a shooting star. The great majority of the meteorites are so small that the heat of their passage through the atmosphere vaporizes them before they reach the ground. If the meteorite is the size of a basketball or larger, its entry into the atmosphere creates a spectacular sight called a fireball.

Prior to the landing on the moon, meteorites were the only samples of extraterrestrial matter available to man. Their examination in the laboratory has provided two results of great importance to students of the origin of the solar system. One relates to the age of the solar system. Measurements of the ages of meteorites by the technique of radioactive dating (Chapter 16) yield results ranging up to 4.6 billion years. Since these 4.6-billion-year-old meteorites are the oldest objects known, their age is taken as a lower limit to the age of the solar system and, therefore, to the age of the earth. This conjecture is strengthened by the dating of the lunar rocks, which also yields ages up to 4.6 billion years (Chapter 17).

A second point of interest is the composition of meteorites. They show no signs of having been worked over by air and water erosion or subjected to a long history of chemical separation, as is the case of rocks on the surface of the earth. For this reason, they are believed to give a better indication of the original state of the materials in the solar nebula than can be obtained on the earth.

The chemical composition of one type of meteorite in particular, called the chondrite, is believed by some geologists to provide an accurate indication of the composition of the materials out of which the earth accumulated as a new planet. Chondrites are often used as a starting point in theoretical studies of the earth's history from the time of its formation.

Comets

The close approach of a comet is one of the most spectacular sights in the heavens. The comet appears first in the telescope as a small, fuzzy, faintly luminous object. At this point it is far from the earth, but is approaching our planet on the inward leg of a long journey from the edge of the solar system. As it comes closer, solar energy warms the head of the comet and vaporizes gases that were frozen in solid crystals during the many years in which the comet was far from the sun. These gases stream out behind the comet's head. Excited to luminescence by the absorption of solar radiation, the stream of gases forms a spectacular, glowing tail, which becomes clearly visible to the naked eye as the comet nears the sun (Figure 15.17).

Comets probably derive their name from the appearance of the comet tail; since the word comes from the Latin *cometes,* which

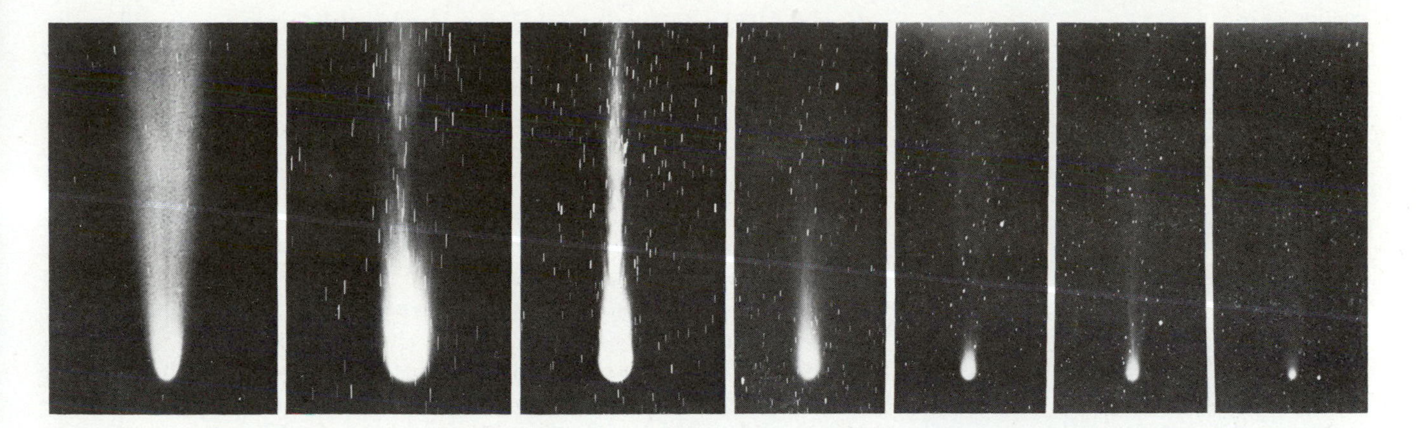

FIGURE 15.17
The gradual disappearance of the tail of Halley's Comet as it recedes from the sun.

means "long-haired." All comet orbits are ellipses. Most, however, are highly elongated, approaching closely to the sun, crossing the orbits of the planets on their approaches to the sun, and then retreating far beyond the orbit of Pluto on the outward leg of their journeys. Many comet orbits cross the orbit of the earth (Figure 15.18). The most elongated orbits are estimated to reach one-fifth of the way to the next nearest star. A comet in one of these orbits requires several million years to complete one circuit of the sun, and most comets are estimated to require 10,000 years or more for the round trip. The vast majority of the comets we observe will not be seen again from the earth for thousands of years. Billions of these long-period comets are thought to reside in the outer fringes of the solar system, in a vast cluster called the Oort cloud, located at a distance of about a light-year from the sun.

Some comets have shorter periods, ranging down to 3.3 years for Enke's Comet. Most of these short-period comets come close to the orbit of Jupiter, suggesting that Jupiter's gravity has deflected them into new orbits that remain within the solar system. The most famous comet in this group is Halley's Comet, named after Edmund Halley, a friend of Newton, who studied the records of comets dating back to 1531 and decided that several of these comets were a single body making a repeated appearance in the sky. Halley predicted that this comet would return in 1758, and it reappeared on Christmas night of that year. The last appearance of Halley's Comet was in 1910, and its next appearance is calculated to be in 1986.[2] Teams of scientists from several countries plan to launch spacecraft in time for a close rendezvous with Halley's comet.

What is a comet? The most widely held theory suggests that the nucleus of the comet is a swarm of rocky and metallic particles coated with frozen ices of water, ammonia, methane, and carbon

[2] The period of Halley's Comet varies from 74 to 79 years from one orbit to the next as the result of changes produced by Jupiter's gravity.

FIGURE 15.18
The current orbit of Halley's Comet, bringing it inside the earth's orbit to a perihelion in 1986. As with all comets, the tail points away from the sun because of the outward pressure created by streams of solar particles and radiation.

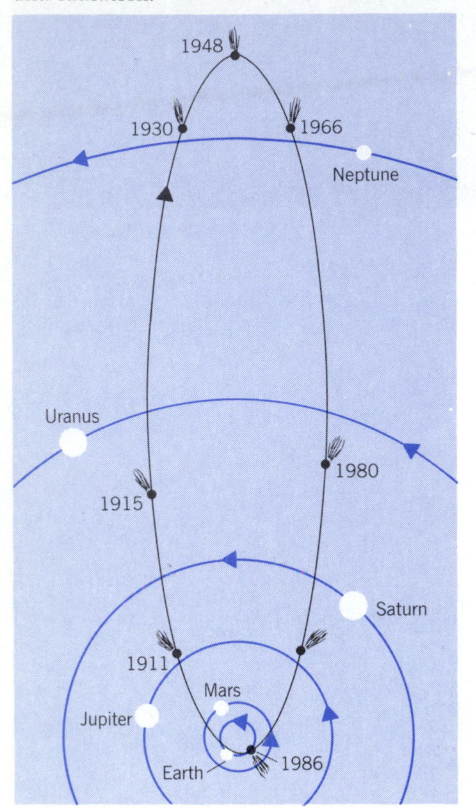

dioxide. This collection of dirty, slushy substances resembles the mixture of ice and rocky materials in the original solar nebula out of which the planets condensed. If this theory is correct, the spacecraft missions to Halley's comet in 1986 should yield new information about the primordial matter of the solar system.

The nucleus of the comet is a small object, a few miles in diameter, probably with a very loose structure and a low average density. The ice crystals vaporize to form the tail on each sweep around the sun. As the comet moves out toward the edge of the solar system once more, the gases condense and freeze again, and the tail disappears (Figure 15.17).

There is evidence that the earth occasionally passes through the tail of a comet, or through a swarm of particles that have become detached from the tail of a comet that passed by previously. On some evenings, thousands of meteor trails are visible in the sky during the course of the night. These displays are known as meteor showers. The trajectories of the meteorite trails can be calculated by photographing them with special cameras. Most coincide with the orbits of known comets, suggesting that these grains of dust are cometary particles entering the atmosphere. The meteor shower occurring each year around October 21 is produced by particles formerly in the tail of Halley's Comet, while the spectacular shower that occurs about August 12 is identified with the anonymously named comet, 1862 III. Table 15.2 lists the major meteor showers and their dates of occurrence.

TABLE 15.2
Meteor Showers.

Name	Date of Maximum	Approximate Limits	Number per Hour at Maximum
Quadrantids	Jan 4	Jan 1–6	110
Lyrids	April 22	April 19–24	12
Eta Aquarids	May 5	May 1–8	20
Delta Aquarids	July 27–28	July 15–Aug 15	35
Perseids	Aug 12	July 25–Aug 18	68
Orionids	Oct 21	Oct 16–26	30
Taurids	Nov 8	Oct 20–Nov 30	12
Leonids	Nov 17	Nov 15–19	10
Geminids	Dec 14	Dec 7–15	58

Main Ideas

1. Formation of the solar system out of the products of earlier generations of stars; probability of other solar systems.
2. Origin of the giant planets.
3. Origin of the earthlike planets; differences between the earthlike and giant planets.

4. Basic properties of the planets and their orbits.
5. Structure of the giant planets.
6. Moons of the giant planets; composition and surface properties.
7. Minor bodies of the solar system: asteroids, meteorites and comets.

Important Terms

Apollo asteroid	greenhouse effect	planet
asteroid	meteor	retrograde rotation
chondrite	meteor shower	rings (of Saturn and Jupiter)
comet	meteorite	Roche limit
ecliptic	micrometeorite	satellite
Galilean satellites	moon	solar nebula
giant planets	nucleus (of comet)	terrestrial planet

Questions

1. What events that took place earlier in the history of the Galaxy dictated the eventual composition of the earth? Explain.
2. What is the difference between a star and a planet? Between the sun and Jupiter?
3. Sketch the solar system. Label the planets in order and give their principal characteristics.
4. Using Table 15.1, list the fundamental differences between the inner planets and the outer planets. Explain the differences in terms of their formation and early history.
5. Suppose that a planet circles Barnard's Star at a distance equal to the radius of the earth's orbit. Which body—the earth or Jupiter—would you expect this planet to resemble more closely? Suppose that a planet circled Deneb at a distance equal to the orbital radius of Jupiter. Would this body resemble a giant planet or a terrestrial planet? Explain your answers.
6. Describe and compare the four Galilean satellites of Jupiter.
7. Explain why the rings of Saturn have not condensed into a moon.
8. How many years remain in the lifetime of the solar system according to the theory of stellar evolution? Describe what you think will happen to the earth and to Jupiter, as the sun approaches the end of its life. Base your response on information in Chapter 8.
9. What are the asteroids? Where are the orbits of most of them located? How do they compare in size with the major planets?
10. What is the difference between a meteor and a comet?
11. What do you think would happen if a comet hit the earth?

THE EARTH

16

The early years of the earth's history are shrouded in mystery. Erosion by wind and running water, and the upheavals that accompany the building of continents and mountain chains – all have combined to erase the record of the earth's past. Although the earth is under our feet, and the stars are far away, writing the history of our planet has turned out to be a far more difficult task than piecing together the life story of the stars, for the skies contain stars of many different ages, and all are available for examination in our telescopes. Through the study of these young, middle-aged and old stars we have learned the story of the red giants and white dwarfs. But planets of different ages are not available for inspection. We have no direct knowledge of the conditions that might exist on an earthlike planet during its lifetime.

Faced with this problem, students of the earth and its history – geologists – have arrived at ingenious methods for reconstructing the past history of the earth. Some of their investigations are concerned with scattered fragments of rock lying on the surface. The means for studying these rocks range from absurdly simple operations – such as tasting or feeling the rock – to highly complex, delicate laboratory analyses in which the rocks are taken apart almost atom by atom. To begin with, the geologist brushes the dust off, looks at the clean surface of the rock carefully, hefts it, and then hits it with a hammer. He grinds a smooth face on one of the fragments and inspects it under a low-powered microscope; then he cuts off a thin slice, grinds that down to a transparent slab one-thousandth of an inch thick and sends polarized light through the slab

The Mount Saint Helens eruption in 1980.

FIGURE 16.1
Types of rock: (a) sedimentary; (b) fine-grained igneous; (c) coarse-grained igneous.

(a)

(b)

(c)

to see what colors are produced. Afterward he bombards another small piece with electrons and X-rays. Finally, he vaporizes the rock and studies it an atom at a time.

Much can be learned about a rock's past by looking at it. Does the rock look like it has been built up layer upon layer? If so, it was probably formed by an accumulation of sediments, perhaps by silt filtering down onto the bottom of a lake or an ocean. Such *sedimentary rocks* (Figure 16.1*a*) are common on the earth, nearly all of whose surface has been covered at one time or another by water. Is the texture of the rock not layered but homogeneous, made up of many fine crystals of different types? If so, it is an *igneous rock*—solidified lava—that came up to the surface of the earth in a molten stream from the deep interior and rapidly cooled there. Is it an igneous rock with large crystals? That means that this rock, again originating in the deep interior, has collected in a pocket under the surface of the earth and cooled and solidified there very slowly, subsequently to be exposed on the surface by a subterranean uplift or by the forces of erosion (Figure 16.1*b* and *c*).

The rocks that lie on the surface of the earth come from depths as great as 60 miles, but no greater. How does the geologist penetrate deeper into the earth to determine the materials of which it is composed and the temperature and pressure at great depths?

The problem is like that of the physician who seeks a knowledge of his patient's interior without taking a cross-sectional cut. Earthquakes provide the geologist with the diagnostic tool equivalent to the X-ray and the electrocardiogram. Vibrations set up by subterranean disturbances travel through the body of the earth at speeds that depend on the properties of the materials through which they travel. By studying the earthquake records—seismograms—from stations located at many points and by comparing the properties and arrival times of the signals from an earthquake, the geologist can deduce a great deal about the earth's interior (Figure 16.2). In this way he has learned that the center of the earth is largely liquid—a *core* of molten iron 1800 miles in radius. Within the molten core lies an inner solid core, whose atoms are forced into the solid state, in spite of the high temperature at the earth's center, by the enormous pressure of the overlying layers. Surrounding the molten outer core is a *mantle* of dense rock 2200 miles thick. The mantle is capped by a rigid *crust* of lighter rocks with an average thickness of 10 miles (Figure 16.3).

The sum of the evidence gives a surprising picture of dynamic change and transformation within the earth's interior and on its surface throughout its history from the moment of its birth. The solid earth, viewed in the span of geologic time, has been the scene of violent activity that belies its seemingly static nature. The pattern of those changes has been assembled painstakingly by geologists through a combination of crude and sophisticated methods, but a key element was missing from earth science until a few short years ago. Prior to the mid-1960s, geologists could see what had happened, but they could not determine *why* it had happened. The missing element, which is connected with the concept of "continental drift" or plate tectonics, is described in the last section of this chapter.

FIGURE 16.2
The seismograph: a heavy weight suspended from a spring remains relatively motionless while the ground shakes beneath it during an earthquake. A pen, attached to the nearly stationary weight, traces the vibrations of the earth on a rotating drum.

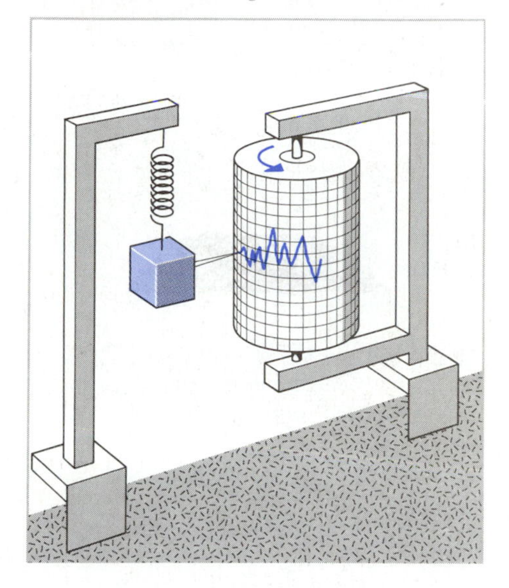

THE EARLY HISTORY OF THE EARTH

In Chapter 15 we described how the earth was probably formed out of small grains of iron and rocky substances. The elements contained in these grains of matter were, in addition to iron, also oxygen and silicon, and smaller amounts of magnesium, nickel, sulfur, aluminum, and calcium. These elements, which make up the bulk of the earth's mass, were the products of 10 billion years of prior element synthesis in the massive stars of the Galaxy.

We do not know exactly how the grains of matter first began to collect to form the earth. However, once that happened, and the nucleus of the earth began to form, other bits of material surrounding the planetary nucleus must have been drawn to its surface by the force of its gravity. In a relatively short period of time the nucleus would have developed into a full-sized planet.

As the earth grew to its final size, the force of its gravitational attraction mounted in proportion to the mass of the accumulated material. Toward the end, the force of gravity on the earth's surface was as strong as it is today.

An extraterrestrial fragment of rock—such as a meteorite—drawn down onto the surface of the earth by gravity, crashes into the planet at a speed of about 25,000 miles an hour. The amount of energy liberated by the impact at that speed is greater—pound for pound—than the energy liberated in the explosion of TNT. If the object is substantial in size, an enormous amount of heat is created when it hits the surface. If we can take our imaginations back in time to the period when the earth was almost fully formed, but there was still a large amount of planetary debris circling around the sun, we can conceive not only of an occasional meteorite hitting the earth, but of a heavy bombardment of rocks of all sizes raining down on the surface.

We can see that the earth must have been heavily scarred by that early bombardment. In fact, calculations indicate that large parts of the earth's outer layers could have melted as a result of the temperature rise caused by the bombardment during the final stages of its birth.

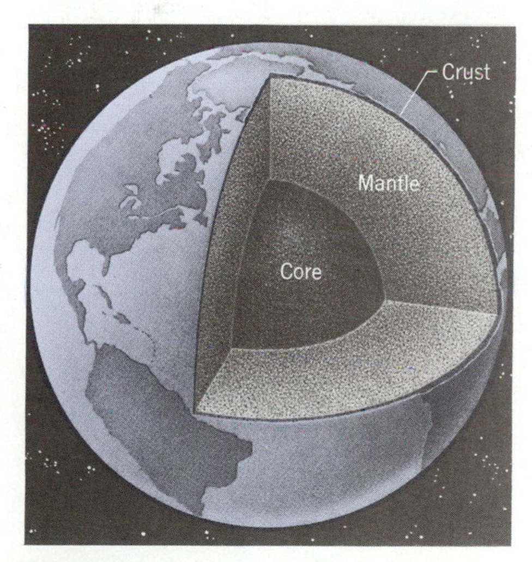

FIGURE 16.3
The structure of the earth.

RADIOACTIVE HEATING OF THE EARTH

Subsequently, this early melting by bombardment was supplemented by an additional source of heat: the energy released in the decay of radioactive elements present in the earth's interior. These elements—uranium, thorium, and potassium—have the special property, unique among all elements found within the earth, of disintegrating without any external stimulus when a sufficient amount of time has passed. In the disintegration, a piece of the nucleus of the radioactive element breaks off and is ejected at high speed, leaving behind a smaller and different nucleus than existed before. The breakup of the nucleus is called radioactive decay.

The fragment of the original nucleus that has broken off, speeding away from the scene of the radioactive decay, crashes through the surrounding atoms of the solid rock in which the radioactive element is located (Figure 16.4). Colliding with these atoms, the nuclear fragment transfers energy to them and heats the rock.

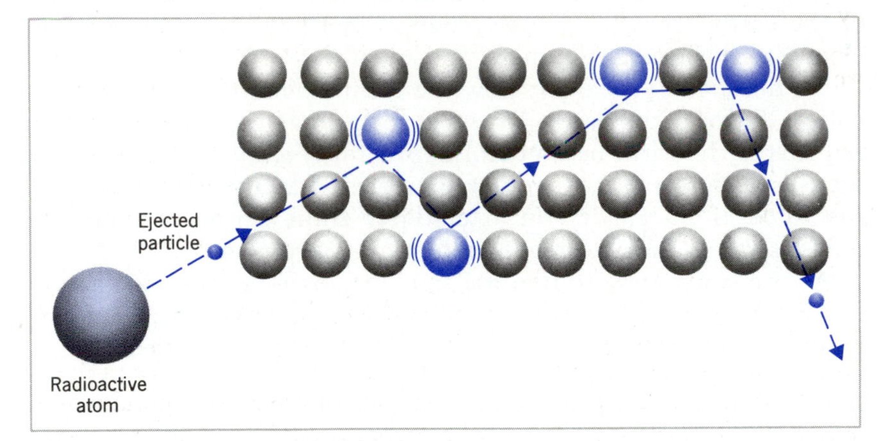

FIGURE 16.4
The heating effect of radioactive elements. A particle ejected from a radioactive atom strikes atoms of the surrounding rock (blue) and transfers energy to them.

Ernest Rutherford—the discoverer of the atomic nucleus—was the first person who thought about the heating effects of the radioactive elements in the earth. He was also the first person to measure the amount of heat, and the first to realize the implications of this heating for the interior of the earth, and for the understanding of the earth's history. Rutherford measured the temperature of a small amount of radium—a radioactive element derived from uranium—and found that if the radium was carefully insulated from all other sources of heat, it steadily became warmer. The rise in temperature was the result of radioactive decays.

Rutherford calculated the heat that would be released by the radioactive decays in ordinary rock using his measurements of the heat released by the decay of radium atoms. He found that the heating effect was very small as measured by ordinary standards. The radioactive heat released inside a fist-sized piece of rock would take a century to raise the temperature of a thimbleful of water by one degree. Yet Rutherford found that the heat from these decays, accumulating inside the earth for a billion years or more, could raise the temperature of the earth by thousands of degrees. This was sufficient to melt the solid rock interior of the earth itself.

Calculations indicate that by the end of the first billion years of the earth's existence, enough radioactive heat had been released to melt the iron in the earth's interior. From that point on, the earth had approximately its present structure—a largely molten iron core extending outward from the center, surrounded by a mantle of warm, yielding, but for the most part solid, rock.

Differentiation of the Earth's Interior

The formation of the iron core produced additional heat as the iron melted and ran to the center of the earth. As the iron flows toward the center through small channels in the rock, its flow is impeded by frictional resistance, which heats the surrounding rock. The pull of the earth's gravity on the iron is the ultimate source of this heat. According to some calculations, the heat produced by the formation of the iron cores must have been sufficient by itself to melt the interior of the earth. If the heat released by the formation of the core did not melt the earth's interior, the release of radioactive heat, continuing year after year, would have led somewhat later to the same result.

Once the rock mantle melted, the only way in which it could become completely solid again would be by solidifying from the interior of the earth outward toward the surface. You might expect the opposite—that a crust would form on the surface of the molten rock first, just as ice forms on the surface of a lake or a pond when the temperature falls. But water is very different from molten rock. When water freezes into ice, the ice weighs less than the water and, therefore, floats on top. That is why ice forms on the surface of a cooling body of water. Solid rock, on the other hand, is denser than liquid rock. If the entire mantle melted, and then a solid crust formed on the surface of the earth, it could not remain there. Being heavy, it must have broken up and sunk into the molten material, melting as it sank. Again and again the crust would have formed, broken up, sunk into the liquid rock beneath, and melted. Final solidification could never occur at the surface; it must have started at the "bottom," that is, at the base of the mantle.

It is believed likely that only a part of the mantle was molten at one time. Therefore the picture presented above is oversimplified. The actual history of the earth's mantle probably involved the repeated melting, solidification, and remelting of local regions. The details of the process are complex, but a thorough analysis leads to the same result that one obtains from the simpler picture.

What happened next? As soon as the interior of the earth began to solidify, a complication entered the story. We have been speaking of the rocks in the interior as if they were made of a simple substance like salt or sugar. In actuality, however, the substance we have been calling "rock" is a mixture of many different compounds or minerals. Common rocks generally contain about one dozen separate minerals in varying proportions. For example, consider the rock *granite.* If you look closely at a block of granite on the face of a building, you will see that it contains many small grains of different colors and shapes. Some grains are ivory white in color. These light grains are called *orthoclase feldspar.* Other grains are almost entirely transparent except for a slight cloudiness. They are the mineral *quartz.* The granite is also apt to contain dark specks—nearly black, which are crystals of the mineral *biotite.*

Another common rock is the dark, fine-grained volcanic material called *basalt.* If you have ever seen solidified lava, you probably have seen basalt. Nearly all rocks brought back from the lunar landings are basalts. Basalt usually contains a mixture of three minerals called *pyroxene, olivine,* and *plagioclase feldspar.*

What does the complex mineral structure of rocks have to do with the story of the earth's past? The answer is that each mineral freezes out of a cooling mass of liquid rock at a different temperature. The different minerals also have different densities. In general, the minerals that freeze at the highest temperatures have the highest densities. Table 16.1 lists the commonest rock minerals with their freezing-point temperatures and densities, that is, temperatures at which a molten mineral becomes solid. The minerals in this table make up 95 percent of the earth's rocks.

TABLE 16.1
Properties of Common Rock Minerals.

Name	Chemical Formula	Freezing Point, °C,[a] of Single Crystal	Density g/cm^3
Quartz	SiO_2	600–900	2.7
Biotite	$K(Mg, Fe)_3(Si_3AlO_{10})(OH, F)_2$	900–1000	2.7–3.3
Hornblende	$Na, Ca_2(Mg, Fe^{II}, Al)_5(Si, Al)_8$	1060–1200	
	$O_{22}(OH, F)_2$	1000–1200	3.0–3.5
Feldspar	$KAl(Si_3O_8)$		
	$NaAl(Si_3O_8)$	1100–1400	2.6
	$CaAl_2Si_2O_8$		
Pyroxene	$Ca(Mg, Fe, Al)(Si_2O_6)$	1200–1400	3.2–3.6
	$CaMg(Si_2O_6)$		
Olivine	$(Mg, Fe)_2SiO$	1400	3.2–4.4
Garnet	Mg_2SiO_4		

[a] Although feldspar has a relatively high melting point, its solid form is less dense than the molten form of other minerals. Therefore, as it formed in the mantle, it floated upward to the crust where it is found today.

As the interior of the earth cooled, beginning at the base of the mantle, the minerals with the highest freezing points were first to freeze. Being also relatively dense, they sank to the base of the mantle.

The table indicates that olivine and garnet have the highest freezing-point temperatures. Therefore these minerals must have been the first to crystallize out of the melt. They are also denser than the other minerals in the table, and would tend to sink to the bottom of the melt. Accordingly, the rocks deep within the mantle must be rich in olivine and garnet. In later stages of cooling pyroxene appears. The last minerals to appear, concentrated near the surface of the earth, would be feldspar and quartz.

Degrees of Differentiation. The process by which the earth was partitioned into core, mantle and crust, and various rock minerals were concentrated at different depths within the planet, is called *differentiation* by geologists (Figure 16.5).

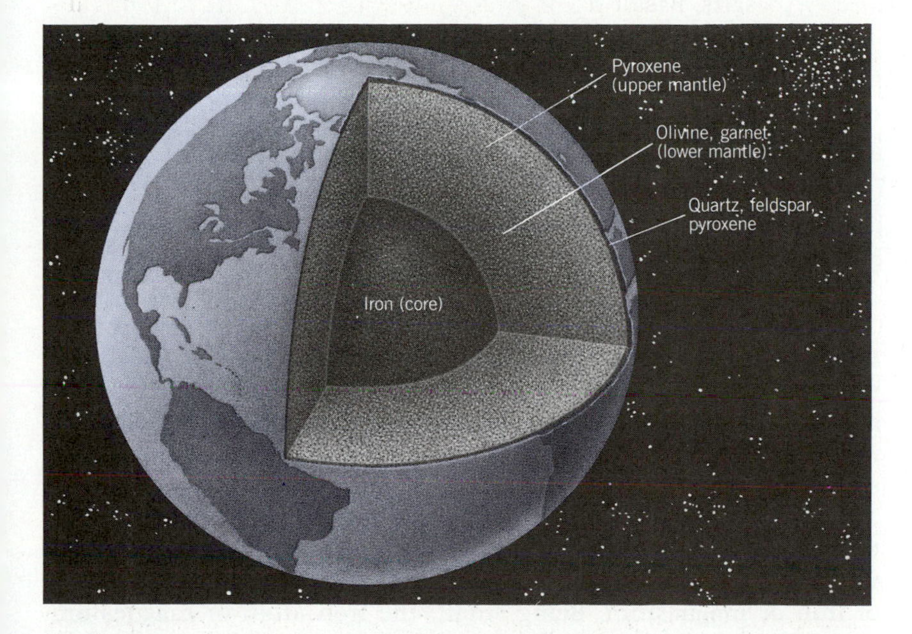

FIGURE 16.5
Major concentrations of minerals in the earth's crust and mantle.

Geologists refer to different degrees of differentiation of a planet. They consider that differentiation has proceeded to an advanced degree within the earth, for example, because of the fact that all or most of the earth's iron has collected at the center to form a nearly pure iron core, while the lightweight minerals have been very strongly concentrated at the surface in the crust.

As we will see in the next chapter, the moon appears to be a partly differentiated planet in which the process of melting and recrystallization has occurred to some degree, but not to the extent that it has occurred on the earth. Mars is probably more differentiated than the moon but less than the earth. If a planet has never been melted or partly melted throughout its history, it will probably have the same composition at the surface as at the center. Such a planet is called *undifferentiated.*

Composition of the Crust. Next to iron, oxygen and silicon are the most abundant substances in the earth. They are also among the most abundant substances in the Universe. They are the principal constituents of all the rock minerals mentioned above. Quartz, for example, is pure oxygen and silicon in the proportions given by the formula SiO_2. Its basic structure is shown in Figure 16.6. A tetrahedron—called the silicic tetrahedron—is formed by four oxygen atoms enclosing the relatively small silicon atom at the center. Since each oxygen atom is shared by two silicon atoms in neighboring tetrahedra, the chemical formula is SiO_2 instead of SiO_4.

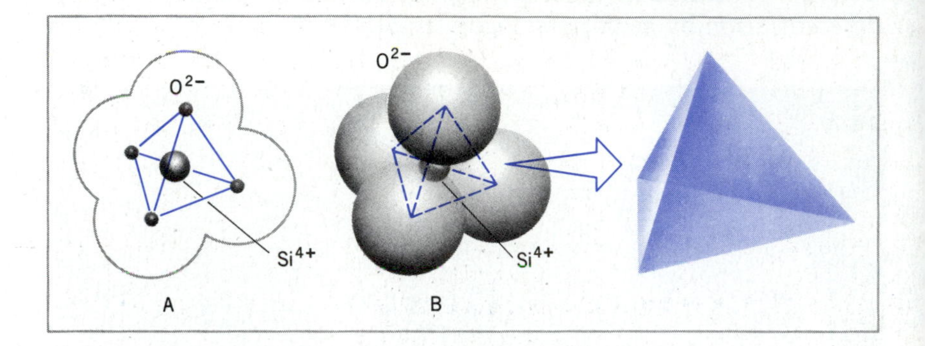

FIGURE 16.6
The silicon tetrahedron.

The silicic tetrahedron is the basic building block for most of the minerals in the earth's mantle and crust. Quartz consists of pure silicic tetrahedra, and other minerals are composed of silicic tetrahedra linked by regularly spaced atoms of iron, magnesium, aluminum, sodium, calcium, or potassium. For example, Figure 16.7 shows the structure of olivine.

The smaller the atom that serves as the link between adjacent tetrahedra, the tighter and more compact is the resultant structure. Olivine, for example, consists of silicon tetrahedra linked by atoms of iron or magnesium. Being small, the iron atom or magnesium atom fits very easily into the spaces between the other atoms that make up the crystal structure of olivine. The result is a mineral with a very compact and, therefore, very dense structure. This is the reason why olivine tends to settle to the bottom of the melt when it crystallizes out of molten rock.

The presence of iron and magnesium atoms in olivine also explains why this mineral is the first to crystallize out of a cooling mass of molten rock—that is, why it has a higher freezing-point temperature than other minerals. Iron and magnesium increase the strength of the bonds in the mineral by sharing electrons with the

II. PLANETS

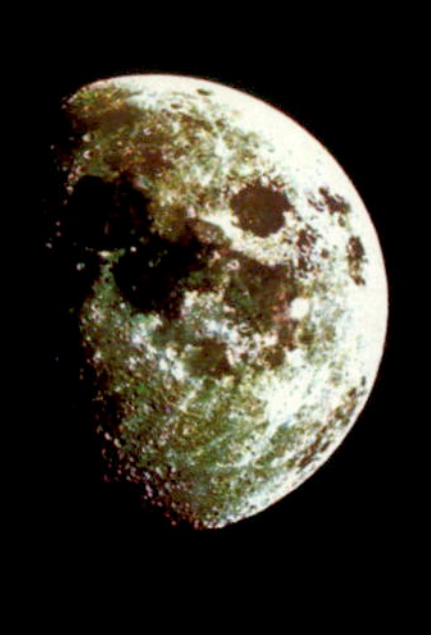

PLATE 15. THE EARTH AND THE MOON. Photographed from space, shown in proportion to true size.

PLATE 16. THE SURFACE OF THE MOON. A scene at the edge of the lunar highlands, photographed during the APOLLO 17 exploration. The Taurus-Littrow Mountains appear in the background. The large rocks in the foreground are highland basalts that probably rolled down the nearby mountainside several hundred million years ago. These rocks are 4 billion years old.

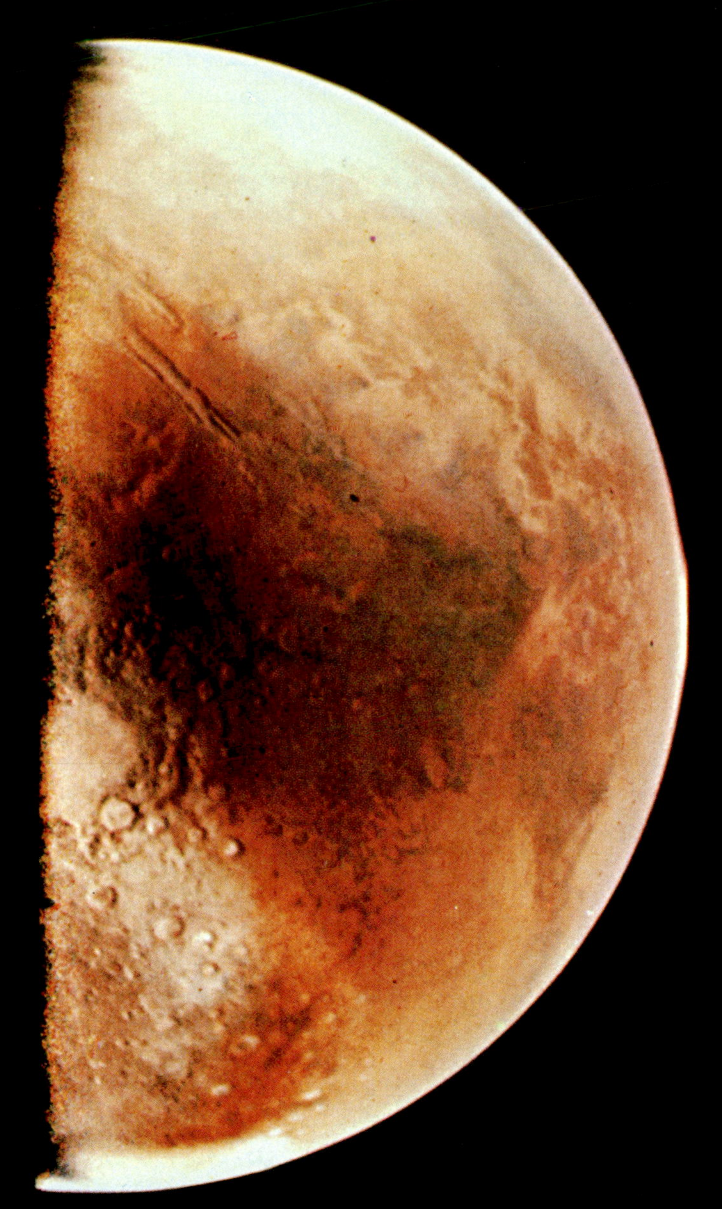

PLATE 17. MARS. Mars photographed from space. The white area at bottom is the polar cap. The cratered surface is old terrain marked by meteorite impacts. The Valley of the Mariner appears at upper left

PLATE 18. THE SURFACE OF MARS. Martian landscape on the Plain of Chryse, 4600 miles to the east, photographed from Viking I. The salmon color of the sky results from particles of red dust (in the atmosphere). The large rock in the foreground is a foot across. The horizon is two miles away. Bright patches of bare rocks toward the horizon are bedrock, probably of volcanic origin. The cause of the orange-red color is not certain. It may be iron oxide stain caused by weathering in an earlier era when oxygen may have been present in greater abundance.

PLATE 19. MOUNT OLYMPUS. The summit of this extinct Martian volcano is 70,000 feet above the surrounding terrain. The large crater at the summit, called a caldera, is caused by the collapse of rock into pools of lava below. The smaller craters on the flanks are probably meteorite impacts.

PLATE 20. JUPITER. This photograph of Jupiter was taken by cameras on the Voyager spacecraft at a distance of 17 million miles. The smallest features that can be seen in the photograph are 300 miles across. The bands circling the planet are belts of moving clouds made of crystals of frozen ammonia. The Red Spot *(lower left)* and several smaller white spots are mammoth storm centers. Other photographs suggest high-speed jet streams flowing to the right and the left above and below the Red Spot.

PLATE 21. THE RED SPOT. The Red Spot is believed to be a gigantic anchored hurricane, rotating counterclockwise every six days. Its diameter is approximately 15,000 miles. The Red Spot has persisted on Jupiter for at least 300 years. The turbulent patterns in its vicinity change with time.

PLATE 22. JUPITER AND THE GALILEAN SATELLITES. Jupiter *(background)* and the Galilean satellites–Io, Europa, Ganymede and Callisto–are seen in this montage of Voyager photographs from a vantage point just beyond Callisto *(lower right)*. Callisto and Ganymede *(lower left)* are probably spheres of water and rock with frozen crusts of rock and ice. Io *(upper left)* and Europa *(center)* have the density of rock and probably are composed largely of rocky materials like our moon and the terrestrial planets.

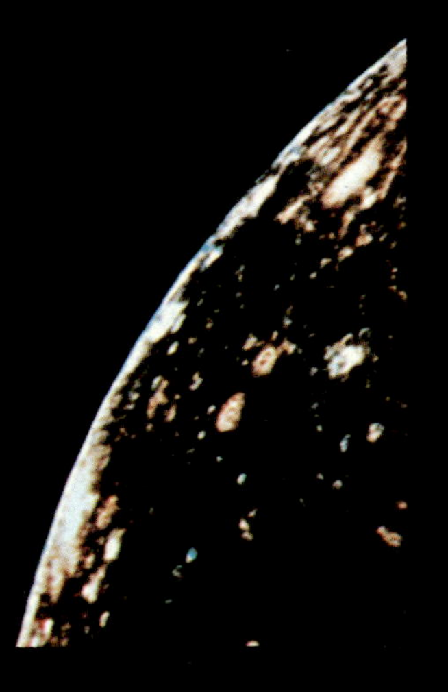

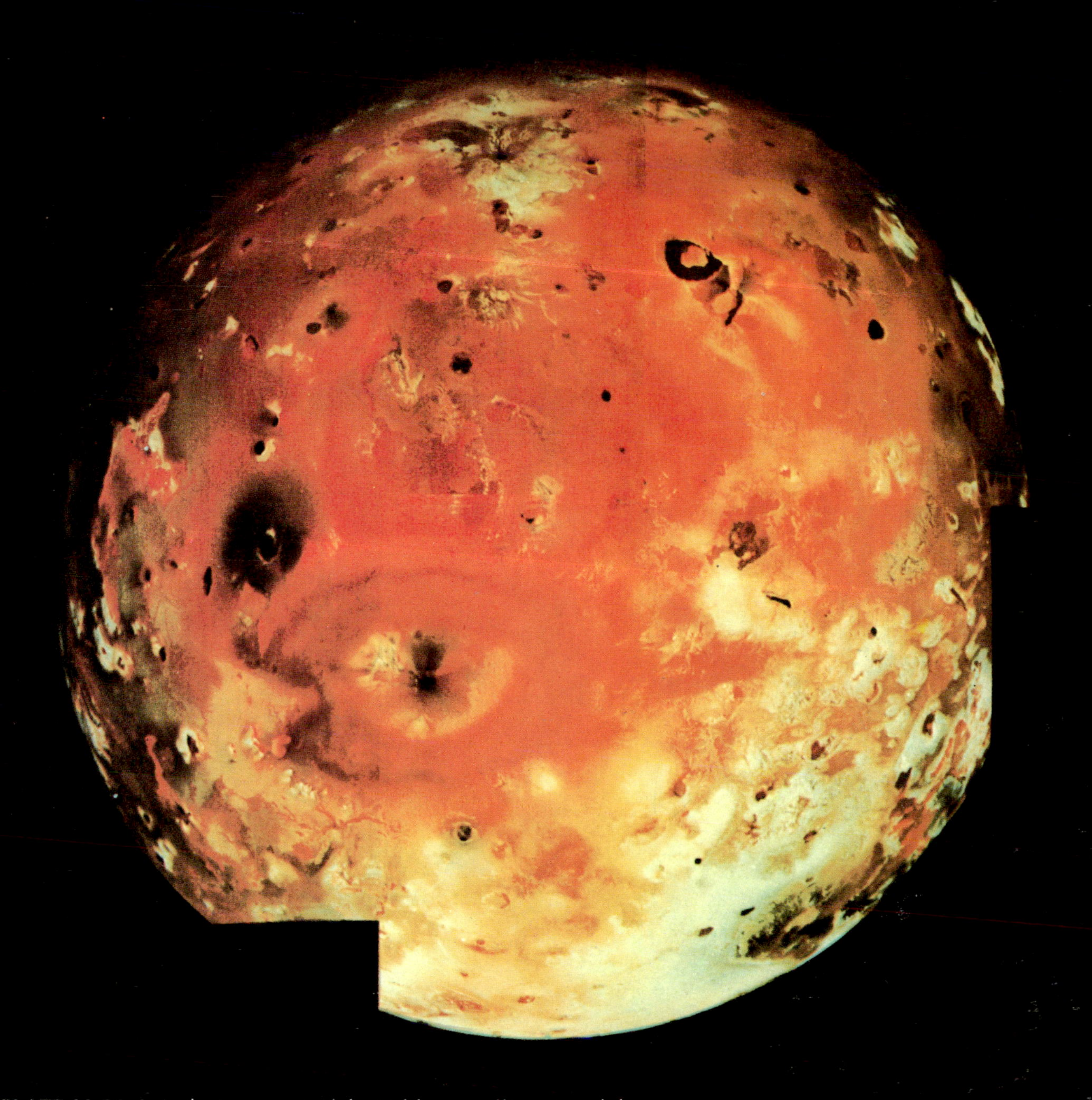

PLATE 23. IO. Io is the innermost of the Galilean satellites. Io and the earth are the only known bodies in the solar system with active volcanoes. The source of energy for Io's volcanoes is not radioactive heat as on the earth, but tidal friction caused by the varying force of Jupiter's gravity over the course of Io's elliptical orbit. The vivid colors are thought to be deposits of sulphur and sulphur dioxide brought to the surface by volcanic eruptions.

PLATE 24. IO's VOLCANOES. This photograph of the crescent Io shows eruptions of the volcanoes Amirani *(top)* and Maui. The plumes rise 60 miles above the surface. They are probably particles of sulphur dioxide snow.

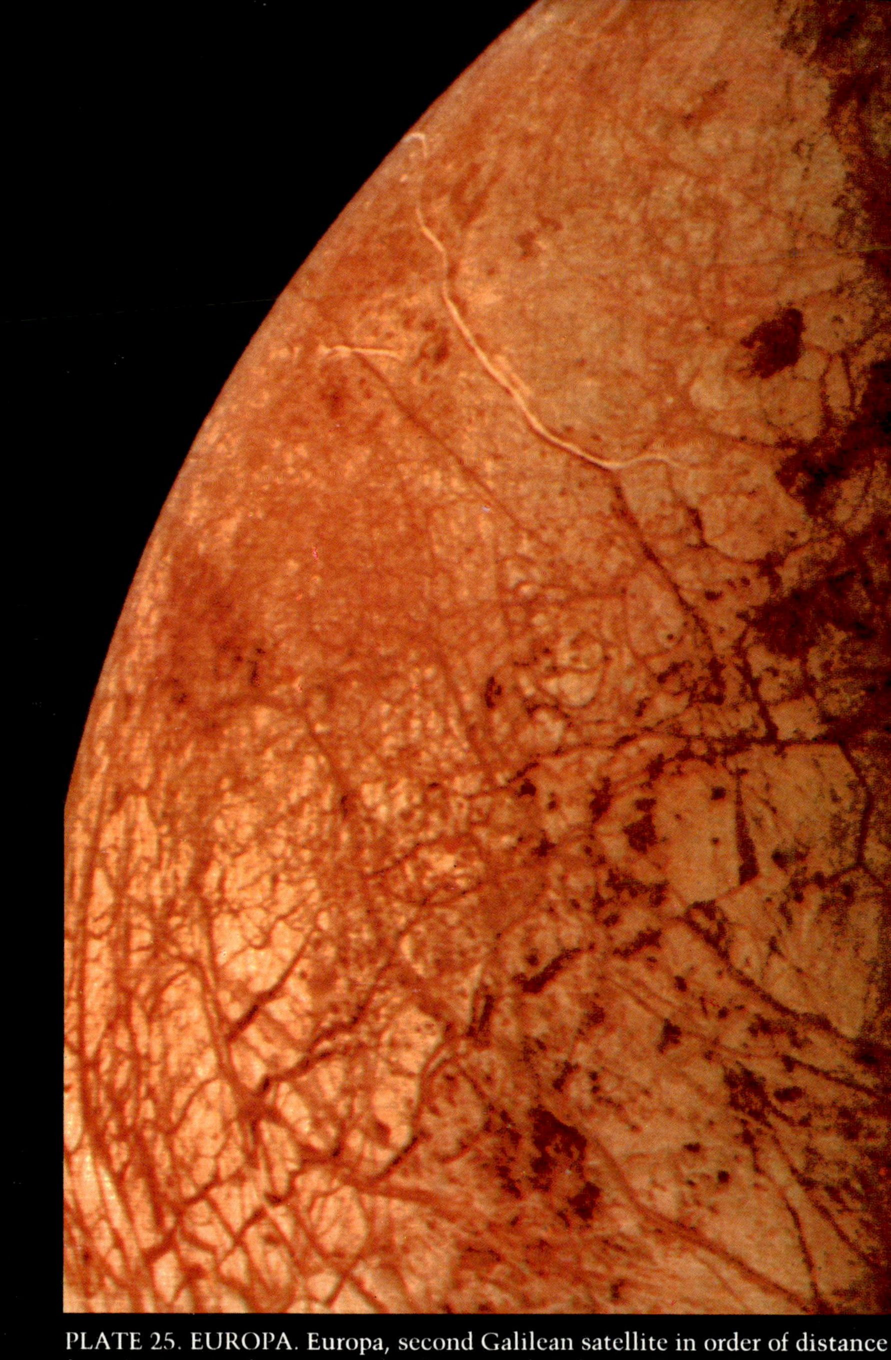

PLATE 25. EUROPA. Europa, second Galilean satellite in order of distance from Jupiter, is almost completely covered with ice. The surface is extraordinarily smooth, with changes in elevation of only some hundreds of feet over the entire moon. The narrow, dark streaks, extending for miles, are of uncertain origin.

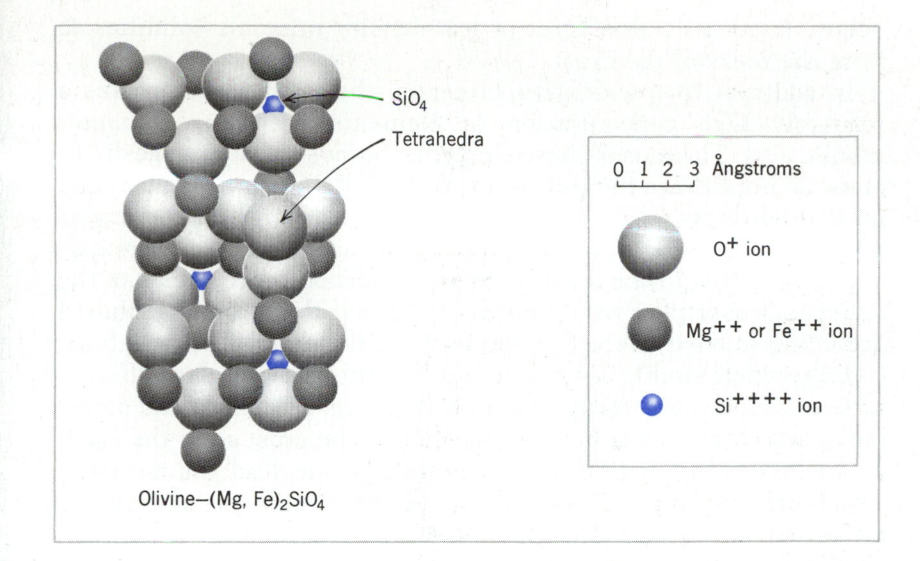

FIGURE 16.7
The structure of olivine.

oxygen atoms in the tetrahedra. Because the iron and magnesium atoms are small, they come into close contact with surrounding oxygen atoms. The electrical forces of attraction that tie atoms together to make a crystal are stronger when the atoms are closer together. This circumstance makes the bonds particularly strong in minerals with iron or magnesium atoms. Suppose now we consider what happens as a molten mass of rock cools down. Collisions occur continually between neighboring atoms in the molten rock, breaking up the bonds that attract one atom to another, and preventing solid crystals from forming. At very high temperatures, these collisions are violent enough to break the bonds of all minerals, including the very strong bonds that tend to tie together the atoms that make up a crystal of olivine.

However, as the temperature drops the violence of the collisions diminishes. When the temperature falls to 1400 degrees, the violence of the collisions is sufficiently diminished so that the atoms that make up a crystal of olivine can lock into place and stay in place in spite of continuing collisions with their neighbors. But at this temperature the collisions are still violent enough to break up the other minerals—with a looser, less compact, and less tightly bound structure—that start to form in the molten mass. These minerals cannot solidify or crystallize out of the melt until the temperature decreases still further.

Thus, as the earth solidifies from the inside out, the small atoms are locked into the structures of the minerals that appear at the base of the mantle, while the large atoms are forced successively upward, layer by layer, because they do not fit into the spaces between other atoms. Finally, at the top there is a residue of the low-density minerals containing large atoms, which float on top of the denser rocks below. Eventually this layer of low-density minerals solidifies to form the rocks of the crust.

In addition to low-density minerals, the crust also contains a relatively high concentration of elements with large-diameter atomic ions. The ions of these elements, because of their large diameters, do not fit into the compact structure of the minerals that exist under high pressure deep in the interior. Such large-diameter ions tend to be squeezed upward into the crust as a result of differentiation. They exist in the crust in great abundance, far in excess of the abundance that they would normally have if they were distributed according to their average proportions in the original solar nebula.

Large-sized atomic ions squeezed up into the crust by differentiation include such fairly common elements as sodium and potassium, which are ten times as abundant in the crust as in the earth as a whole. They also include elements such as lead and mercury, which are rare in the Cosmos, and therefore relatively rare in the crust, but are to be found there in much greater abundance than their average abundance in the earth as a whole.

Finally, all the radioactive substances—radium, thorium, potassium, and uranium—have atoms with very large diameters. These radioactive substances are also to be found in far greater abundance in the crust than their average concentration in the earth. Differentiation is estimated to have removed more than half of the radioactive elements from the interior of the earth and to have concentrated them in the relatively thin crust during the course of the earth's history.

If the radioactive substances were still distributed throughout the mantle, their radioactive heat would be sufficient to keep a large part of the mantle in a molten condition. Probably a solid and relatively permanent crust could not exist in that case. In summary, heat released by radioactive substances led to differentiation in the mantle, and the differentiation led to the removal of radioactive substances from the mantle, and hence the removal of the heat source. Differentiation has thus introduced inherent stability into the temperature history of the earth.

THE FLOATING CRUST

The lightweight rocks that form the crust of the earth are the familiar *granites* that are found on the continents, and the somewhat denser *basalts* that make up the rocks underlying the oceans. Like the granites, basalts are also found in substantial amounts on the continents. Granites and basalts together form the crust that floats on the denser mantle underneath.

The word "floats" is used with care. The rocks of the earth's interior—while solid—are nonetheless warm enough so that they have a kind of plasticity. They yield under pressures, provided the pressure is applied for a long enough time. They are very much like Silly Putty, which seems as hard as steel when hit sharply, but yields and flows like a viscous liquid under a steadily applied force.

As the photographs show (Figure 16.8), a slab of Silly Putty yields in 30 minutes under the pressure of a weight placed on top. How much time elapses before the solid rocks of the earth's interior yield under the weight of a mountain on the surface? The answer is approximately one million years. If a large mass—the size of a mountain—is placed on the surface of the earth, during the course of a very long time the underlying rock of the mantle yields under its weight, and slowly but steadily the mountain sinks into the earth's interior. But several million years must elapse before this occurs. During appreciably shorter periods of time—such as a year, a century, or even 10 centuries—very little happens. The atoms of the mantle rock remain locked in fixed positions, each atom bound to its neighbor by an electrical force.

But if solid rock is subjected to a large force, now and then a small layer of atoms somewhere within the rock will slip over an adjoining layer of atoms as a consequence of the pressure on its surface. The sliding motion of one layer of atoms over another occurs only in scattered places in the solid rock, and only at rare intervals. No change can be seen in the rock if you look at it for a short time. But over a sufficiently long period, the accumulation of many tiny displacements adds up to a "flow" of one part of this seemingly rigid, solid body over another part.

The slow movement of a solid body—such as rock—under heavy pressure is called *creep*. The phenomenon of creep is responsible for the fact that the solid rocks of the earth's interior yield and flow under the weight of mountain ranges on the surface.

Mountains are made of crustal rocks—mostly granite—which are lighter than the rocks of the mantle as a consequence of differentiation. The light rocks of the crust are buoyant; they float in the yielding mantle like a block of wood in a tank of water.

Only a small part of a floating block of wood rides above the surface of the water. Similarly, when you see a mountain range—such as the Rockies or the Himalayas—you are only viewing a small part of the complete mass of rock. The remainder is "submerged" below ground. The Himalayas extend upward 4 to 5 miles above sea level and downward about 25 miles. Continents also float on the mantle and have deeply submerged roots that extend far into the underlying rock. In fact, the entire crust of the earth—including the rocks on the floor of the ocean—floats on the mantle, with the larger part submerged and hidden from view (Figure 16.9) (p. 408).

FIGURE 16.8
Silly Putty yields under steady pressure.

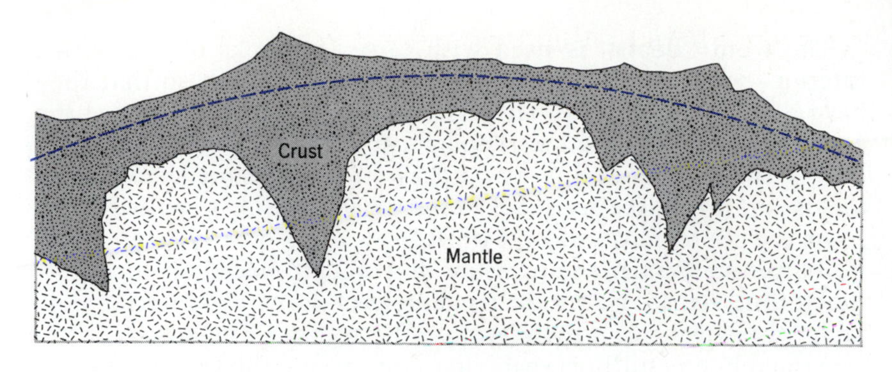

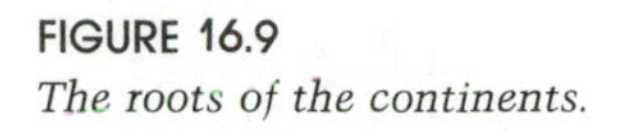

FIGURE 16.9
The roots of the continents.

PLATE TECTONICS

In recent years evidence has accumulated that makes it indisputable that the continents have not always been located in the positions they now occupy on the globe. Antarctica was once located in a pleasant climate far from the South Pole; North America was once joined to Europe; South America was once joined to Africa; and all these continental masses were far from their present locations.

The Zone of Weakness

At first thought it seems that these results must be incorrect. How can a continent slide about like a cake of soap in the bathtub? The answer is that a *zone of weakness* exists in the earth's interior at a depth of some 60 miles, well below the deepest continental roots. At this particular depth the mantle rocks—while still below the melting-point temperature—come nearer to melting than at any other place in the interior of the earth. The rocks at a depth of 60 miles are still solid, but only barely so. They are so close to melting that they are as soft as warm butter. The mantle above this deep layer of soft and yielding rock is colder and more rigid. This rigid 60-mile layer is called the *lithosphere.*

The lithosphere is broken up into a number of slabs, called *plates* by earth scientists. The term conveys a thinness that seems inconsistent with a slab of rock 60 miles thick. However, the plates average thousands of miles in size, or more than 20 times their thickness (Figure 16.10). The movement of the earth's plates, and the disturbances of the crust that these movements produce, are the subject of *plate tectonics.*

If a continent is embedded in a particular plate, the continent moves along with the plate as it slides over the zone of weakness. The movement is slow—usually no more than one inch per year—but over an appreciable period of time in the earth's history (e.g.,

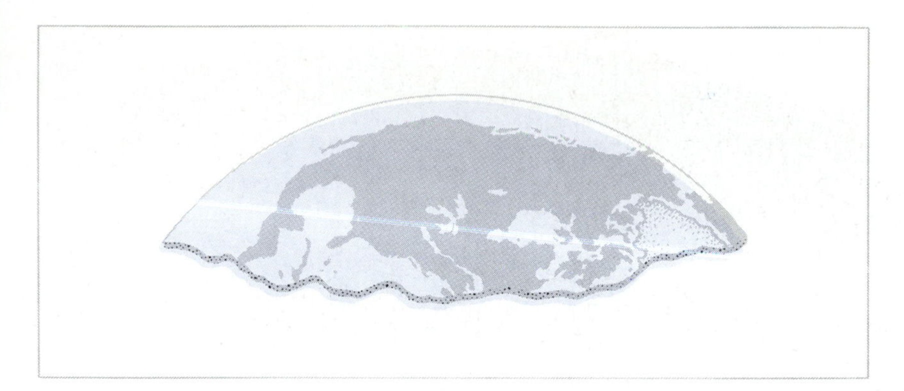

FIGURE 16.10
The dimensions of a plate.

100 million years or more) these slow movements of the plates add up to displacements of thousands of miles. Plate tectonics provides the explanation of the drifting continents.

A Map of the Earth's Plates

Between 1965 and 1970, geologists succeeded in mapping the boundaries of most of the great plates into which the earth's surface is divided. The boundaries of the plates are shown in Figure 16.11 (p. 410). Arrows drawn on the plates indicate the direction of the movements. All movements are relative to the Eurasian landmass, which is regarded as fixed in this map. Locations of all earthquakes occurring during the last 10 years are also marked by small circles or dots. The zones of intensive earthquake activity played a major role in defining the plate boundaries. The most striking geological consequences of the map are explained below.

The African Plate. In some parts of the globe, two plates appear to be colliding head-on, crumpling the material of the earth's crust in great folds. The map shows that a great plate containing the African continent and part of the Atlantic Ocean is moving northward and plowing into a second plate containing the Eurasian landmass. The Alps are giant wrinkles in the edge of the Eurasian landmass produced by a collision between the two plates at the rim of the Mediterranean. The Mediterranean itself is gradually disappearing as the northern movement of the African plate closes the gap between the two continents. The earthquakes that plague Greece and Turkey, of which the most recent occurred in Turkey in 1975, are signs of the subterranean adjustments occurring as one mammoth mass of rock pushes its way into another.

The India-Australia Plate. A crustal block in the Indian and Pacific oceans, carrying the subcontinent of India and the continent of Australia, is currently moving to the north, again at about the

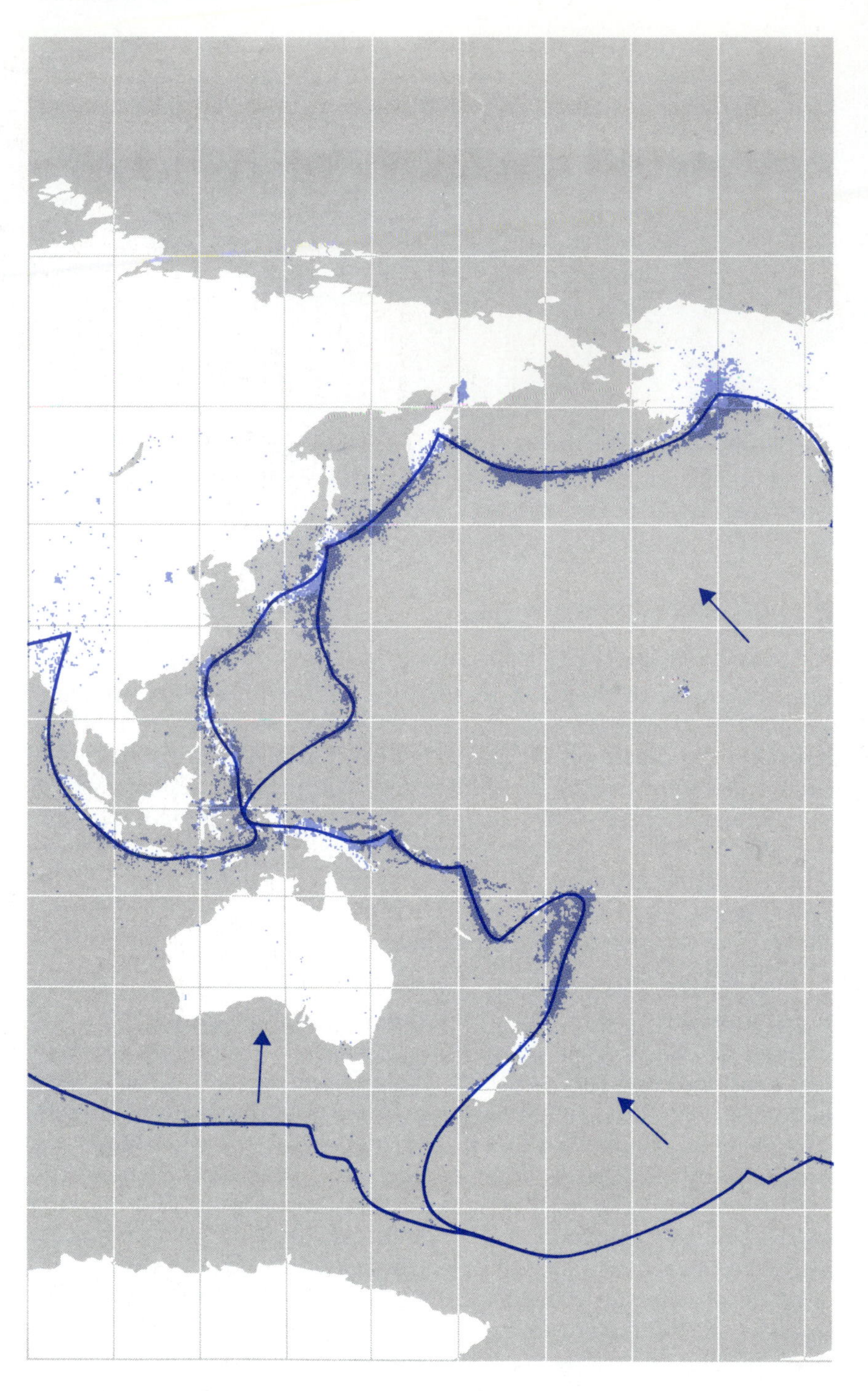

FIGURE 16.11
The boundaries of the earth's plates. Arrows indicate plate movements relative to the Eurasian landmass. Shading indicates zones of earthquake occurrence.

rate of one inch a year, crumpling up the lower part of the Eurasian landmass and creating the Himalayan Mountain system. The map indicates a zone of earthquake activity in Asia near the plate boundary, produced again by the subterranean impact of the collision.

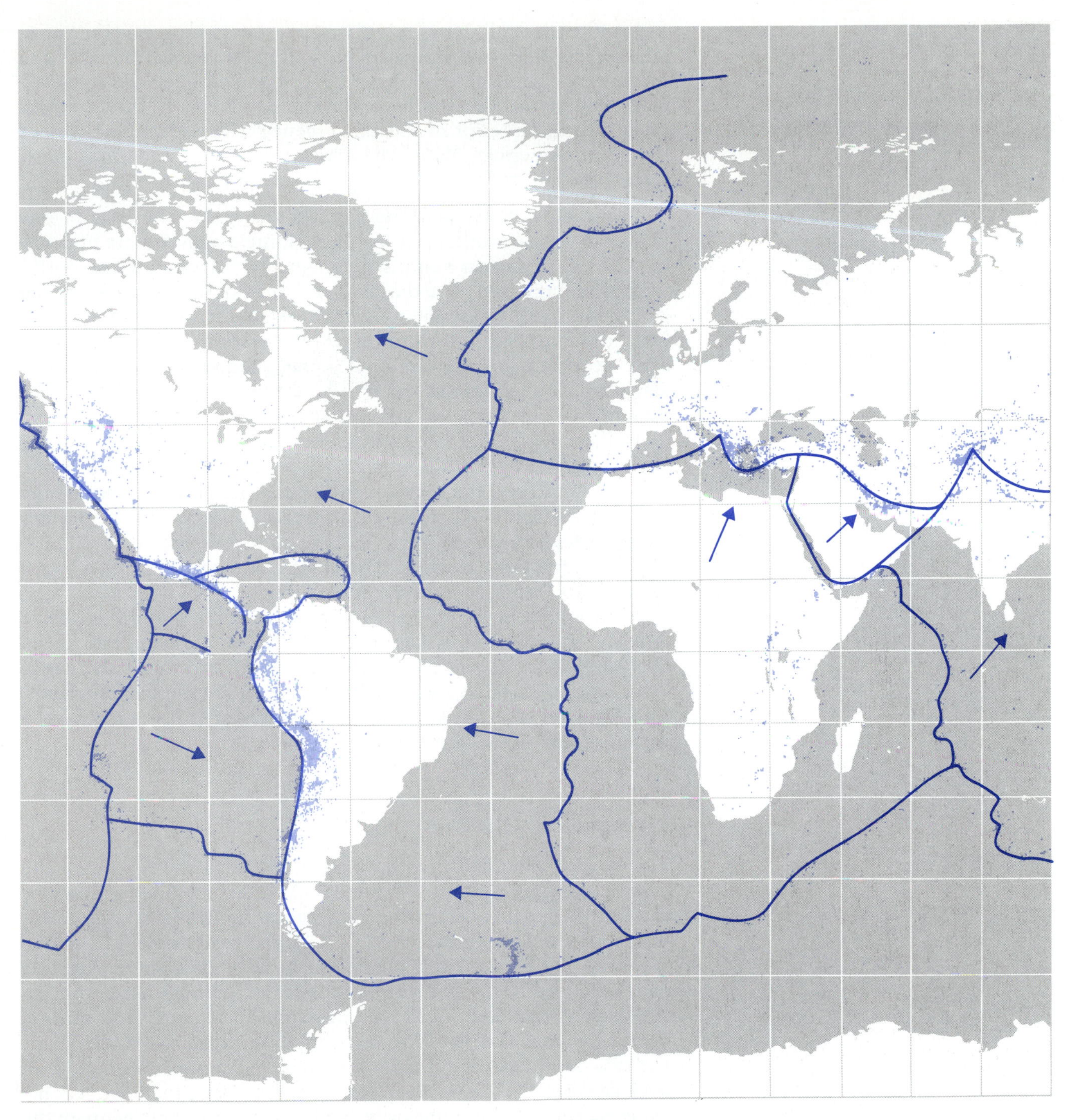

The South American Plate. The westward movement of the plate containing the South American continent is clearly indicated on the map. The western edge of this plate also marks the western coast of South America. When a plate containing a continental landmass on its leading edge collides with a plate underlying the ocean, the continental plate rides over the ocean plate. Deep-seated earthquakes and intensive volcanic activity occur in Peru and Chile as the adjoining plate underlying the southeastern Pacific Ocean is thrust downward into the interior of the earth, beneath the advancing edge of the South American plate. The edge of the South American plate, forced upward, became the Andes Mountain Range.

The Pacific Plate. In some parts of the globe, the edge of one plate slides past another like a ship scraping a pier. A plate underlying the Pacific Ocean basin is moving slowly upward to the northwest, scraping past a plate that includes most of the North American continent. In this case the movement is about 1.5 inches per year. Over the course of a century or so, the eastern rim of the Pacific plate—carrying Los Angeles—will move about 15 feet to the north. San Francisco lies on the other side of the plate boundary. At the present rate, Los Angeles will enter the suburbs of San Francisco in 10 million years. The San Andreas Fault marks the boundary of the plate running up the California coastline. The motion of the Pacific plate slowly bends the rocks of the crust where they run across the San Andreas Fault. When the displacement across the fault reaches about 15 feet, the crust breaks in two and the broken ends vibrate, causing earthquake tremors. The crust last snapped in 1906, causing the San Francisco earthquake and fire.

At its northern edges, the Pacific Ocean plate collides with the Eurasian plate near the Aleutian islands. The collision has forced the Pacific Ocean plate downward into the interior of the earth, creating an intensive zone of earthquakes and volcanic eruptions. The Aleutian island chain is the accumulation of lava produced by the eruptions—called an *island arc.*

Evidence for Continental Drift

The bulging Brazilian coastline fits nicely into the hollow of the Ivory Coast in West Africa (Figure 16.12). This coincidence provided the first suggestion of drifting continents. Impressed by the agreement between the coastlines, a German meteorologist, Alfred Wegener, put forward the suggestion in 1910 that South America and Africa had once been part of a single landmass. According to Wegener, Africa and South America split apart about 150 million years ago. Since the matching coastlines are about 3000 miles apart, the two continents must have drifted away from each other at an average speed of one inch per year. A speed of one inch per year became accepted as the standard velocity for drifting continents in the Wegener theory.

Wegener's proposal was greeted with scorn by most of the world's geologists, although he pointed to other evidence, such as similari-

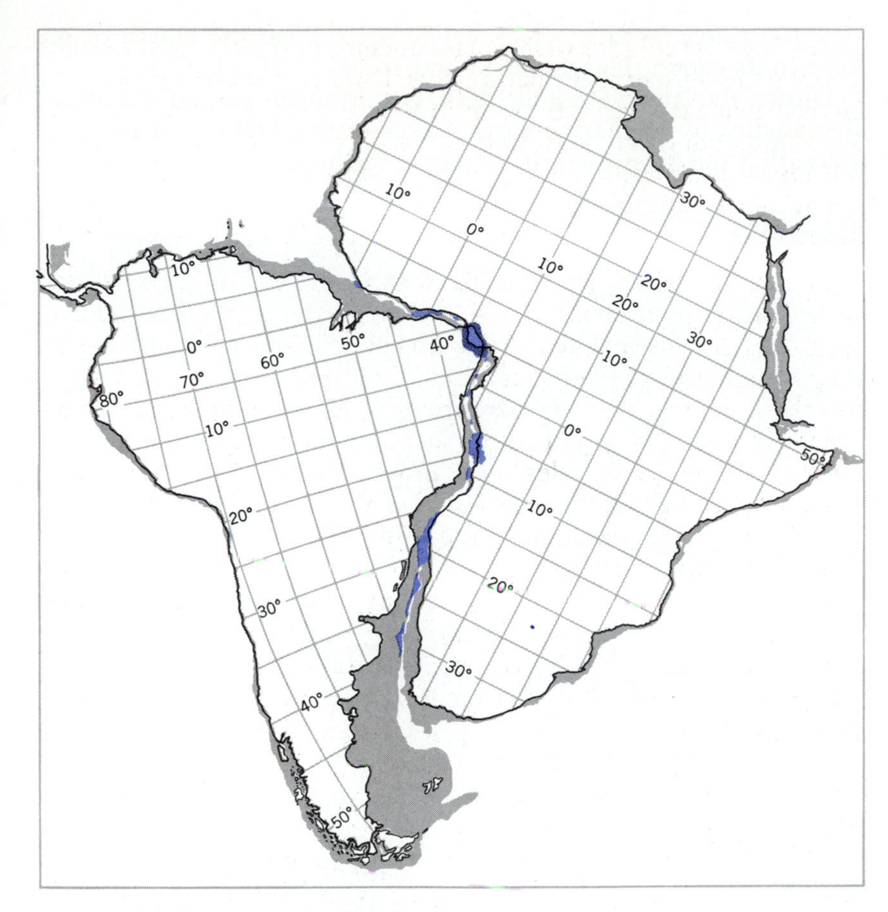

FIGURE 16.12
The fit of the African and South American coastlines.

ties in fossil plants and animals discovered on the two continents. In spite of circumstantial evidence accumulating in its favor, the theory of continental drift continued to be greeted with skepticism.

Interest in continental drift was revived after World War II by a series of geophysical discoveries relating to the floor of the Atlantic Ocean. Each discovery by itself seemed to have little bearing on the Wegener theory until 1966, when everything clicked into place and it became clear to geologists everywhere that the theory of continental drift—long ignored or dismissed—was valid.

The Mid-Atlantic Ridge. The first important discovery was made by scientists of the Lamont-Doherty Geological Observatory, who measured the topography of the ocean bottom during a series of cruises of the oceanographic research vessel *Vema.* They found that an underwater chain of mountains ran along the bottom of the Atlantic Ocean from north of Scandinavia to the latitude of Cape Horn at the tip of South America. The subterranean mountain chain was a nearly continuous ridge rather than a series of separate peaks. Because it ran along the middle of the ocean floor, dividing the Atlantic into two nearly equal parts, it became known as the *Mid-*

Atlantic Ridge. A large crack called the *rift* ran the length of the ridge in its center (Figures 16.13 and 16.14).

During the late 1950s and 1960s, geophysicists carried out intensive studies of submarine earthquakes in the Atlantic Ocean and found that the great majority of these earthquakes were located on

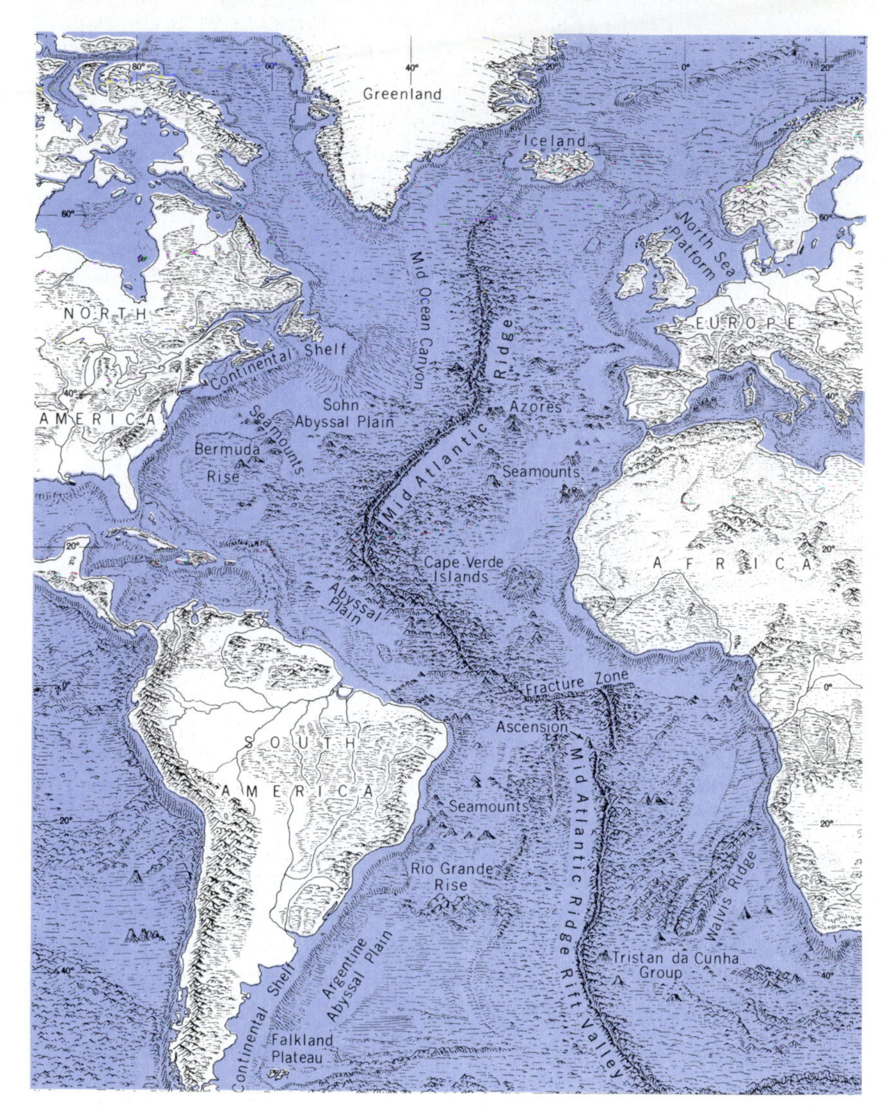

FIGURE 16.13
Map of the Mid-Atlantic Ridge.

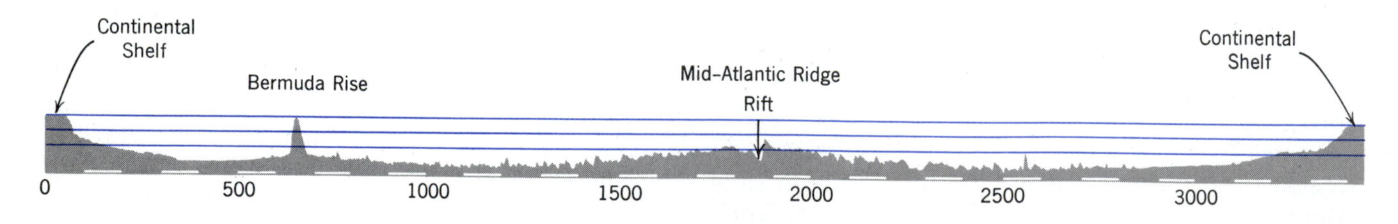

FIGURE 16.14
Profile across the North Atlantic from Cape Henry to Rio d'Ora. The horizontal scale is in nautical miles and the vertical scale in thousands of fathoms (after Heezen et al., 1959).

or near the Mid-Atlantic Ridge. During the same period, evidence was found of extensive volcanic activity on the Mid-Atlantic Ridge. Finally, measurements of the flow of heat through the floor of the Atlantic Ocean revealed an anomalously large amount of heat emerging from the Central rift in the ridge.

The flow of heat through the rift suggested a crack in the earth's crust at that point, connecting the surface to the warm interior of the planet. With this thought born, all the facts regarding the Mid-Atlantic Ridge began to fit together. Suppose a large amount of heat appeared in the rift because hot, molten rock was emerging through this crack in the earth's crust and flowing out to either side along the ocean bottom. The material would create fresh rocks continually on the floor of the ocean as it came up from the interior.

The theory could be tested by measuring the ages of the ocean-bottom rocks in the Atlantic. If the ocean floor were continually renewed by fresh lava flowing out of the rift, the ocean-bottom rocks in the vicinity of the ridge should be younger than rocks at a distance from the ridge. The rocks near the continental boundaries should be the oldest rocks in the Atlantic Ocean.

The ages were measured by collecting samples of ocean-bottom sediments and by determining the ages of fossil animals in the deepest layer of sediments lying directly on the ocean floor. The results fully confirmed the theory: the rocks making up the floor of the Atlantic Ocean were very young in the neighborhood of the Mid-Atlantic Ridge, and they became progressively older as the distance from the Mid-Atlantic Ridge increased.

Moreover, no rocks older than 150 million years were found anywhere in the Atlantic between the South American and African continents, indicating that this part of the ocean did not exist prior to that time. Those rocks that were as old as 150 million years were all located very close to the continental shorelines.

No proof could be more convincing. Africa and South America had been a single landmass with no ocean between them until 150 million years ago, when a buried crack—the forerunner of the rift—appeared in the crust, and the landmass broke in two.

Pangaea

Accumulating evidence indicated that 225 million years ago all the earth's continents—and not only South America and Africa—were collected into a single land mass called Pangaea that stretched nearly from pole to pole in a broad belt of land about 5000 miles wide and 10,000 miles long. Approximately 200 million years ago, a rift developed at the equator between the northern and southern halves of Pangaea, separating it into two supercontinents called Gondwanaland and Laurasia. Gondwanaland, located largely in the

FIGURE 16.15
Gondwanaland and Laurasia.

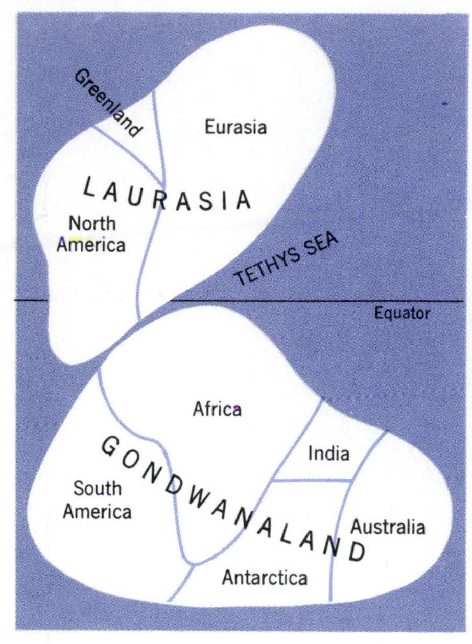

Southern Hemisphere, contained in a single mass the blocks of rock that broke off subsequently to form the separate continents of Africa, India, South America, Australia, and Antarctica. The other supercontinent, called Laurasia, located largely in the Northern Hemisphere, contained the blocks that broke off and drifted apart to become North America, Eurasia, and Greenland.

Painstaking correlation of a mass of items of information has produced reconstructions of the supercontinents of Gondwanaland and Laurasia (Figure 16.15). The solid lines in the diagram show the boundaries along which the supercontinents came apart into the separate pieces that form the continents as they are today.

A body of water called the Tethys Sea originally separated the supercontinents. When Gondwanaland and Laurasia broke up, the fragment containing Africa and India moved northward, pushing up into Eurasia and closing up the Tethys Sea. These events took place between 130 and 150 million years ago. The northward movement of India into Eurasia compressed and buckled the sediments that formerly covered the bottom of the Tethys Sea. The folded layers of ocean-bottom sediment became the Himalayas. Marine fossils, found on the summit of Mount Everest during the first ascent, had been lifted to the roof of the world from the bottom of the Tethys Sea by the events that followed the breakup of Gondwanaland.

In the same period North America divided from Eurasia, opening up the North Atlantic, and South America divided from Africa, opening up the South Atlantic. The Atlantic continued to widen, and, by 50 million years ago, the arrangement of the continents and the oceans had become recognizably similar to the one that exists at present, although the Atlantic Ocean was still about 1000 miles narrower than it is today. The plates continued to move apart at the rate of approximately one inch per year or 20 miles every million years. Indirect evidence indicates that the movement of the plates and their continents is still going on at the present time.

What Moves the Earth's Plates?

With many features of the earth's surface and history explained, one mystery still confronts the geologist: What forces move the earth's plates? As yet, no one knows. One theory proposes that the warm rocks of the mantle move up to the surface of the earth from the deep interior and down again, in steadily circulating currents of solid rock. Similar currents—called convection currents—are set up in any container of fluid when it is heated from below. In the case of the earth, the moving material is a solid rather than a fluid, but because of the creep the solid material moves like a viscous fluid. According to this theory, sluggishness of the flow will cause each cycle of convection to last for millions of years, but the mechanism is the same one that causes convection in a pot of heated porridge.

Imagine a current of warm rock ascending toward the surface (Figure 16.16). Near the surface, just under the lithosphere, the current of warm rock meets the resistance of the cold, rigid material above, and divides into two separate currents. These currents proceed in opposite directions. Eventually they descend into the interior of the earth again.

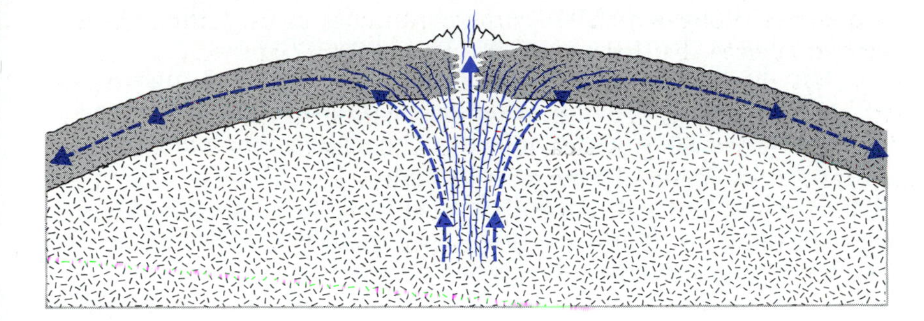

FIGURE 16.16
The intrusion of magma into the edges of the plates.

In the lithosphere, above the place where the current separates into two streams, the solid rock is pulled in two directions. It breaks apart; a crack appears; and the lithosphere separates into two plates. The crack manifests itself on the surface of the earth as a rift running down the center of the midocean ridge. The hot but solid rock in the upward-moving convection current is exposed to a lower pressure in the crack or rift. As a result, it melts the rock. The molten rock, which has a lower density, is buoyant and therefore rises. Near the surface it cools and solidifies, filling the crack. In this way, as the plates move away from the rift, fresh material is continually added to the edge of each plate. In effect, fresh ocean floor is being created steadily.

The midocean ridge is a bulge resulting from the fact that the freshly made plate material under the rift is warm and therefore expanded. Moving away from the rift, the plate material becomes colder and therefore denser, and the ridge subsides in height.

As each sluggish current of rock in the mantle moves horizontally under the lithosphere, it drags the overlying plate with it. Several currents or convection cells exist in the earth's mantle, each carrying its own plate. As the plates move apart, they collide with other plates, carried along by the motions of their own convection cells.

This relatively simple picture accounts for many features of the boundaries, but leaves some unexplained. Other plate mechanisms have been proposed. According to one view, a factor that contributes to plate movements is the fact the oldest part of a plate, farthest from the ridge, is cold and therefore dense. Being dense, it tends to descend into the mantle, dragging the rest of the plate with it. It is possible that it will be necessary to invoke these mechanisms, and others as well, to obtain a complete explanation of the movements of the earth's plates.

Main Ideas

1. Early history of the earth; heating of the earth's surface by meteorite bombardment.
2. Heating of the earth's interior by radioactivity.
3. The meaning of differentiation; effect of repeated melting on minerals with different densities and melting points; formation of mantle, crust and core; effect on radioactive elements and other elements possessing large ions.
4. The floating crust; phenomenon of creep in the earth's interior.
5. The zone of weakness and the lithosphere; division of the earth's surface into lithospheric plates.
6. Explanation of mountains and other features of the earth's surface in terms of collisions of plates.
7. Evidence for plate tectonics; fit of the continents; the midocean ridges; ages of ocean-bottom rocks.
8. Pangaea; the Tethys Sea; opening of the Atlantic Ocean.
9. The forces that drive the motion of the plates.

Important Terms

basalt
continental drift
convection currents
core
creep
crust
differentiation
granite
igneous rock
lithosphere
mantle
midocean ridge
mineral
plate
plate tectonics
radioactive heating
rift
rock
sedimentary rock
seismograph
zone of weakness

Questions

1. What was the major source of the heat that melted the interior of the earth? What element melted first? Where did this element concentrate?
2. The most abundant elements in the earth are iron (35 percent),

oxygen (30 percent) and silicon (15 percent). Compare these with the three most abundant elements in the sun after the volatile elements H, He, C, N and Ne have been removed. Base your answer on the graph of cosmic abundances in Chapter 9. Are the three most abundant (non-volatile) elements in the sun also the three most abundant elements in the earth. Would you expect this to be so? Why? How do the percentages compare? Can you explain the differences?

3. Explain why some minerals have higher melting points than others.
4. Why did some minerals begin to crystallize deep in the interior of the earth, while others did not do so until the final crust formed? Why is uranium, a heavy element, concentrated in the earth's crust instead of sinking with iron to the earth's core?
5. What is meant by differentiation of the earth's interior? Describe the probable composition and physical nature of the earth's crust, mantle and core.
6. Describe the phenomenon of rock creep. How does it explain the fact that the roots of the continents go deep into the mantle?
7. If a two-billion-year old bedrock were found in the Atlantic Ocean floor, what impact would such a find have on the plate tectonics theory? Why?
8. Draw the boundaries of the earth's plates on a map of the world. Indicate the motions of the plates relative to the Eurasian land mass. Explain three mountain ranges or other important features of the earth's surface in terms of these motions.
9. Where is most of the new crustal material made on the earth today? How is it made?
10. What strikes you as the most significant evidence for plate tectonics? Why?
11. Mountains generally are formed at the edges of plates in plate collisions. What explanation can you suggest for the origin of a mountain range such as the Urals, which are located in the middle of the Eurasian landmass far from any existing plate boundary?
12. Suppose an earth-sized planet is formed in another solar system, in which the abundance of the radioactive elements is negligible in comparison to their abundance in the solar nebula out of which the earth condensed. What would the internal structure of this earth-sized planet be after 4.6 billion years of geological history? What would its surface look like? Explain. What existing celestial bodies would it resemble? Suppose the radioactive elements were considerably more abundant in this earth-sized planet than in the earth itself. What would its surface look like after 4.6 billion years? Explain.

THE MOON

17

When Galileo, the first man to look at the moon through a telescope, turned his primitive instrument on that body in 1609, he saw large, dark areas resembling the earth's oceans, and mountainous light-colored areas that seemed to resemble the continents. This pattern of light and dark regions, visible to the naked eye, makes up the face of the man-in-the-moon (Figure 17.1) (p. 422). Galileo thought the dark areas were actually oceans, and called them *maria,* or seas. The light-colored, mountainous regions came to be known as the lunar *highlands.*

Today we know that these similarities to the surface of the earth are illusory. The lunar seas contain no water; no storms rage across the dark plains; no streams flow down from the highlands. And the lunar highlands are not similar in any way to the earth's continents; they do not resemble continental rocks chemically, and the forces that created the lunar mountains were entirely different from the forces that thrust up the great mountain ranges of the earth.

Moreover, a casual inspection of the moon through a telescope reveals that the texture of the moon's surface is completely different from that of the earth. Photographs of the moon taken through a large telescope show that both the maria and the highlands are pitted by innumerable craters of all sizes. Most of these craters have been produced by the impact of meteorites that have been raining down on the moon's surface for billions of years. Many craters are circled by ramparts ranging up to 10 thousand feet in height. Some of these ramparts must be more than one billion years old, yet photographs taken with a telescope clearly indicate that

Geologist Harrison Schmitt studies a house-sized rock on the moon.

Sea of
Showers
Sea of
Serenity
Sea of
Crisis
Sea of
Tranquility
Ocean of
Storms
Sea of
Fertility
Sea of
Clouds

they have been preserved almost unchanged, with little of the original material worn away (Figure 17.2).

FIGURE 17.1 (opposite)
Photo of the moon taken from the earth, made by joining two half-moon photos together to provide maximum contrast.

FIGURE 17.2
The ancient surface of the moon.

Meteorites have collided with the earth throughout its history, just as they have collided with the moon, and they have produced similar craters; but all traces of the older craters are gone. Only the scars of the most recent collisions, such as the Arizona meteorite crater, formed about 30,000 years ago, are still visible on the earth (Figure 17.3) (p. 424).

Air and water—the elements that make our planet livable—have worn down the oldest rocks and washed away their remains into the oceans, while the movement of the plates, and associated mountain-building activity and volcanic eruptions, have churned the surface and flooded it repeatedly with fresh lava. These natural forces have entirely removed the materials that lay on the earth's surface when it was first formed. But on the moon there are no oceans and

FIGURE 17.3
The Arizona meteorite crater.

FIGURE 17.4
Small craters on the moon. The crater in the center is 500 feet in diameter.

atmosphere to destroy the surface, and there is little or none of the plate movement and mountain-building activity that rapidly change the face of the earth. Over large areas, the materials of the moon's surface are as well preserved as if they had been in cold storage.

THE LUNAR SURFACE

Pictures of the moon taken by spacecraft provide further proof that the surface of the moon changes very slowly. Many craters that had never been seen before in photographs taken with telescopes on the earth were visible, ranging in size from a few feet up to hundreds of feet (Figure 17.4). These small craters must have existed on the earth as well, but were wiped out almost immediately by the wearing effect of winds and running water. The Arizona Crater probably will last no longer than 10 million years, which is a blink of an eye in the scale of geological time. On the moon the shallow footprints of the Apollo astronauts, six inches deep, will last at least for that long (Figure 17.5).

FIGURE 17.5
Neil Armstrong's footprint on the surface of the moon.

The Small Rate of Lunar Erosion

Why should the footprints of the astronauts not last forever? The principal force of erosion on the earth is running water; why should the moon, lacking water and even an atmosphere, suffer any erosion at all?

A part of the explanation is connected with the very thinness of the moon's air, which allows a continuous hail of extremely small meteorites—called *micrometeorites*—to reach the surface of the moon. Micrometeorites bombard both the earth and the moon at all times, but in the case of our planet they are burned up in the outer layers of the atmosphere and never reach the earth's surface.

These tiny grains of rock and metal contribute appreciably to the erosion on the moon because, although very small, they are present in enormous numbers, and collide with the moon at very high speeds. The sizes of typical micrometeorites range from one ten-thousandth to one-thousandth of an inch and they travel at speeds ranging up to 70,000 miles per hour. A micrometeorite moving at 70,000 miles an hour possesses one hundred times as much energy as an equivalent mass of TNT, and it can do one hundred times as much damage. Figure 17.6 shows a pit blasted out of an iron meteorite—picked up on the surface of the moon by the Apollo 11 crew—by a microscopic grain of meteoritic matter one ten-thousandth of an inch in diameter.

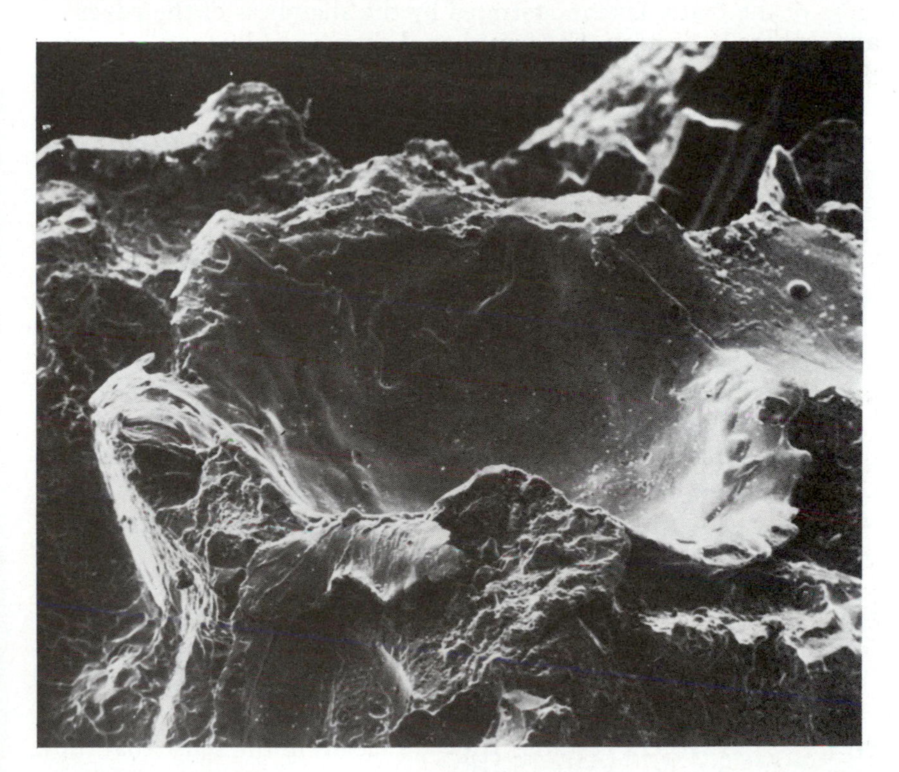

FIGURE 17.6
A lunar micrometeorite crater one-hundredth of an inch in diameter.

The moon's surface is continually fragmented and churned by the impact of these tiny particles, as well as the larger meteorites in the range of sizes that also reach the surface of the earth. Whether the meteorite is small or large, the effect of its impact is to pulverize the surface of the moon, ejecting particles of rock in a spray of fine dust, and, in the case of large meteorites, rock fragments ranging up to 30 or 40 feet in size.

Meteorite Erosion. Each time a meteorite hits the surface, it disturbs the layer of rock dust created by previous impacts, and as the dust particles are shifted about under the succession of impacts, they always tend to slide downhill, filling in all the depressions on the surface. Also, the continuing bombardment steadily wears away the high points of the lunar surface. Thus, the crater edges become steadily more rounded, and the craters themselves are gradually filled in.

The evidence of lunar erosion can be seen clearly in Figure 17.4. This photograph was taken by a lunar satellite orbiting the moon. It shows an area of 100 acres located in the ocean of Storms near the western limb of the moon. The crater in the middle of the area is 500 feet in diameter. It looks freshly made and probably was formed recently, within the last 200 million years, by the impact of a meteorite about 10 feet in diameter and weighing 500 tons. The freshness of the crater is demonstrated by the sharpness of the edges of the crater wall, and by the fact that the surrounding terrain is still littered with blocks of rock ranging up to 10 or 15 feet in size.

Two craters, indicated by arrows at the right of the fresh crater, demonstrate the effects of lunar erosion by micrometeorites. Their edges are rounded and the craters are partly filled in. These craters are clearly older than the fresh crater. They were formed by a meteorite impact approximately one billion years ago.

A study of many lunar photographs shows that frequently small craters are worn away and filled in, but large craters never are entirely obliterated. The dividing line between the two groups of craters occurs at a crater size of 200 feet and a crater depth of 50 feet. That is, in the 4.6-billion-year history of the moon, meteorite and micrometeorite bombardment has moved the top 50 feet of the moon's crust from place to place, wearing down the high points and filling in the hollows in the moon's surface. Fifty feet of erosion in 4.6 billion years represents a rate of erosion one ten-thousandth of the erosion rate on the earth.

The Unchanging Surface of the Moon

We referred above to events occurring on the moon 200 million years ago as "recent" events. They are recent in the sense that they occurred only a short time ago in comparison to the age of the moon and the age of the solar system. But consider how the surface of the earth has been transformed in the last 200 million years. Two hundred million years ago the continents of the earth were located in entirely different places than they are today: Africa and South America probably were joined in a supercontinent; the Rocky Mountains, the Alps, and the Himalayas did not exist; the Appalachians were just being formed; and the eastern seaboard and most of the southwest region of the United States were at the bottom of a shallow sea. Yet the moon has scarcely changed at all in this period. Fragments of rock, hurled out of the fresh crater in Figure 17.4 when it was formed, still lie on the surface, precisely where they fell 200 million years ago.

The excellent preservation of these materials on the moon's surface—and the low level of lunar seismic activity described below—suggest that much of the moon's interior is cold and rigid. We would expect more of the plate movements and mountain building that have molded the earth, transformed its surface, and continue to transform the surface today. Because of the relatively modest level of geological activity on the moon, this body has preserved the record of its past for an exceptionally long time; it holds clues to the early history of the earth which are unavailable on our own planet.

THE APOLLO FINDINGS

These circumstances lent considerable scientific interest to the exploration of the moon. During the Apollo landings 836 pounds of rocks, hundreds of miles of scientific records, and many thousands of photographs were collected. This mountain of facts will be sifted for nuggets of information for years to come. Thus far, lunar scientists have established the following major features of the Apollo results.

Chemistry of the Moon

The chemical ingredients of the moon's surface were analyzed with the aid of instruments carried around the moon in orbit during the Apollo flights, and by studying rock samples in the laboratory. The results show that the entire surface of the moon—including the maria and the highlands—appears to be covered with basalt—a type of rock formed by the cooling of molten lava.

Two kinds of lunar basalt were discovered. One, called mare basalt, came from the maria and was composed mainly of the same minerals that make up the bulk of the basalts found in ocean crust on the earth. These minerals are pyroxene and plagioclase feldspar. In addition, the mare basalts also contained small amounts of olivine and a much rarer mineral called ilmenite, which is an oxide of iron and titanium. While olivine is present in about the same concentration as in terrestrial basalts, ilmenite is much more abundant in the lunar basalts than on the earth.

The other type of basalt was collected mainly during the landings on the highlands, and is called highland basalt.[1] The predominant mineral in highland basalt is plagioclase. Some pyroxene was found, but much less than in the mare basalts. Only a trace of ilmenite was detected.

The abundance of ilmenite—a dark mineral—in the mare basalts, and its scarcity in the highland basalts, largely account for the blackness of the lunar seas in comparison to the highlands.

Dryness of the Rocks. In addition to analyzing the composition of the rocks, the Apollo investigators also made a very careful search for water. They looked for traces of actual moisture in the rocks, and also for water in the form of molecules locked up in the crystalline structure of the minerals. The laboratory analyses showed the lunar rocks were bone-dry in both respects; they contained neither free moisture, nor minerals with water molecules included in their structure.

Since water is an essential ingredient for life as we know it, the dryness of the moon rocks suggests that life does not exist on the moon at the present time. Moreover, the measurements of the ages of the rocks, discussed below, indicate that they have been in their present state about three billion years. Therefore, we can be confident that no life has existed on the moon for that long interval of time.

The lunar rocks also showed no traces of organic matter, such as would have been left by biological organisms. Laboratory tests failed to detect any fossil organisms or residual molecular building blocks of living matter, such as amino acids and nucleotides.

Evidence for the Early Melting of the Moon

The discovery that the entire surface of the moon—both maria and highlands—is covered with basalt was one of the most important results of the entire Apollo program. On the earth, basalts are created only by the cooling of molten lava. Thus, this single finding from Apollo indicates that the surface of the moon was entirely molten at some point in its past.

What melted the moon's surface? Only two possible answers are known. One is that the moon was melted by an intense meteorite bombardment during a short period at the beginning of its life. This explanation fits in with the currently favored theory on the origin of the solar system, which proposes that all the planets, and their

[1] Most highland rocks are actually *breccias*—rocks made up of fragments from preexisting rocks. These rocks were first broken apart by intense meteorite bombardments, and later packed together again by the force of further impacts.

moons, condensed out of particles of gas, dust, and fragments of rock of various sizes. As our moon grew to final size in the last stages of this birth process, gravity pulled the material around it down onto the surface with great force. Each rock generated some heat as it crashed into the surface. The heat would be radiated away slowly to space, but if enough impacts occurred in a short period of time, the total accumulation of heat could be sufficient to melt the moon to a considerable depth.

The alternative explanation is that the moon was melted later by radioactive energy released in the decay of uranium and other radioactive elements scattered throughout the moon's interior. The Apollo measurements showed that the moon rocks contain a substantial amount of radioactivity, suggesting that this could be an explanation for the melting of the moon.

However, radioactive elements release their heat very slowly. In fact, calculations show that it takes about one billion years of steady radioactive heating to bring the interior of a planet like the moon or the earth to the melting point. If the rocks on the surface of the moon were melted when the moon was at least one billion years old, that fact would suggest that radioactive heat was the likely cause of the melting; but if the rocks were melted when the moon was considerably younger than one billion years, it would be necessary to look to another factor—presumably meteorite bombardment—for the explanation.

The Ages of the Lunar Rocks

These ideas indicate that the times at which the moon rocks were melted could provide the clue to the cause of their melting. With this remark we come to the second critical Apollo result, which is the measurement of the ages of the moon rocks. The ages of the rocks are measured by the technique of radioactive decay, which tells how long they have been in their present crystalline form (see Chapter 16). In other words, it tells how much time has elapsed since they were last melted. Thus, the age measurements provide precisely the information needed to distinguish between the two causes of the moon's melting.

But nature rarely gives up her secrets without a struggle. When the results of the age measurements became available, they yielded the ambiguous answer that *both* causes of melting probably had played roles in the moon's history. The lunar highlands probably were melted by meteorite bombardment early in the moon's life, while the lunar seas were melted later by internal radioactive heat.

This fact did not become clear until the astronauts were in true highlands terrain. All the highland rocks collected during the Apollo missions turned out to be at least 4 billion years old, indicating that these rocks crystallized when the moon had existed for no more than 600 million years. Several highland rocks—the oldest rocks found on the moon thus far—are 4.2 billion years old. Many small fragments of rock 4.6 billion years old also have been found. From the age of these fragments it follows that the surface of the moon must have been melted and resolidified when the moon was less than 100 million years old. One hundred million years is probably too short a time for the moon to have been melted by the slow process of radioactive heating. Therefore, meteorite bombardment must have melted the highland rocks.

The highlands are known to be older than the lunar seas, and are thought to be derived from the moon's original crust, parts of which were later covered over by the darker materials of the lunar seas. Thus, it appears that the original surface of the moon was entirely melted by meteorite bombardment during the moon's birth, or shortly after. This inference from the Apollo findings agrees with the general expectation that an intense meteorite bombardment must accompany the formation of every planetary body such as the moon or the earth.

Evidence for a Second Melting

But the story of the moon's melting does not end there, because other moon rocks—those collected from the lunar seas—have younger ages, ranging from 3.1 to 3.9 billion years. Since the moon was formed 4.6 billion years ago, these ages imply that the materials of the maria began to melt when the moon was 700 million years old. This time period would correspond to the time required for radioactive heating of the moon to melt the rocks in its interior, creating floods of molten lava on the surface.

Presumably, the lunar seas are pools of solidified lava that accumulated in the basins of the original highland crust as a result of repeated volcanic eruptions and lava flooding, during this later period of radioactive heating. Photographs of the walls of the Hadley Rille, taken during Apollo landings, clearly reveal several layers of the kind that would result from repeated flooding of the moon's surface by lava (Figure 17.7).

Another important conclusion follows from the fact that the oldest mare rocks are 3.9 billion years old and the youngest go back 3.1 billion years. This spread of ages suggests that the moon experienced an episode of sustained volcanism lasting about 800 million years. The 800-million-year interval of melting and lava flooding on the moon was undoubtedly accompanied by volcanic eruptions, moonquakes, and all the other manifestations of internal heat within a planet that are so familiar to us through our experience with volcanic activity on the earth. When the volcanism ended, the

FIGURE 17.7
Layers in the surface of the moon exposed in the wall of the Hadley Rille, and photographed from the far side by the Apollo astronauts. The clearly defined layer in the middle consists of four distinct strata, each about 10 feet thick.

moon subsided into geological lifelessness. For the last 3 billion years, the moon has been geologically dead. Today there seems to be scarcely any volcanism on the moon, apart from a few wisps of gas vented at the surface now and then.

The Seismometer Results

Is it possible that volcanoes have erupted more recently, but only in places that were not sampled during the lunar landings? Another basic Apollo finding—the information obtained from the lunar seismometers—excludes this possibility.

Five seismometers were placed on the moon during the Apollo program (Figure 17.8) (p. 432). According to the seismic data, the energy released by moonquakes per year is only a billionth to a trillionth as much as the energy released in one year by quakes on the earth. The most powerful moonquakes detected by the Apollo seismometers have a rating of 2 on the Richter scale. If you were standing directly over a quake of this size on the earth, it would not produce a perceptible vibration in your feet.

The weakness of the moonquakes detected by the Apollo instruments strongly suggests that there cannot be intense volcanic activity or lava flooding anywhere on the moon at the present time. Volcanoes and lava flows would be accompanied by movements of material within the moon, which would create vibrations detectable by the seismometers left behind on the lunar surface. These vibrations would show up in the records as major moonquakes.

FIGURE 17.8
Emplacement of a lunar seismometer. The instrument is housed in the drum-shaped container. The side panels are photocells converting sunlight to electric power.

The seismometer signals have provided other details regarding the moon's internal structure. The signals indicated that the quakes occurred at a surprisingly great depth, most originating at depths between 500 and 600 miles, whereas quakes on the earth usually originate within 60 miles of the surface.

The fact that the moonquakes occur at depths of 500 miles or more suggests that little, if any, molten rock or warm and plastic rock exists in the outer 500 miles of the moon. In other words, the moon is capped by a 500-mile thick shell of strong and rigid rock.

Beneath the rigid shell, however, there appears to be a layer of warmer rock, analogous to the earth's zone of weakness. This layer is the source of the moonquakes.

THE HISTORY OF THE MOON

The Early Bombardment

Armed with the facts from Apollo, we can now reconstruct the full life story of the moon. Our satellite condensed out of cold matter, but as it grew its surface was bombarded by a vast quantity of meteorites which rained down on the surface with ever increasing speed. The rocks in each collision area had no chance to cool off from the heat of one impact before they were warmed up by the next. The temperature rose as the intense bombardment continued. Eventually, the whole outer layer of the moon melted, and remained molten until the bombardment subsided. As the holocaust ended, the molten outer layer cooled and solidified. The moon became a geologically quiet planet—a sphere of solid rock from its center to its surface.

By the time the moon was several hundred million years old, the intensity of the bombardment must have fallen off from the earliest years. However, some meteorite collisions still occurred at a reduced rate. These collisions produced the craters visible today in the highlands. Once in a great while a meteorite of exceptional size, perhaps as large as a large asteroid, hit the moon. These mammoth chunks of rock, up to 60 miles in diameter, presumably blasted out the basins of the circular maria.

The 800-Million-Year Episode of Volcanism

As the number of meteorite impacts diminished, simultaneously the interior of the moon began to heat up as a result of the decay of radioactive elements below the surface. When the moon was nearly a billion years old, the radioactive heating raised the temperature of parts of the interior to the melting point. Occasionally a flood of lava poured across the surface, as molten rock forced its way upward through natural channels provided by fractures in the crust.

The impact of a very large meteorite would produce such fractures in the crust. There must have been numerous fractures under the circular basins that had been the sites of the biggest impacts. This explains how the lunar seas came to be formed; molten rock rose up repeatedly under the surface along these numerous fractures, and flooded the basins, creating the seas as we know them today.

Volcanic activity and lava flooding—signifying a partly molten interior—continued for about 800 million years, and then subsided, leaving the moon quiet once more. The last major lava flow appears to have occurred 3.1 billion years ago. The next three billion years in the history of the moon were geologically uneventful.

The Moon Today

Why did lava flows dry up on the moon three billion years ago? The answer probably is that the moon, being a small planet, lost its heat to space rapidly. As the moon's heat disappeared, and its temperature dropped, the extent of the warm, molten region diminished, and the outer zone of cold, strong rock became thicker. According to the Apollo seismometer data, this cold outer zone, which is the moon's lithosphere (page 408), today extends from the surface to a depth of 500 miles. Beneath the thick layer, there exists a warmer zone of rock, analogous to the earth's zone of weakness. Below a depth of 600 miles the moon may be partly molten.

It is extremely unlikely that molten rock lying at such great depths could force its way upward through the thick layer of solid rock that caps the moon. This casing of cold rock probably explains the absence of volcanic eruptions on the surface of the moon during the last three billion years. Magma from this depth might reach the surface through an intervening layer of 500 miles, but only very infrequently and with great difficulty. A 500-mile outer casing of rigid rock is also too thick to break up into several plates that move against each other, as the earth's plates do, producing earthquakes, volcanoes, and mountain buildings (Chapter 16). These circumstances account for much of the difference between the geology of the moon and the geology of the earth, and explain why the appearance of the moon today is so different from that of our own planet.

What the Moon's History Reveals About the Earth

Scientists had always hoped that they would recapture some of the earth's missing past as they unraveled the history of the moon. Now that the Apollo facts are in, what, if anything, can they tell us about our own planet? Consider first the evidence for the early melting of the moon by meteorite bombardment. If the moon and the earth had similar bombardment histories, the outer layers of the earth would have been melted by bombardment at about the same time that the first episode of melting occurred on the moon.

After the initial bombardment subsided, the earth, like the moon, would have cooled and been a nearly solid body from the center to the surface. It would than have remained solid, for a period of about 500 million to a billion years, while the radioactive heat within it slowly built up, and its internal temperature rose.

If the abundance of radioactive elements is about the same in the interiors of both the earth and the moon, then, after about a billion years, the earth's interior, as in the case of the moon, would have reached the melting point of rock. At that time, on the earth as well as the moon, molten rock would have started to rise to the surface.

Where did the molten rock first rise? We believe that on the moon it rose in the places where the crust had been punctured or damaged by giant meteorites. The evidence for this conclusion is to be found in the ages of the materials filling the lunar seas. The ages, ranging from 3.1 to 3.8 billion years, indicate that the mammoth craters left by the impact of those meteorites were filled in by lava at just about the point in the moon's history when radioactive melting would have occurred. On the earth, lava must also have filled similar basins left by the largest meteorite collisions, producing terrestrial equivalents of the lunar seas at the same time that they were formed on the moon.

Today, the lunar seas, dating back to that early time when the moon was about one billion years old, are still as well preserved as if the entire sequence of events had happened yesterday. They are preserved because three billion years ago the moon became geologically inactive, and has remained inactive right down to the present. But the "seas" of lava on the earth were not preserved, and are no longer visible. They have been obliterated by the volcanic activity and mountain building that have reworked the surface of the earth continuously for nearly four billion years.

THE ORIGIN OF THE MOON

The history of the moon has been carried from its first melting by meteorite bombardment through its second melting by internal heating, and then through the three billion years of geological inactivity that followed, down to the structure of the moon as it is today. Left unexplained is the question of the moon's origin: How and where was the moon formed?

The subject has always been one of the most vigorously debated questions in science. In 1898 the physicist George Darwin, son of Charles Darwin, proposed that the moon had been wrenched from the earth by centrifugal forces, when the earth was a young and rapidly spinning planet. This idea came to be called the fission theory. According to Darwin, the Pacific Ocean Basin was the scar left by the violent separation that created the moon.

Recently, a variation of the Darwin theory—called the spin-off theory—has been proposed. This theory holds that the moon was spun off as a circular ring of gaseous matter from the equator of the young, rapidly rotating earth, like a stream of mud leaving the rim of a rapidly turning bicycle wheel.

A third theory suggests that the moon was formed independently of the earth, as a small planet in another part of the solar system, and was captured subsequently by the earth's gravity as it hurtled past in a near-encounter. This idea is known as the capture theory.

Very few scientists support Darwin's fission theory today. One reason is that enormous frictional forces would be created when the moon started to separate from the earth, and calculations show that the frictional forces would slow down the rotation of the earth to such an extent that the moon would never get away.

The spin-off theory also suffers from basic difficulties. One relates to the plane of the moon's orbit. According to the spin-off theory, the material of the moon should have been spun off from the earth's equator, since the equator is the place on the earth where the centrifugal force is greatest. The material that was originally on the earth's equator would continue to circle in the earth's equatorial plane after it had been spun off. The moon, accumulated out of this material, would move in an orbit that is also in the earth's equatorial plane. However, the moon's orbit is actually inclined at a rather large angle to the earth's equatorial plane, averaging 23.5 degrees.[2] Calculations show that the moon's orbital plane could never have drifted this far from the equatorial plane of the earth, if it had started out in that plane.

The second difficulty with the spin-off theory is that the earth probably was not spinning fast enough to produce the moon in this way. The earth had to turn on its axis once every two and one-half hours, in order to produce sufficient centrifugal force to spin off the moon, but from the rate at which the earth and the moon are rotating today, it can be proved that the earth would have been spinning on its axis once every five hours. That is, it was rotating only half as fast as would have been necessary to spin off the moon.

This and the previous objection also apply to the fission theory. Both objections must be rejected if it happened that another planet flew off at the same time as the moon. It has been suggested that Mars was created in this way, spinning off the equator of the earth along with the moon, in one dramatic episode at the beginning of the solar system. There is no theoretical objection to this hypothesis. However, the observed facts about Mars and the moon seem to work against it. The theory implies that the moon and Mars are made of exactly the same materials: namely, the materials that made up the outer layers of the earth at the time they were formed. This conclusion works well enough for the moon, whose density is about the same as the density of the earth's mantle, but it does not work for Mars, whose density is considerably greater than that of the mantle.

The third theory—formation of the moon in a different part of the solar system, followed by capture by the earth—has no obvious flaws except that it constitutes an extremely improbable event. For capture to occur, the moon must have moved past the earth at exactly the right distance. If too far away, it would have been whipped around the earth by the force of our gravity, without dropping into orbit; if too close, it would have collided with us. Calculations indicate that the range of conditions that can lead to capture is very narrow, and the probability of a capture occurring is correspondingly small.

A fourth theory suggests that the materials of the moon were

[2] The angle varies between 17° and 28° in an 18-year cycle.

captured, but not in one piece. This idea avoids some of the defects in the original capture theories. It proposes that the earth's gravity captured planetesimals and bits of debris from the solar nebula, building up a ring of material that orbited around it. The moon aggregated out of this ring of finely divided material by a process not clearly understood, but presumably similar to the process by which the earth itself condensed out of planetesimals. This theory of condensation in orbit is speculative, but it has the advantage over competing ideas that no obvious objections to it have yet been raised.

Main Ideas

1. Description of the lunar surface: meteorite bombardment; low level of surface erosion; antiquity of the lunar surface.
2. The Apollo findings; dryness of the moon, absence of life, ages of the highland rocks and maria rocks, evidence for early melting of the surface, evidence for a second melting in the interior, seismic evidence for low level of geological activity.
3. History of the moon deduced from the Apollo results; implications for the history of the earth.
4. Theories of the origin of the moon.

Important Terms

basalt
highlands
maria
meteorite bombardment
meteorite crater
meteorite erosion
micrometeorite
seismometer

Questions

1. What are the lunar highlands? What are the lunar seas? How were they formed? Why is the surface of the moon different from the surface of the earth?
2. What is the origin of most lunar craters?
3. What are the principal items of evidence indicating that the moon is geologically lifeless? Describe the evidence indicating when the moon's geologic activity commenced and when it terminated.
4. What bearing does the presence of a substantial abundance of radioactive elements in the lunar rocks have on the thermal history of the moon? How would the Apollo findings have been altered if the moon had double the abundance of radioactive elements it actually possesses?

5. How does the history of the moon illuminate the history of the earth? How would you expect the history of a planet to compare with that of the moon if that planet were half the moon's diameter? Twice the moon's diameter? Four times the moon's diameter? What would you expect the surfaces of such planets to look like? Base your answer on the information in this and the preceding chapter. Are there bodies in the solar system whose diameters are twice and four times the moon's diameter? How do their properties compare with your predictions?
6. Cite evidence for and against the current theories of the moon's origin.

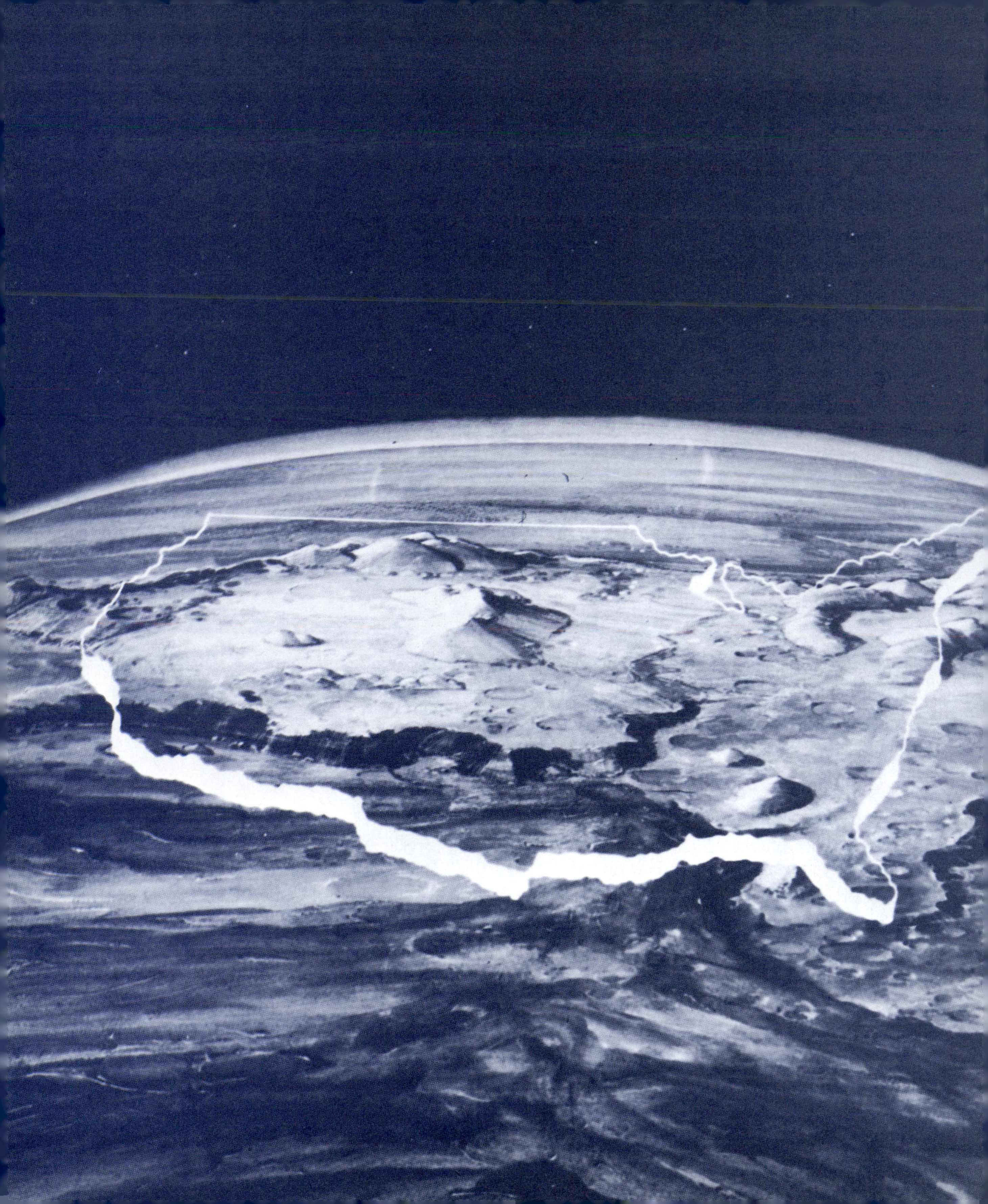

VENUS

18

Venus and the earth are sister planets, closely similar in size and weight, and situated at distances from the sun that are not very different. Conditions on the surface of Venus have always been an enigma because the planet is completely covered by clouds; yet hope has flourished throughout the last 300 years—since Galileo first turned his telescope on Venus in 1610 and discovered that the planet was a round body like the earth and the moon—that beneath the clouds on Venus lie teeming masses of exotic flora and fauna. In 1686 de Fontenelle, in his book *Conversation on the Plurality of Worlds,* described the characteristics he expected to find in the people on Venus:

> *"I can tell from here . . . what the inhabitants of Venus are like; they resemble the Moors of Granada; a small black people, burned by the sun, full of wit and fire. . . ."*

THE GEOLOGY OF VENUS

The density of Venus is very close to the density of the earth, indicating that it probably has a similar bulk composition, dominated by iron and rocky materials. As a consequence, it can be assumed to have the same concentration of radioactive elements. Since Venus is close to the earth in size, it should contain its internal heat about as well as our planet, and should have a similar geology,

Ishtar Terra, a mountainous plateau on Venus. The outline of the continental United States is superimposed.

with a partial melting of the mantle at some depth below the surface, a lithosphere, and plate movements. There should also be collisions at plate boundaries, with zones of earthquakes, volcanoes, and "midocean" ridges.

However, it has turned out that the geology of Venus is rather different from that of the earth. Radar signals reflected from Venus, which reveal the large-scale topography of the planet, show little evidence of the plate movements that generate the large-scale features of the earth's surface. These features may be present but, since they are blurred by the limited resolution of the radar images, they do not stand out clearly. It is also possible that Venus resembles the earth in an earlier stage of its evolution when the lithosphere and crust of our planet were thinner. According to this view, the geology of Venus will come to resemble that of the earth in several billion years, when its internal temperature has diminished and the lithosphere has thickened to the point where plates can move about largely intact, except for collisions at their boundaries.

The radar images do reveal two large areas that stand out above the surrounding terrain and resemble continents on the earth (Figure 18.1). One, called Ishtar Terra, is comparable in size to the United States. Ishtar Terra, also shown in the illustration facing page 440,

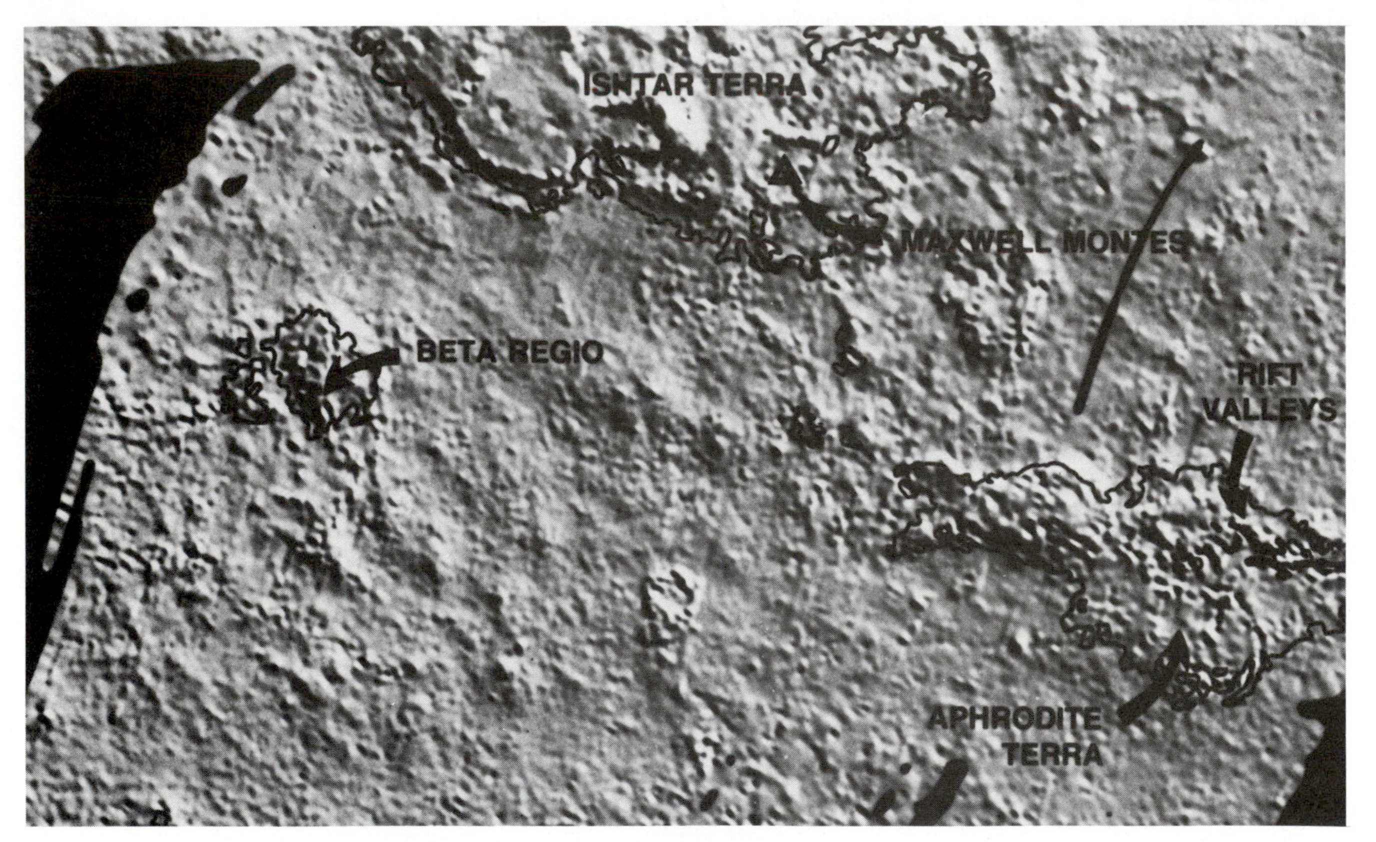

FIGURE 18.1
A relief map of Venus based on radar observations.

contains a mountain massif called Maxwell Montes. Maxwell Montes is higher than Everest, its summit being 35,400 feet above "sea level" on Venus—that is, above the average elevation of the Venus surface. The second highland region, called Aphrodite Terra, sits on the equator, and is about the size of the northern half of Africa. Just off the eastern boundary of Aphrodite lies the lowest point on Venus, about 9500 feet below the mean level of the Venus surface. This feature bears some resemblance to the Valley of the Mariner on Mars, and also to rift valleys on the earth which mark the incipient stages of plate breakup and movement. It is the only feature found on Venus thus far that suggests the existence of lithospheric plates.

An elevation to the south of Ishtar Terra, called Beta Regio (Figure 18.2), consists of huge volcanic mountains similar in appearance to the volcanic islands of the Hawaiian chain, but larger. Beta Regio shows signs of recent volcanic activity, and may still be erupting. Presumably it sits on top of an upwelling plume or hot spot in the Venus mantle, similar to the hot spot that lies under the island of Hawaii.

Most of the remainder of Venus is covered by a fairly flat, rolling plain. Numerous circular dark features in the radar images appear to be large meteorite craters, several hundred miles in diameter. As on the moon, the presence of these craters suggests that much of the original crust of the planet is still preserved.

FIGURE 18.2
The Beta Regio elevation: two volcanic mountains resembling the largest Hawaiian islands.

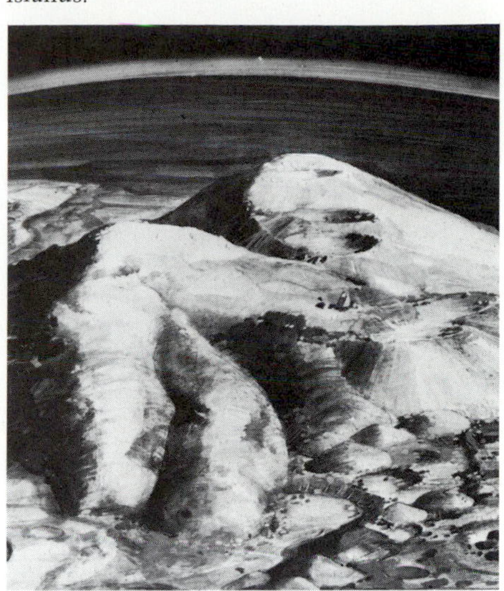

SURFACE CONDITIONS

The planetary properties of Venus suggests that it should provide an agreeable climate for living organisms. The planet is 67 million miles from the sun, while the distance of the earth is 93 million miles. Because Venus is closer it receives 1.9 times, or approximately twice, the intensity of sunlight falling on the earth. This is true because the intensity of sunlight falls off in proportion to one over the square of the distance from a planet to the sun. Therefore, the intensity of the sunlight received by Venus in comparison to the sunlight received by the earth is equal to:

$$(93/67)^2 = 1.9$$

The heavy cloud cover on Venus keeps out 80 percent of this excess of solar energy, but the predicted average temperature on the planet is, nonetheless, slightly higher than the average temperature on the earth, and very comfortable by terrestrial standards.

In 1956 radio astronomers obtained evidence which suggested that the climate on Venus might be far from balmy. In fact, the measurements indicated that Venus is a very hot planet. The results obtained by the radio astronomers depended on the fact that a heated object radiates energy at all wavelengths, although the most intense radiation occurs at one particular wavelength that is determined by its temperature. Figures 7.6 and 7.7 in Chapter 7 show radiation curves for several different temperatures. An object that is only hot to the touch, but not visibly glowing, such as a heated iron, radiates most of its energy in the infrared region; but, like every heated object, it also radiates energy at all other wavelengths from the shortest to the longest. For example, the relative amount of energy radiated by a hot iron in the short ultraviolet wavelengths, or at very long radio wavelengths, is very small compared to the amount emitted at the peak intensity in the infrared; but nevertheless, the radiation at these wavelengths is present.

A planet heated by the rays of the sun to a temperature of a few hundred degrees is like a hot iron. Most of the energy radiated from its surface is in wavelengths in the far infrared region, around 100,000 Å, or 10 microns,[1] but some energy is radiated at shorter wavelengths and some is also radiated at very long wavelengths.

The long-wave radiation includes waves ranging from a fraction of an inch in length up to many miles. These waves fall into the parts of the electromagnetic spectrum that are called the *radar* or the *microwave* region, and the *radio* region. This radiation is of great interest to astronomers because of the fact that after it has been emitted from the surface of the planet, it passes through the atmosphere of the planet relatively unhindered, and escapes freely to space. Furthermore, that part of it which reaches the earth also passes through the atmosphere of the earth relatively unhindered and thus can be detected by radio telescopes on the earth's surface (see Chapter 5).

Especially when the surface of the planet is covered by clouds, as is the case of Venus, the long-wave radiation, which can penetrate through these clouds, is the only means available to the earthbound astronomer of obtaining information about the surface conditions on the planet.

Motivated by this fact, radio astronomers at the Naval Research Laboratory pointed their antenna in the direction of Venus and tried to pick up long-wave radiation coming from the direction of that planet. They were immediately successful in detecting the signal from Venus. In fact, it was considerably stronger than their calculations led them to expect.

The calculations were based on the assumption that the temperature on Venus was about the same as the temperature on the earth. The fact that the signal from Venus was more intense than expected indicated that the temperature on Venus must be considerably

[1] A micron, which is a convenient unit for radiation in the far infrared, is one-millionth of a meter. The visible band of wavelengths extends from 0.4 to 0.7 microns.

higher than the temperature on the surface of the earth. Fitting their data to a radiation curve, the Naval Research Laboratory astronomers deduced that the temperature on Venus was a sizzling 800°F.

This temperature is hot enough to melt lead. It is also high enough to break apart all the delicate molecules that make up the essential ingredients of a living cell. It is unlikely that no organisms remotely resembling any form of life on earth can exist on the surface of Venus.

Yet hope lingered on for the discovery of a green world on Venus. Some astronomers argued that the intense radiation might come from the atmosphere of Venus and not from its surface. Others suggested that life might be supported at the north and south poles, which should be cooler. Another suggestion was that organisms might have developed on Venus with a very specialized form, consisting of gas-filled bladders, something like beach balls, whose buoyancy would cause them to float high in the atmosphere, where the temperature should be considerably cooler than it is on the ground.

THE VENUS ATMOSPHERE

In recent years several Soviet and American spacecraft have reached the planet and carried out measurements that show the temperature of the surface is, in fact, extremely high, the average value being about 900°F.

These measurements remove the last doubts regarding the temperature deduced by the radio astronomers. Venus is hot enough to melt lead, and there is no reasonable chance of finding life on its surface.

The American and Soviet measurements also indicate that the atmospheric pressure at the surface is nearly one ton per square inch—100 times the pressure on the surface of the earth.

Finally, the experiments show that carbon dioxide makes up about 96 percent of the atmosphere, compared to 0.03 percent in the earth's atmosphere. Nitrogen, which constitutes 78 percent of the earth's atmosphere, is only 2 to 4 percent of the Venus atmosphere. With allowance for the greater density of the Venus atmosphere, the absolute amount of nitrogen is approximately the same on Venus and the earth. This result supports the theory that the atmosphere of an earthlike planet consists of gases from the planet's interior, which are released at the surface in volcanic eruptions. If Venus and the earth experienced the same amount of volcanic activity during their lifetimes, they should have similar amounts of nitrogen in their atmospheres. They should also have similar amounts of other volcanic gases, such as water vapor. The experi-

TABLE 18.1
The Venus Atmosphere

Carbon dioxide	about 96%
Nitrogen	2.5% to 4.5%
Water vapor	About 0.2%
Sulfur dioxide	100 to 350 ppm[a]
Argon	20 to 150 ppm
Neon-20	10 to 40 ppm
Helium	10 to 40 ppm
Oxygen	Less than 30 ppm

[a] Parts per million.

ments also suggest the presence of a small amount of water, sufficient, if condensed to liquid form, to cover the surface of the planet up to a depth of one foot. The composition of the Venus atmosphere is shown in Table 18.1.

The Venus Clouds

Soviet and American spacecraft have recorded conditions beneath the clouds and down to the surface. At an altitude of about 40 miles, a smog layer of sulfur particles covers the planet. Below the smog lies a thick cloud layer composed of sulfuric acid droplets. The clouds thin out below 20 miles and the atmosphere becomes entirely clear. The sun's light grows weaker as the atmosphere approaches the density of soup. At the surface the intensity of sunlight corresponds to a heavily overcast day on the earth, and the light has a strong orange-brown cast.

The Greenhouse Effect

Why is Venus so hot? What determines the temperature on the surface of this planet, or any planet? A planet's temperature is determined primarily by its distance from the sun, which controls the intensity of the solar heat falling on the planet's surface. Most of this heat reaches the planet in the form of visible light. A part of the light is reflected back to space by clouds and scattered in the atmosphere; the remainder passes through the atmosphere to the surface, which is warmed by the absorption of this solar energy. The warm surface radiates heat back to space in the form of infrared radiation. Over a long period of time a planet must radiate as much heat back to space as it receives in the form of visible sunlight.

At first the surface of the planet will rise steadily in temperature as it receives the solar energy, but as it becomes hotter it radiates more heat to space, until eventually the outgoing heat just balances the inflow of sunlight. For any planet, the temperature can be calculated at which the radiated heat equals the influx of sunlight. In the case of the earth, we find in this way that the average ground temperature should be −20°F.

But the actual average temperature of the earth is 60°F, and not −20°F. The increase in temperature is produced by our atmosphere, which acts as an insulating blanket, trapping a part of the radiation from the ground and returning it to the planet, where it adds to the heat provided by the absorption of sunlight. In this way the average temperature of the earth is raised from the theoretical value of −20°F to the level of 60°F, which is actually observed.

The outgoing heat is trapped by the trace amounts of water

vapor, carbon dioxide, and ozone in the atmosphere. Although these constituents together make up less than one percent of the earth's atmosphere, they absorb 90 percent of the heat radiated into the atmosphere from the surface.

The heat insulation provided by the atmosphere is called the "greenhouse effect." A greenhouse has a glass cover that is transparent to the sun's visible radiation (just as the earth's atmosphere is transparent), but blocks the heat radiated from the plants within. Thus the heat is trapped within the greenhouse and warms the interior, just as water vapor, carbon dioxide, and ozone retain heat in the earth's atmosphere and warm the surface of our planet (Figure 18.3).

The greenhouse effect provides the answer to the mystery of Venus' high temperature. Carbon dioxide, of which only a trace exists in the earth's atmosphere, is present in the atmosphere of Venus in such massive amounts—100,000 times more than in the earth's atmosphere—that it makes the Venus atmosphere almost impervious to infrared radiation. Nearly all the heat radiated from the surface of Venus is trapped by the insulating atmospheric blanket composed of carbon dioxide, and smaller amounts of water vapor and sulfur dioxide. The remainder is trapped by clouds and haze. The trapped heat causes the temperature of the surface to rise far above the value it would have if the atmosphere of Venus were similar to ours.

FIGURE 18.3
The Greenhouse effect. Visible light from the sun passes freely through the atmosphere to the surface of the planet. The surface, heated by the sun's rays, sends out infrared radiation, which does not pass as freely through the atmosphere. Some is absorbed by the atmosphere, heating it, and the heated atmosphere, in turn, warms the planet to a higher temperature than it would have in the absence of an atmosphere.

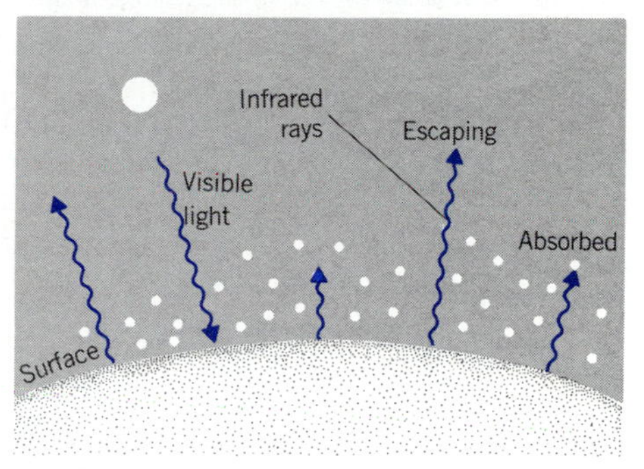

ATMOSPHERIC EVOLUTION ON THE EARTH AND VENUS

The abundance of carbon dioxide on Venus explains its high temperature. However, this explanation leads to another question: Why does Venus have so much CO_2? Why did the earth and Venus, formed out of similar materials, develop different atmospheres? The answer emerges from an examination of the factors that govern the evolution of a planet's atmosphere.

In the beginning the earth and Venus must have been airless bodies of rock. Today each planet is blanketed by a substantial atmosphere. Where did these atmospheres come from? They could not have been drawn in from the surrounding space by the gravitational attraction of the planet, because if that were the case, the atmosphere would still contain substantial amounts of such chemically inert gases as neon. However, these so-called rare gases are virtually absent. The only remaining possibility is that the present atmospheres were exhaled from the interiors of the planet as volcanic vapors during the course of their geologic evolution. There is evidence in support of this theory in the case of the earth. As a result, geologists are generally agreed that the oceans and atmosphere of the earth have accumulated during the lifetime of our planet by the emission of volcanic gases from the interior through cracks and fissures in the crust.

Accumulation of CO_2 in the Atmosphere

Carbon dioxide is an abundant gas in volcanic eruptions on the earth, second only to water vapor. Since most of the water vapor condenses into the oceans, the carbon dioxide, which remains in the atmosphere, should be the dominant atmospheric gas on the earth. In fact, the earth should be covered with a dense blanket of CO_2, exerting a pressure of half a ton per square inch at the surface.

Assuming that the earth and Venus have had similar histories, that should also be true of the Venus atmosphere. The atmosphere of Venus is observed to have just the amount of carbon dioxide predicted by this reasoning. However, the earth has 100,000 times less. Now we see that the problem is not to explain why Venus has so much CO_2 in its atmosphere, but why the earth has so little.

Removal of CO_2 from the Earth's Atmosphere

The answer to the question of the missing CO_2 is connected with the fact that the rocks of the earth's crust are able to absorb carbon dioxide. The absorption takes place in a chemical reaction in which atmospheric carbon dioxide combines with rocks to form carbonates, somewhat as oxygen in the atmosphere combines with iron to form rust. Geochemists have long been familiar with chemical reactions taking place at the surface of the earth which demonstrate this process. An example of such reactions is

$$CO_2 + CaSiO_3 \rightarrow SiO_2 + CaCO_3$$

in which carbon dioxide combines with calcium silicate, a variety of rock, to form quartz plus calcium carbonate. Sand is an example of nearly pure quartz, and marble and limestone are examples of calcium carbonate.

This and similar reactions, going on slowly but steadily at the surface of the earth over millions of years, presumably removed carbon dioxide from the atmosphere and converted it into solid compounds nearly as fast as it entered the atmosphere through volcanic eruptions.

A partial confirmation of this theory is found in the fact that the amount of carbon locked up in the sedimentary rocks in the earth's crust in the form of carbonates is approximately equal to the amount of carbon contained on Venus in the form of atmospheric carbon dioxide.

Perhaps the removal of carbon dioxide by the rocks of the earth's crust was the reason why the temperature on the earth stayed at a comfortable level long enough for life to take hold here. As soon as animal life appeared in abundance, its own chemistry would have further reduced the level of carbon dioxide in the earth's atmosphere. Marine animals absorb carbon dioxide from seawater and convert it within their bodies to the solid substances called carbonates.[2] For every molecule of carbon dioxide removed from the water by this process, one molecule eventually enters the ocean from the atmosphere. Thus, the absorption of carbon dioxide from seawater by marine animals is equivalent to its removal from the atmosphere.

The Runaway Greenhouse Effect

The rocks on the surface of Venus must contain the same silicates that removed carbon dioxide from the earth's atmosphere. Why was CO_2 not absorbed on Venus?

The answer is that the reactions that remove carbon dioxide depend strongly on temperature. The higher the temperature of the rock, the less effective it is in removing carbon dioxide. When the earth was a young planet—old enough to have cooled from its birth process, but still too young to have accumulated its insulating blanket of atmospheric gases—its average temperature was about −20°F. At this low temperature the rocks of the earth's crust would absorb nearly all carbon dioxide from the atmosphere. Thus no appreciable greenhouse effect developed, and the earth stayed cool.

When Venus, on the other hand, was a young planet without an atmosphere, its average temperature was higher—about 140°F—because of its closeness to the sun. At this higher temperature, the rocks on its surface absorbed far less carbon dioxide than was absorbed by rocks on the earth in the same period. According to the measurements on the temperature dependence of the reactions, the amount of carbon dioxide remaining in the atmosphere of Venus must have been 10 to 100 times greater than the amount in the earth's atmosphere.

This blanket of carbon dioxide, although still thin at that early time in comparison with the heavy CO_2 atmosphere on Venus today, would still have been great enough to produce an appreciable greenhouse effect. The greenhouse effect, raising the temperature of the surface of Venus still higher, would further diminish the effectiveness of the Venus rocks in absorbing the CO_2 that continued to pour into the atmosphere through volcanoes. As more CO_2 accumulated in the atmosphere, the greenhouse effect became

[2] Seashells, for example, are nearly pure calcium carbonate.

stronger, the temperature of the rocks rose higher, their effectiveness in absorbing CO_2 was still further diminished, the greenhouse effect was further enhanced, and so on. In this way, a runaway greenhouse effect started, leading to the massive carbon dioxide atmosphere and ovenlike conditions that characterize Venus today.

Presence of Oxygen on the Earth

These ideas explain the difference in carbon dioxide on the earth and Venus. They do not account for the presence of free oxygen, the second most abundant constituent of our modern atmosphere. This gas should not exist in the atmosphere in appreciable concentrations despite its high abundance in the Cosmos, because its strong tendency to enter into chemical reactions removes it promptly from the atmosphere and deposits it on the surface in the form of solid oxides. The explanation lies in the presence of life on the earth. The main source of atmospheric oxygen is generally believed to be plant life. Plants—especially marine algae—absorb carbon dioxide from the atmosphere, convert the carbon into compounds that enter the structures of their bodies, and return the residual oxygen atoms to the atmosphere. These biological activities continually renew the supply of free oxygen in the atmosphere, which would otherwise be depleted rapidly—in 2000 years or so—by chemical reactions with the materials of the earth's crust.

Why Venus Has Less Water than the Earth

Water, like carbon dioxide, was one of the gases that must have been trapped in the interiors of both the earth and Venus when they first condensed out of the solar nebula. The release of trapped water vapor through cracks and fissures in the crust of the earth has led to the accumulation of oceans covering three-quarters of the surface of the planet to a mean global thickness of 8000 feet.

A comparable amount of water should exist on the surface of Venus. Because of the planet's high temperature, this water should be in the form of vapor in the atmosphere, rather than a liquid filling the natural basins in the topography of the crust. But it is known from the results of the measurements performed by the Soviet spacecraft that the equivalent of an ocean of water is not present on Venus. In place of the 8000 feet of liquid water that cover the face of the earth, the atmosphere of Venus has no more than a few feet of liquid water.

Recent American spacecraft measurements may have solved the problem of the missing Venus water. A package of instruments dropped into the Venus atmosphere included a device for making accurate measurements of the masses of atoms and molecules in the atmosphere. The measurements yielded a proportion of heavy water molecules to ordinary water molecules in the Venus atmosphere. The heavy water molecule is one in which one or both of the hydrogen atoms are replaced by atoms of deuterium or heavy hydrogen. During the history of the atmosphere of a planet such as the earth or Venus, water is lost continually from the upper levels of the atmosphere because solar ultraviolet radiation breaks up the water molecules into their basic hydrogen and oxygen atoms, and the hydrogen atoms, being relatively light, evaporate to space. The oxygen atoms remain behind, because they are heavier and are held by the planet's gravity, and enter into chemical combinations with elements in the crust. As a result of this process, water is removed steadily from the surface and atmosphere of an earthlike planet.

When the atmosphere of a planet contains a mixture of ordinary water molecules and heavy water molecules, the ordinary water is removed more rapidly than heavy water, because the hydrogen it contains is a lighter atom and escapes more rapidly to space. Thus, as time goes by, both ordinary water and heavy water escape, but since the ordinary water escapes more rapidly, the proportion of heavy water to ordinary water gradually rises.

A measurement of the relative amounts of heavy water and ordinary water on Venus should enable us to infer how much water has escaped from the planet during its history. The measurements show that a great deal has escaped; in fact, the amount now present in the atmosphere of the planet is only a hundredth of what was originally present. This means that at one time Venus had a global ocean on its surface at least several hundred feet deep, and possibly thousands of feet deep.

Circumstances That Preserve Water on the Earth. Why did the photodissociation process that removed water from Venus not also remove water from the earth? The answer probably depends on two factors. *First,* the temperature of the air drops with increasing height above the surface of the earth, reaching a low point of about −80°F at a height of approximately 30,000 feet. As the temperature falls, the water vapor condenses into droplets of liquid water or tiny crystals of ice, forming clouds. The water droplets or ice crystals are immune to destruction by solar ultraviolet radiation, which can only destroy individual molecules of water. Only one-tenth of one percent of the water in the atmosphere remains in the form of vapor at high altitudes, as a result of the so-called "cold trap" at 30,000 feet. *Second,* solar ultraviolet radiation coming from the sun is absorbed rapidly as it enters the atmosphere from above, partly by photodissociation and partly by reactions in which ultraviolet photons eject electrons from atoms of air. Oxygen atoms, which are abundant at great heights in the earth's atmosphere, are particularly effective in absorbing the ultraviolet wavelengths that break up water molecules. By the time the solar radiation has penetrated to 30,000 feet, its ultraviolet component has been almost entirely removed.

Thus, in the atmosphere below 30,000 feet, where an abundance of water vapor exists, the ultraviolet intensity is negligible because it has been screened out by oxygen; and in the upper atmosphere, where the ultraviolet intensity is still strong, the amount of water vapor is negligible because of the cold trap. This combination of circumstances has protected the water on our planet. Consequently, water has accumulated steadily on the surface of the earth throughout its history, gradually filling up the empty basins of the earth's crust to form the oceans.

Disappearance of Water from Venus. What are the corresponding circumstances on Venus? Temperatures in the atmosphere of Venus are higher than at the comparable levels in the earth's atmosphere. The consequence of this fact is that 200 times more water can remain in the atmosphere in the form of vapor than in the case of the earth. In addition, Venus lacks the shielding screen of oxygen molecules that make up 20 percent of the terrestrial atmosphere and protect the water vapor at lower levels from photodissociation. The amount of solar ultraviolet radiation penetrating to the lower levels of the Venus atmosphere is 1000 times greater than in the case of the earth.

The product of the two factors—200 times more water vapor, and 1000 times more solar ultraviolet radiation—corresponds to an increase by a factor of 200,000 in the rate of removal of water molecules from the Venus atmosphere by photodissociation, in comparison with the rate of removal of water from the earth by the same process. This factor is adequate to account for the scarcity of water on Venus and its abundance on the earth.

CONDITIONS FOR LIFE

How far from the sun must a planet be to maintain a congenial climate for life? We do not know. The exploration of Venus has told us only that the planet was too close, and that was its undoing. If it had been nearly as far away as the earth is, the temperature of its surface might have climbed slowly enough to permit life to gain a toehold, and once life began, it could have held the abundance of carbon dioxide in check, and might have prevented the temperature from climbing out of bounds subsequently. But Venus, having lost its chance to harbor life when the solar system was first formed, could never again recapture the opportunity.

The Zone of Life

This comparative study of the histories of the atmospheres of the earth and Venus demonstrates that an earthlike planet must not be too close to its sun if conditions are to remain favorable for the evolution of terrestrial forms of life. In the case of our solar system, the critical threshold lies somewhere between the orbits of Venus and the earth.

How far on the other side of the earth's orbit can a planet be, and yet maintain a comfortable climate for the support of life? The answer depends mainly on the size of the planet. It may be quite far from its sun, and the warming effect of the sun's rays correspondingly small, but if the planet is large enough to seal in its radioactive heat and build up a high level of internal temperatures, volcanic eruptions will be frequent, and copious emissions of gases can be expected from the planet's interior. Thus, its atmosphere will accumulate to a high density, and a significant greenhouse effect will develop. The greenhouse effect will trap the heat of the sun and create a comfortable temperature on the planet's surface, in spite of its distance from the sun and the weakness of the solar heat.

If, however, the planet is relatively small—the size of the moon, for example—its internal temperatures will not rise to a sufficiently high level to produce substantial volcanic activity with copious emission of gases. Moreover, the relatively weak gravitational field of the planet will render it less effective in preventing the lighter gases from escaping to space, further diminishing the density of the atmospheric blanket. For these two reasons, the greenhouse effect will be too small to increase the temperature of the planet's surface appreciably. The scarcity of water—as a further consequence of the diminished emission of gases from the interior of the planet—will also tend to act against the onset of biological evolution.

Mars is an example of a planet intermediate between the two sets of circumstances. It is one-tenth as massive as the earth, but ten times more massive than the moon. Consequently, Mars may have developed internal temperatures high enough to create volcanic eruptions, accompanied by the emission of water and atmospheric gases. According to the spacecraft evidence, these events actually took place earlier in the history of Mars, but ceased roughly one billion years ago when the planet became geologically inactive (page 458). As long as the volcanism persisted, the planet probably possessed a congenial environment for life, but later the gases disappeared to space or were locked up in frozen form in the rocks of the crust, and Mars became a less agreeable place.

An interesting conclusion follows: Mars is quite cold and dry, and less congenial for living organisms than the earth, not primarily because it is farther from the sun, but because it is smaller and geologically less active. If a giant hand moved the earth out to the orbit of Mars, the average temperature of our planet would drop. However, its insulating blanket of air, continually replenished by volcanic gases for some billions of years longer, would maintain a comfortable temperature over large areas. The difference would be noticeable, but life could continue.

Main Ideas

1. The geology of Venus; radar images of large-scale continental features and volcanic mountains; Ishtar Terra, Maxwell Montes and Beta Regio.
2. Surface conditions and cloud cover on Venus.
3. Properties of the Venus atmosphere; abundance of carbon dioxide.
4. Explanation of high surface temperatures on Venus in terms of the CO_2 greenhouse effect.
5. Volcanic origin of water and atmospheric gases on the earth and Venus; explanation for the abundance of CO_2 in the Venus atmosphere and scarcity of CO_2 in the terrestrial atmosphere.
6. The runaway greenhouse effect.
7. Explanation for the large amount of water on the earth's surface and disappearance of water from the Venus surface.
8. Conditions for life on an earthlike planet; the zone of life around a star.

Important Terms

carbon dioxide
carbonate
cold trap
greenhouse effect
photodissociation
runaway greenhouse effect
zone of life

Questions

1. Why would you expect the geology and surface features of Venus to be similar to those of the earth? In what ways have the surfaces of the two planets turned out to be different? Are there any resemblances?
2. How has the abundance of CO_2 in the Venus atmosphere produced the high temperature on its surface?
3. Venus is roughly twice as far away from the sun as Mercury. Why is the surface temperature on Venus greater than on Mercury?

4. What is believed to be the source of the gases now present in the atmospheres of the earth and Venus? What is the source of the water in the earth's oceans?
5. How was CO_2 removed from the earth's atmosphere? Why did CO_2 accumulate in the Venus atmosphere?
6. Explain the runaway greenhouse effect.
7. If Venus were at the earth's distance from the sun, what would be the conditions on its surface? Explain.
8. What effects do animal and plant life have on the earth's atmosphere and its surface temperature?
9. Use the properties of water to define the inner and outer boundaries of a "zone of life" around a star.
10. List and explain the principal factors that, in your opinion, determine the conditions on a planet to be favorable for life as we know it. According to your ideas, how would the luminosity of a star affect the position of the "zone of life" for earthlike planets circling it? Where would you expect the "zone of life" to be for such planets circling a Main Sequence blue giant? A Main Sequence red dwarf? How would the position of the zone depend on the size and composition of an earthlike planet?

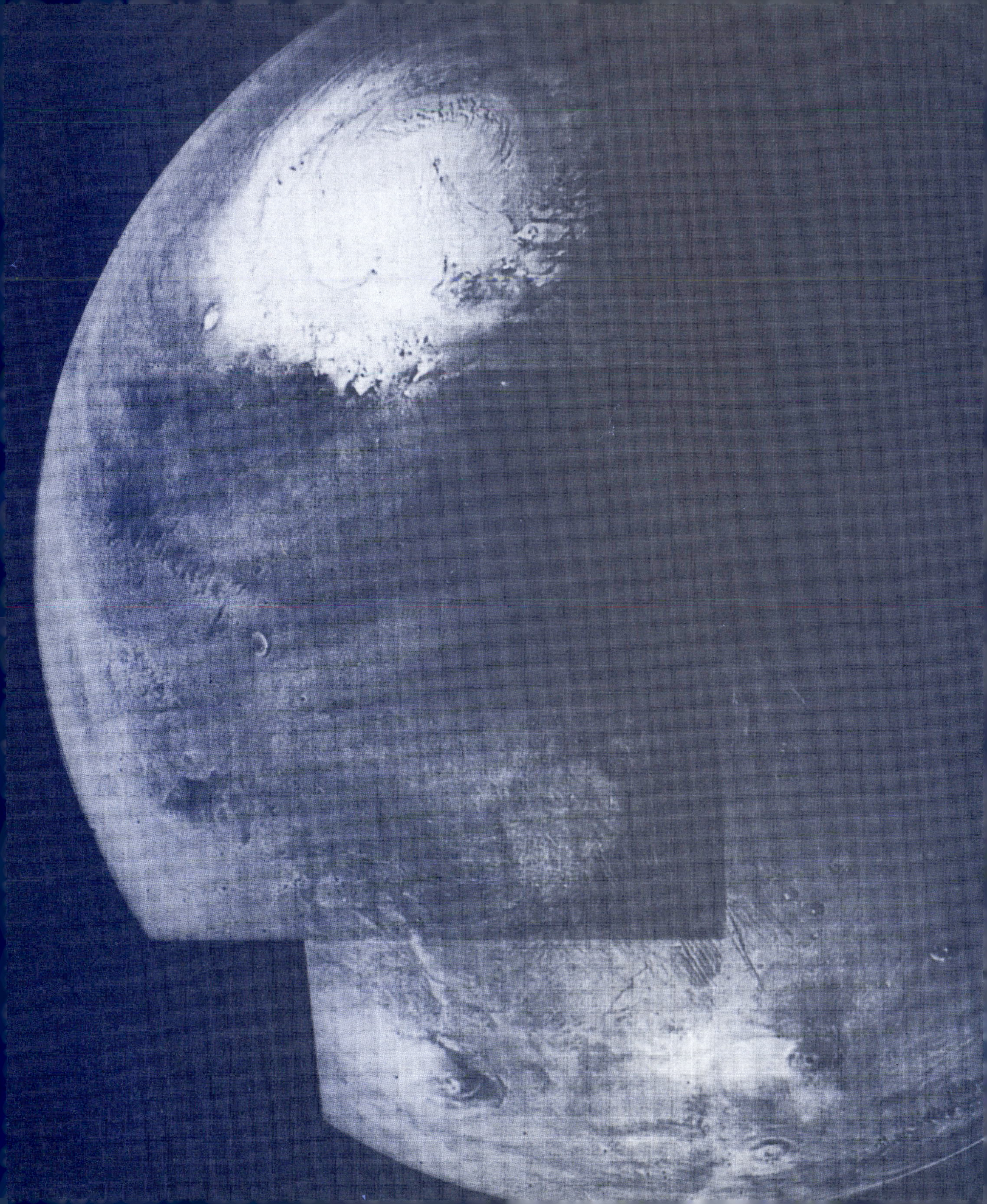

MARS

19

Mars has generated more speculation regarding extraterrestrial life than any other planet in the solar system. Several characteristics of the planet have contributed to the growth of these speculations. Its surface, free of clouds, reveals changes during the Martian year that resemble the march of the seasons on the earth. In each hemisphere a polar cap grows larger in the fall and winter and diminishes in the spring and summer. Dark regions appear each spring that are suggestive of the seasonal growth of vegetation. Toward the end of the nineteenth century, some observers reported a planetary network of canals presumably engineered by intelligent life.

Twentieth-century studies of Mars have failed to confirm the existence of the canals. However, earth-based and spacecraft observations made in recent years have contributed a great deal of new information that bears directly on the prospects for Martian life. Since Mars is nearly free of clouds, conditions of its surface can be observed readily from the earth through telescopes. These earth-based observations reveal that the planet has a more agreeable climate than Venus. A trace of moisture exists in the atmosphere, equivalent to a film about a thousandth of an inch thick. Although this small amount of water is not adequate to permit life to originate, it could support life that had evolved earlier, in a wetter and more favorable age, and had then adapted slowly to drier conditions. The suggestion that Mars once had an abundance of water has received support from spacecraft observations carried out in orbit above Mars and on its surface. All in all, as an abode of life Mars is considerably more promising than Venus.

A composite of pictures of Mars obtained from a spacecraft. The north polar cap, Valley of the Mariner (bottom right), and six large volcanoes appear clearly.

PLANETARY PROPERTIES

Mars stands midway between the moon and the earth as a planetary body. It is approximately twice as large as the moon, and one-half the size of the earth; its mass is approximately 10 times the moon's mass, and one-tenth the mass of the earth. These facts lead to a prediction regarding volcanism on Mars. We know from the study of the earth that volcanism results from the release of radioactive heat within the body of an earthlike planet. The source of the heat is the radioactive decay of the elements uranium, thorium, and potassium, which exist in the interior of the earth in small concentrations of a few parts per million. The Apollo findings indicate that these elements exist in very roughly the same concentrations in the moon, and presumably they also exist in the interior of Mars.

The heat that the radioactive elements create must have gradually increased the temperature in the interior of the moon and Mars, as well as the earth, during their early years. Volcanism may have commenced on the earth, the moon and Mars at about the same time, when the three planets were about one billion years old, as the product of this chain of circumstances.

Although the volcanism probably began at about the same time on each planet, it did not last for the same length of time on each, because of the differences in their sizes. The reason is that the heat generated in the interior of a small planet has to travel only a short distance to reach the surface. Heat is lost from the surface of a small planet at a relatively rapid rate because of this fact. If the planet is very small, its temperature will never reach the melting point of rock, and volcanism will never commence.

If the planet is somewhat larger, its internal temperature may reach the melting point of rock, but will not remain there for more than a short time. This appears to have been the case for the moon, whose episode of volcanism lasted only 700 million years.

If the planet is very large, its internal heat must travel through a relatively thick layer to reach the surface. This thick layer acts as an insulating blanket, bottling up the radioactive heat, and causing the temperature in the interior to remain at a high level for a relatively long period of time. This is the case for the earth, which has remained volcanically active throughout most of its lifetime, and is still active.

Mars, intermediate in size between the moon and the earth, must have retained its radioactive heat longer than the moon, but not as long as the earth. Therefore, Mars should have been volcanically active for longer than 700 million years, and may have been active for as long as several billion years. This remark implies that Mars may have been the scene of active emission of volcanic gases until a relatively recent period in its geologic history.

Water vapor and other gases released by internal heating are the probable sources of water on the earth's surface, and gaseous elements in its atmosphere. If volcanic activity persisted on Mars for two or three billion years, and the relative amount of water trapped in the interior of Mars was roughly the same as the amount of water trapped within the earth, the surface of Mars may once have been covered by water to a considerable depth. A substantial atmosphere also may have existed during this long period of volcanic activity.

Evidence for Geological Activity

The suggestion of early volcanism on Mars was confirmed when television cameras, mounted on a satellite placed in orbit around Mars, obtained excellent pictures of the entire planet. The televised pictures were taken at altitudes varying between 775 and 10,500 miles. When the first pictures were obtained, the surface of the planet was obscured by a violent dust storm that persisted for nearly two months. Eventually the dust settled, and the television cameras revealed details of the surface of Mars that had never before been seen by man.

The most conspicuous feature was the huge volcanic mountain, Mount Olympus, shown in Figure 19.1 (p. 460). The crater at the summit of this mountain is about 40 miles in diameter. The entire mountain is 300 miles across at its base, and rises at least 70,000 feet above the surrounding floor.

The roughly conical shape of the mountain and the presence of a crater at its summit clearly establish it as a volcano. The summit crater, called a caldera, is a familiar feature of terrestrial volcanoes. It is formed by the collapse of surrounding material into the molten pit of lava at the top of the volcano (Figure 19.2) (p. 461).

In every respect the Mars mountain resembled the mounds of congealed lava that form on the earth, when many successive outpourings of molten rock occur at a single spot over a period of millions of years. If the Pacific Ocean basin could be emptied, the Hawaiian Islands would be revealed as similar but smaller mounds of lava, rising out of the floor of the ocean basin in the same way that the Mars mountain rises out of the surrounding terrain.

About a dozen large volcanic mountains have been discovered on Mars. Six can be seen clearly in the photograph on page 456. There is evidence that the volcanoes have been extinct for a long time. The photographs show many modest-sized meteorite craters on the flanks of the volcanoes. These relatively small craters would have been obliterated if fresh floods of lava had run down the sides of the volcanic mountains in recent times. From the number of craters present, it has been estimated that the great Martian volcanoes may have been active as recently as a few hundred million years ago.

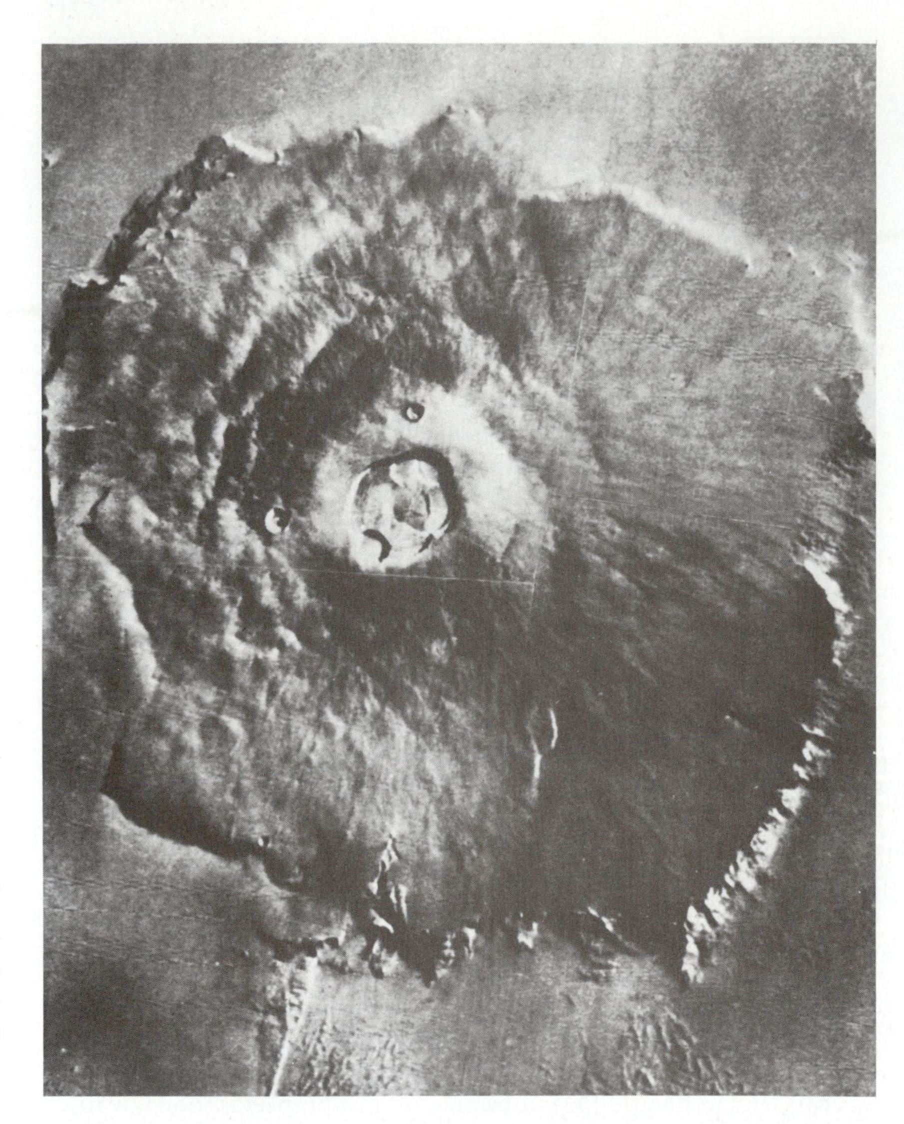

FIGURE 19.1
Mount Olympus, a large volcano on Mars. The base is 300 miles across, and the summit rises at least 70,000 feet above the surrounding terrain. The volcanic mound terminates at its base in a scarp or sharp cliff that is the origin of the circular feature called Nix Olympica by earlier Mars observers. The large crater at the summit is shown in a detailed view in Figure 19.2.

The Valley of the Mariner. On the earth, volcanism is associated with the movement of large slabs of rock, generally thousands of miles in size, that slide over underlying layers of warm and yielding material. These slabs of rock are the earth's plates (Chapter 16). When the edge of one plate slides past another, cracks such as the San Andreas fault appear in the surface of the planet. When two plates collide head-on, their edges crumple, and huge masses of rock are thrust upward to form mountain ranges such as the Himalayas. When two plates move apart, a shallow cleft, called a rift valley, appears between them. The Great Rift Valley, which runs through the Jordan Valley, the Dead Sea, and the Red Sea, and continues into East Africa, is an example.

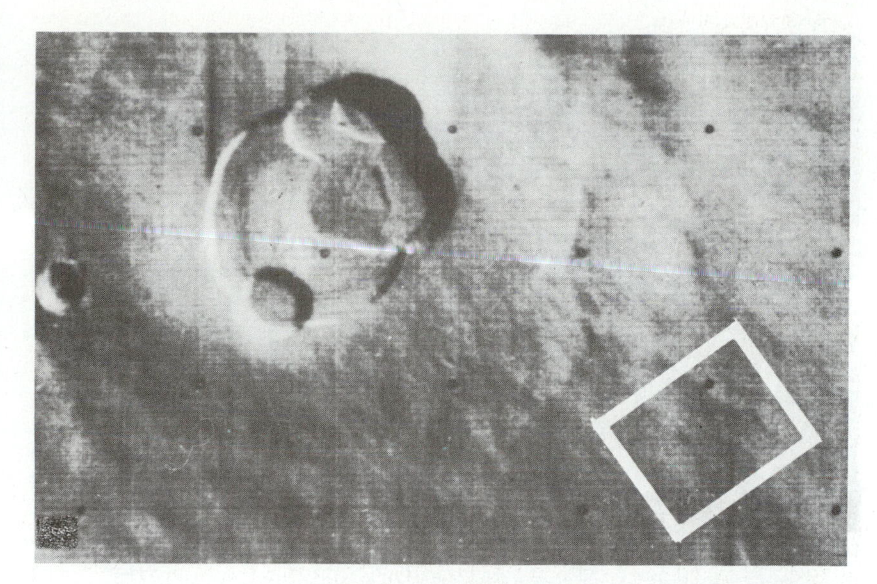

FIGURE 19.2
The crater at the summit of the Mars volcano shown in Figure 19.1.

Are there also signs of the movement of plates on Mars, to accompany the clear evidence of Martian volcanism? Thus far, no evidence has been found of features like the San Andreas fault or the Himalayan Mountains, but a feature that does resemble a long, straight crack or fault has been found. This is the Valley of the Mariner, a broad canyon 3000 miles long, roughly 75 miles across, and 15,000 feet deep at its lowest point. It is visible in the composite photograph opposite page 457 as a jagged crack at the lower right, near the edge of the planet. An overhead view of a 300-mile segment of the Valley of the Mariner is shown in Figure 19.3*a* (p. 462). The region enclosed in the rectangle at the right side of Figure 19.3*a* is shown in an enlarged view in Figure 19.3*b*. This photograph is a mosaic of high-resolution images obtained from a spacecraft in orbit around Mars.

The arrows in Figure 19.3*b* point to jumbled terraces of rock, which seem to be giant landslides formed by the slumping of material from the north rim of the Valley. Their presence provides a clue to the process by which the Valley of the Mariner may have been formed. At some point in the geological evolution of Mars, forces deep within the planet must have pulled the crust apart along a line roughly coinciding with the present Valley. Numerous small, parallel cracks appeared in the crust as a result of these forces, forming the beginning of the Valley of the Mariner. Repeated slumping and landslides at the rims of the cracks gradually widened the Valley. Some of the original cracks are still visible in Figure 19.3*b* (page 462), running parallel to the axis of the Valley above the north rim.

A similar process is believed to occur on the earth, when two plates in the lithosphere begin to break up and move apart (Chapter 16). The separation of the plates containing the South American and

FIGURE 19.3
(a) The Valley of the Mariner, photographed from a spacecraft. (b) A detailed view of the area within the rectangle in (a). The arrows point to giant landslides.

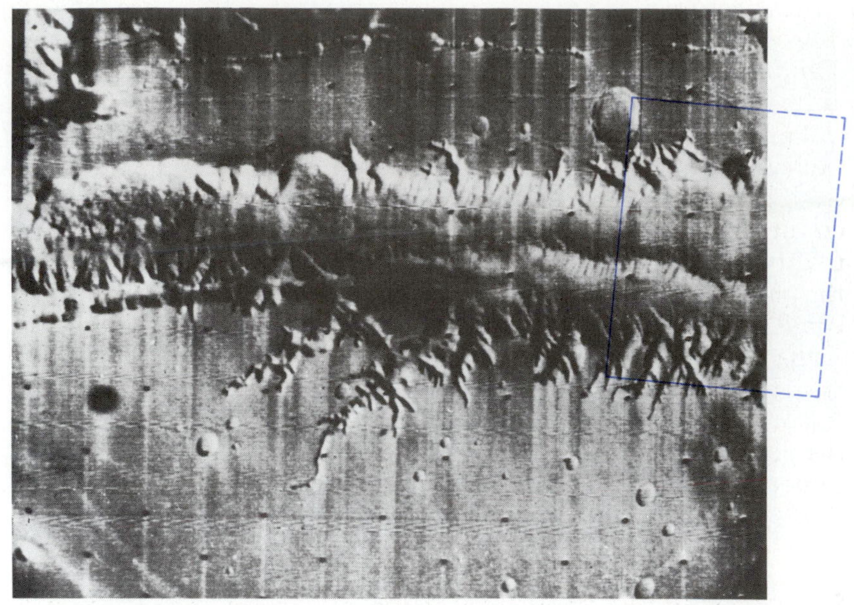
(a)

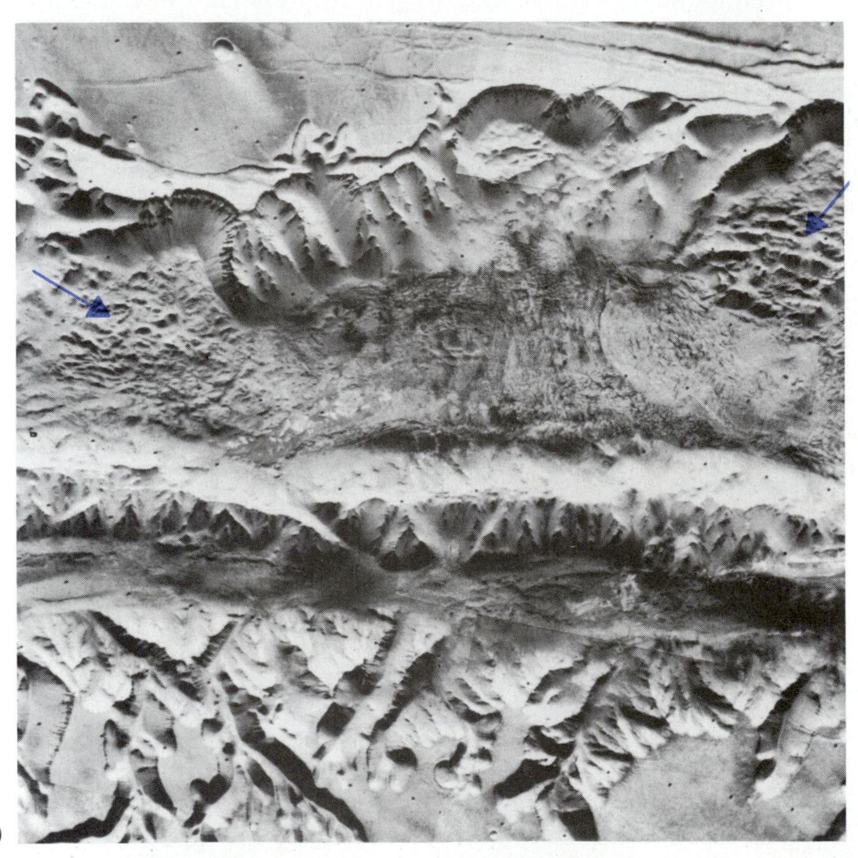
(b)

African continents is the clearest example (page 413). Since there is, however, no sign of major plate movements on Mars, it must be assumed that Martian tectonic activity never progressed beyond the incipient stage.

The absence of plate movements on Mars provides a likely explanation for the large size of such Martian volcanoes as Mount Olympus. On the earth, when a pile of congealed lava accumulates on the surface to form a volcanic island, as in the Hawaiian Islands, the accumulation of lava never grows to a very great height because the movement of the earth's plates—in the case of Hawaii, the Pacific plate—carries the volcano away from the location of the rising lava. It is because of this circumstance that an entire chain of islands has developed in Hawaii; each one represents the accumulation of lava at the surface between these intermittent motions of the plates.[1] If, however, there is no plate movement, the lava continues to pile up at the same place throughout geologic periods of time, growing eventually to very great heights and covering a very large area.

Photographic Evidence for an Early Abundance of Water

Figure 19.4, taken from the Viking spacecraft in 1976, shows a clear pattern of converging channels, resembling the channels cut by large volumes of water when flash floods occur in desert regions on the earth.

Other indications of an early abundance of water appeared in the spacecraft pictures. Figure 19.5 shows particularly clear evidence for a Martian riverbed. This mosaic of pictures shows a braided pattern of channels difficult to explain in any way other than by the flow of large volumes of water. Figure 19.6 (465) shows a winding channel resembling a riverbed, with tributaries feeding in at the top, and a distinct pattern of meanders in one section. From the angle at which the tributaries enter the main channel, it can be concluded that the direction of the flow of water or other fluid was from top to bottom in the photograph.

The Martian Polar Caps. Spacecraft pictures of the Martian channels provide evidence for the existence of a large amount of water on Mars in the past. However, the evidence is circumstantial; there is no direct proof of the presence of water. Direct evidence comes from measurements of the temperature of the Martian north polar cap, made from an orbiting spacecraft when the northern hemisphere

[1] Reference to the map of the earth's plates and the direction of the plate's movements on pages 410–411 indicates that the line of the Hawaiian Islands parallels the direction of motion of the Pacific plate, in confirmation of this view of the Islands' formation.

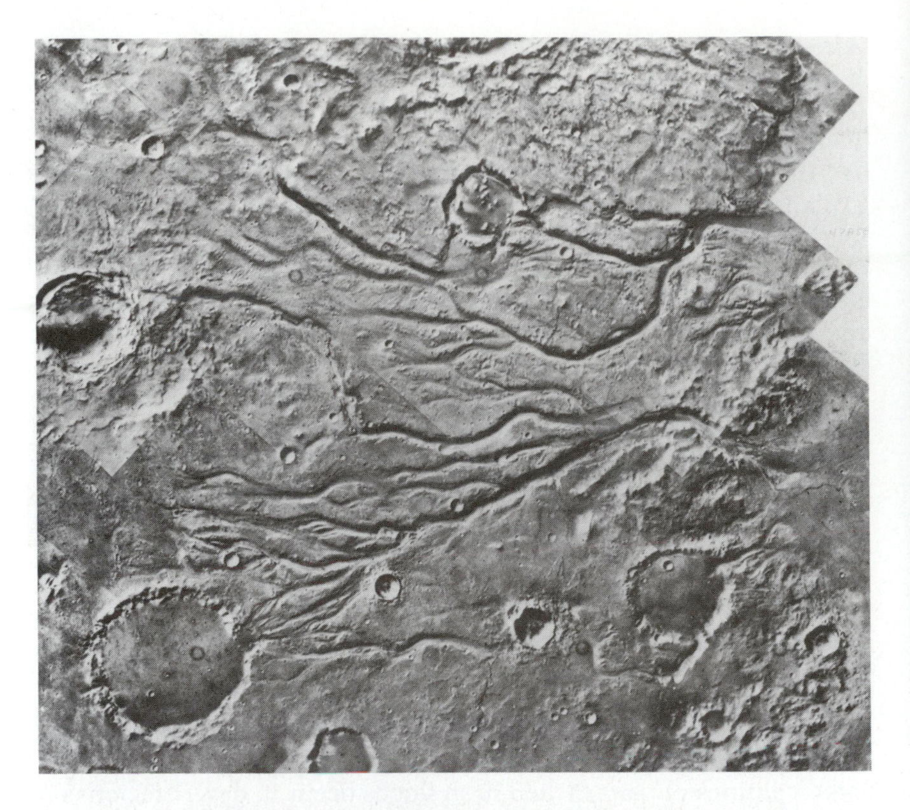

FIGURE 19.4
Converging channels on the surface of Mars, resembling channels formed by water erosion in desert regions on the earth.

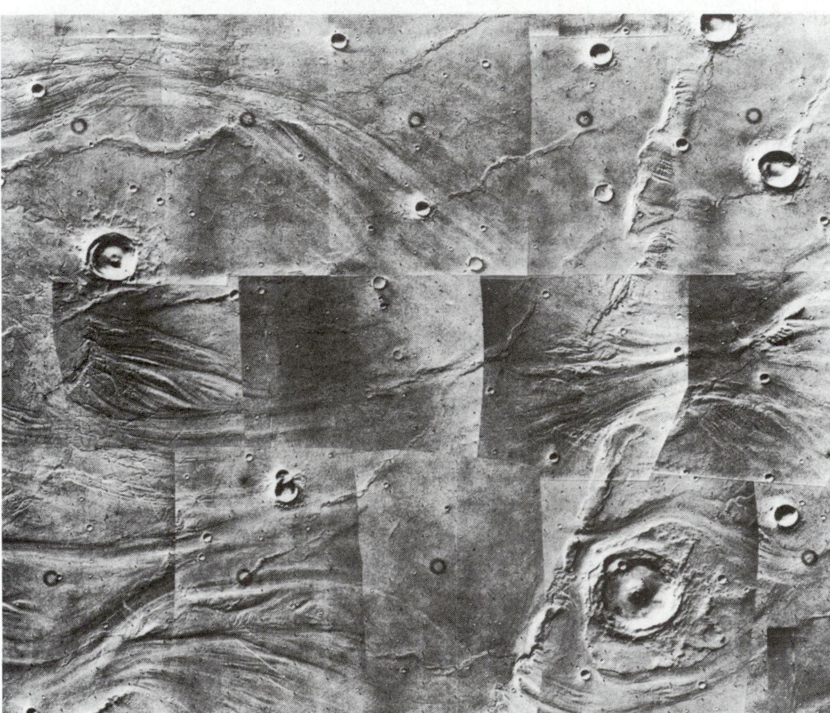

FIGURE 19.5
A braided pattern in a Martian channel, suggesting a flow of water.

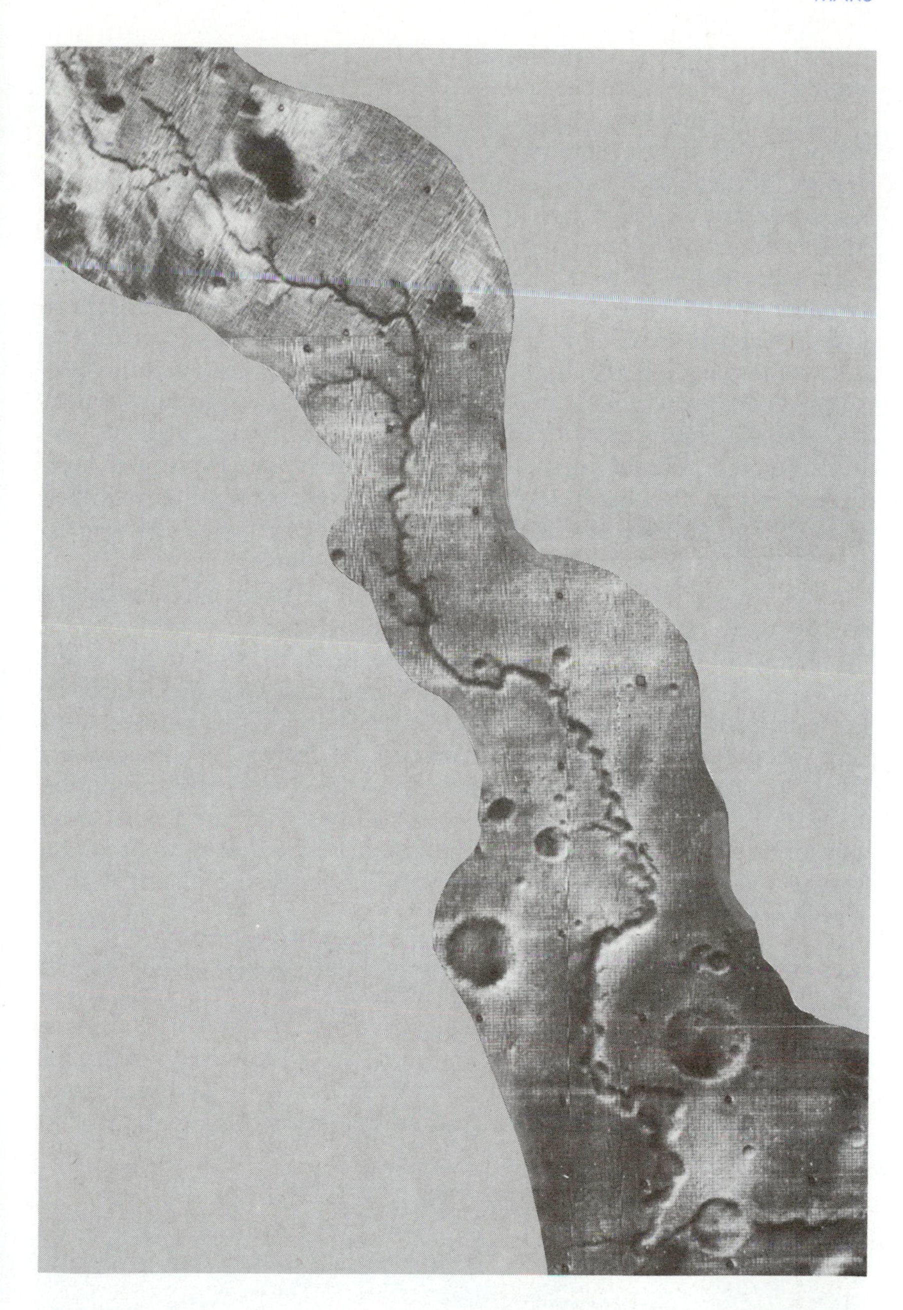

FIGURE 19.6
A Martian feature about 200 miles long, resembling a dried riverbed with tributaries feeding into the main channel and a characteristic pattern of meanders.

was in its summer season. The polar cap appears in the photograph facing page 457. Prior to the spacecraft measurements, the Martian polar caps were considered to be mainly composed of dry ice or frozen carbon dioxide, with a smaller admixture of ordinary ice or frozen water. However, under Martian atmospheric conditions the temperature of a cake of dry ice would be about −190°F. The measurements revealed the temperature of the northern polar cap to be

−90°F, which is too warm for dry ice but consistent with the interpretation that the caps are composed of frozen water.

The inference that the polar caps are composed of water ice is supported by measurements of the humidity of the Martian atmosphere, also made from an orbiting spacecraft. These show that the air above the north polar cap is somewhat more humid than the atmosphere over the rest of the planet. That would be expected if the caps were made of water ice, since water molecules leave an ice surface by the process of evaporation and enter the atmosphere directly above, increasing its humidity. Water molecules enter the atmosphere even more rapidly from a surface of *liquid* water; thus, at the edge of the polar cap, where ice is melting continuously during the summer season and liquid water is always present, the humidity should be particularly high. This prediction was confirmed by the Viking water vapor measurements, which showed that the air directly above the edge of the north polar cap was 10 times more humid than the average for the planet.

This evidence firmly establishes that much of the water released to the atmosphere during earlier volcanic eruptions is still present on Mars, but locked up at the poles in frozen form. How much water is present? A tentative answer comes from the fact that relatively large meteorite craters can be seen lying underneath the caps with only their rims visible. From the size of the craters, and the general relation between the diameter of a meteorite crater on Mars and the height of its rim above the plain, it follows that the ice is some hundreds of feet thick. If this huge amount of ice were melted and spread over the surface of Mars, it would cover the planet with a shallow sea several feet deep.

Climate Changes on Mars. A detailed photograph of the edge of the polar cap (Figure 19.7) shows terraced slopes or layers. These layers are believed to be evidence of repeated cycles of climate change on Mars, alternating from warm to cold periods, each perhaps 100,000 to one million years in duration. At the present time, Mars appears to be experiencing an ice age in its climate cycle (Figure 19.8) (p. 467).

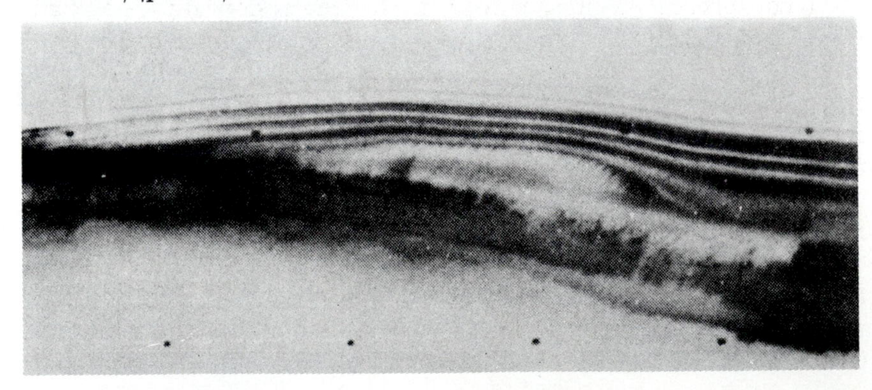

FIGURE 19.7
A detailed view of the edge of the polar cap, showing layers that suggest the alternation of ice ages with warmer periods.

Recent advances in understanding of climate changes on the earth suggest a plausible explanation for these episodes of Martian climate change. Primarily because of the gravitational pull of Jupiter, the shape of the earth's orbit varies every 100,000 years or so. Therefore, the earth's distance of closest approach to the sun also varies. In addition, the tilt of the earth's axis varies in a cycle with a duration of about 40,000 years. Finally, the direction of the earth's axis of rotation varies in a cycle of about 26,000 years (see page 29). Because of these variations, the northern hemisphere of the earth will, on some occasions, be tilted sharply toward the sun at the same time that the planet is closest to the sun in its orbit. When this happens, summers in the northern hemisphere will be very warm and winters will be correspondingly cold. When the northern hemisphere is tilted away from the sun during the time of closest approach to the sun, northern-hemisphere summers will be cool and winters will be mild. These circumstances are known to affect the rate of accumulation of ice, and are one of the major causes of the ice ages on our planet.

Similar variations occur in the orbit of Mars, again because of Jupiter's gravitational pull. The shape of the orbit of Mars varies in a cycle of two million years; the tilt of the axis of rotation varies in a cycle of one million years; and the direction of the axis changes in a cycle of 50,000 years. The combined effect of these orbit variations can easily explain the climate changes that seem to have occurred on Mars.

Still other evidence suggests that large deposits of ice lie under the surface of Mars in other places besides the polar regions. In some instances, these buried ice deposits seem to have melted suddenly, creating massive runoffs of ice water. Figure 19.9 shows an area about 300 miles across, just north of the Martian equator, photographed by the Viking spacecraft from orbit in 1976. The most conspicuous feature in the photograph is a triangular-shaped depression, with numerous channels running out of the depression toward the east, or to the right in the photograph. This is also the direction of the regional downhill slope of the terrain. The chaotic, jumbled appearance of the floor of the depression suggests that buried ice or permafrost melted suddenly, releasing huge quantities of water that flowed eastward, that is, following the downhill slope. It is possible that some of the Martian channels were carved by flash floods of melted ice from regions such as that shown in Figure 19.9, during warm periods in the history of Mars. The pattern of riverlike channels at the right side of Figure 19.9 can be followed for hundreds of miles with the aid of other Viking photographs. The signs of the great flood peter out near the site on which the first Viking spacecraft landed. From the number of ancient meteorite craters in the area, geologists infer that the melting and flooding occurred 2 or 3 billion years ago.

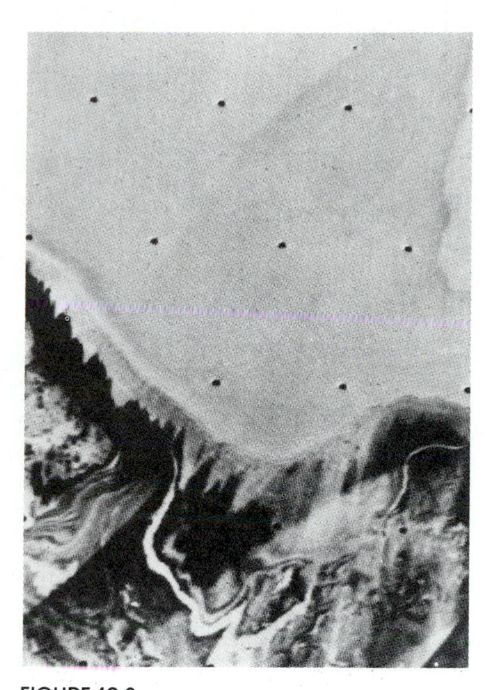

FIGURE 19.8
Another detailed view of the Martian polar cap showing tongues of ice—Martian glaciers—several miles long, advancing over the edge of the cap and into the region beyond. This photograph lends support to the suggestion that Mars is a planet experiencing an ice age.

FIGURE 19.9
A collapsed region 150 miles in extent, possibly caused by sudden melting of buried ice or permafrost.

A new and more complete picture of Mars emerges from this evidence, as a planet characterized by climatic intervals of relative warmth and moisture, during which conditions may have been more congenial for the chemical evolution of life. Combining our knowledge of the present scarcity of water on Mars with the evidence for its abundance in the past, we are led to a tentative picture of the history of water on the planet. At one time, Mars may have had a substantial amount of water, which accumulated during a relatively long period of volcanism, possibly lasting as long as a billion years. During that period its surface may have been traversed by streams and rivers, but when the volcanism subsided, the source of the water disappeared. Individual molecules of water, leaking steadily into space, gradually diminished the amount of available moisture, and the streams and rivers dried up. Finally the planet reached the conditio in which we find it today, with only a trace of moisture in the atmosphere, and no liquid water visible on the surface.

PROSPECTS FOR LIFE ON MARS

Does life exist on Mars? Possibly. Mars is dry, cold, and less favorable than the earth for the support of life, but not implacably hostile. Experiments have shown that some terrestrial plants could exist in the Martian climate, although they would not flourish there. Cosmic-ray and ultraviolet radiation bombard Mars with an intensity that would be lethal to terrestrial life, but biologists have suggested ways in which Martian organisms could have evolved natural shields against these deadly rays.

The Relationship Between Water and Life. Water is the key element in a discussion of the prospects for Martian life. Water is an essential ingredient for the development of the kinds of life with which we are familiar, because it provides a fluid medium in which the complex molecules of the cell can move freely. This movement leads to frequent collisions between neighboring molecules and, as a consequence of these collisions, to chemical reactions that make up the ongoing process of life. The basic building blocks of living matter—immersed in the shallow Martian seas—would have collided ceaselessly; now and then the collisions would have linked them into the large molecules—proteins, DNA, and RNA—which are the essence of the living organisms. The linking of small molecules to form large ones would have marked the first step along the path from nonlife to life.

The relationship between water and life lends a special interest to the spacecraft evidence for large amounts of water on Mars at an earlier time. If life exists on Mars today, it probably can be traced back to a golden age on the planet, when emission of gases from volcanoes maintained a substantial amount of liquid water on the surface, as well as a denser atmosphere, and the climate rivaled the climate of the earth. The transition to the drier climate of today may have occurred very slowly, over a period of millions of years and a very large number of generations. In this case, Martian life could have adapted progressively to the gradual onset of severe conditions. There seems no reason to doubt that life could exist on Mars today as a result of this long-continued process of natural selection *if* the planet once had an abundance of water.

The Search for Martian Life

Two American spacecraft have been landed on the surface of Mars, equipped with a variety of instruments intended for the study of Martian geology and biology. Each spacecraft contained a pair of television cameras providing stereoscopic photographs of the nearby terrain, partly to provide geological information and partly to search for fossils and scan the horizon for signs of existing life. The cameras recorded a scene reminiscent of some of the desert areas of the southwest, with rust-colored soil, dunes, and other windblown formations, and scattered blocks of volcanic debris (Figure 19.10 and Color Plate 18). Each spacecraft also contained a small, automated biochemical laboratory designed to carry out a series of tests for Martian life. The laboratory was operated by remote control from earth and included instruments for the detection of simple forms of life as well as the molecular building blocks of living matter. All the tests used samples of Martian soil that had been scraped from the ground near the spacecraft by a small shovel mounted on the end of a boom (Figure 19.11). The samples were deposited in

FIGURE 19.10
(Pages 470 and 471) A Martian landscape photographed by the Viking lander. The large rock at left is about 3 feet high. The horizon is 2 miles away.

FIGURE 19.11
At left, the Viking sample collector pushes aside a rock to get at the underlying soil, protected against the destructive effects of solar ultraviolet radiation. At right, the collector has scooped up the soil and is transferring it to the spacecraft for biological and chemical tests.

hoppers leading to separate chambers. The entire spacecraft had been sterilized before the flight to ensure that any organisms the instruments detected would be indigenous to Mars, and not carried there from the earth.

The first test was based on the reasoning that all organisms known on the earth give off gases as waste products. For example, plants release oxygen as a waste product, and animals and most kinds of microbes release carbon dioxide. If the Martian soil contained organisms resembling plants, a pinch of soil placed in a chamber would produce a small amount of oxygen. If the soil contained microbes or animals, carbon dioxide would be produced. An instrument adjacent to the chamber could detect this and other gases.

When the test for gases was carried out, carbon dioxide was released from the soil, indicating that small animals or microbes might live in it. At the same time, a substantial amount of oxygen appeared, suggesting that the soil might contain plant life. The carbon dioxide was released slowly and steadily over a period of several days, as would be expected from a population of microbes in the soil. The oxygen, however, came off in a burst in the first few hours of the experiment. This was a surprise, because if plants were producing the oxygen, it would be released at a slow, steady rate, like the carbon dioxide.

The very rapid release of oxygen was more consistent with a chemical reaction in the soil, rather than a biological process. One kind of chemical reaction that could release a burst of oxygen is known from laboratory experiments. Solar ultraviolet radiation—falling on a surface of soil—can produce molecules of hydrogen peroxide, or H_2O_2, which then adhere to the grains of rock in the soil. If the grains of rock are moistened, the hydrogen peroxide molecules break up very quickly into water and oxygen. Since the initial step in the Viking experiment was the exposure of the soil to moisture, the release of oxygen could be explained readily by the chemical process just described, without recourse to life processes.

Another test was designed primarily for microbes. In this test, a pinch of soil was placed in the chamber and moistened with a nutrient broth containing amino acids and other food substances. These substances, like all foods, were made from atoms of carbon, oxygen, nitrogen, and other chemical elements. If microbes existed in the soil, they would consume the food and digest it. Most of the atoms in the food would be incorporated into the bodies of the microbes as they grew and reproduced, but some carbon atoms would be released into the atmosphere in the form of molecules of carbon dioxide. All animals, and most microbes, exhale carbon dioxide in this way as a by-product of their metabolism.

How could scientists on the earth find out whether that complicated process was taking place in a chamber on Mars, more than 200 million miles away? The answer is ingenious. Before the flight to Mars, a special kind of food had been prepared, in which some of the carbon atoms were the radioactive isotope of carbon, C^{14}. If Martian microbes were present, they would digest the C^{14} atoms and exhale radioactive carbon dioxide. To find out if this were happening, a detector sensitive to radioactivity was placed in an adjoining chamber, located above the first one and separated from it by a thin tube. If the detector signaled the arrival of radioactive carbon dioxide, that would suggest that Martian microbes existed in the chamber below.

When the test was carried out, the detector indicated that a large amount of radioactive carbon dioxide had been produced in the pinch of soil below, and had passed through the tube into the upper chamber. Thus, this test suggested the presence of Martian life.

But the test for microbes, like the first test, also could be explained by a chemical reaction not involving organisms. Suppose Martian soil contains hydrogen peroxide, as seemed to be indicated by the release of a burst of oxygen in the previous experiment. When the soil was moistened by the nutrient broth, the peroxide compounds would decompose the particles of food in the broth, breaking them up into smaller molecules, including molecules of carbon dioxide. The carbon dioxide would be released to the atmosphere. Some of this carbon dioxide would be radioactive. In that way, chemical reactions could simulate the presence of microbes in the second test. However, repetitions of the microbe test under different conditions yielded results that are difficult to explain by chemical reactions, unless the reactions are exceedingly complicated.

Additional information came from still another experiment, not designed specifically as a test for life, which tested the soil for complex compounds containing carbon. These compounds, known as organic molecules, are the building blocks of all life on the earth. If life exists on Mars and is similar to terrestrial life, there should be an abundance of organic molecules in the Martian soil. But the test failed to reveal any organic molecules within the limits of sensitivity of the instrument. That result suggested that life does *not* exist on Mars, and tended to support the chemical explanation for the Viking results. On the other hand, the absence of organic molecules could also be explained as being due to the same peroxides or other strongly oxidizing compounds that appear to be present in the Martian soil. These compounds would rapidly decompose all organic debris in the soil, but still-living organisms might have evolved defenses against the oxidizing action. The sensitivity of the organic molecule test was such that a substantial population of living microbes, equal to the population of a rich garden soil on the earth, could exist without being detected by it. The final resolution of the question of Martian life may have to await more elaborate tests, in future landings on that planet's surface.

The Significance of the Search for Martian Life

The discovery of Martian life—even in as simple a form as a microbe—would have portentous implications for the probability of life on other earthlike planets in the Cosmos. At the present time no one knows what that probability is. Perhaps the evolution of life out of inanimate atoms requires the simultaneous occurrence of many special circumstances, whose combined probability is so low that this event has happened only once in the Universe. How can we determine whether this is so?

The discovery of life on a second planet in our solar system would go far toward settling that question. Life on *one* planet—the earth—tells us nothing about the probability of life in the Universe, but life on *two* planets in one solar system—the earth and Mars—would tell us nearly everything. For if life has arisen independently on two planets in a single solar system, it cannot be a rare accident, but must be a fairly probable event. If life is found on Mars, we will know that countless inhabited planets probably surround us in the Milky Way Galaxy and other galaxies. No scientific discovery more significant in its implications can be imagined.

Main Ideas

1. Surface conditions on Mars; composition of the atmosphere.
2. Relation between size and geological activity on a planet; prediction regarding volcanic activity on Mars; implications for Martian water and atmosphere.
3. Evidence for an early abundance of water on Mars.
4. The Martian polar caps.
5. Prospects for life on Mars; results of the Viking tests for life on the Martian surface.

Important Terms

biological and chemical reactions
caldera
metabolism
microbe
Mount Olympus
organic molecule
polar cap
radioactive isotope
Viking (spacecraft)
volcanism

Questions

1. How does the size of a terrestrial planet influence its geological history? Describe and compare the surfaces of the moon and Mars and explain the similarities and differences in terms of the relationship between planetary size and geological activity.
2. What features on the flanks of the Martian volcanoes indicate that they have been extinct for a long time? Explain your reasoning.
3. Draw up a table for Mars, Venus and the earth listing, for each planet, the total atmospheric pressure at the surface, principal constituents in the atmosphere, and percent abundance of each constituent. Explain the important features of the table, including: differences in surface pressure; presence or absence of particular gases, such as oxygen; and relative abundances of such gases as carbon dioxide. Use information from this and previous chapters on the source of an earthlike planet's atmosphere and size of an earthlike planet and its level of geologic and volcanic activity.

4. What is the evidence for water on Mars in the past? Today? In what form(s)?
5. Why is the evidence for water on Mars, past and present, considered to be of special interest?
6. Describe in detail one Viking biological experiment. Explain the reasoning involved in the interpretation of the combined results from the experiments.
7. Can you devise one or more experiments designed to search for life on an alien planet? Your experiments should be different from those described in the text. Consider the practicality and probable cost of your proposal in framing your answer.

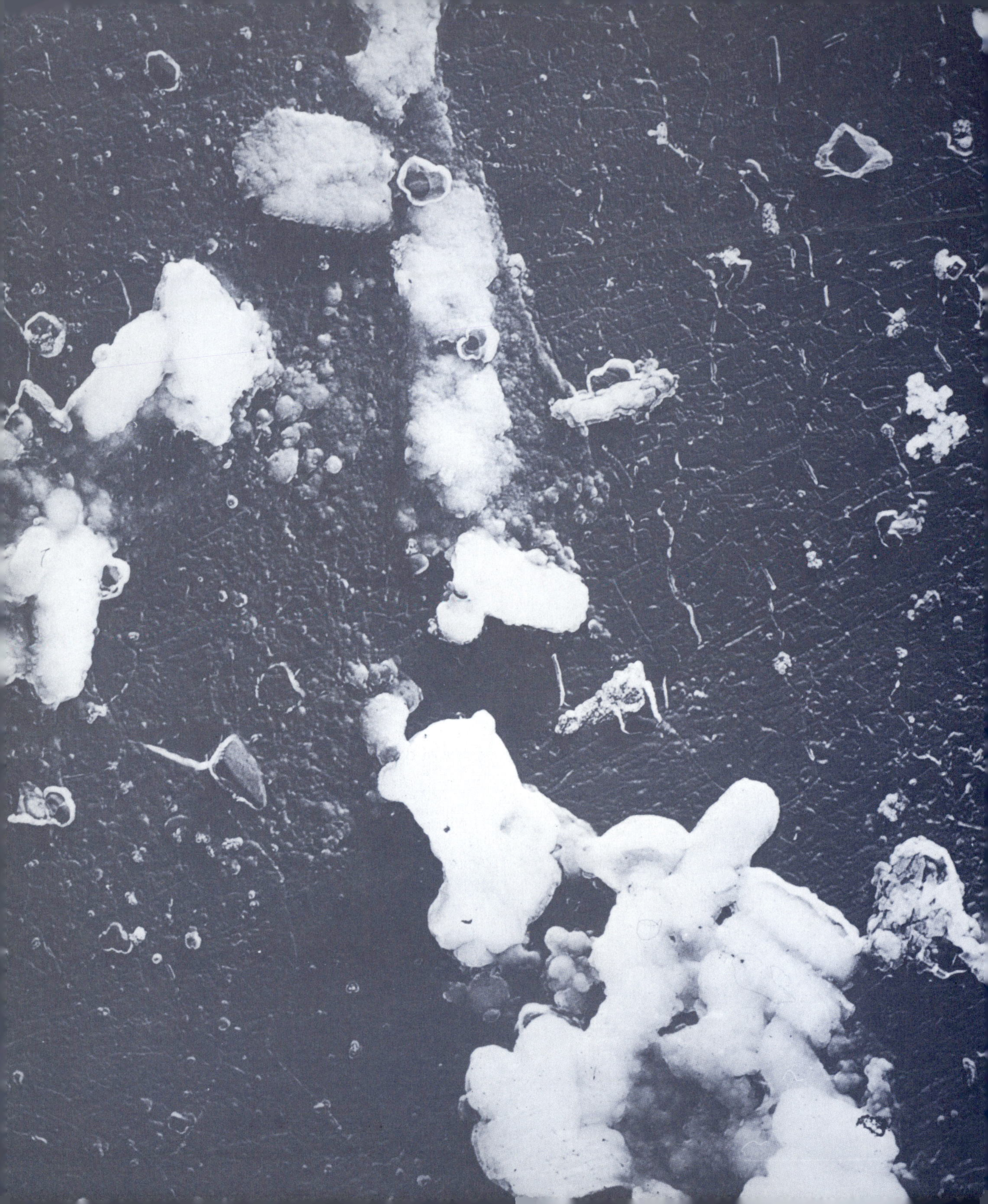

20 LIFE IN THE COSMOS

Discoveries in recent decades have linked the Universe of the stars to the world of life in a chain of cause and effect that extends over billions of years, commencing with the formation of stars in our galaxy and ending with the appearance of man on the earth. At the beginning of this history there existed only atoms of the primeval element, hydrogen, which swirled through outer space in vast clouds. These clouds were the raw stuff out of which stars and planets were made. Occasionally the atoms of a cloud were drawn together by the attractive forces of gravity; with the passage of time the cloud contracted to a small, dense globe of gas; heated by self-compression, it rose in temperature until, at a level of some millions of degrees, its center burst into nuclear flame. Out of such events, stars were born.

Within the newborn star a series of nuclear reactions set in, in which all the other elements of the Universe were manufactured out of the basic ingredient, hydrogen. Eventually these nuclear reactions died out, and the star's life came to an end. Deprived of its resources of nuclear energy, it collapsed under its own weight, and in the aftermath of the collapse an explosion occurred, spraying out to space all the materials that had been created within the star during its lifetime.

In the course of time, new stars, some with planets around them, condensed out of these materials. The sun and the earth were formed in this way, four and a half billion years ago, out of materials that were manufactured in the bodies of other stars earlier in the

Fossil bacteria several billion years old.

life of the Galaxy, and then dispersed to space when those stars exploded.

When the earth was formed, it must have been barren, but within one billion years or so life appeared on its surface. How can we explain this fact? If we continue the inquiry within the boundaries of science, discoveries made during the last few decades suggest a tentative scientific explanation for the presence of life on the earth.

THE ORIGIN OF LIFE

A sound scientific basis exists for the view that life on the earth evolved spontaneously out of chemicals that filled the atmosphere and oceans of the planet in the early years of its existence. The following facts suggest this hypothesis.

First, biologists have shown that all living organisms on the face of the earth depend on two kinds of molecules—amino acids and nucleotides—which are the basic building blocks of life—just as the physicists have shown that all matter in the Universe is constructed out of three building blocks—the neutron, proton, and electron.

Second, chemists have manufactured these molecular building blocks of life in the laboratory out of the gases that are believed to have filled the atmosphere of the earth when it was a young planet.

These experimental results provide a plausible scientific basis for the view that life can evolve out of nonliving matter.

The Building Blocks of Life

The basic building blocks of living matter are more complicated than the building blocks of the physical world. Approximately 20 different kinds of amino acids play a critical role in living creatures, and five different kinds of nucleotides. Furthermore, each amino acid or nucleotide is, itself, a rather complex molecule made up of approximately 30 atoms of hydrogen, nitrogen, oxygen, and carbon, bound together by electrical forces of attraction. An example of the structure of a typical amino acid is shown in Figure 20.1.

The amino acid and the nucleotide have very different functions in the chemistry of life. Within the cell the amino acids are linked together into very large molecules called *proteins.* One class of proteins, called structural proteins, makes up the structural elements of the living organism—the walls of the cell, hair, muscles, and bone. The structural proteins are like the steel framework and walls of a building. The other type of protein is called the enzyme. Many kinds of enzymes exist; each kind controls one of the many

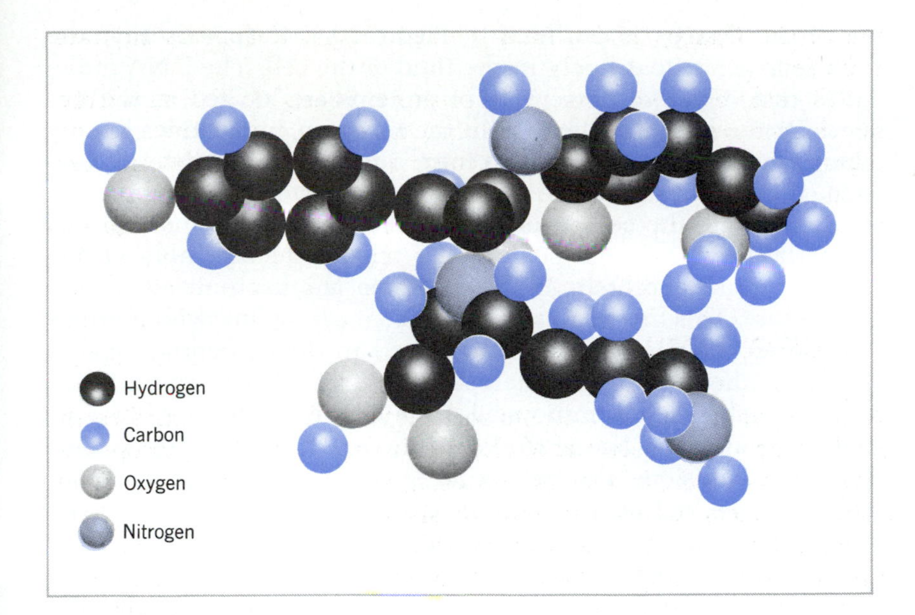

FIGURE 20.1
Three amino acids linked into a short segment of a typical protein.

chemical reactions that are necessary to sustain the life of the organism.

All proteins, in all forms of life, plant and animal, are constructed out of the same basic set of 20 amino acids. One protein differs from another only in the way in which their constituent amino acids are linked together. However, these differences are all-important. The distinction between a man and a mouse, both in appearance and personality, depends entirely on the differences between the proteins contained in the cells of their bodies.

Proteins are assembled within living organisms by the second set of building blocks—the nucleotides. Nucleotides are joined together within the cell to form very long chains, called *nucleic acids*. The most important type of nucleic acid is called deoxyribonucleic acid, or DNA for short. DNA is the largest molecule known, containing, in advanced organisms such as man, as many as 10 billion separate atoms. The size of the DNA molecule is understandable when we consider the complexity and importance of its functions in the living cell. The DNA molecule is the most important molecule in every living organism, even more important than the protein, because it determines *which* proteins will be assembled; the DNA molecule has the master plan for the organism.

The DNA molecule has the shape of a twisted ladder or double helix. The nucleotides are the rungs of the ladder. The double structure of DNA is shown in Figure 20.2.

How does DNA control the assembly of proteins in the cell? The general features of the process began to emerge some years ago, although many details are still not clearly understood. The

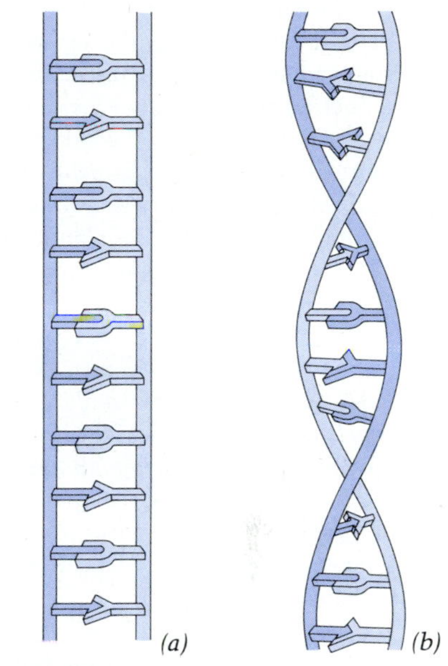

FIGURE 20.2
The structure of DNA. Watson and Crick discovered the structure of the DNA molecule in 1952. DNA resembles a ladder (a) twisted into a double spiral or helix (b). Each rung of the ladder is a linked pair of nucleotides, represented by the symbols

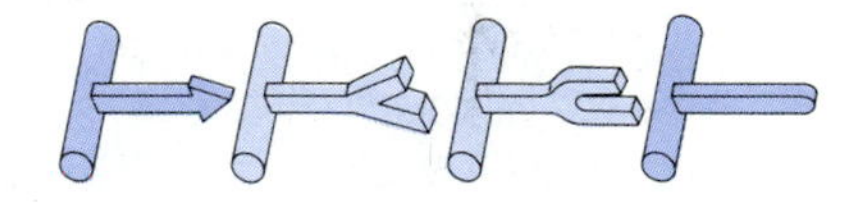

FIGURE 20.3 (opposite)
How DNA directs the assembly of amino acids into proteins. The DNA molecules reside in a nucleus at the center of the cell, enclosed by a membrane. When the molecule is about to make a protein, a segment of the DNA—one gene—splits into two strands, and a series of nucleotides assembles along one of the strands (1), copying the master sequence on the DNA.

The assembled string of nucleotides, slightly different chemically from DNA, is called RNA. As the RNA molecule assembles, it peels off and goes out into the cell through an opening (2) in the membrane surrounding the nucleus. This RNA molecule is called "messenger RNA" because it carries the message for building a protein from the DNA to the rest of the cell.

The cell contains another kind of RNA molecule resembling a stick (upper left) with an attachment for an amino acid on one end, and attachments for a set of three nucleotides on the other end. This stick-like molecule is called transfer RNA. There are 20 kinds of transfer RNA molecules—one for each kind of amino acid. Each kind of transfer RNA can pick up only one particular kind of amino acid. The other end of this transfer RNA, in turn, fits one particular set of three nucleotides on the messenger RNA. When a transfer RNA molecule is in action, it picks up an amino acid (3) and transfers it to the proper spot on the messenger RNA (4), where it adds to the growing chain of amino acids that makes up the complete protein (5). After the transfer RNA deposits its amino acid, it goes off into the cell (6) to find another one of its kind of amino acid and do its work again.

sequence of steps is outlined in Figure 20.3. It appears that the 20 amino acids float freely in the fluid of the cell. The DNA molecules that direct the assembly of proteins are located in the nucleus, at the center of the cell. In the first step, nucleotides line up alongside the DNA segment to form a replica of it. In the second step, the replica detaches itself from the master DNA chain, and moves off into the cell; it is a messenger that carries instructions from the DNA into the body of the cell for the assembly of one particular kind of protein. In the third step, another molecule enters the picture. This molecule serves as a connecting link which brings the amino acids in the fluid of the cell to the appropriate places alongside the messenger. There are 20 kinds of connecting links, one for each kind of amino acid. Each of the connecting links attracts one and only one of the 20 amino acids. When, in the course of chance collisions, the right kind of amino acid comes into contact with the end of the connecting link designed for that particular amino acid, it is held fast there. At the other end of the connecting link is another set of molecules, making up a surface of nooks and crannies so constructed that it can only fit into the appropriate place along the length of the messenger. When the connecting link takes its place along the messenger, it adds the amino acid to the chain of amino acids that has already been built up. When the assembly of the chain is completed, the finished protein then detaches itself from the messenger and drifts off into the fluid of the cell (Figure 20.3).

By this rather complicated process, the essential proteins are built up within an organism in accordance with the order of the nucleotides in its DNA molecules. The segments of the DNA molecule are "read" like the words of a book. Each DNA segment, controlling the assembly of one or more proteins, is like a sentence in a book. The order of the nucleotides in the DNA segment is like the order of the letters in the words of the sentence. The full set of DNA molecules contained within a cell is the library of genetic information for the organism. The DNA molecules in the cells of a human being are a library of information for the assembly of the amino acids in the human body into human proteins; the different DNA molecules in the cells of a mouse are a library of genetic information for the assembly of its amino acids into mouse proteins.

The Mechanism of Inheritance of DNA

How is the plan for the assembly of the right kind of proteins passed from one generation to the next? How do progeny inherit their characteristics from their parents? The answer lies in a most extraordinary property of the DNA molecule—the ability to make a copy of itself. The mechanism by which DNA copies itself was discov-

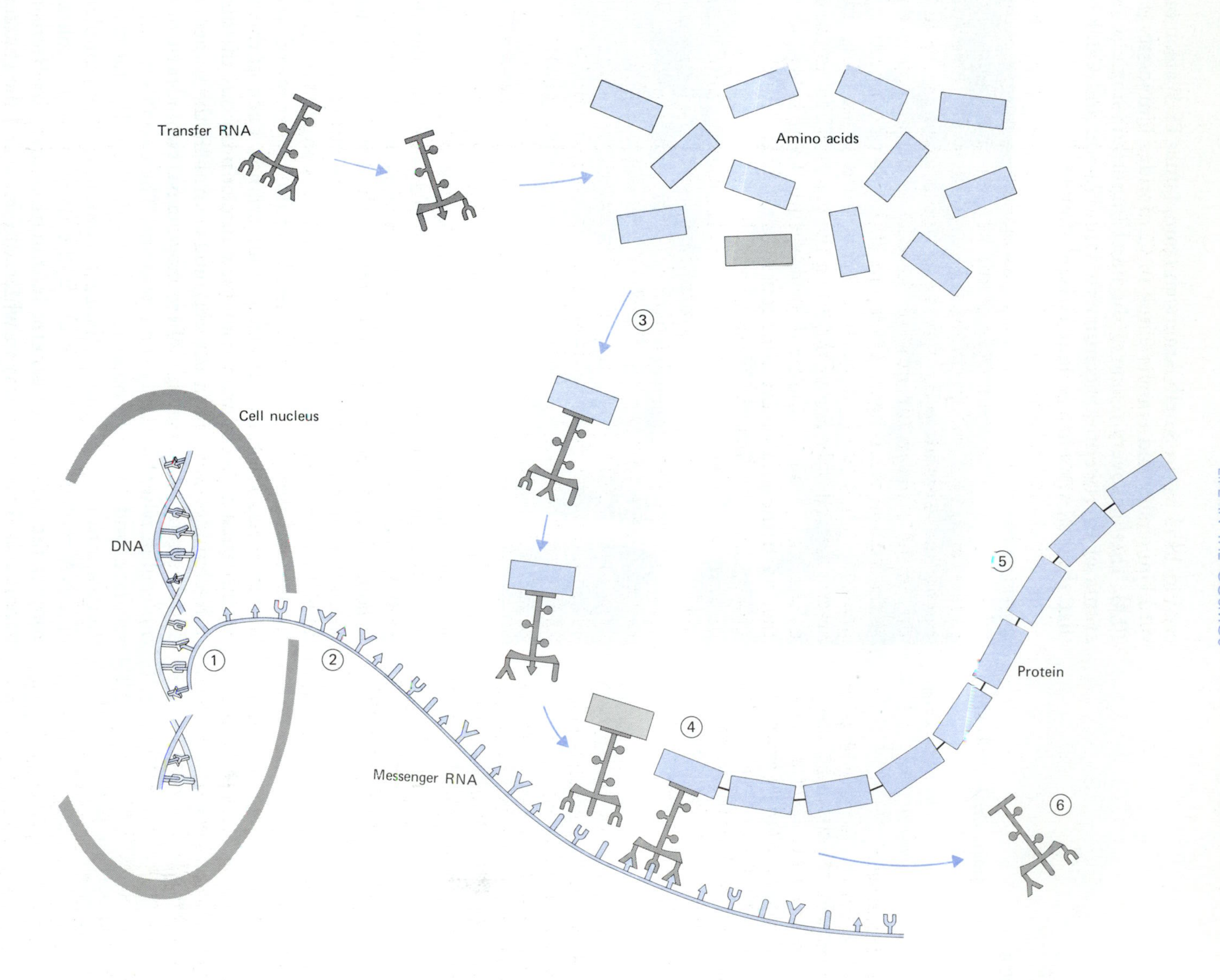
Transfer RNA
Amino acids
Cell nucleus
DNA
Messenger RNA
Protein
1
2
3
4
5
6

ered in 1952 by an Anglo-American team, James D. Watson of Harvard University and Francis Crick of Cambridge University (Figure 20.4). This discovery is one of the most important single scientific events of the twentieth century to date. Watson and Crick found that the DNA molecule is a twisted double chain of molecules,

FIGURE 20.4
James Watson and Francis Crick at the time of their discovery of the double-helix structure of the DNA molecule. Their first DNA model is shown in the background.

or double helix. The two chains are joined at regular intervals by molecules that run between them like rungs of a ladder (Figure 20.5). In each rung of the ladder there is a weak spot, which is easily broken. During the early history of a cell the two strands remain connected, but when the cell has attained its full growth, and the division into two daughter cells is about to commence, the weak connections running down the middle of the ladder break, and the double strand separates into two single strands. Each of the single strands then collects unto itself new nucleotides out of the pool of nucleotides floating in the cell, and assembles them into a new double-stranded molecule. There now exist two identical DNA molecules, where formerly there was one. The two DNA molecules separate, and move to opposite corners of the cell; the cell then divides into two daughter cells, each containing one complete set of DNA molecules. Thus, each of the daughter cells contains a copy of the volumes of genetic information that has been in the parent cell. This is the way in which the shape and the character of a plant or an animal are transmitted from generation to generation.

In summary, the DNA molecule controls the assembly of proteins, and the proteins determine the nature of the organism. Each

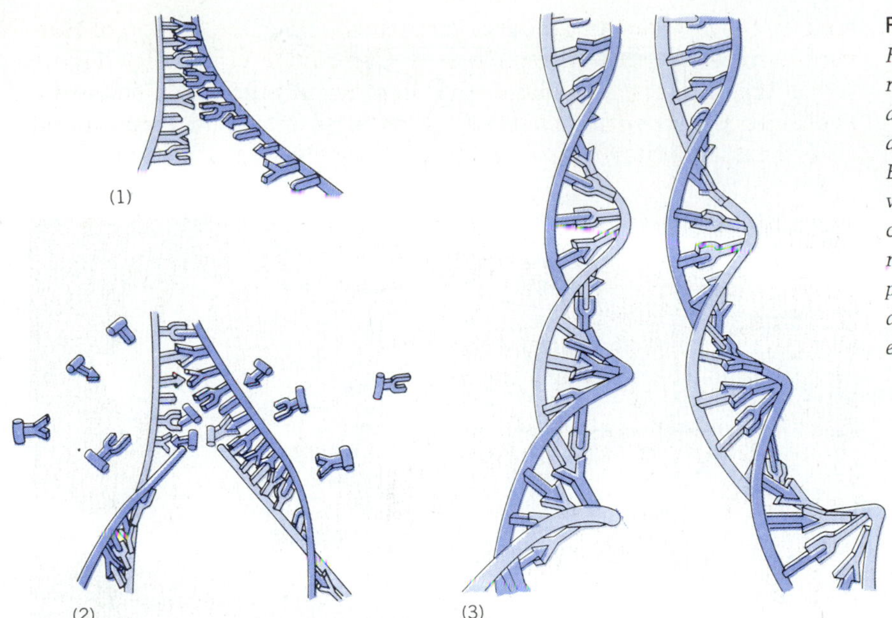

FIGURE 20.5
How the DNA molecule copies itself. DNA resembles a twisted ladder (p. 481). (1) When a cell is about to divide, the ladder untwists and the rungs break at their midpoints. (2) Each strand of "broken" DNA collects individual nucleotides floating nearby in the cell and begins to form a new ladder. (3) The result is two DNA molecules, replicas of the parent DNA. In this way the traits of the individual are passed from generation to generation.

living organism has its own special set of DNA molecules; no two organisms have the same set unless they are identical twins. However, *the basic nucleotides and amino acids are the same in every living creature on the face of the earth, whether bacterium, mollusc, or man.*

Formation of the Building Blocks Out of Atmospheric Gases

With this fundamental property of living creatures in mind, one can appreciate the importance of a critical experiment performed in 1952 by Stanley Miller, at that time a graduate student working on his Ph.D. dissertation under Harold Urey. At the suggestion of Urey, Miller mixed together the gases—ammonia, methane, water vapor, and hydrogen—which were abundant in the parent cloud of the earth, and which were probably abundant in the atmosphere of the young earth. He circulated the mixture through an electric discharge. At the end of a week, Miller found that the water contained several types of amino acids. Figure 20.6 is a schematic diagram of Miller's apparatus.

Subsequently, nucleotides were created in the laboratory under similar conditions. In some experiments, amino acids and nucleotides have been manufactured out of a variety of gas mixtures, using various sources of energy—bombardment by alpha particles, irradiation with ultraviolet light, and simple heating of the ingredients. The results of all these experiments, taken together, demonstrate that the molecular building blocks of life could have been

FIGURE 20.6
The apparatus in which Miller first produced amino acids from gases of the primitive atmosphere.

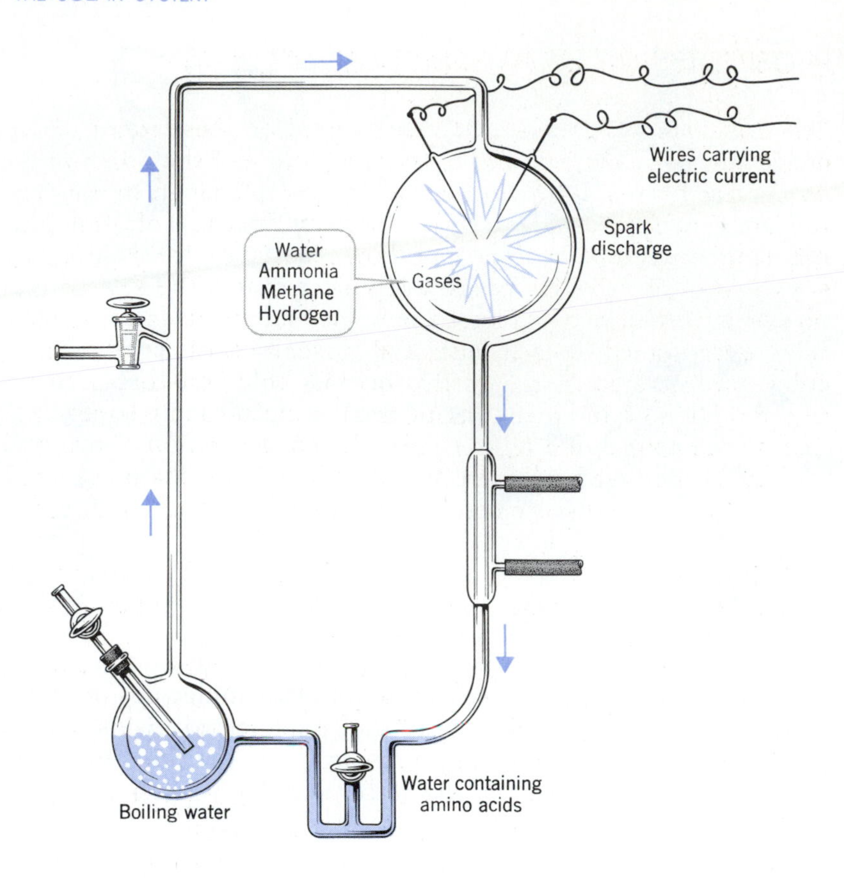

created in any one of many different ways during the early history of the earth.

Amino acids and nucleotides might have been formed on the earth out of these same ingredients four billion years ago, by the discharge of lightning in primitive thunderstorms, or by the action of solar ultraviolet rays from the sun. We can guess what happened thereafter. Gradually the critical molecules drained out of the atmosphere into the oceans, building up a nutrient broth of continuously increasing strength. Over a long period of time the concentration of amino acids and nucleotides built up, until eventually a chance combination of building blocks produced still more complex molecules—primitive proteins and nucleic acids. With the further passage of time, cells developed, many-celled organisms appeared, and living organisms were started down the long road toward the complexity of the creatures that exist today.

PROSPECTS FOR EXTRATERRESTRIAL LIFE

The imagination of the scientist has seized on these items of evidence and has fashioned out of them a picture of the origin of life on the earth. No living form existed on our planet in its infancy; the atmosphere was filled with a noxious mixture of ammonia, methane, water, and hydrogen; peals of thunder rumbled across the sky; flashes of lightning occasionally illuminated the surface, but no eye perceived them; minute amounts of amino acids and nucleotides were formed in each flash, and gradually these critical molecules accumulated in the earth's oceans; collisions occurred between them now and then, linking small molecules into larger ones. During the course of a billion years the concentration of complex molecules increased; eventually a complete DNA chain appeared. Thus, the threshold was crossed from inorganic matter to the living organism.

According to this story, life can appear spontaneously in any favorable planetary environment, and evolve into complex beings, *provided vast amounts of time are available.*

How much time is needed? Studies of the fossil record suggest that life appeared sometime during the first billion years of the earth's history. Apparently a billion years or so is the length of time required.

Our knowledge of the life cycle of a star indicates that the necessary period of a billion years or more will be available for the chemical evolution of life on any planets that circle around a star similar to our sun. Stars larger than the sun burn out too quickly to provide the needed time. Stars smaller than the sun are suitable, provided that they have planets close enough to them to raise their temperatures to a comfortable range. All stars the size of the sun, surrounded by one or more planets that are approximately the same distance from them as the earth is from its star, should certainly provide very favorable circumstances for the development of living organisms.

There are 100 billion stars belonging to the cluster we call our galaxy. Ten billion other galaxies, each with 100 billion stars—and probably a like number of planets—are within the range of the largest telescopes. Perhaps only a small fraction of them are earthlike planets, but that would mean millions of earthlike planets in our galaxy alone.

Can we maintain our belief in the uniqueness of life on the planet earth in the face of these numbers? We can on the astronomical evidence alone, because all planets except ours could be dead bodies of rock; but the biological discoveries described in this chapter suggest that this may not be the case. *First,* all life on earth depends on a few basic molecules, and these molecules have been created out of simple atoms in the laboratory; *second,* the atoms that compose the basic ingredients of life are the same as the atoms that exist on every other star and planet in the Universe; *third*, there is reason to believe that the same laws of physics apply in every corner of the Cosmos. If a chain of physical and chemical reactions led to the appearance of life on the earth, a similar chain could also have occurred on other planets.

These circumstances make it plausible that if sufficient time is available, a kind of life resembling ours can develop on many planets in our galaxy, and on planets in other galaxies as well.

THE HISTORY OF LIFE ON THE EARTH

Throughout the present discussion of stars, galaxies, and planets, our viewpoint has been that of the physical scientist, seeking to understand the essence of the world around him in terms of a few simple principles—the laws of physics and chemistry—which are the distillation of man's knowledge of the physical world acquired during thousands of years of observation. These laws have carried us from the origin of the Universe through the formation of stars and planets to the birth of the sun and the earth, and—with speculation and uncertainty—across the threshold of life on the earth.

Now we come to the explanation of the subsequent course of events in the history of life, leading from the first simple organisms to man. Here, for the first time, the principles of physics are no longer helpful. The stars and planets have yielded the secrets of their history to the physicist; the molecular foundations of living organisms are beginning to be understood; but the higher forms of life are incalculably more complicated than any star, or planet, or giant molecule. New insights are needed for the understanding of the circumstances that led to their evolution. A new law must be found.

FIGURE 20.7
Charles Darwin in 1881.

The new law was discovered by Charles Darwin more than a century ago (Figure 20.7). Darwin showed that evolution is controlled by a "force" that acts on plants and animals slowly, over the course of many generations, to produce changes in their forms. This force has no mathematical description; it is not to be found in any textbook of physics, listed alongside the basic forces that control the world of nonliving matter; but, nonetheless, it guides the course of evolution from simple cells to advanced creatures—on this planet and on all planets on which life has arisen—as firmly and as surely as gravity controls the motions of the stars and the planets.

Darwin began with an almost self-evident set of observations on the nature of life: All living things reproduce themselves; reproduction is the essence of life; *but the process of reproduction is never perfect.* The offspring in each generation are not perfect copies of their parents; brothers and sisters differ from one another; no two individuals in the world are exactly alike.

Usually the variations are small; brothers and sisters resemble one another, all human beings look more or less alike, and all elephants look more or less like other elephants.

Yet, Darwin asserted, these small variations are critically important; for, in the struggle for existence, the creature that is distinguished from its brethren by a special trait, giving it an advantage in the competition for food, or in the struggle against the rigors of the climate, or in the fight against the natural enemies of its species —that creature is the one most likely to survive, to reach maturity, and to *reproduce its kind.* The offspring of the favored individual will inherit the characteristic that has given it its advantage. Some among them will possess the desirable characteristic to a greater degree than others. These individuals, doubly favored, have a still greater chance to survive to maturity and to produce offspring of their own.

Thus, through successive generations, the advantageous trait appears with ever-increasing strength in the descendants of the individual who first possessed it.

Not only does the strength of the trait increase with the passage of time, but the number of individual animals possessing it also increases. For these individuals have slightly larger families than the average for the species; in each generation they leave behind a greater number of offspring than their less-favored neighbors; their descendants multiply more rapidly than the rest of the population, and in the course of many generations, their progeny replace the progeny of the animals that lack the desirable trait.

In *The Origin of Species,* Darwin gave this process the name by which it is known today:

> *"This principle of preservation or the survival of the fittest, I have called* Natural Selection."

Through the action of Natural Selection, a trait that first appeared as an accidental variation in a single individual will, with the passage of sufficient time, become a pronounced characteristic of the entire species. So the deer became fleet of foot, for the deer that ran fastest in each generation escaped their predators and lived to produce the greatest number of progeny for the next generation. So did man become more intelligent, for superior intelligence was of premium value: the intelligent and resourceful hunter was the one most likely to secure food. Thus developed the brain of man; thus, too, in response to other pressures and opportunities in their environments, developed the trunk of the elephant and the neck of the giraffe.

Of course, the incorporation of one new trait does not create an entirely new animal. But if we count all the births that occur to a single species over the face of the globe in one year, an enormous number of variations will appear in this multitude of young creatures. On all these variations the same process of selection works steadily, preserving for future generations the new traits that give strength to the species, and eliminating those that lend weakness. The changes are imperceptible from one generation to the next, but over the course of many generations the accumulation of these many favorable variations, each slight in itself, completely transforms the animal. According to Darwin,

> *"Natural selection is daily and hourly scrutinising, throughout the world, the slightest variations, rejecting those that are bad, preserving and adding up all that are good; silently and insensibly working at the improvement of each organic being in relation to its . . . conditions of life."*

Natural selection is the force that molds the forms of living creatures. Under the continuing action of the law of natural selection the qualities and shapes of animals change with time; old species disappear in response to changing conditions, and new ones arise. Few of the species of animals that roamed the face of the earth 100 million years ago still exist today, and few of those existing today will exist 100 million years hence. From *The Origin of Species:*

> *". . . Not one living species will transmit its unaltered likeness to a distant futurity."*

But natural selection works its effects slowly. Its influence is not felt in one individual or in his immediate descendants. Ten thousand generations may elapse before a change becomes noticeable; in man that amounts to a quarter of a million years.

Darwin knew that this concept of infinitesimally slow change was bewildering when first encountered. He wrote in the *Origin:*

> *"The mind cannot grasp the full meaning of the term of even a million years; it cannot add up and perceive the full effects of many slight variations accumulated during an almost infinite number of generations. . . . We see nothing of these slow changes in progress, until the hand of time has marked the lapse of ages, and then . . . we see only that the forms of life are now different from what they formerly were."*

A vast amount of time was needed for natural selection to produce the highest forms of life out of these simple beginnings. Yet, ever since Rutherford measured the age of the earth, we have known that enough time is available. Our planet has existed for billions of years; that is the secret strength of Darwin's theory.

The Evolution of Intelligence on the Earth

Certain highlights in the history of life—revealed by the fossil record—supply clues to the origin of intelligence on the earth. Innumerable skeletons and fossil remains mark the path by which life climbed upward from its crude beginnings. The first steps along the path are not known, for those early forms were very fragile and no trace of them remains. The earliest signs of life appearing in the record, already far advanced beyond the molecule lying on the threshold between life and nonlife, are the deposits of simple one-celled plants called algae, and the shells of rod-shaped organisms resembling bacteria. These are found in rocks solidified three billion years ago, when the earth was approximately one billion years old.

Little else appears in the fossil record during the first three billion years. One of the mysteries in the study of life is the fact that suddenly, in rocks 600 million years old, the record explodes in a profusion of living forms. A great variety of animals appears in the record at that time. Perhaps the forms of life were nearly as numerous and populous just prior to this ciritical date, but left no trace of their existence because they lacked the body armor that is most easily preserved.

Somewhat more than 400 million years ago an event occurred that is of great consequence for the development of man. There appeared, for the first time, a new kind of creature—one with an internal skeleton and a backbone. This animal—the vertebrate—evolved out of a wormlike ancestor resembling the modern lancelot, a small, translucent creature, lacking fins and jaws, but possessing gills and, most important, a primitive version of the backbone.

Among the descendants of the first vertebrates were the fishes. Some of the early fishes contained crude lungs for gulping air at the surface of the water, as well as gills. These lungs were lost or converted to other uses in most instances, but in some forms of fish, perhaps those living in small bodies of water such as ponds and tidepools, the lungs came into frequent use. Whenever a drought developed and the water level in the ponds dropped, the fish with the best lung capacity survived where others perished. They lived to produce progeny that inherited their superior capacity for breathing air. In this way, an efficient lung evolved gradually among the fish that inhabited shallow bodies of water.

Some of the air-breathing fish were doubly favored in possessing strong fins that enabled them to waddle over the land from one pond to another in search of water. By a slow accumulation of favorable mutations, the muscle and bone of the fin gradually changed into a form suitable for walking on land. In this way, the fin evolved into the leg. The metamorphosis took place over a period of perhaps

50 million years, and a like number of generations. The result was a four-legged, air-breathing animal known as the amphibian.

The amphibians were born in water, lived most of their adult lives near water, and almost always returned to the water to lay their eggs. For 50 million years they flourished on shores and river banks. Some became large, aggressive carnivores as much as 10 feet in length, fearing no other animals of their time. The amphibians attained the peak of their size 250 million years ago, and thereafter they declined. Today their common descendants are the diminutive frog, toad, and salamander.

In the course of time, some of the ancient amphibians, again by the chance occurrence of a succession of favorable mutations, developed the ability to lay their eggs on land. These eggs were encased in a firm, leathery shell, which retained moisture and provided the embryo with its own private pool of fluid. Other mutations led to a tough, leathery hide, which preserved the water in their bodies without the need for continual immersion. Such creatures were completely emancipated from the water. They were the first reptiles.

The reptiles marked a very successful step in evolution, for they had access to rich resources of food previously denied to the fishes and the amphibians. The reptiles flourished and developed into a great variety of forms, including the ancestors of every land animal with a backbone now on this earth. They reached their evolutionary zenith in the dinosaurs. These animals ruled the earth for 100 million years. They displayed an extraordinary vigor, evolving into such extreme forms as the giant vegetarian swamp-dweller, *Brontosaurus,* 70 feet long and weighing 30 tons; and the meat-eating *Tyrannosaurus rex,* 40 feet high, with a 4-foot skull filled with dagger-like teeth—unquestionably the fiercest land-living predator the world had ever seen.

Two hundred million years ago, somewhat before the appearance of the first dinosaurs, another branch of the reptile class veered off on an entirely different course. This particular group may have lived in places on the edge of the temperate zone where the weather was relatively severe. Through the action of natural selection on chance variations, the new branch of the reptiles acquired a set of traits that fitted them uniquely for survival in a rigorous climate. They developed the rudimentary characteristics of a warm-blooded animal. The naked scaly skin of the reptile was replaced in these animals by insulating coats of hair and fur that kept them warm in low temperatures, while sweat glands under the skin, controlled by an internal thermostat, cooled the body by evaporation when the temperature rose too high.

In spite of this and other advantages, the mammals remained subordinate to the dinosaurs for more than 100 million years—

small, furry animals, inconspicuous, keeping out of sight of the rapacious reptiles by living in the trees or in the grasses.

But some 80 million years ago the dinosaurs began to decline, and by 65 million years ago, they were all gone. The reasons for their disappearance are still obscure. According to one view, their downfall was the consequence of a worldwide change in climate, which they were ill-equipped to survive. Dinosaurs, like all reptiles, were cold-blooded animals; that is, they lacked the internal heat controls that could maintain the temperature of the body at a constant level regardless of the rigors of the climate. We know that the period during which they disappeared was marked by repeated upheavals of the earth's crust in which many new mountain ranges were formed. The Rocky Mountains were among the ranges created in these upheavals. Most probably, the upward thrust of huge masses of rock disrupted the flow of currents of air around the globe; perhaps the climate of the temperate zone was changed in this manner from one of uniform warmth and humidity, agreeable to a cold-blooded animal, to a climate marked by major changes of temperature from season to season.

As the population of dinosaurs dwindled, the mammals came down from the trees and up from their burrows in the ground, and they inherited the earth. Quickly they spread out across all the continents. Within 20 million years, the basic mammalian stock evolved into the forebears of most of the mammals with which we are familiar today—bats, elephants, horses, whales, and many others.

But one group of mammals remained in the trees. These mammals—the primates—were small, insect-eating animals, the size of a squirrel, and similar in appearance to the modern tree shrew of Borneo. Man owes his remarkable brain to the fact that these animals required two physical attributes for survival in their arboreal habitat: first, they needed hands and an opposable thumb for securing a tight hold on branches; and second, they needed sharp binocular vision to judge the distance to nearby branches. In the competition for survival among primitive tree-dwelling mammals, 100 million years ago, those who possessed these characteristics in the highest degree were favored. They were the individuals most likely to survive and to produce offspring. Through successive generations the desirable traits of a well-developed hand and keen vision, passed on from parents to offspring, were steadily refined and strengthened. Fifty million years ago they existed in advanced form in the animals from which the modern tree shrew, lemur, and tarsier are descended. They became even better developed in some of the immediate descendants of these animals, under the continued pressure of the struggle for survival in the trees. Gradually, the evolutionary trends established by the requirements of life in the trees transformed some of these early primates into animals resembling the monkey.

Animals with hands also had the potential capacity to exercise rudimentary manual skills; when this potential was combined with the development of the associated brain centers, such animals had, almost by accident, the ability to use tools. For those who had this ability, great value became attached to the mental capacity for the remembrance of the usage of tools in the past, and for the planning of their use in the future; thus, by the action of natural selection on a succession of chance mutations, those centers of the brain developed and expanded in which past experiences were stored and future actions were contemplated. These mental qualities proved to be of great value in meeting the general problems of survival. As a result, the brain evolved and expanded under the continued pressure of the struggle for existence. It doubled in size in 10 million years, and nearly doubled again in the next two million. These circumstances established the line of ascent leading to intelligent life.

Prospects for Intelligent Life

Throughout this long history, the most intelligent form of life present on earth in each era has been the rootstock out of which new and still more intelligent forms have evolved. The line of increasing intelligence stretches unbroken from the reptiles to the mammals, the primates, and then to man. Apparently, intelligence —which permits a flexible response to changing conditions—has exceptionally great survival value.

These circumstances suggest that intelligent life may evolve on planets in other solar systems, provided those planets and their stars have existed for a sufficiently long time. Judging from the history of life on the earth, several billion years is the time required. How many solar systems have existed for that length of time? Astronomy provides a tentative answer.

In round numbers, the age of the Universe is about 15 billion years; stars began to form when the Universe was about one billion years old; the first stars to form were made entirely of hydrogen and helium, and could not have earthlike planets or forms of life as we know it. By the time the Universe was several billion years old, appreciable amounts of the heavier elements—including the ingredients of earthlike planets and life—had been made in stars, dispersed to space in supernovas, and become available for the formation of new solar systems. Solar systems formed in that time or later will have the necessary ingredients for the evolution of life. However, they must not have been formed too recently, or there will not have been time for intelligence to evolve on their planets. Stars with ages in the neighborhood of 5 to 10 billion years would appear to offer the best prospects. This line of reasoning suggests that roughly a third of the stars in the Universe may be candidates for the evolution of intelligent life. Of the stars in our close neighborhood, several fall into this favorable category (Table 20.1).

How many of these stars actually bear life? Some scientists feel the conditions for life, and especially intelligent life, are very restrictive, involving a concatenation of many special circumstances. The product of this chain of probability is vanishingly small in their view, and the prospects for life on nearby stars are correspondingly remote. It seems unlikely that either theoretical reasoning or laboratory experiments will settle the question of life's probability. The only way to resolve the issue is to set up an antenna and start to listen.

TABLE 20.1
The Dozen Nearest Stars Outside Our Solar System, with an Estimate of the Chance of Life on Each

Name	Age	Chance for Life
A. ALPHA CENTAURI	Approx. the same as the sun (4.6 billion years)	Good.[a] This triple star is about as old as the sun, thus was formed when the Universe had large amounts of carbon, oxygen, and other elements essential for life.
B. BARNARD'S STAR	Old as the Universe (15 billion years)	Poor. Too old; no carbon, etc. available when this star formed.
C. WOLF 359	15 billion years	Same as above.
D. LAL 21185	15 billion years	Same as above.
E. SIRIUS	300 million years	Uncertain. It is a young star; any life would be primitive.
F. UV CETI	Uncertain	Poor. It is a flare star, emitting bursts of lethal, ionizing radiation.
G. ROSS 154	Younger than the sun (?)	Fair, if the star is not too young. The red color means this star may not emit sufficient light of suitable wavelengths for photosynthesis
H. ROSS 248	20 billion years	Poor. Too old. See note on Barnard's Star.
I. EPSILON ERIDANI	Approx. the same as the sun	Good chance for planets and good chance for life.
J. ROSS 128	Same as the sun	Good chance for planets; fair chance for life. See Ross 154.
K. L 789-6	20 billion years	Poor chance for life; too old.
L. 61 CYGNI	20 billion years	Same as above.

[a] Double or triple stars such as Alpha Centauri are often classified as poor for life because any planets around these stars would be thrown out of the system by the changing force of gravity. However, recent calculations show that if a planet is close to one of the stars (less than one-quarter of the distance between them), its orbit is stable.

COSMIC EVOLUTION

In Chapter 3 of this book we suggested that the study of astronomy can illuminate basic problems of interest to every person, regardless of the degree of his professional interest in science: What am I? How did I get here? What is my relationship to the rest of the Universe? These questions acquire a new meaning when viewed in the billion-year perspective of stellar lifetimes. The human life span seems so short in comparison with astronomical time scales that the student of these questions cannot help but feel a new humility as he contemplates the 4½ billion years that have elapsed in the history of our solar system and the 6 billion years that lie ahead before the sun becomes a red giant.

The evidence provided by astrophysics and cosmology suggests that we owe our physical existence to events that took place billions of years ago in stars that lived and died long before our solar system was formed. According to the results of astrophysical investigations, the atoms in our bodies—and the atoms that make up the body of the earth—are drawn from the interiors of countless stars that have ended their own lives in supernova explosions earlier in the history of the Galaxy. The substance of each of those stars in turn was condensed out of an interstellar medium that had been enriched by innumerable supernova explosions that occurred still earlier in the history of the Universe. Step by step, astronomy traces the origin of our material substances back through time and into the parent cloud of hydrogen. The philosophical implication in this union of astronomical, geological, and biological concepts is that man has appeared on the scene as the product of an unbroken sequence of events, extending over more than 10 billion years, in which the Universe expands and cools, stars are born and die, the sun and earth are formed, and life arises on the earth.

"We are brothers of the boulders, cousins of the clouds."
Harlow Shapley

Main Ideas

1. Theories of the origin of life; two items of evidence.
2. Basic molecules of living matter: amino acids and nucleotides.
3. Structure of proteins; two functions of proteins.
4. Structure and functions of DNA; how DNA controls the assembly of proteins; DNA and the mechanism of inheritance.
5. The Urey-Miller experiment; formation of the building blocks of living matter in the earth's primitive atmosphere.
6. A chain of reasoning suggesting that inhabited earthlike planets may be common in the Universe.

7. Circumstances that would influence the course of evolution on an inhabited earthlike planet; Darwin's theory of natural selection.
8. History of life on the earth; the evolution of intelligence.
9. Prospects for intelligent life in other solar systems; relation to the age of a star; chances for life on the stars closest to the sun.
10. Cosmic evolution.

Important Terms

amino acid
Darwin
DNA
enzyme
extraterrestrial life
molecule
natural selection
nucleic acid
nucleotide
protein
RNA

Questions

1. Describe the chain of events leading from the beginning of the Universe to the threshold of life on the earth. Include the major points of relevance in cosmology and stellar evolution. Limit your answer to not more than six critically important steps in the chain of events. Briefly describe the conditions that existed at the time of each critical event.
2. What roles do proteins play in living organisms? What is the structure of a protein?
3. What roles does DNA play in living organisms? What is its structure?
4. How have the building blocks of life been produced in the laboratory? What gases were used? Are these gases present in the earth's atmosphere today? Are they present in the atmosphere of any planet today? Why is it probable or possible that they were present in the earth's atmosphere when it was a young planet?
5. Give your own definition of a living organism. For each property that you consider essential to the property of "life" try to think of a nonliving object which shares this definition with living organisms. Is it possible to isolate one characteristic that is unique to living organisms? Explain your answer.
6. Briefly describe the scientific picture of the origin of life on the earth.
7. Do you think that there is life elsewhere in the Universe? Why?

8. Given your knowledge of stellar evolution and stellar populations, where in the Galaxy would you consider searching for life? Why?
9. Summarize in 5 or 10 sentences the reasoning that led Darwin to his theory of evolution by natural selection.
10. Briefly describe the major events that have taken place in the history of life on the earth during the sun's last two turns around the center of the Galaxy.
11. Describe how environmental pressures produced a number of critical evolutionary advances in the chain of events leading from the threshold of life to man. Pick one of these important advances, and invent a different set of circumstances in the environment that might have deflected terrestrial life into a different line of evolution. Apply the Darwinian reasoning on natural selection to predict the alternative path along which life would have evolved on the earth in these alternate circumstances. Is your alternative line of evolution likely to lead to intelligent life? Why?

METHODS OF MEASUREMENT

APPENDIX A

In numerous places in the text, references have been made to stellar distances and masses without an indication of the way in which these important properties are measured. The determination of the distances and masses of astronomical objects presents a formidable problem that has taxed the ingenuity of astronomers for centuries. The solutions are a triumph of observational astronomy, and provide the empirical foundation for the description of stellar evolution in Chapters 8 and 9.

MEASUREMENT OF ASTRONOMICAL DISTANCES

The determination of stellar and galactic distances depends on several distinct methods, beginning with the method of trigonometric parallax and the moving-cluster method for nearby stars, and extending outward to greater distances by a succession of other methods.

Each method is calibrated by applying it to objects whose distances have already been measured by a previous method. Ultimately, the entire system of distance determinations rests on a set of known distances to nearby stars, which have been measured with the aid of the trigonometric parallax and moving-cluster methods. This scheme of distance determination may be compared to an inverted pyramid, with the parallax method and the moving-cluster method at the bottom of the pyramid, bearing the weight of the entire structure.

The Method of Trigonometric Parallax

The trigonometric parallax method is often called simply the method of parallax. It consists in observing the position of a nearby star against the background formed by the distant stars in the sky, and then measuring the position again six months later, when the earth is on the other side of its orbit. Because the star being studied is close to the solar system, its position seems to shift against the background of the distant stars when the position of the earth changes (Figure A.1).

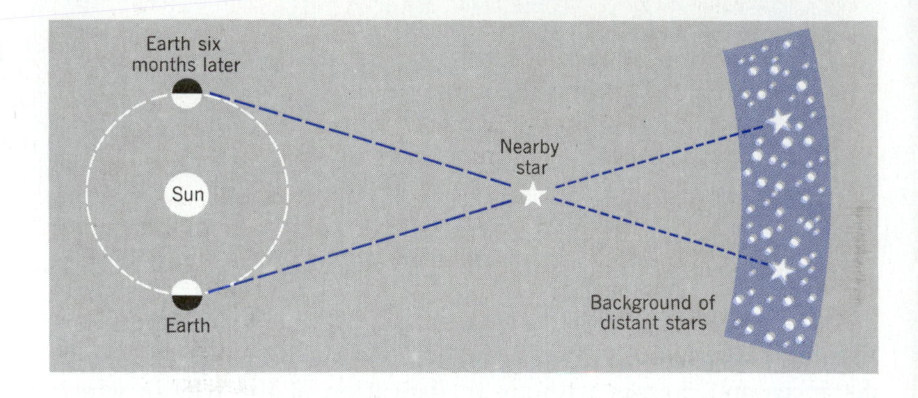

FIGURE A.1
Shift in the apparent position of a nearby star against the background of distant stars.

The observer can demonstrate the parallax effect by holding up a pencil one foot away and closing first one eye and then the other as he views the pencil against a background of more distant objects, such as the edge of the blackboard or a tree across the street. This is the parallax effect. If the sighting is repeated with the pencil held farther away from him at arm's length, the shift in apparent position is not as great as before. In other words, the amount of parallax depends on the distance to the pencil.

Figure A.2 illustrates the change in parallax when the pencil is held close (*a*) and at arm's length (*b*).

The shift in apparent position of a star, as observed from the earth on opposite sides of its orbit, is called *stellar parallax.* The stellar parallax is defined as one-half of the apparent shift in angular position observed when the earth moves across its orbit (Figure A.3). It is measured in seconds of arc.

The largest stellar parallax is 0.763″, determined for Proxima Centauri (the companion of Alpha Centauri), the closest star to the sun.

The best parallax measurements are uncertain by ±0.005″. Because of this uncertainty, the method is limited to a useful range of roughly 100 light-years. A star at that distance has a parallax of 0.03″, and a probable error of 15 percent or 15 light-years in the distance derived from its parallax.

Because the observation of parallax is the basis of all distance-measuring techniques, astronomers often use the term "parallax" as

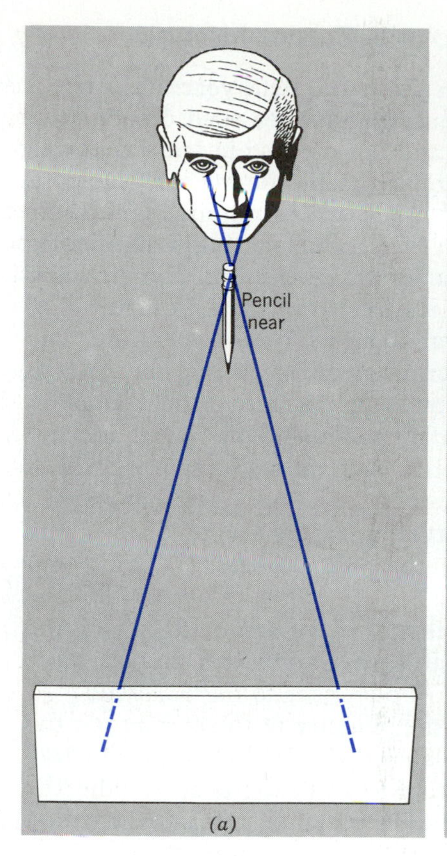

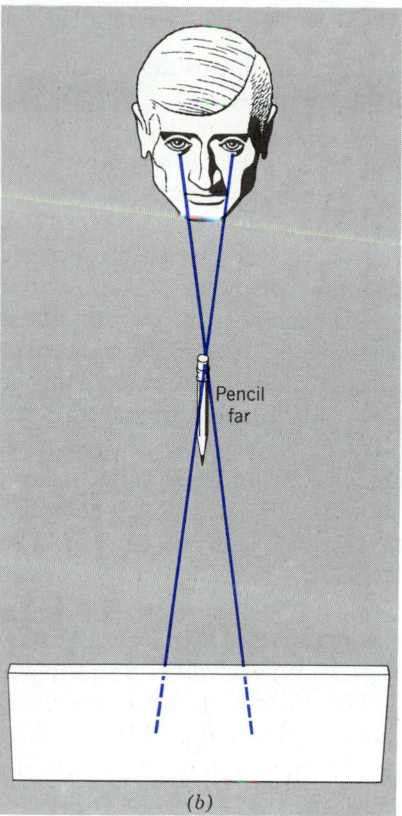

FIGURE A.2
Illustration of the parallax effect.

a synonym for distance to an astronomical object, regardless of how that distance has been measured.

Another unit of stellar distance, frequently used in place of the light-year, is derived from the parallax method. This unit, called the parsec (abbreviated pc) is the distance at which a star has a parallax of one second of arc. One parsec = 3.26 light-years. The distance of Proxima Centauri, for example, is 1.3 pc.

For very large distances, units of the kiloparsec (10^3 pc, written kpc) and the megaparsec (10^6 pc, written Mpc) are employed.

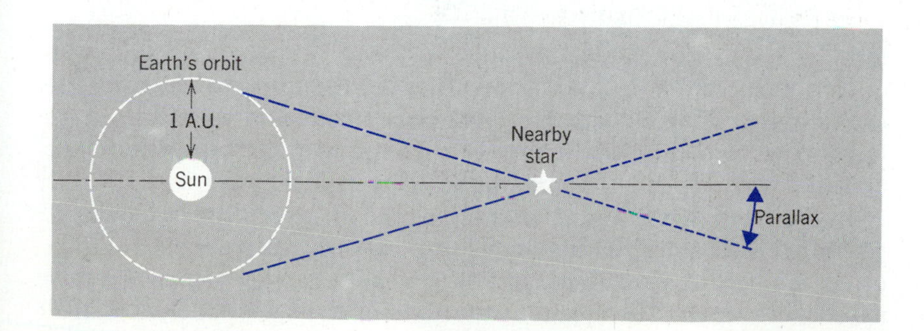

FIGURE A.3
Stellar parallax.

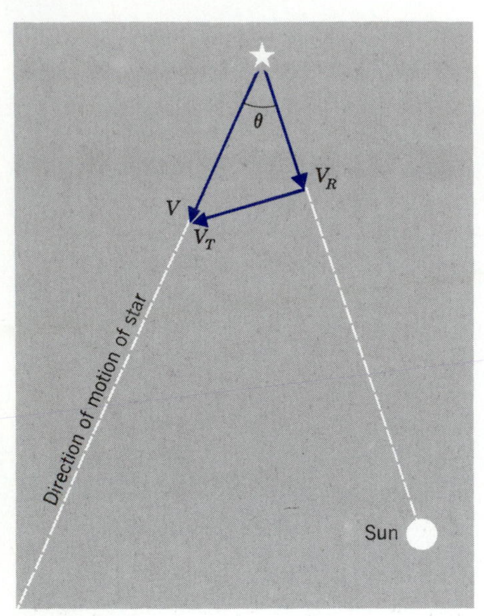

FIGURE A.4
The arrow labeled V is the velocity of the star relative to the sun. The arrows V_R and V_T are the components of v along (radial) and across (tangential to) the line of sight, respectively.

Beyond the Parallax: The Moving-Cluster Method

As noted, a star more distant than about 100 light-years has a trigonometric parallax too small to be accurately measured. However, if the star is a member of a cluster, another method of direct distance determination can be employed to determine the mean distance to the stars in the cluster. This technique, called the moving cluster method, is more complicated than the parallax method and involves a series of steps. The first step is to determine the direction in which the star is moving through space relative to the sun. In Figure A.4, the direction of motion of the star is shown by a dashed line. The method for determining the direction of the star's motion through space is explained below. The second step is to measure the Doppler shift in the spectrum of the star. From the Doppler shift and the formula on page 532, the speed with which the star is approaching or receding from the sun can be calculated. This means the speed along the line of sight to the observer. In Figure A.4, the speed along the line of sight is indicated by the arrow labeled V_R (for radial velocity).

It is important to note that V_R is not the velocity with which the star is traveling through space relative to the sun. It is only the *component* of that velocity *along the line of sight to the observer.*

The third step is to calculate the velocity of the star *across* the line of sight. In Figure A.4, this is labeled V_T (for tangential velocity). If θ is the angle between the line of sight to the observer and the direction of the star's motion, V_T is obtained from the formula (Figure A.4).

$$V_T = V_R \tan\theta$$

The fourth step is to measure the shift in the apparent position of the star against the background of the fixed stars during the course of an extended period of time. Usually a number of years is needed for an accurate measurement of the shift in a star's position. This measurement yields the *rate* at which the position of the star is changing. The rate of change of a star's position is called its *proper motion,* usually denoted by μ.

Finally, the distance of d to the star follows from the relation

$$d = \frac{V_T}{\mu}$$

The most difficult observational part of the procedure occurs in the first step. How do we determine the direction in which the star is moving through space relative to the sun? The answer can be clarified with the aid of an example.

If you have driven along a straight road at night with telephone poles on either side, you may have noticed how the pairs of poles on either side of the road ahead of you appear to move apart as the car approaches them. The apparent motions of the telephone poles are such that they appear to diverge from a point in the distance. This point represents the direction on the horizon toward which the car is traveling.

A second example is the flight of a flock of birds coming toward the observer. As they approach the observer and fly overhead, their flight paths also appear to diverge from a point in the distance, because of the same perspective effect. The line drawn from the observer to that point is the direction of travel of the flock.

To apply these ideas to the direction of motion of stars, consider a cluster of stars, such as the galactic cluster discussed in Chapter 11. All the stars in the cluster move with about the same speed relative to the sun, as a part of the general motion of the entire cluster through space. In this respect, they resemble the flock of birds. When the proper motions for the members of the cluster are plotted, they will be seen to converge toward or diverge from a single point, for the same reason that the flight paths of the birds appear to originate in a single point.

The chart (Figure A.5), showing the proper motions of stars in the Hyades Cluster, provides an example of the convergence to a point.

The proper motions converge in this example because the relative motion of the Hyades is away from the sun. If the relative motion is toward the sun, the proper motions diverge from a point in the sky.

Why is a cluster of stars necessary for this method? In principle, one star is sufficient if its direction of motion relative to the sun is known. In general, however, the direction is not known. The proper

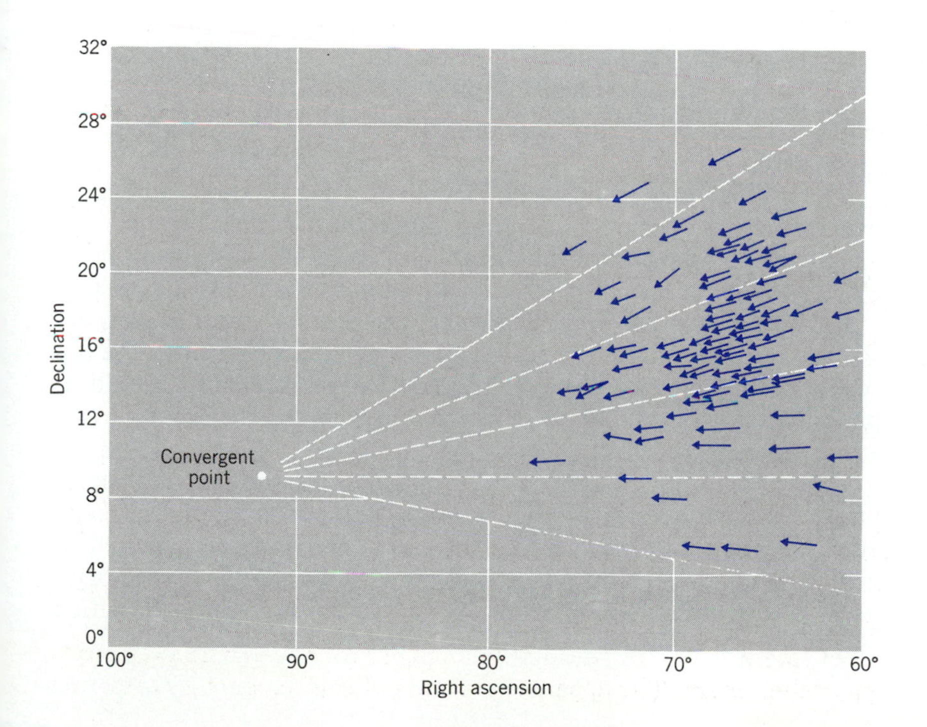

FIGURE A.5
The moving-cluster method applied to the Hyades Cluster.

motions for the entire cluster are needed to give the direction of relative motion, as explained.

The moving-cluster method breaks down at great distances because the proper motions become too small to be measured accurately in a reasonable interval of years. The effective limit on accurate distance determinations using this method is about 500 light-years.

Actually, the moving-cluster method is used primarily for two particular stellar groups—the Hyades cluster and the Scorpio-Centaurus association, at distances of 120 light-years and 500 light-years, respectively. These are the only groups of stars that are close enough to permit an accurate measurement of the proper motions of their members and, at the same time, contain a sufficiently large number of stars so that the directions of motion can be located with high precision.

The Method of Spectroscopic Parallax

The method of spectroscopic parallax depends on using the spectrum of a star to locate its place on the H-R diagram. When the star has been located on the H-R diagram, its absolute luminosity or magnitude can be read off the diagram. At the same time, the apparent magnitude of the star is measured by photometry. Knowing the star's absolute magnitude and apparent magnitude, the distance to it can be calculated from the formula on page 167. Solving for distance:

$$d \text{ (parsecs)} = 10^{\frac{m - M + 5}{5}}$$

where m and M are the apparent and absolute magnitudes, respectively.

The critical step in the spectroscopic parallax method is the location of the star on the H-R diagram. Once that is done, the absolute magnitude is known, and the distance follows. The key question is: How is a star's spectrum used to locate its place on the H-R diagram?

If the star is on the Main Sequence, the measurement of its spectral type or temperature is sufficient for this purpose, since the observed Main Sequence "line" (although really a band of finite width) provides a fairly well-defined relationship between spectral type and absolute magnitude.

If the star is not on the Main Sequence, the same remarks apply, but the precision of the method is not as great as for Main Sequence stars. Suppose, for example, that the star is known to be a giant. The giants also lie in a definite region of the H-R diagram—a broad band with luminosities ranging from $10^2 L_\odot$ to $10^4{}_\odot$. If a star is classified as a giant and its spectral type is also known, we have a rough knowledge of its position on the diagram and its absolute luminosity.

But how do we know whether or not a star is on the Main Sequence? The star's spectrum provides the answer. The absorption lines in the spectrum of a giant are much narrower than the same lines in the spectrum of a Main Sequence star, because giants are distended stars with tenuous atmospheres, in which collisions are relatively infrequent. Therefore, collisional broadening (see page 530) is much less pronounced in the spectral lines of giants than it is in Main Sequence spectra. In Figure B-7, which compares the spectra of a giant and a Main Sequence star of the same spectral type, the difference in the widths of the lines is clearly evident to the eye.

The classification of stars according to the widths of their spectral lines has been refined into a system of luminosity classes. All stars can be assigned to one of these classes on the basis of spectral line widths and related properties of their spectra. The luminosity class is denoted by a Roman numeral between I and VI. Stars in luminosity class I have the highest absolute luminosities, in the range from 10^4 to 10^6 solar luminosities. These are the supergiants. The most luminous supergiants, such as Deneb and Rigel ($L \sim 2 \times 10^5 L_\odot$), are called la, and the less luminous supergiants, like Antares ($L \sim 4 \times 10^4 L_\odot$), are designated lb. Bright giants like Epsilon Canis Majoris are placed in luminosity class II, which includes luminosities between 10^3 and 10^4. Normal giants such as Arcturus ($L \sim 100 L_\odot$) are in class III. Stars lying between the giants and the Main Sequence are placed in class IV, with luminosities ranging from $10 L_\odot$ to $10^2 L_\odot$. Main Sequence stars comprise luminosity class V. Stars less luminous than Main Sequence stars — the subdwarfs — sometimes are placed in a luminosity class VI. The white dwarfs — the least luminous stars in a given spectral class — are given a special classification wd or D.

Figure A.6 shows luminosity classes superimposed on an H-R diagram of the form used in Chapters 6 and 7. Normally a star is labeled by its spectral type and luminosity class in that order, for example, A2 I for Deneb or G2 V for the sun. This specification in terms of the spectral type and luminosity class is a two-dimensional system of classifying stellar spectra, known as the Morgan-Keenan or MK system. Classification of a star's spectrum in the MK system permits the position of the star to be located approximately on the H-R diagram, fixing its absolute luminosity, and, therefore, its absolute magnitude. When the apparent magnitude is measured, the distance follows from the formula on page 503. The distance to a star determined in this way is called the star's *spectroscopic parallax.*

Distances determined by the method of spectroscopic parallax are accurate to about 15 percent. This fractional error is roughly independent of the distance of the star. In the method of trigonometric

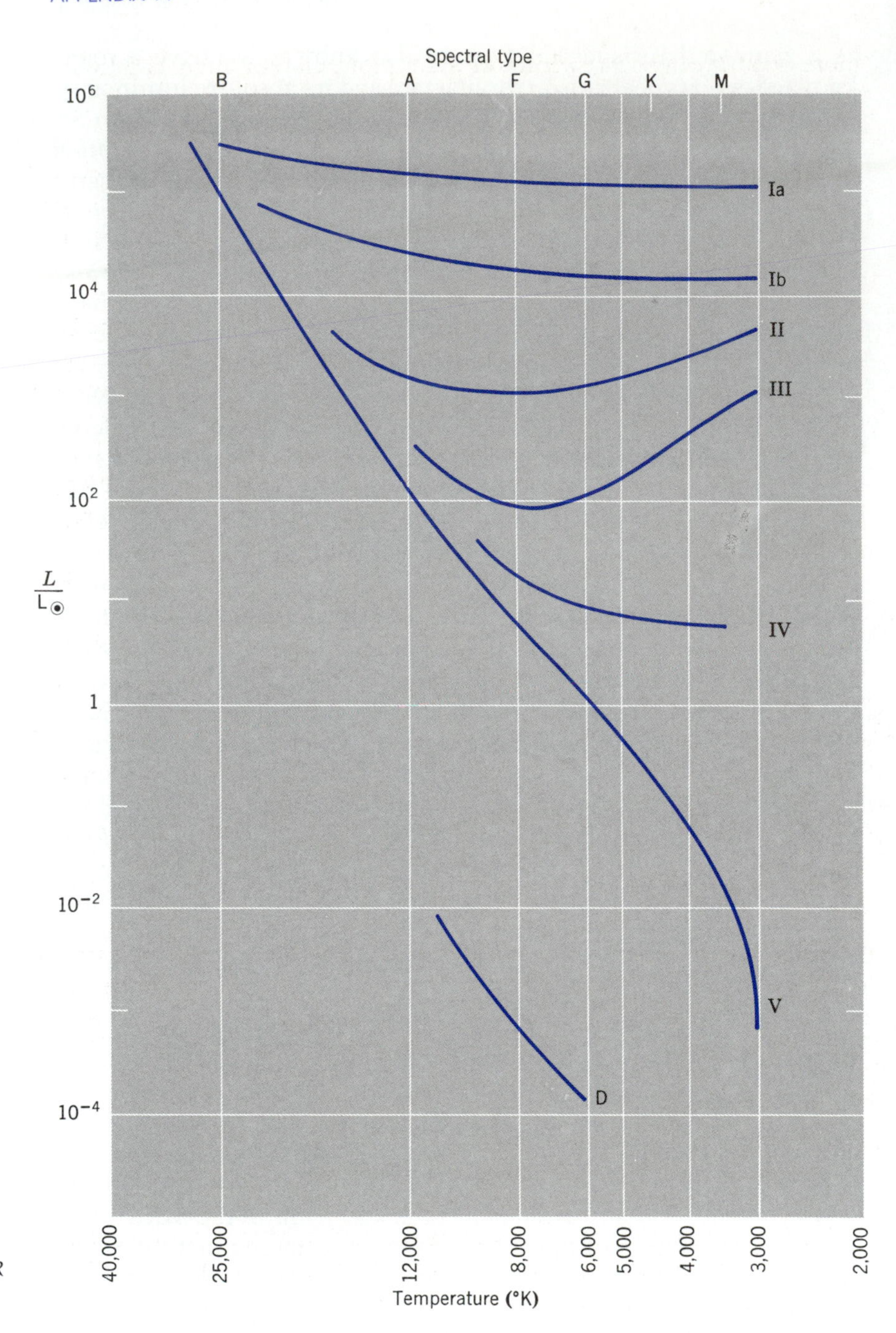

FIGURE A.6
Location of luminosity classes on the H-R diagram.

parallax, on the other hand, the error has a fixed absolute value, and, expressed as a fraction, is much smaller than 15 percent for nearby stars, but larger for distant stars. The two methods yield the same error for stars at a distance of about 100 light-years. If a star is closer than 100 light-years, the method of trigonometric parallax yields more accurate results, while for stars more distant than 100 light-years, the method of spectroscopic parallax is superior.

Main Sequence Fitting. In principle, the method of spectroscopic parallax is applicable to single stars. In practice, the method achieves its greatest accuracy when applied to a cluster of stars, rather than to a single star, because it is nearly always possible to identify the Main Sequence stars in a cluster unambiguously. As noted, Main Sequence stars are the most suitable candidates for accurate distance determinations with this method. The application of the method of spectroscopic parallax to clusters is known as Main Sequence fitting.

As an illustration of the use of the method of Main Sequence fitting consider first one of the relatively young clusters, such as the Praesaepe or Pleiades clusters, which are discussed in Chapter 11. In a young cluster, many of the stars are still on the Main Sequence; few have yet evolved off the Main Sequence into the red giant or red supergiant regions.

The first step in the method is to plot the directly observed H-R diagram for the stars in the cluster, giving their *apparent* magnitudes versus spectral types. The stars will be strung out along a line on the diagram, which is clearly the Main Sequence line.

The next step is to compare this H-R diagram, plotted in terms of apparent magnitude, with a calibrated H-R diagram plotted in terms of *absolute* magnitude. This diagram is made up by using stars whose distances and, therefore, absolute luminosities have been determined previously by the trigonometric parallax or the moving cluster method. This is the first instance of the procedure described in the introductory paragraph, whereby each successive method of distance measurement is based on a previous one.

Since all the stars in the cluster are at approximately the same distance, the difference between the two H-R diagrams, one in terms of absolute magnitudes and the other in terms of apparent magnitudes, will only be a shift in the vertical scale. This can be seen from the formula,

$$M = m + 5 - 5 \log d$$

in which M and m are the absolute and apparent magnitudes, respectively, and d is the distance in parsecs (Chapter 7).

The determination of the distance is equivalent to laying one diagram over the other and sliding it vertically until the two Main Sequences fit. The distance is calculated from the difference in the two magnitude scales, using the formula above.

An example of Main Sequence fitting is shown in Figure A.7.[1]

All the Main Sequence lines for the clusters in Figure A.7 are anchored to the Main Sequence line for the Hyades Cluster. The ab-

[1] Adapted from *Basic Astronomical Data,* edited by Strand, Volume III, p. 407.

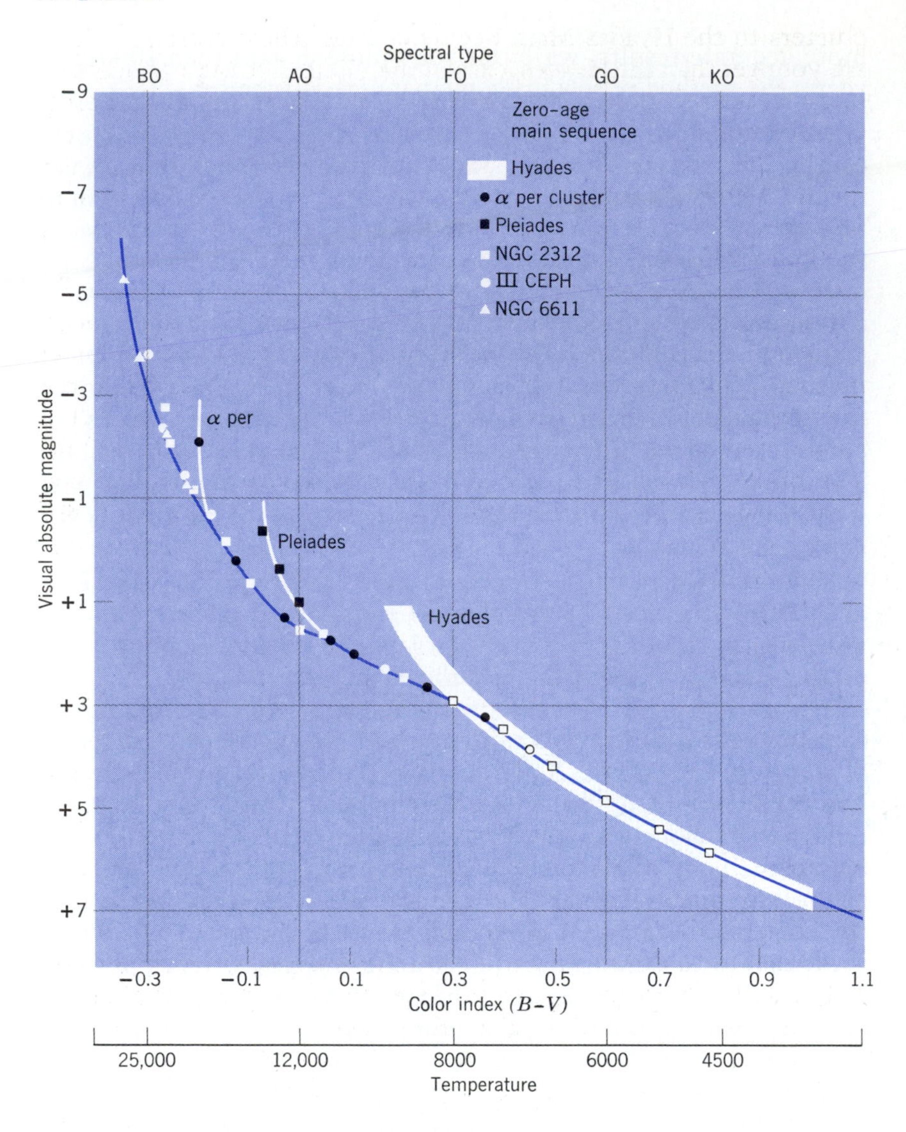

FIGURE A.7
Main-Sequence fitting.

solute magnitudes for the Hyades Cluster are known with great accuracy because this cluster happens to be particularly suitable for the application of the moving-cluster method of distance determination. Therefore it is used as the anchor line for the other clusters. The solid line in the diagram gives a calibrated "Zero-Age Main Sequence" line determined by overlapping the Main Sequence lines for the six clusters listed in the diagram.

If the Hyades' distances are known so well, why use the other clusters in addition? The reason is that while the Hyades Cluster calibrates the lower Main Sequence, it does not provide all the needed information for the upper Main Sequence, because its more massive stars have evolved away from that region. The upper Main Sequence line is established by fitting the five other overlapping clusters to the Hyades Main Sequence line. These clusters, which are younger than the Hyades, extend the "Zero-Age Main Sequence" line successively upward from the K, G, and F stars into the region of O and B stars.

The precision of the Main Sequence fitting method is typically ±15%. As in the case of the method of spectroscopic parallax, Main Sequence fitting is more accurate than the trigonometric parallax method for distances greater than 100 light-years.

It will be seen below that most of the methods for measuring extragalactic distances rest on the spectroscopic parallax or Main Sequence fitting methods. These, in turn, rest on the moving-cluster method applied to one particular cluster—the Hyades. Thus, the moving-cluster method carries a greater weight than any other single method in the scheme of cosmic distance determinations. The other independent methods—trigonometric parallax and statistical parallax (described below)—also play important roles as primary yardsticks, but neither is quite as important as the moving-cluster method.

Globular Clusters as Standard Candles

The methods discussed thus far are not usable for measuring distances to objects outside the Galaxy.[2] Globular clusters provide the first method of determining extragalactic distances.

Globular clusters are plentiful in our galaxy, approximately 150 having been observed thus far. They are also plentiful in some other galaxies. If we assume that the extragalactic globular clusters have the same absolute magnitudes as those in our galaxy, and also assume that the absolute magnitudes of the brightest globular clusters in our galaxy are known, the distances to these extragalactic clusters can be determined by comparing their apparent and absolute magnitudes.

The first step in translating this idea into practice is the determination of the distances to globular clusters in our galaxy. The method of Main Sequence fitting is one way of measuring these distances. However, a complication arises through the fact that globular clusters are composed of stars that were formed early in the history of the Galaxy, when relatively little conversion of hydrogen and helium to heavier elements had occurred. When the absolute magnitudes of these stars are plotted against spectral type, they form a Main Sequence that is somewhat different from the Main Sequence line of clusters of stars formed more recently, such as the Hyades Cluster. The stars formed early in the history of the Galaxy are called population II stars, and stars formed more recently are

[2] Except the Magellanic clouds, whose brightest stars fall within the range of the method of spectroscopic parallax.

described as belonging to population I (see Chapter 11). The calibrated Main Sequence line described in the previous section is based on population I stars, and therefore cannot be used for globular clusters. However, a separate Main Sequence line for population II stars can be constructed by observing the stars of this type that happen to be near the sun. These stars are identified as belonging to population II either by their large velocities or by direct spectroscopic evidence for a low abundance of heavier elements. Some of the population II stars are close enough to permit a determination of distances by trigonometric parallax and thus can be used to calibrate the Main Sequence line for population II stars. The calibration is not very accurate because only a few population II stars exist in the neighborhood of the sun.

When the calibrated Main Sequence line for population II stars has been constructed, distances and therefore absolute magnitudes can be determined for globular clusters in our galaxy. Assuming the same range of absolute magnitudes for globular clusters in other galaxies, the distance d to these galaxies can be calculated from the formula on page 503.

Globular clusters provide a method for measuring distances up to 30 million light-years. Beyond that distance they are too faint to be useful.

RR Lyrae Stars as Distance Indicators (The Method of Statistical Parallax)

One method for distance determination that is independent of Main Sequence fitting involves the use of the RR Lyrae variable stars as standard candles. As will be seen, the method based on the RR Lyrae stars also leads to a determination of the distances to the globular clusters in our galaxy and, thus, to an independent calibration of these globular clusters as standard candles for determining distances to other galaxies.

The RR Lyrae method depends on a comparison between the true velocity of a star through space, called its peculiar motion, and its rate of change of angular position, or proper motion. It is clear that the peculiar motion, proper motion, and distance of a star are related. If two stars are moving through space with the same speed, but one star is much farther away than the other, the more distant star will change its apparent position more slowly than the nearer one; that is, its proper motion will be less. If both the peculiar motion and the proper motion of a star can be measured, its distance can be determined through this relationship, and its absolute magnitude can be deduced again from the formula on page 507. The determination of the peculiar motion involves (1) measuring the radial velocity by the star's Doppler shift and (2) assuming that in a statistical average over space velocities, the radial component is one-third of the total space velocity. Because of this dependence on a statistical average over space velocities, the method is called the statistical parallax.

When the absolute magnitudes of a number of RR Lyrae stars are measured in this way, they turn out to be close to 0.6. Apparently, all the members of this particular class of variables have approximately the same absolute magnitude. This fact permits the RR Lyrae stars to be used as standard candles.

The RR Lyrae stars are not bright enough to be visible in external galaxies other than the Magellanic clouds and a few small galaxies in the Local Group (page 80) and therefore cannot be used as standard candles for measuring extragalactic distances in general. However, they can be used extensively within the Milky Way Galaxy. In particular, since RR Lyrae stars exist in the globular clusters in the Milky Way Galaxy, they can be used to determine the distances to the globular clusters, and therefore the absolute magnitudes of these clusters. As discussed above, a knowledge of the absolute magnitude of the globular clusters provides an important means for measuring extragalactic distances. Thus, the RR Lyrae stars help indirectly to expand the range of distance measurements beyond the boundary of the Galaxy.

Cepheid Variables as Distance Indicators

A relationship between periods of pulsation and absolute luminosity is one of the characteristic properties of the Cepheid Variables. This relationship permits the absolute magnitude of a Cepheid Variable, and therefore its distance, to be determined from its observed pulsation period.

Since the Cepheids are considerably brighter than RR Lyrae stars, with absolute luminosities exceeding those of the RR Lyrae stars by more than a factor of a thousand in the case of the brightest Cepheids, they are visible in many external galaxies. Thus, the Cepheids, unlike the RR Lyrae stars, provide a direct method for crossing the void between the Milky Way Galaxy and neighboring galaxies, in the scheme of distance measurements.

Two factors complicate the use of Cepheids as distance indicators. First, two distinct types of Cepheids exist with a separate period-luminosity curve for each. One type is called the Classical Cepheids because they resemble the prototype Cepheid, Delta Cephei. The classical Cepheids are mostly yellow supergiants of population I and are relatively rare. The other type, called the Type-Two Cepheids, are found in globular clusters in the galactic halo and in the center of the Galaxy, and are also relatively rare. The Type-Two Cepheids are population II stars.

The separate period-luminosity lines for the two groups of Ce-

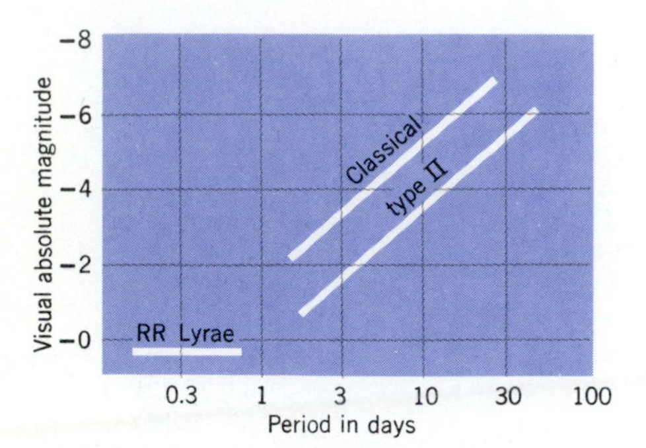

FIGURE A.8
Period-luminosity law for Cepheid and RR Lyrae variables.

pheids are shown in Figure A.8. Either type of Cepheid can be used for extragalactic distance determinations; but it is important to know which type is being used in each case.

When the Cepheids were first used for distance determinations, astronomers were not aware of the existence of two separate period-luminosity curves and misapplied the Type-Two curve to Classical Cepheids in their basic calibration of extragalactic distances. The error, which amounted to 1.5 on the absolute magnitude scale, was discovered by Baade in 1952. The resultant correction doubled the size of the Universe.

The second complication is a dependence of the period of pulsation on the spectral type, or temperature, of the star. As a result of this dependence, the line representing the period-luminosity relationship on the graph should be broadened to a finite band, with a vertical extent of about one magnitude.

Distance determinations based on Cepheids have both advantages and disadvantages, relative to the use of globular clusters as standard candles, for extragalactic distance measurements. An important advantage is their greater accuracy; a disadvantage is the lower luminosity of the Cepheids, which limits the range of the method to smaller distances than the globular-cluster method. Whereas the globular-cluster method extends to about 30 million light-years, the range of the Cepheids for this purpose is only about 10 million light-years.

Bright Blue Stars as Standard Candles[3]

A third technique for estimating extragalactic distances depends on the assumption that the brightest, hottest blue stars in any one galaxy such as our own have the same luminosities as the brightest stars in any other galaxy.

The brightest blue stars are the blue supergiants. The assumption of a known luminosity for all blue supergiants is a reasonable one, since these are believed to be the most massive stars that can be formed in a galaxy, and the upper limit of masses in star formation is probably nearly the same in all galaxies.

If the distances and, therefore, the absolute magnitudes can be determined for a number of blue supergiants in galaxies of known distance, the blue supergiants can serve as standard candles for determination of distances to other galaxies.

Since the absolute luminosities of the blue supergiants range up to $10^6 L_\odot$, or approximately the same as the luminosities of the brightest globular clusters, the range of effectiveness in distance determinations is about the same for the two methods.

The absolute magnitudes of the blue supergiants are determined

[3] Bright red stars are now also being used by Sandage as distance indicators.

by applying the Main Sequence fitting method to a few relative young clusters in our galaxy. The clusters must be young because the blue supergiants are massive stars that only live about 3 million years, and will have completely evolved in any cluster whose age is greater than this.

The Size of H II Regions as a Distance Indicator

Bright blue stars, especially of O type, are intense sources of ultraviolet radiation. The radiation ionizes nearby interstellar hydrogen, producing a sphere of ionized matter around the star with a radius of 150 light-years or more in the case of the hottest O stars. The sphere of ionized hydrogen is known as an H II Region.

Some of the ionized hydrogen atoms in the H II Region recapture electrons, and the electrons cascade down to the ground state in several steps, emitting a part of their recombination energy in the form of photons with energies in the visible part of the spectrum. This radiation creates a luminous sphere around the star, which makes the H II Region detectable.

The sizes of H II Regions in our galaxy have been measured by determining the distances to O-type stars in clusters, using the method of spectroscopic parallax. H II Regions are seen in some other galaxies also, and their angular dimensions can be measured. Assuming that the correlation between the size of an H II Region and the hot star at its center is the same for H II Regions in all galaxies, the measurement of the angular diameter of an H II Region in another galaxy permits the distance to this galaxy to be calculated. If Θ is the angular diameter, measured in radians, of an extragalactic H II Region and L is the linear diameter of a standard H II Region in galaxy of known distance, the distance to the external galaxy is $d = L/\Theta$. This method can be used out to roughly 50 million light-years.

Supernovas as Standard Candles

None of the methods described thus far is useful for distances greater than 50 million light-years. In general, the methods are limited by the luminosities of the astronomical objects involved. All these objects have been stars or clusters with luminosities no greater than roughly $10^6 L_\odot$. To extend the range of distance measurements beyond 50 million light-years, it is necessary to find considerably brighter objects that can be calibrated and used as standard candles.

Only two objects are known that have the necessary brightness. These are supernovas and entire galaxies. The calibration of galaxies as standard candles is discussed below. The use of supernovas for this purpose depends on the fact that the peak luminosity of a supernova is about equal to that of a bright galaxy. If we assume that the supernovas in our galaxy have the same peak brightness as supernovas in other galaxies, an observation of the peak brightness of a supernova in another galaxy immediately supplies a measure of the distance to the galaxy.

Galaxies as Standard Candles

Following the familiar reasoning, this method assumes that the brightest galaxies in our neighborhood are typical of very bright galaxies anywhere in the Universe. That is, a maximum brightness is assumed for galaxies, just as a maximum brightness was assumed for stars. This assumption is supported by observational evidence. Measurements of the absolute luminosities of galaxies, in the various clusters of galaxies around us for which distances have been determined, show that within each cluster of galaxies there is approximately the same cutoff in maximum luminosity. The most luminous galaxy in each cluster is almost invariably a giant elliptical galaxy (page 246) and luminosities of the largest giant elliptical galaxiex are roughly the same (within a factor of two o so) from one cluster to another. These luminosities of the largest giant elliptical galaxies are roughly 10^{45} ergs/second, which is about 10 times higher than the luminosity of the Milky Way Galaxy and other large spirals.

The observation of a uniform maximum luminosity for giant elliptical galaxies makes these objects useful standard candles for distance determination. The method is effective out to the range of visibility of the giant elliptical galaxies, which is two billion light-years. A major cause of uncertainty is the possibility that the evolution of the distant ellipticals has produced large changes in their intrinsic luminosities over the course of the one or two billion years required for their light to reach us. This effect would invalidate the calibration of the distant galaxies in terms of the close ones.

The Red Shift

How can distances greater than two billion light-years be determined? The only method available at the present time is based on the red shifts of galaxies. As discussed in Chapter 13, distant galaxies show a shift to the red in their spectra, which has been found to be proportional to distance for all galaxies whose distances have been determined with reasonable precision. The limit on such distance determinations for galaxies is about three billion light-years. If it is assumed that the same relationship between red shift and distance holds for distances greater than three billion light-years, a measurement of the red shift for a galaxy may be used to determine its distance.

Quasars (See Chapter 12) are the only visible objects detected thus far—with a few exceptions such as 3C295—that have red shifts suggesting a distance in excess of three billion light-years. The red shifts measured for these peculiar objects have been interpreted as an indication that the radius of the observable universe is roughly 20 billion light-years.

However, the use of the red shift in determining quasar distances rests on two assumptions. The first is the validity of the Hubble Law beyond three billion light-years. The second is that quasars are similar to galaxies.

The Hierarchy of Astronomical Yardsticks

The relationships among the methods of distance determination are shown schematically in Figure A.9. Spectroscopic parallax and Main

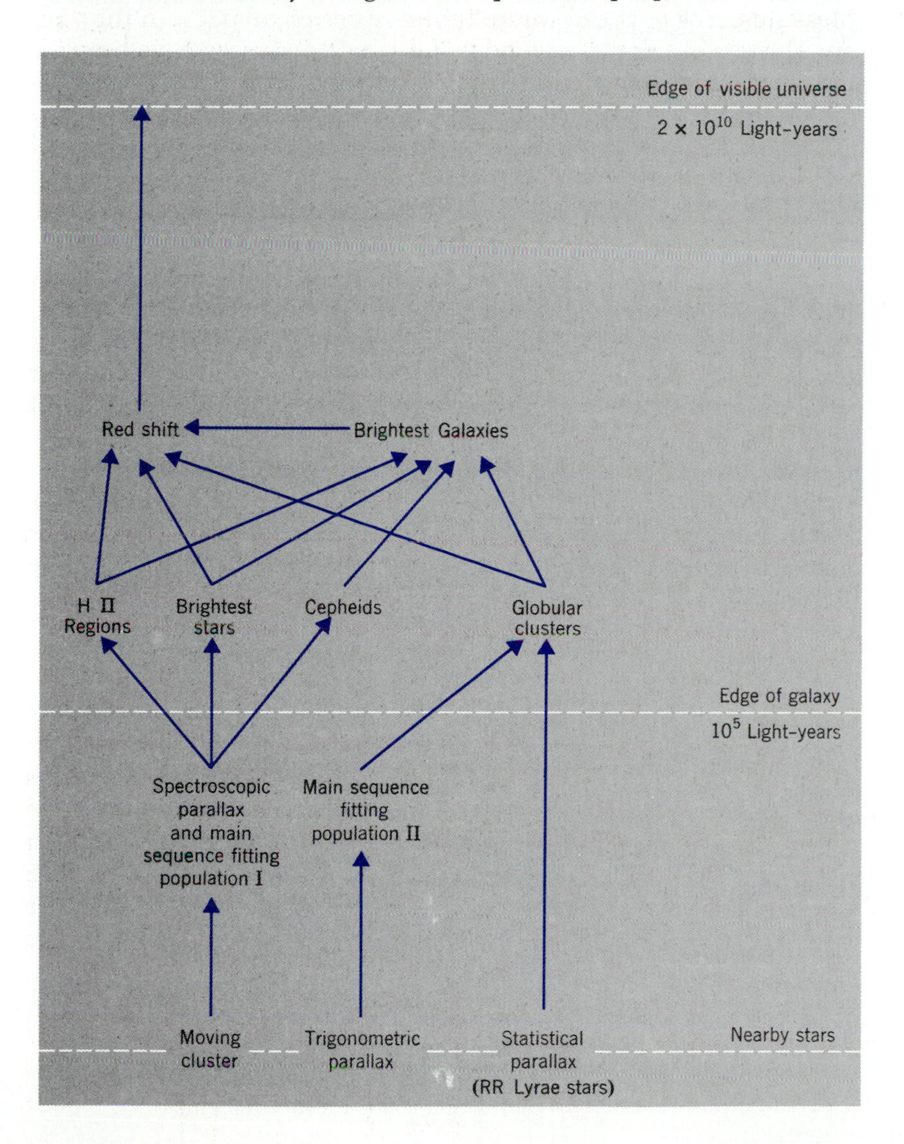

FIGURE A.9
The hierarchy of astronomical yardsticks.

Sequence fitting, both resting on the moving-cluster method, provide the most important base for measuring distances beyond our galaxy. These methods involve population I stars, which are the most common stars in galaxies like ours. Trigonometric parallax and the RR Lyrae stars provide useful independent checks on the measurement of extragalactic distances because they permit Main Sequence fitting to be carried out for the population II stars that make up the globular clusters. In this way, these latter two methods lead to the calibration of the globular clusters in the Galaxy as standard candles. All the methods for distance determination outside the Galaxy, extending from neighboring galaxies to the boundary of the observable Universe, depend on the three fundamental methods – the moving-cluster method, trigonometric parallax and the RR Lyrae stars – used for relatively close stars. Thus, the observed dimensions of the Universe are determined by measurements of the distances to a handful of stars in the neighborhood of the sun.

MEASUREMENT OF APPARENT MAGNITUDES[4]

In the discussion of methods for determining distance or absolute magnitude, we assumed a knowledge of the apparent magnitude. Measuring the apparent magnitudes that provide the basis for this discussion would seem to present no difficulties. One need only observe the brightness of a star with the eye, or record it on a photographic plate, to secure a value for m. However, the measurement of m actually involves serious problems, because this quantity means the energy reaching earth *summed over all wavelengths.* The eye cannot make this measurement because it is sensitive only to a limited band of wavelengths. Also, the earth's atmosphere does not transmit fully in the infrared and the ultraviolet. In the case of very cool or very hot stars, radiating most of their energy in the infrared or ultraviolet, the atmospheric effect is large. Thus, the *visual apparent magnitude,* that is, the apparent magnitude as perceived by the eye, may be very different from the true apparent magnitude.

If a photographic plate is used to record the apparent magnitudes in place of the eye, the situation is improved for hot stars because photographic emulsions are sensitive to ultraviolet radiation. However, plates are less sensitive to red and infrared light. Also, the photographic measurement of apparent magnitudes is equally affected by absorption in the earth's atmosphere.

It is not possible to obtain the missing information about the energy radiated from stars in the ultraviolet and infrared regions unless detectors sensitive to all wavelengths are sent up above the atmo-

[4] Adapted from *A Survey of the Universe,* by D. H. Menzel, F. Whipple and G. deVaucouleurs, pages 438–441. By permission of Prentice-Hall, Inc.

sphere in rockets or satellites. Measurements of this kind have been carried out, but they are still too few in number and too recent to have had a major impact on astronomy, although they will undoubtedly make enormous contributions to future work.

As far as ground-based astronomy is concerned, no purely observational solution exists to the problem of determining apparent magnitudes, and it is necessary to rely on theoretical estimates of the energy radiated from stars in the missing bands of wavelengths. These estimates are described and tabulated below under the heading of Bolometric Corrections.

Visual and Photographic Magnitudes

Before the advent of photography, the eye was the only practical instrument for measuring the brightness of stars. But we know that the brightness of an object depends on the color, or color range, we use to observe it. A red-hot poker, for example, may appear bright to the eye and extremely bright to an infrared detector. However, it is a very weak emitter in the ultraviolet. The human eye is most sensitive to yellow-green light, and thus is best suited for observation in this range of radiation. We employ the term *visual magnitude* to denote the brightness of a star as estimated by the human eye in the yellow-green, or the region around $\lambda = 5500$ Å.

The early photographic emulsions, although sensitive to violet and ultraviolet light, did not respond to radiation in the visible region with wavelengths greater than about 5000 Å. Some of the sensitivity in the ultraviolet was wasted, because most glass lenses absorb light of wavelengths shorter than about 3700 Å, and silvered mirrors lose reflectivity at about 3400 Å. Hence, the early photographic plates recorded the brightness of the stars in the blue-violet region, and the adopted term, *photographic magnitude,* defined a system of magnitudes for wavelengths between roughly 4000 Å and 5000 Å.

Photographic and visual magnitudes differ by an amount that depends on the color of the star. Astronomers have arbitrarily set the zero points of the two magnitude scales to be equal for white stars of spectral class A, such as Sirius. Since the eye is relatively more sensitive to red light than the photographic emulsion, a red star, which radiates relatively more energy in the red than in the blue-violet, appears brighter visually than it does photographically. Stars bluer than Sirius, on the other hand, are brighter photographically than visually.

Photovisual Magnitudes. Ordinary photographic plates are always sensitive in the blue and ultraviolet regions. They can also be made sensitive to other wavelengths by treatment with dyes that absorb certain colors. Thus, it is now possible to take photographs in yellow and red light, and even in the infrared beyond the range of the eye. If an "orthochromatic" emulsion, sensitive to green and yellow light (as well as blue, violet, and ultraviolet), is combined with a yellow filter (that blocks the violet and ultraviolet) the combination closely simulates the color sensitivity of the eye. Magnitudes measured with this arrangement are known as *photovisual* magnitudes.

Color Index

The *color index* of a star is defined as the photographic magnitude minus the visual or photovisual magnitude. The color index is a quantitative measure of the color of a star. For greater accuracy, the photographic magnitude is replaced by measurement with a photometer and filter constructed to transmit radiation with wavelengths in the blue region between 3800 and 4800 angstroms. The intensity in this spectral band is known as the blue magnitude, B, as well as the photographic magnitude. The visual magnitude, V, is precisely defined as the intensity of radiation in the spectral band between 4900 and 5900 angstroms. The color index is the quantity, $B - V$. Because the magnitude scale is such that smaller numbers denote brighter stars, the color index is *positive* for all stars redder than Sirius and *negative* for stars bluer than Sirius. The blue supergiant Rigel has a visual magnitude 0.14, a photographic magnitude -0.03 and, therefore, a color index of -0.17; the red supergiant Betelgeuse has visual magnitude 0.70, photographic magnitude 2.14, and color index $+1.44$.

A star's color index depends on the temperature of its surface. The index provides a useful measure of temperature as long as passage through interstellar dust has not reddened the starlight.

The photographic magnitudes of large numbers of stars can be measured on one photograph. In practice, the stellar images are compared with those of a few stars on the same photograph whose magnitudes have been carefully measured beforehand. The comparison can be made most simply through visual estimates of the relative sizes of the star images (the brighter the star, the larger the blackened area on the plate); with a little training a good observer can easily intercompare stars to within 0.1 mag., or about 10 percent in relative luminosity.

Precise measurements require a more objective device than the eye, such as a photoelectric photometer, to measure the blackness or density of the photographic images. Here again the magnitudes must be referred to a series of stars of known magnitudes on the photographs, which have been carefully determined beforehand. Photoelectric photometers today provide the most precise measurements of individual stellar magnitudes, but the photographic plate is still unsurpassed for the rapid, efficient measurement of the magnitudes of large numbers of stars.

The U, B, V System. Photographic magnitudes are accurate to within a few hundredths of magnitude at best, but photoelectric photometry is capable, in principle, of a precision of about a thousandth of a magnitude. The most commonly used photometric magnitudes have been established in three colors: *near-ultraviolet (U); blue (B); and yellow or visible (V).* This *U, B, V* system of standard magnitudes, as it is called, and the relations between the color indices $B-V$ and $U-B$, have provided precise apparent magnitudes of stars and valuable data on stellar evolution. Approximate shapes of the *U, B,* and *V* bands are shown in Figure A.10.

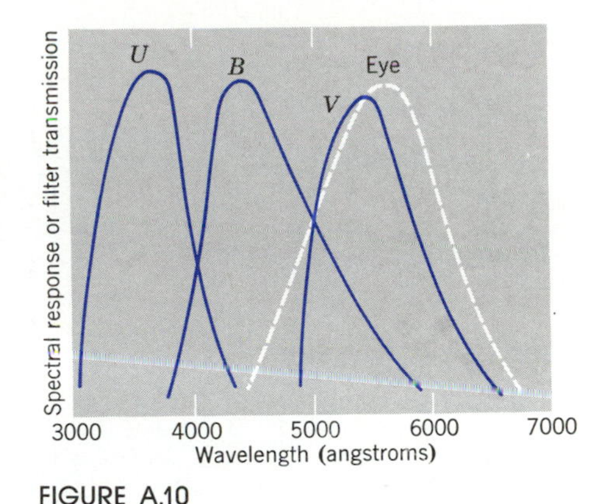

FIGURE A.10
U, B, and V bands compared with the response of the eye (from Clayton, Principles of Stellar Evolution and Nucleosynthesis, *McGraw-Hill, 1968.*

Relation Between Color Index and Spectral Type or Temperature. Table A.1 on page 520 shows the relation between color index and spectral type for stars of various spectral and luminosity classes. It is not surprising that such a relationship exists, since the spectral type of a star depends on its temperature, and the temperature has a strong influence on the star's color.

However, conditions other than surface temperature can also affect the color. If, for example, the starlight has passed through a region of space that contains considerable interstellar dust, scattering of the stellar photons by the dust may reduce their energy and redden the light appreciably.

Recently, observations of the colors of stars have been greatly refined by the precise techniques of photoelectric photometry. If, instead of determining a simple color index by comparing the star's light at two wavelengths only, we use the three-color *U, B, V* system or similar systems, the effect of interstellar reddening often can be detected.

Bolometric Magnitude

Some detectors of electromagnetic radiation respond to all wavelengths. These detectors measure the total rate of energy radiated by a star, that is, its absolute luminosity. Expressed in magnitudes, the total rate of energy radiated by a star at all wavelengths, including wavelengths shorter and longer than those in the visible region, is called the *bolometric magnitude.* Instruments that detect radiation over a large range of wavelengths are frequently called bolometers, from the Greek word *bole,* meaning ray.

The bolometric magnitude is substantially different from the visual absolute magnitude for stars that radiate most of their energy in the blue or red, outside the visible band of wavelengths. However, it is close to the visual magnitude for stars like the sun, whose peak radiation is in the middle of the visible region. The difference between the visual magnitude and the bolometric magnitude is called the *bolometric correction* (B.C.). It is large and negative for stars distinctly more blue or red than the sun, and negligible for stars with temperatures in the neighborhood of the sun's temperature (Table A.1).

A reference to the bolometric magnitude of a star implies that the observations of the radiation from the star have been corrected for the part of the energy curve cut out by absorption in the earth's atmosphere. Most of the radiation of hot stars is in the ultraviolet part of the spectrum, which the earth's atmosphere does not transmit. For these stars we cannot measure the bolometric magnitude directly, except from space vehicles.

The earth's atmosphere introduces many inaccuracies into the measurement of stellar magnitudes by any method. Haze, dust, clouds, molecular absorption, and scattering in the atmosphere all contribute their share of the uncertainty. An observatory, to be useful for photometric work, must possess both a uniform atmosphere and high atmospheric transparency. An apparently clear sky can sometimes be almost useless photometrically because of variable haze too faint to be detected visually. The best photometric sites are generally on top of mountains, in arid, dry regions, as in the southwest of the United States or in the Andes.

TABLE A.1
The visual and bolometric magnitudes, bolometric correction (B.C.), and color index (B-V) for Main Sequence stars, giants, supergiants and for a range of spectral types.

	T_E	Spectral Type	M_{VIS}	M_{BOL}	B.C.	B − V
Main Sequence	35000	05	−6	−10.6	−4.6	−0.45
	21000	BO	−3.7	− 6.7	−3.0	−0.31
	9700	AO	+0.7	0.0	−0.68	0.00
	7200	FO	+2.8	+ 2.7	−0.10	+0.30
	6000	GO	+4.6	+ 4.6	−0.03	+0.57
	4700	KO	+6.0	+ 5.8	−0.20	+0.84
	3300	MO	+8.9	+ 7.6	−1.20	+1.39
Giants, Luminosity Class, III	5400	GO	+1.8	+ 1.7	−0.1	+0.65
	4100	KO	+0.8	+ 0.2	−0.6	+1.06
	2900	MO	−0.3	− 2.0	−1.7	+1.65
Supergiants, Luminosity Class, I		BO	−6.4	− 9.4	−3	− .21
		AO	−6.0	− 6.7	−0.7	0.00
	6400	FO	−5.6	− 5.8	−0.2	+0.30
	5400	GO	−4.4	− 4.7	−0.3	+0.76
	4000	KO	−4.4	− 5.4	−1.0	+1.42
	2800	MO	−4.4	− 6.9	−2.5	+1.94

METHODS FOR MASS DETERMINATION

Five methods are widely used for the determination of stellar masses, in addition to the binary-star method (Chapter 7). They vary in importance, depending on the number of stars whose masses can be measured with a given method, and the accuracy of the measurement.

Empirical Mass-Luminosity Relationship. Mass and luminosity have been measured independently for many stars extending over a broad range of masses. Figure A.11 shows measurements obtained from these stars—visual and spectroscopic binaries—and a dashed line representing an approximate fit to the data. The scatter in the data reflects the observational uncertainties in the mass and luminosity values, and also the effect of variations in chemical composition from star to star. If a star's distance and, therefore, absolute luminosity have been determined, its mass can be read off this curve.

Theoretical Mass-Luminosity Relationship. Computations on stellar structure provide a theoretical relationship between mass and luminosity for stars on the Main Sequence, which can be used to determine the mass of any star whose absolute luminosity is known.

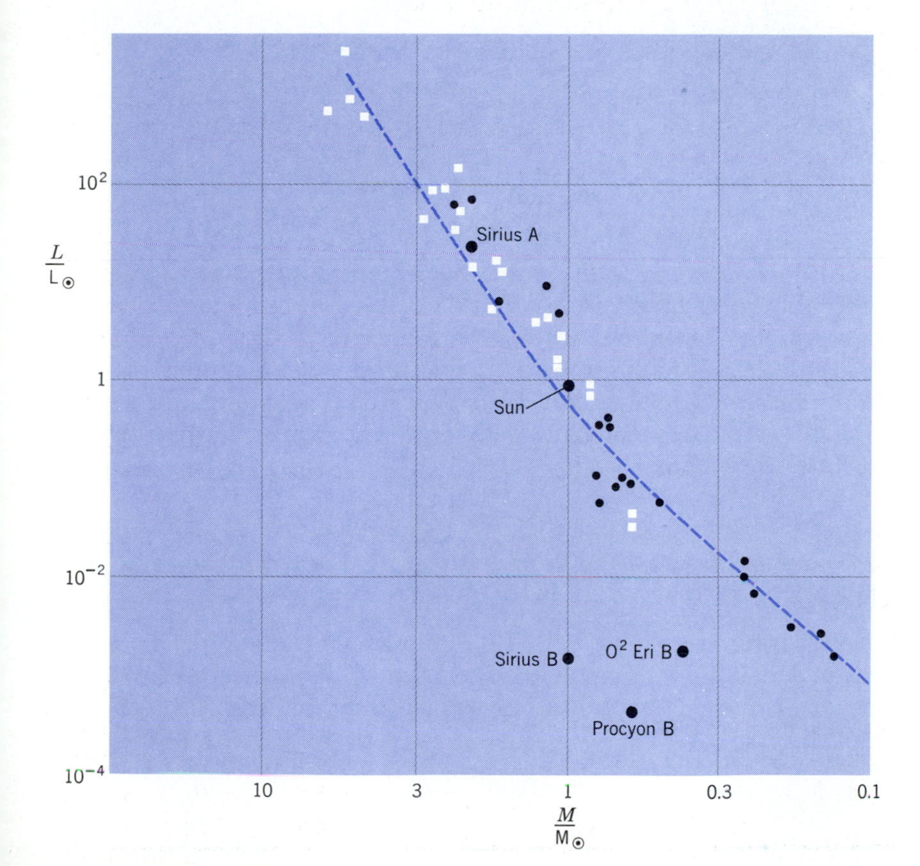

FIGURE A.11
Empirical mass-luminosity curve. Visual and spectroscopic binaries are indicated by circles and squares, respectively.

The theoretical relationship is shown in Figure A.12. The width of shaded area represents the effect of compositional differences in the population of stars with respect to helium and metals abundance.

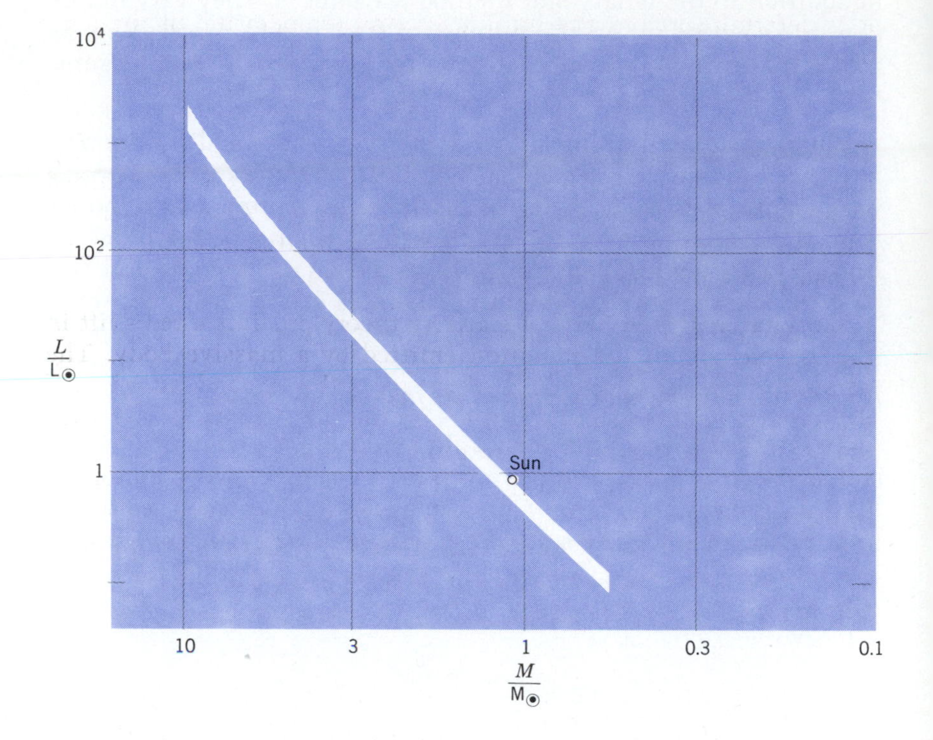

FIGURE A.12
Theoretical mass-luminosity curve.

Stellar Spectra. The analysis of the line intensities in a stellar spectrum yields the star's temperature, and an analysis of the spectral line widths yields the surface acceleration of gravity. From this information an observational value for M/L can be obtained. We use the expressions for the acceleration of gravity at the surface of a star, $g = MG/R^2$, and its luminosity, $L = 4\pi R^2 \sigma T^4$, where M and R are the star's mass and radius, G is the gravitational constant, and σ is the Stefan-Boltzmann constant. Dividing the first relation by the second, we obtain

$$M/L = g/4\pi G \sigma T^4$$

For stars of known distance and, therefore, known L, the value of M follows:

Pulsating Stars. The period P of a pulsating star depends on the mean density of the star, just as the period of vibration of a spring depends on the mass per unit length of the spring. The observed relationship is

$$P = 0.06\sqrt{\bar{\rho}_{\odot}/\bar{\rho}} \text{ days}$$

where $\overline{\rho}_\odot$ and $\overline{\rho}$ are the mean densities of the sun and the pulsating star, respectively. The mean density, $\overline{\rho}$, is defined by the formula,

$$M = (4\pi/3)R^3\,\overline{\rho}.$$

R is determined from the luminosity and temperature of the star, using the relation

$$L = 4\pi R^2 \sigma T^4 \qquad \text{or} \qquad R = \sqrt{L/4\pi\sigma T^4}$$

T is known from the spectral type. If the variable is a cepheid, L is known from the period-luminosity law. If not, L must be determined by an independent measurement of the distance. Assuming that L has been determined, the mass is finally computed from these relationships after R has been eliminated.

Gravitational Red Shift. Relativity theory predicts a red shift in the wavelength of the radiation emitted by a massive body. The amount of the shift depends on the mass of the star and its radius. If the gravitational red shift can be measured, and the radius is known, the mass of the star can be determined.

The relation between the gravitational red shift and the mass can be derived heuristically as follows. Suppose that a photon of energy E is emitted from an atom at the surface of a star. An amount of mass, m_{ph}, can be assigned to the photon according to Einstein's relation,

$$m_{ph} = \frac{E}{c^2}$$

In escaping to space against the pull of the star's gravity, the photon does work against the gravitational pull of the star and loses energy. The photon's energy is related to its wavelength by an inverse proportion: the smaller the energy, the longer the wavelength. Thus, when the photon is pulled back by the star's gravity and loses energy, its wavelength increases. That is, it is shifted toward the red. From Newton's law of gravity it can be shown that the formula for the fractional red shift is

$$\frac{\Delta\lambda}{\lambda} = \frac{MG}{Rc^2}$$

where M is the mass of the star, G is the universal constant of gravity, R is the star's radius, and c is the velocity of light.

For a star of solar mass and radius, the fractional red shift is

$$\frac{\Delta\lambda}{\lambda} = \frac{M_\odot G}{R_\odot c^2} = \frac{2 \times 10^{33} \times 6.7 \times 10^{-8}}{7 \times 10^{10} \times (3 \times 10^{10})^2} = 2 \times 10^{-6}$$

Which amounts to 10^{-2} Å for a line in the middle of the visible spectrum at 5000 Å. A red shift of roughly the predicted magnitude has been measured in the sun's spectrum, and it is also barely detectable in other Main Sequence stellar spectra. In white dwarfs, however, the gravitational red shift is much larger, because R is smaller. For a white dwarf of solar mass, with a typical radius of 20,000 km,

$$\frac{\Delta\lambda}{\lambda} = 7 \times 10^{-5}$$

which corresponds to a shift of 0.35 Å for a 5000 Å line. A shift of this magnitude can be measured with good accuracy in white dwarf spectra. Thus, the gravitational red shift method of mass determination is useful for white dwarfs, but not for most other stars.

In applying the method to white dwarfs, an obvious question arises: How can the gravitational red shift be separated from the Doppler shift caused by the motion of the white dwarf relative to the sun? The two effects cannot be separated for a single white dwarf, but if measurements are made on several white dwarfs, a statistical analysis can be performed to eliminate the effect of the Doppler shift. The idea of the statistical analysis is that if many stars are considered, on the average half will be moving toward the sun, and half away from it. The two groups of stars produce Doppler shifts of opposite signs. Therefore, if the wavelength shifts are averaged for several white dwarfs, the Doppler effect should cancel out, leaving only the gravitational shift.

On this basis the average mass of white dwarfs has been determined to be about $0.7M_\odot$. Most white dwarf masses are believed to be close to this value, although some, determined by independent methods, have turned out to be as low as $0.4M_\odot$.

INTERPRETATION OF STELLAR SPECTRA

APPENDIX B

The detailed shape of the lines in a star's spectrum is one of the main sources of information regarding the surface properties of stars. If you measure the degree of blackness of an absorption line on a photographic plate, you will find that usually the line is blackest, or most intense, at its center, and shades off gradually to either side of the center, that is, to longer and shorter wavelengths. The shape of the line refers to the precise manner in which the intensity falls off to either side of the center.

The photograph below (Fig. B.1) provides a clear example of the gradual decrease in the intensity of the absorption on the short-wavelength and long-wavelength sides of an absorption line. This photograph is an enlargement of a section of Figure 6.14. If we could measure the degree of blackening of the photograph at various

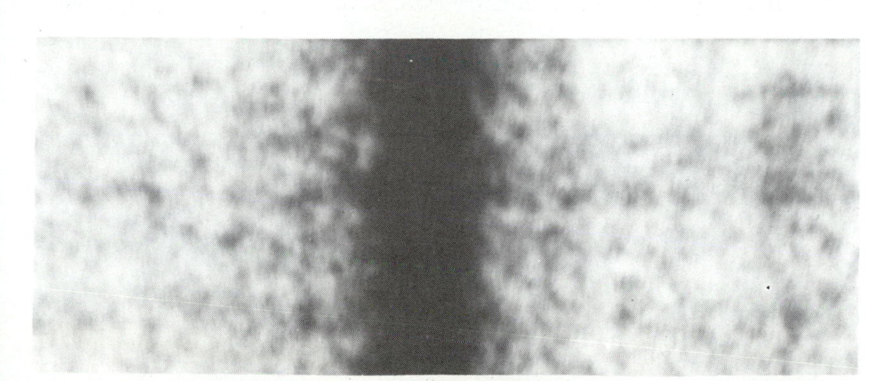

FIGURE B.1
Photograph of 4101 Å absorption line of hydrogen in the spectrum of an A-type star (Altair).

wavelengths in the vicinity of this line, we would obtain a graph similar to Figure B.2. This figure shows the percent of light transmitted through the star's atmosphere at wavelengths near 4101 Å.

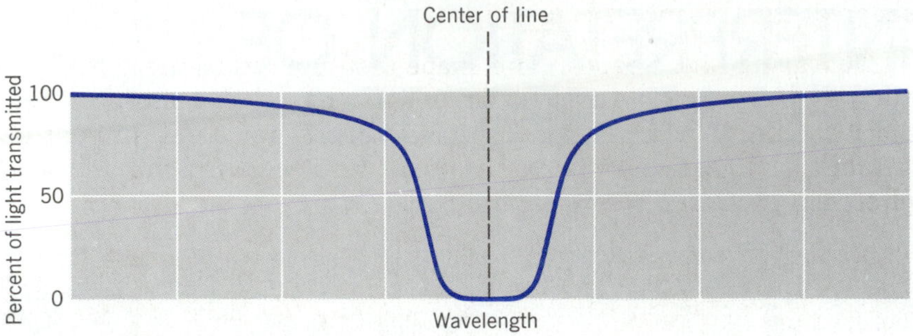

Center of line
Percent of light transmitted
100
50
0
Wavelength

FIGURE B.2
Variation of absorption with wavelength for the intense 4101 Å line in Figure B.1 (approximate).

Figure B.2 is the *shape* of this particular line. Because the hydrogen spectrum is relatively intense in an A-type star, its shape is characterized by a flat bottom, indicating nearly complete absorption over a substantial range in wavelength. Weaker lines have the bell-shaped appearance shown in Figure B.3.

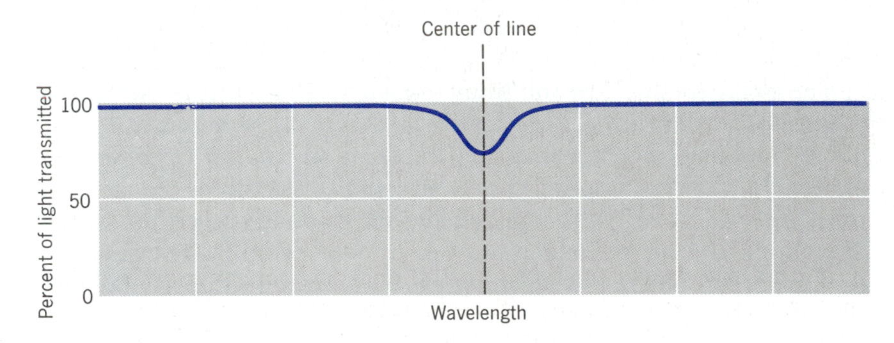

FIGURE B.3
Inverted bell-shape characteristic of a relatively weak absorption line.

Figure B.4 shows shapes of several lines of varying intensities ranging from very weak (*a*) to moderately strong (*b*) and very strong (*c*). These line shapes were calculated for the absorption line of singly ionized calcium located in the ultraviolet region of the solar spectrum at 3934 Å. This is the first calcium line in the solar spec-

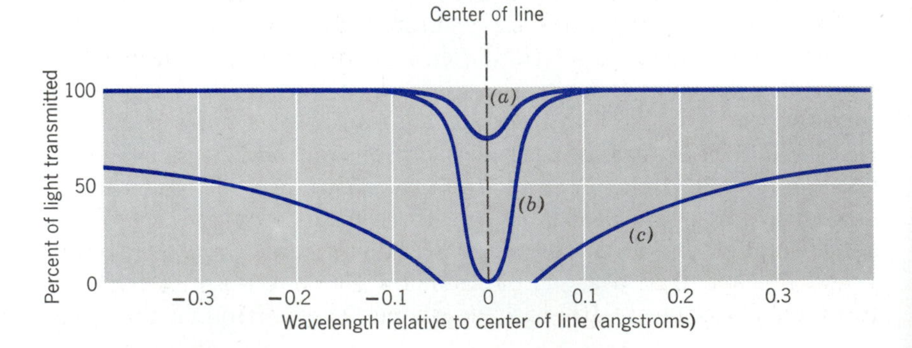

FIGURE B.4
Shapes of lines of varying intensity: (a) weak; (b) strong; (c) very strong, calculated for the line of singly-ionized calcium at 3934 Å in the solar spectrum.

trum facing page 133. The computed line shapes shown in the graph correspond to calcium ion abundances of approximately 3×10^{11}, 3×10^{14}, and 3×10^{16} ions per square centimeter in the sun's atmosphere. Line shape (*c*) agrees approximately with the observed shape of the Ca line in Figure B.1.

Measuring Line Shapes. The shape of a line can be measured accurately by passing a narrow beam of light through the photographic plate on which the star's spectrum is recorded, and determining the amount of light transmitted through the plate with the aid of a photocell mounted below (Figure B.5).

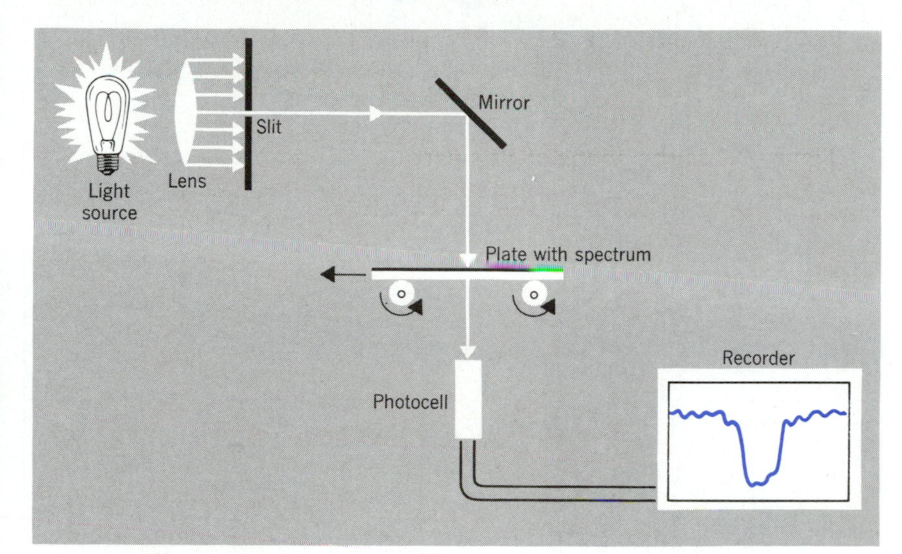

FIGURE B.5
Essential elements of a densitometer.

The narrow beam of light and the photocell are stationary. The photographic plate bearing the spectrum is mounted on a sliding track controlled by an accurately machined gear that moves it slowly through the light beam. Whenever the beam passes through an absorption line in the course of the plate's motion, the photocell records an increase in the intensity of the light passing through the plate. (The intensity increases because the plate is a photographic negative.) The photocell readings, plotted on a wavelength scale, yield a graph similar to Figures B.2 and B.3, on which the intensities and shapes of spectral lines are accurately displayed.

Devices of this kind are basic items of equipment in every astronomical observatory. They are called *densitometers* because they measure the density or degree of blackening of the photographic emulsion. The beam is as narrow as a thousandth of a centimeter in the most precise instruments, which are called microdensitometers.

Figure B.6 shows the result of a microdensitometer scan of the spectrum of Zeta Ophiuchi, a star at a distance of about 500 light-years. The spectrum was photographed with a one-hour exposure at the Lick Observatory 100-inch telescope. The portion of the spec-

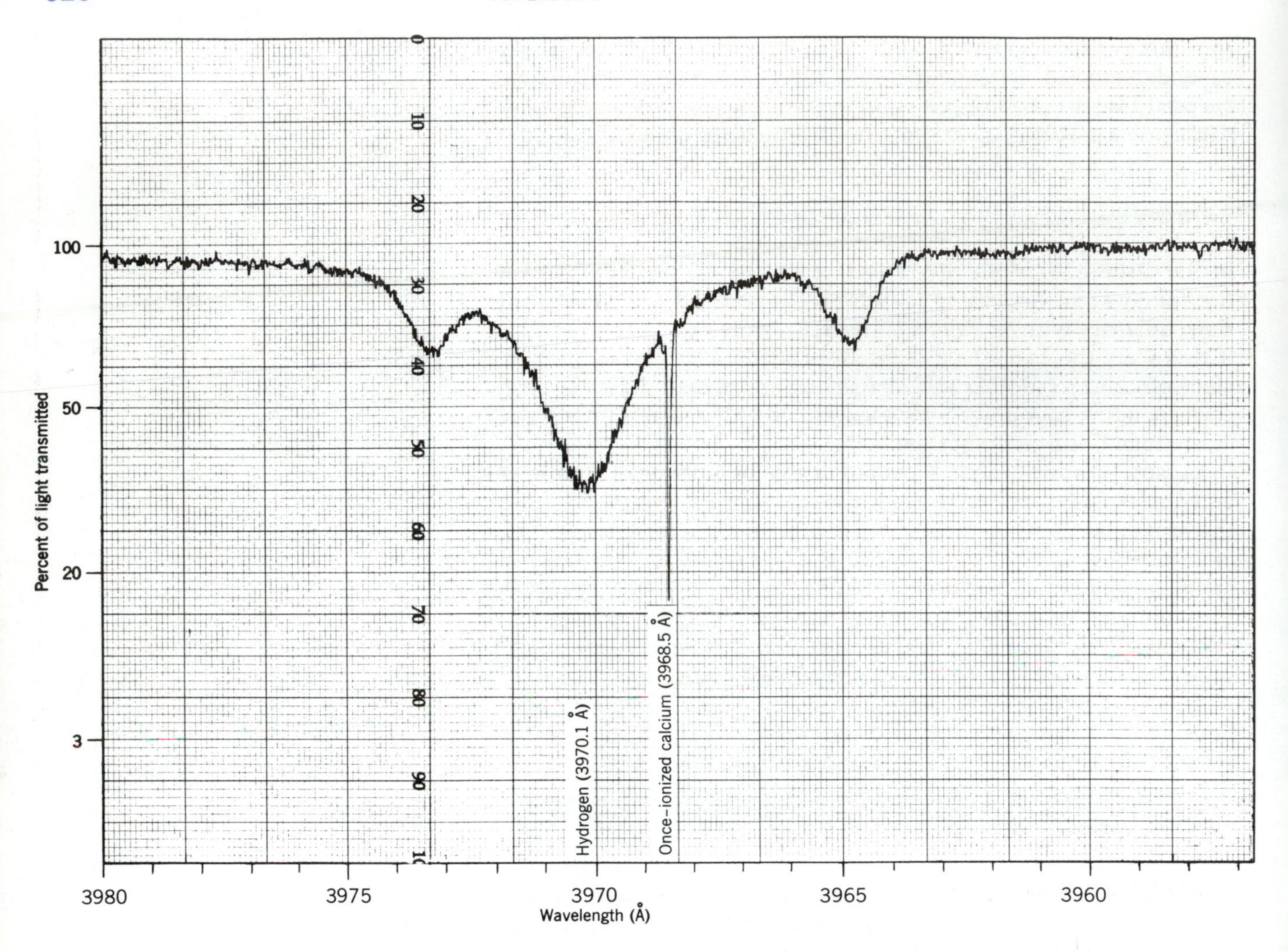

FIGURE B.6
The spectrum of Zeta Ophiuchi measured with a microdensitometer, showing broad absorption lines produced in the atmosphere of the star and narrow lines produced by absorption in the low-density interstellar gas.

trum shown in the figure spans the interval from 3957 Å to 3980 Å. The record shows a broad, relatively intense line centered at 3970.07 Å with a width of about 20 Å. This is a line in the Balmer series of hydrogen produced by absorption in the atmosphere at Zeta Ophiuchi. Broad but somewhat weaker absorption lines appear at 3964.8 and 3973.4 Å. These lines are probably due to iron.

An exceedingly narrow line, less than an angstrom wide, is also visible at 3968.5 Å. This line has been identified as an absorption line produced by atoms of singly ionized calcium, located in the clouds of interstellar gas and dust that occupy a part of the space between our solar system and Zeta Ophiuchi.

Why should the absorption lines produced by hydrogen and iron atoms in the atmosphere of Zeta Ophiuchi be very broad, while the absorption line produced by atoms of calcium in the space between the stars is extremely narrow? To answer this question, we must try to understand all the major factors that influence the shape of a spectral line. In acquiring this understanding, we will also gain the means of measuring many important properties of stars.

Factors Influencing the Shape of a Spectral Line

In the discussion of atomic energy states given on pages 134–138, the energy of an electron in its orbit was described as a precisely defined quantity with no variation in energy from one atom to another. If this were the case, the photon absorbed in a transition from one electron orbit to another would also have a precisely defined energy, corresponding to a unique wavelength, and all spectral lines would be sharp and infinitely narrow.

In fact, however, there is a small degree of blurring of atomic energy states during the transition from one orbit to another, because the atom's structure is disturbed slightly by the absorption of the photon. As a result, the absorption lines produced in the transition are also blurred to a slight degree. That is, the lines are broadened. This blurring of a spectral line, which occurs in the spectrum of every atom, is called the *natural width* of the line. It is very small, being a hundredth of an angstrom in typical cases.

The natural width is the smallest source of broadening of spectral lines. In the sections below, we discuss more important factors that contribute to the broadening of spectral lines and influence their shapes; and we go on to consider the properties of stars that can be deduced from a detailed study of line shapes.

Collisional Broadening. When two atoms collide, the electrons on one atom repel the electrons on the other atom, distorting the shapes of their orbits and changing the energies of the atomic states. If one of the atoms happens to absorb a photon during the collision, the energy of the absorbed photon will be affected by this distortion. If the density of the gas is low, collisions between atoms are infrequent, and the absorption of a photon usually takes place when the atom is in an undisturbed state. In this case, collisions will have only a small effect on the width of a spectral line. If the density of the gas is high, an atom will frequently be involved in a collision at the same time that it absorbs a photon, and the energy of the absorbed photons will vary over a considerable range. As a result, the absorption line will be substantially broadened.

This effect on the shape of a spectral line is known as *collisional broadening*. The width of the broadened line is determined by the frequency of collisions between atoms. The collision frequency depends on the gas density and temperature. Thus, if the temperature

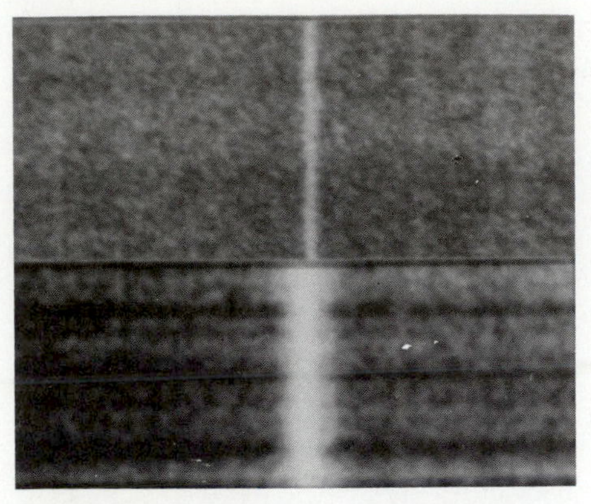

FIGURE B.7
Comparison between the spectrum of a red giant (above) and a Main-Sequence star (below) with the same spectral type and surface temperature.

is known, the width of the spectral lines indicates the density of the gas.

Chapter 7 classifies the population of stars into broad categories, including Main-Sequence stars and red giants. A Main-Sequence star is a "normal" hydrogen-burning star, while a red giant is a star in a later stage of its evolution, in which the outer layer of the star has expanded into an enormous, tenuous sphere of low-density gas. A red giant can be distinguished clearly from a Main-Sequence star through the difference in the widths of their spectral lines, even if the surface temperatures of the two stars are identical.

For a Main-Sequence star, the width of a collision-broadened line is typically a few tenths of an angstrom. In the outermost layers of a red giant, the density is much less than the density at the surface of a Main-Sequence star, and the absorption lines are correspondingly narrower. Figure B.7 compares the spectrum of a red giant with a Main-Sequence star. The narrowness of the line in the red-giant spectrum is apparent. Theoretical investigations of collisional broadening have led to an accurate relationship between spectral line-width and density for a given temperature.

Doppler Broadening or Thermal Broadening. The lines in stellar spectra are also broadened through the combination of their thermal motion and the effect of the Doppler shift. Because of the random thermal velocities of the atoms in the star's atmosphere, some atoms move toward us and others move away from us when they emit their photons. These motions produce a Doppler shift in the spectrum. The result is a broadening of the line, called *Doppler broadening,* which adds to the broadening produced by collisions. Since the degree of broadening is determined by the average thermal velocity of the atoms, that is, by their temperature, the effect is also known as *thermal broadening.*[1]

The effects of Doppler broadening or thermal broadening are somewhat smaller than those of collisional broadening in Main-Sequence stars, the width of a thermally broadened line being 0.05 to 0.1 Å in typical cases. However, the shapes of the lines produced by the two types of broadening are quite different. In a Doppler-broadened line, the absorption falls off sharply to either side of the center of the line, while in a collisionally broadened line, the absorption diminishes more gradually, forming two broad wings to either side of the center.

Figure B.8 shows the theoretically computed shape of the absorption line of ionized calcium in the spectrum of the sun at 3934 Å. The difference between the two types of broadening is clearly evident in the figure.

[1] Random turbulent motions of the gas in the atmosphere of the star also broaden the lines as a result of the Doppler effect. This *turbulence broadening* produces line shapes similar to those of Doppler-broadened lines.

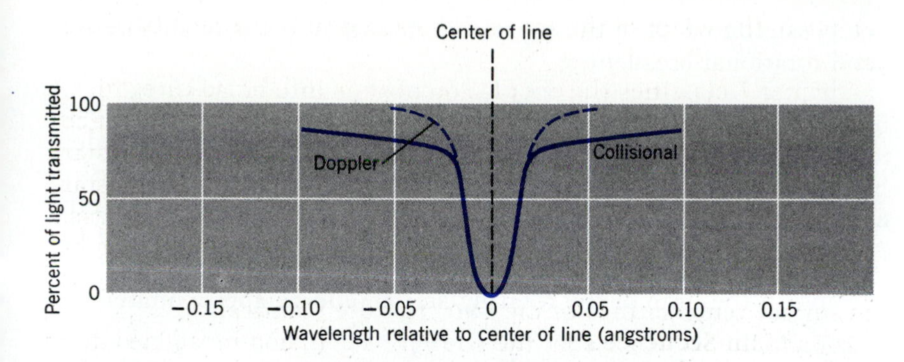

FIGURE B.8
Line shapes for Doppler broadening and collisional broadening.

Rotational Broadening. If a star is rotating, the atoms on its surface move toward or away from the observer during the course of each rotation, unless the axis of rotation is pointed directly at the observer. Let us assume for simplicity that the axis is perpendicular to our line of sight. Assume also that the star is rotating counterclockwise as seen from above (Figure B.9). Then all the atoms on the left half of the star's disk will be moving toward the earth and all the atoms on the right half of the disk will be receding. These motions produce a Doppler shift in each line of the star's spectrum.

The atoms on the left limb of the star and on its equator will have the largest velocity toward the earth and, therefore, the contribution of these atoms to the absorption spectrum of the star will be shifted by the greatest amount toward the blue end of the spectrum. The atoms on the right limb of the star and on the equator will have the greatest velocity away from us and, therefore, their contribution will be shifted by the maximum amount to the red. Atoms at other points on the star will give rise to intermediate values of the Doppler shift. The superposition of these many absorptions will give rise to a broadened line. The extent of the broadening will depend on the rate of rotation and the angle of inclination of the axis of rotation to our line of sight. This effect is known as *rotational broadening.*

How can rotational broadening be distinguished from collisional broadening and thermal broadening? An absorption line looks approximately the same to the eye, regardless of the type of broadening that has produced it. However, theoretical calculations of the shapes of rotationally broadened lines reveal that they differ significantly from other line shapes. The differences are not sharp enough to be evident to the eye when a spectrum is examined, but they can be detected if a densitometer tracing is made of the line shape, as in Figure B.6. A theoretical formula, representing a combination of the three broadening effects acting simultaneously, can be fitted to the observed line shape as given by the densitometer tracing. By adjusting the relative contributions to the theoretical formula from each of the three broadening effects in order to secure the best possible agreement between the formula and the measured line shape, it may

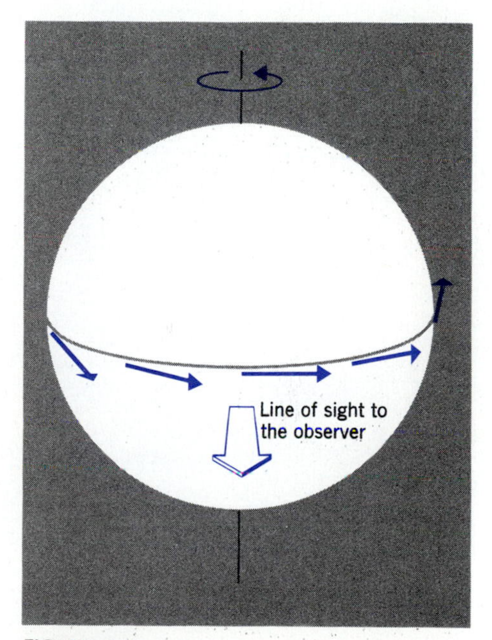

FIGURE B.9
Motions of atoms on the equator of a rotating star giving rise to rotational broadening.

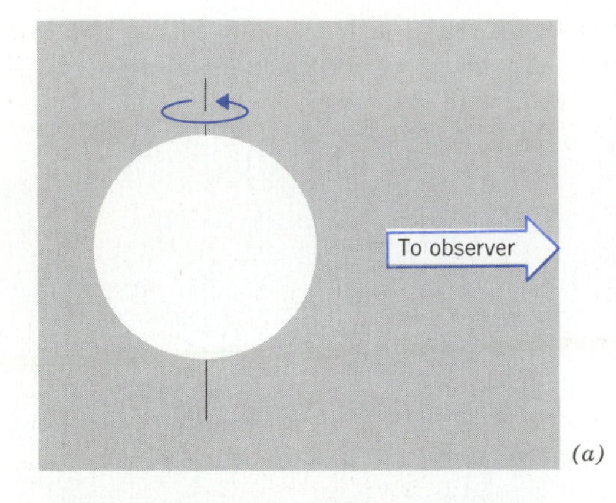

(a)

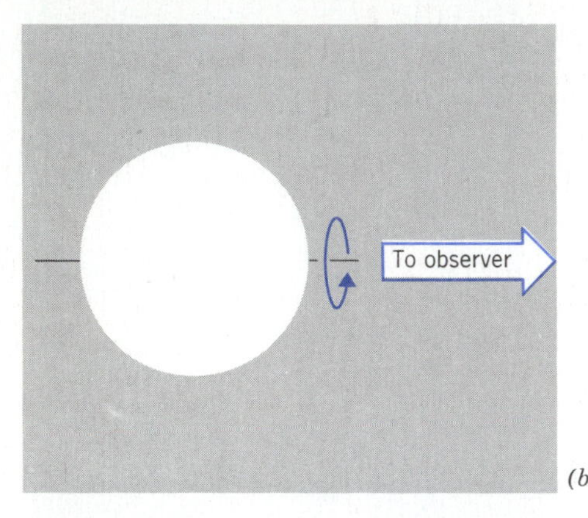

(b)

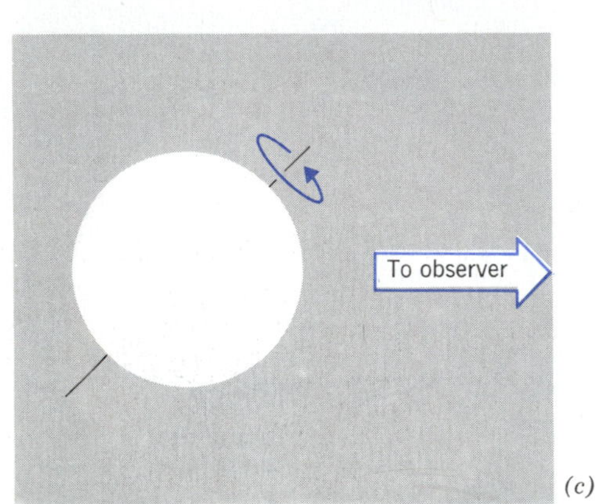

(c)

FIGURE B.10
Angle between the axis of a rotating star and the line of sight to the observer: (a) perpendicular; (b) parallel; (c) intermediate.

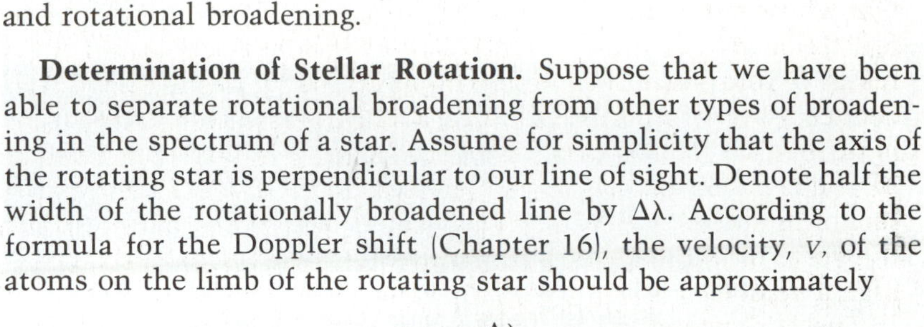

be possible to deduce the separate amounts of collisional, thermal, and rotational broadening.

Determination of Stellar Rotation. Suppose that we have been able to separate rotational broadening from other types of broadening in the spectrum of a star. Assume for simplicity that the axis of the rotating star is perpendicular to our line of sight. Denote half the width of the rotationally broadened line by $\Delta\lambda$. According to the formula for the Doppler shift (Chapter 16), the velocity, v, of the atoms on the limb of the rotating star should be approximately

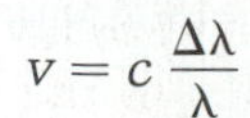

$$v = c\,\frac{\Delta\lambda}{\lambda}$$

In this formula, v is the speed that the atoms on the limbs of the rotating star have toward or away from us as a result of the rotation, and c is the velocity of light. If the radius, R, is known, the period of rotation, T, that is, the time in which the star completes one turn on its axis, can be calculated from the relation

$$T = \frac{2\pi R}{v}$$

At the beginning of the discussion we assumed for simplicity that the star's axis of rotation was perpendicular to the observer's line of sight (Figure B.10*a*). Usually the axis is tilted at an angle to our line of sight, diminishing the amount of rotational broadening. For example, if a star is rotating very rapidly, but the axis of rotation is pointed directly toward the earth (Figure B.10*b*), atoms on the surface of the rotating star will have no velocity toward or away from the observer, and the star's spectral lines will show no rotational broadening.

In general the axis of rotation is tilted at an intermediate angle (Figure B.10*c*).

There is no way of determining the inclination of the axis of rotation separately from the speed of rotation. Therefore, the astronomer's only recourse is to assume that the axis of rotation is perpendicular to the line of sight and apply the formulas given above. The speed of rotation calculated in this way will be correct if the axis of rotation happens to be perpendicular to our line of sight. In other cases, the calculated value will be less than the true speed of rotation. Thus, the formulas for rotational broadening (above) will always yield a lower limit to the true rate of a star's rotation.

Using this method, it has been found that the hottest stars, Type O and B, rotate most rapidly in general. A-type and F-type stars rotate somewhat less rapidly. A sharp decrease occurs in the rate of rotation in the neighborhood of F-type stars. Nearly all G-type stars, including the sun, rotate relatively slowly.

The difference between the rapidly rotating and slowly rotating stars is very sharp. Many O and B-type stars rotate on their axes as rapidly as once every 4 hours. This is the case for Zeta Ophiuchi, the star whose spectrum was shown in Figure B.6. The width of the very broad line at 3970 Å in the spectrum of Zeta Ophiuchi is entirely the result of rotational broadening. From the width of the line it can be deduced, by the methods described above, that this massive object is turning on its axis every four hours, and possibly even more rapidly, depending on the angle between its axis of rotation and the line of sight to earth. A star rotating on its axis every four hours is on the verge of flying apart, like a centrifuge that has been spun at too high a velocity.

The sun rotates on its axis today at the leisurely pace of once every 27 days. Some theories of the origin of the solar system assumed that the sun rotated as rapidly as once every three or four hours in its youth. According to these theories, the birth pangs of the planets slowed down the rotation of the sun in some way. Carrying this line of thought further, some astronomers believe that if any star is rotating slowly, a possible reason is that a family of planets was formed around it, braking its rapid rotation during the process of planetary birth. If this theory is correct, it provides indirect evidence of a multitude of solar systems in the Cosmos around us, since the majority of stars rotate at the relatively slow rate of the sun.

The Zeeman Effect and Magnetic Fields of Stars. When an atom is placed in a magnetic field, each of its energy levels may be split into several levels by the action of the field on the magnetic properties of the atom. If an energy level is split into three separate levels, for example, the photons emitted in transitions involving these levels can have three distinct energies and wavelengths. Consequently, a single line in the original spectrum will be split into three separate lines when the magnetic field is present.

The splitting of a line by a magnetic field is called the *Zeeman Effect.* It is described in more detail in Chapter 13. Figure B. 11 shows the splitting of a line by the Zeeman effect in a laboratory experiment.

FIGURE B.11
Magnetic splitting of a line in the laboratory.

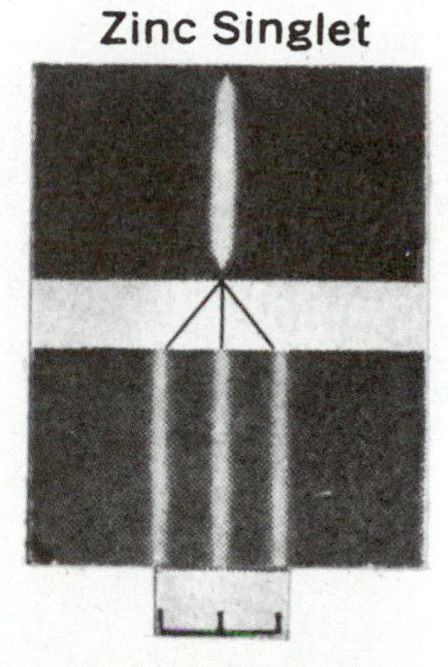

A clear splitting of the lines in stellar spectra by the Zeeman effect can usually be observed only in the spectra of sunspots, which contain intense magnetic fields of thousands of gauss. In the spectra of stars more distant than the sun, the light from the entire disk of the star passes through the slit of the spectrograph, and many different intensities of magnetic fields, from different areas on the surface of the star, contribute to the observed spectrum. As a result, each line is a blurred average over varying degrees of magnetic splitting. The resultant *magnetic broadening* is an indication of the average strength of the magnetic field on the surface of the star.

The average magnetic field on the surface of a star must be several hundred gauss in order to produce an observable degree of magnetic broadening. Clear indications of magnetic broadening and magnetic field strengths of this magnitude have been found in roughly 150 stars. The largest magnetic field observed in a star thus far is 34,000 gauss. Figure B.12 shows the spectrum of this star, named HD215441. In this exceptional case the magnetic splitting into distinct lines is discernible because of the very strong field.

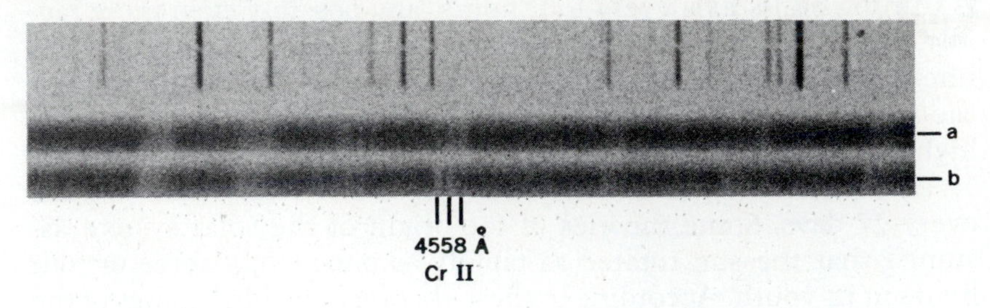

FIGURE B.12
Magnetic splitting of lines in the spectrum of a star. The absorption lines appear light against a dark background because the spectrum is shown as a photographic negative. The sharp lines above and below the faint stellar spectrum are a known spectrum exposed on the same plate in the observatory in order to determine the wavelengths of the stellar lines accurately.

GLOSSARY

absolute luminosity This indicates the amount of energy flowing out of an object per unit time.

absolute magnitude The absolute magnitude of an object is defined to be the *apparent magnitude* of the object if it were moved to a distance of 10 *parsecs* from the observer. Larger numerical values for the *magnitude* represent smaller amounts of radiation (light) emitted by the object.

absorption spectrum A spectrum in which dark *absorption lines* partially block the light of an otherwise *continuous spectrum.*

amino acids Molecule that serves as the building blocks of *proteins.*

Angstrom A distance that is 10^{-10} of a meter. For example, the *wavelength* of blue light is 4000 Å (angstroms).

annular eclipse A *solar* eclipse for which the moon does not appear to completely cover the bright surface of the sun. Thus the sun's surface peeks around the moon's edge and produces a complete circle of bright light.

antielectron Same as *positron.*

antimatter The complement to ordinary matter. The charge of an antimatter particle is the opposite of its matter complement. Matter and antimatter may, on collision, annihilate each other, leaving only energy.

antiproton The negatively charged *antimatter* complement to a *proton.*

apparent magnitude A value that indicates the amount of radiation (light) received by an observer from an object. It is a logarithmic measure of the *apparent luminosity.* An increase by one in the value of the magnitude corresponds to a decrease by about 2.5 in the apparent luminosity. Thus, the larger the value of the magnitude the less radiation received. The value of the apparent magnitude is determined both by the intrinsic radiation energy output of the object and its distance.

apparent luminosity A measure of the amount of radiation (light) received by an observer from an object. Its value depends both on the object's intrinsic radiation energy output and its distance from the observer.

asteroids Small solid bodies whose orbits about the sun are typically between Mars and Jupiter.

astronomical unit The distance between the earth and the sun. 1.5×10^{11} meters or 93 million miles. Abbreviated as A. U.

atmosphere (stellar) Gaseous outer layer of a star from which light escapes into space. This layer is the origin of the typical *absorption spectrum* of the stars.

atmospheric blurring Blurring of the images of astronomical objects caused by refraction in the earth's turbulent atmosphere.

atom A tiny object that has a central *nucleus* and one or more orbiting *electrons.* The number of protons in the nucleus of the atom determines which element that particular atom represents.

atomic spectral line An *absorption* or an *emission line* caused by the absorption or emission of a *photon* by an *atom.*

atomic spectrum The *spectrum* arising from a particular *element.* (The spectrum from a particular *molecule* is usually termed a molecular spectrum.) The *wavelength* locations of the *atomic spectral lines* indicate which *element* produced that spectrum.

aurora Glowing nighttime displays that are produced by the interaction of the *solar wind* with the particles of the earth's atmosphere.

Balmer line A hydrogen *spectral line* for which the first *excited state* is the lower level of the transition of the electron. The wavelengths of these lines are in the *visible* region of the *spectrum.*

Balmer series The collection of all Balmer *spectral lines.*

band spectrum A collection of *spectral lines* that are closely clustered at neighboring wavelengths. They arise from *transitions* in *molecules.*

barred spiral galaxy A *galaxy* with an apparent bar running through the center (the *nucleus*). Each end of the bar serves as the origin of a curved *spiral arm.*

basalt A dark fine-grained volcanic rock. Solidified lava.

Big-Bang cosmology A theory of the origin of the *Universe* in which the entire *Universe* originated in one hot, explosive instant—termed the Big Bang.

billion 1,000,000,000.

binary stars Two stars in orbit about each other due to their mutual gravitational attraction.

black body A hypothetical object that perfectly absorbs and emits *photons* at all *wavelengths.* The characteristics of the *spectrum* of the emitted *photons* (called black-body radiation) depend only on the *temperature* of the object. All liquids, solids, and high-density gases behave as almost perfect black-body objects.

black hole An object whose surface gravity is so high that nothing, not even light, can escape.

Bohr atom A model of an *atom* developed by Niels Bohr in which *electrons* are constrained to orbit the *nucleus* at specified distances.

blue giant A large, hot star whose position is plotted in the upper left-hand corner of the *H-R diagram.*

bolometric correction The numerical difference between the *visual* and the *bolometric magnitudes* of an object.

bolometric magnitude The term "bolometric" indicates that this *magnitude* is a measure of the radiation at all *wavelengths.* The *magnitude* may be either an *absolute* or an *apparent magnitude.*
carbon cycle The conversion of hydrogen into helium, using carbon and nitrogen in intermediate steps in the series of *nuclear reactions.*
carbonates Rocks formed from carbon and oxygen ($CO_3^=$), combined with a metal ion, usually calcium or magnesium. Typical examples are limestone and dolomites.
Cassegrain reflector A reflecting telescope in which the *image* is brought to focus behind the *primary mirror* of the telescope. This is accomplished with a second (secondary) mirror in front of the primary mirror and a hole in the center of the primary mirror.
celestial equator The outward projection of the earth's *equator* onto the *celestial sphere.*
celestial sphere The apparent distant sphere of the sky centered at the earth. The Greeks thought that the stars were placed on it.
centrifugal force A fictitious force that appears to push an object away from the center of a circular motion. The effect is real, but is not due to a force. An object, unless constantly deflected into a curved path by a force, will move in a straight line, and hence away from the center of the circular motion.
cepheid variable A *pulsating star* with a regular period of pulsation. The cepheid variables serve as useful cosmic yardsticks because their pulsational periods specify their average *luminosities.*
Charon The moon of Pluto.
chemical reaction Chemical reactions alter the pairings of *atoms* through modifications of the *electron's* orbits while leaving the *nuclei* of the atoms unchanged.
chondrite A stony meteorite whose texture includes small rounded bodies known as chondrules.
chromosphere A layer of the sun's *atmosphere* that is just above the *photosphere.* Atoms in this low-density layer produce many of the observed solar *absorption lines.*
chromosphere network An irregular pattern of dark linear markings visible at the depth of the *chromosphere* in the sun. The network is believed to be caused by the subsurface solar *convection zone.*
cluster of galaxies A physical clumping of *galaxies* that may contain a few to a few thousand galaxies.
collisional broadening Broadening of *spectral lines* due to the frequent collisions between *atoms* in the gas. The greater the density of the gas, the greater the number of collisions, and hence the broader the lines.
color A sensation due to the eye's and mind's discrimination of the *wavelengths* of photons.
color index A number defined as the numerical difference between the *photographic* and *visual magnitudes* of an object. The color index is a measure of the redness (or blueness) of an object.
comet A small, icy, dusty object that becomes visible as a glowing, diffuse nucleus with a long tail during its near approaches to the sun.
comparison spectrum An *emission line spectrum* of a gas that is exposed

adjacent to the astronomical *spectrum.* It serves as a standard to assist the astronomer in his determination of the *wavelengths* of the lines in the astronomical spectrum.

composition The specification of the relative amounts of the different *elements* or, sometimes, *molecules* in a substance.

concave mirror The reflecting surface of a concave mirror is curved inward.

constellation An apparent pattern of stars in the sky.

continental drift The theory that the continents move across the face of the earth. The modern version of this theory is known as *plate tectonics.*

continuous spectrum A smooth spectrum of *light* without *absorption* or *emission lines.*

convection A flow of mass that moves hot material into cooler regions and vice versa. This flow transports heat from one location to another.

convection zone A region or a layer in which *convection* is occurring. In the case of the sun, the convection zone is a shell of material just beneath the surface *(photosphere)* of the sun.

convex mirror The reflecting surface of a convex mirror is curved outward.

core (earth's) The central nickel-iron region of the earth.

core (stellar) The central region of a star. The term often denotes that central region in which *nuclear burning* is occurring.

corona The outermost gaseous layer of the sun that starts at the top of the *chromosphere* and extends through the *Solar System.*

coronagraph A telescope that produces artificial *solar eclipses* for use in the investigation of the *corona* and other phenomena occurring above the surface of the sun.

cosmic abundance Relative abundances of the *elements* as determined as an average for the *Universe* as a whole.

cosmic rays High energy (high speed) atomic *nuclei* that strike the earth's atmosphere.

cosmological red shift A *red shift* due to the expansion of the *Universe.*

cosmology The study of the *Universe* as a whole—its past, present, and future.

creep The slow movement of a solid body such as the creep of the rock in the earth's *mantle.*

crescent moon A *phase* of the moon for which the observer can see less than one-half of the illuminated hemisphere of the moon.

crust (earth's) The outermost solid layer of the earth.

cyclotron A device used by physicists to accelerate small particles such as *electrons* to speeds just under the speed of light. Collisions of these high velocity particles on special targets yield information about the structure of the *atoms* and their constituent particles.

declination The *celestial sphere* analogue of *latitude.* Measurements are made in degrees from the *equator.*

densitometer An astronomical instrument that produces a graph of the information contained on a photographic *plate* (or a spectrum, for example) by measuring the density (blackness) of the *plate.*

deuterium An *atom* with a *nucleus* that contains one *proton* and one *neutron.*

deuteron The *nucleus* of a *deuterium* atom; a paired *proton* and *neutron.*
differential rotation A body rotating at different angular rates at different distances from the equator is said to exhibit differential rotation. Solid bodies rotate uniformly, not differentially.
differentiation A separation of materials at differing depths in a body, usually caused by prolonged internal melting. The earth is a differentiated planet.
diffraction The bending of light as it passes the edge of an object.
diffraction disc The broad central image of a point of light (such as a star); the image of the point of light is broadened by *diffraction* in the telescope itself.
diffraction grating An optical surface with many fine parallel scratches that produces a spectrum by reflection or transmission.
dish (radio) The large surface of a *radio telescope* that reflects *radio waves* and focuses them at the point of the *radio wave* detector.
dispersion The bending of light to a degree depending on wavelength. For example, a *prism* is said to disperse light in the production of a spectrum.
dispersive power The degree to which a *grating* or a *prism* spreads out the *spectrum.*
DNA (deoxyribonucleic acid) The long-chained *molecule,* residing in cells, which contains the genetic code and has the ability to reproduce itself.
Doppler broadening A broadening of *spectral lines* due to the superposition of the random (*temperature* induced) thermal velocities of each of the absorbing or emitting atoms.
Doppler shift The change in the *wavelength* of radiation due to the relative velocities of the source and the observer. Only the velocity along the line of sight between the observer and the source contributes to the displacement of the wavelength.
dust clouds Clouds of small dust-sized particles in interstellar space that absorb visible light. The composition of these dust-sized particles is poorly known.
dwarf (star) A *Main Sequence* star of average or small diameter and luminosity.
dwarf galaxy A smaller-than-average *galaxy.*
eclipse season A semiannual period during which *solar* or *lunar eclipses* may occur.
earthlike planets See terrestrial planets.
eclipsing binary stars A binary system whose orbital plane is sufficiently close to our line of sight that one star periodically partially, or totally, blocks (eclipses) the light from the other star.
ecliptic The apparent yearly path of the sun on the *celestial sphere.*
effective temperature The effective temperature is defined to be the temperature that a perfect *black body* of the same size must have to produce the same total *luminosity* as the object in question.
electrical field A region of space in which the *electrical force* may be detected.
electrical force A force between two electrically charged objects.
electromagnetic force A force originating from electrical charge, current (electrical charge in motion), or a magnetized body.

electromagnetic spectrum The distribution of *electromagnetic radiation* over all possible wavelengths.

electromagnetic radiation Waves produced by vibrating charges that travel through a vacuum at the speed of light.

electron A small negatively charged particle that may normally be found in orbit around the *nucleus* of an *atom.*

electron shell In the *Bohr atom* the *electrons* are constrained to orbit at only certain specified distances from the *nucleus;* these locations are termed shells.

element The specification of the type of *atom,* based on the number of *protons* in the *nucleus.*

ellipse A precisely defined geometrical figure that appears as a squashed circle. It has many variations. Two extreme examples of ellipses are circles and straight lines.

elliptical galaxy A *galaxy* whose outline forms the shape of an *ellipse.*

emission line A single bright *spectral line.* Its *wavelength* corresponds to a unique downward *transition* in a particular ionized state of a particular *element.*

emission nebula A gaseous *nebula* that is visible through the *emission lines* that it produces. The *excited-state atoms* that produce the emission lines are usually excited by the *ultraviolet* photons from a nearby star.

emission spectrum A spectrum containing *emission lines.*

envelope (stellar) That portion of a star between the *core* and the *atmosphere.*

equator A line circling a spherical body that is always at an equal distance from the rotation poles of the body.

equatorial bulge A bulge around the *equator* of a rotating planet or other spinning object.

equatorial plane A plane containing the *equator* of the body. The equatorial plane is perpendicular to the axis of rotation of the body.

eruptive prominence A solar *prominence* that seems to eject a cloud of glowing matter.

excited state In the *Bohr atom* model, an *electron* is said to be in an excited state (orbit) if there is at least one lower orbit that it may descend to by releasing a photon; also, an *atom* with one or more of its *electrons* in an excited state is itself termed an excited state atom.

filament (solar) Dark streaks visible in the sun's *chromosphere* associated with regions of a strong *magnetic* field.

fireball An impressive *meteor* that is often bright enough to be visible in the daytime.

first-quarter moon A phase of the moon in which one half of the visible hemisphere, or one quarter of the complete surface, is illuminated. This *phase* occurs approximately one week after the *phase* of *new moon.*

fission (atomic) The splitting of a nucleus.

flare An explosive brightening visible on the solar surface.

flash spectrum An *emission line spectrum* obtained from the sun's *chromosphere.* It may be obtained for a few seconds before and after the time of totality of a *solar eclipse.*

fluorescence The absorption of light at one *wavelength* and the subsequent reemission of that energy at other *wavelengths.* A common example is the absorption of invisible *ultraviolet photons* and the re-emission of that energy as *visible photons.*

focus The point to which an optical device converges light from a distant source.

Fraunhofer line The name given to the prominent solar *absorption lines.* Named after an early solar spectroscopist.

frequency The number of vibrations in a certain time interval; the number of vibrations per second.

full moon The *phase* of the moon in which the observer on the earth sees the moon's entire illuminated hemisphere.

fusion The joining of two *nuclei.*

galactic cluster Between 10 and 10,000 stars found in a loose clumping. The galactic clusters are found in the *disc* of the *Galaxy.*

galactic disc The highly flattened spherical distribution of stars in a *galaxy.*

galactic halo A spherical volume surrounding a spiral galaxy, with radius roughly equal to the radius of the galactic disc, with numerous individual Population II stars and globular clusters scattered irregularly throughout the volume.

galaxy A collection of stars (and often gas and dust) held together by mutual gravitation. While the actual number of stars may vary over a wide range, 100 billion is a typical number.

Galaxy The *Milky Way Galaxy.*

Galilean satellites Four large satellites of Jupiter first observed through a telescope by Galileo. Their names in order of increasing distance from Jupiter are Io, Europa, Ganymede and Callisto. Ganymede is larger than the planet Mercury.

gamma rays The most energetic photons. Their wavelengths are shorter than those of *X-ray photons.*

gauss A measure of a magnetic field's strength. On the earth's surface the magnetic field is approximately one-half gauss.

giant A star whose location is plotted above the *Main Sequence* on the *H-R diagram.* Blue Main Sequence stars are also often termed giants.

giant planets Jupiter, Saturn, Uranus, and Neptune; planets composed mainly of hydrogen and helium, in contrast to the smaller *terrestrial planets.*

gibbous moon A *phase* of the moon between first and third quarters. During the gibbous phase more than half the moon's illuminated surface is visible.

globular cluster Approximately one million stars found in a compact clumping. The stars are old. These clusters are found in the *halo* of our *galaxy* and other *galaxies.*

globule A dark spot seen against a bright *nebula.* Believed to be a fragment of an interstellar cloud that is in the process of collapsing prior to star formation.

granite A coarse-grained intrusive *igneous rock* composed of quartz, orthoclase feldspar, plagioclase feldspar, and mica.

granules "Specks" across the surface of the sun that give it an oatmeal appearance. Each "speck" is approximately 1000 miles in diameter. They are believed to be caused by upward- and downward-moving columns of gas in the *convection zone.*

gravitational redshift An increase of the *wavelength* of *photons* caused by a high gravitational field.

gravity The attraction of one mass to another; one of the basic forces in the physical Universe.

greenhouse effect A planetary atmosphere that transmits incoming visible solar radiation and blocks outgoing *infrared radiation* is said to exhibit the *greenhouse effect.* By sealing in the infrared radiation, the atmosphere acts as a blanket to increase the surface temperature of the planet.

Greenwich meridian The *meridian* that passes through Greenwich, England. It is defined to be zero degrees *longitude.*

ground state The lowest energy orbit that an *electron* may occupy in an *atom.* Also, an atom that has all its electrons in their respective lowest-energy orbits is said to be a ground-state atom.

HI region An *interstellar* gas cloud composed of neutral *hydrogen.*

HII region An *interstellar* gas cloud composed of *ionized hydrogen.*

halo The outer region of a *galaxy,* roughly spherical in shape, mainly containing old stars and a small amount of gas.

heavy elements In astronomical usage heavy elements are all elements other than *hydrogen* and *helium.*

helium An *element* having two *protons* in its nucleus.

helium flash An explosive ignition of a *helium-burning core.*

helium burning The *fusion* process that converts *helium* into *heavier elements* with the attendant release of energy.

Hertzsprung-Russell diagram A plot of the *luminosity* (or *absolute magnitude)* against the *spectral type* (or surface temperature or *color index*) of a collection of stars. The plot is named after the two independent creators of the diagram.

hidden mass Matter present in the Universe in a nonluminous, relatively invisible form, i.e., distinct from stars and galaxies. The hidden mass may compose most of the matter in the Universe.

highlands Refers to those regions on the moon that are not flooded by lava.

H-R diagram Same as the *Hertzsprung-Russell diagram.*

Hubble Constant The proportionality constant in *Hubble's law.*

Hubble's law The (linear) relation between the distance of an astronomical object and its *red shift* due to the expansion of the *Universe.*

hydrogen An *atom* with only one *proton* in the *nucleus.*

hydrogen burning The *fusion* of *hydrogen* to produce *helium* with the release of energy.

igneous rock A rock formed by the cooling of molten silicate minerals, usually from volcanic processes.

image Optical duplication of the appearance of an object, by light focused through the use of lenses or mirrors.

image intensifier Same as *image tube.*

image tube A device similar to a television camera. It intensifies the images obtained through a telescope.

infrared radiation *Electromagnetic radiation* of a *wavelength* just longer than *visible radiation.*

interstellar (dust, medium, molecules) Materials found in the space between the stars.

inverse square law Something is said to follow the inverse square law if its strength decreases inversely with the square of the distance ($1/d^2$) from its source. If the distance is increased by a factor of 3, then its strength would decrease by a factor of 9.

ion An *atom* whose number of *electrons* is not equal to its number of *protons.* (It is not neutral.)

ionization The act of removing one or more *electrons* from an *atom.*

irregular galaxy A galaxy whose appearance does not satisfy the criteria for a *spiral, barred spiral,* or *elliptical galaxy.* A *galaxy* without symmetry.

isotope *Elements* with the same number of *protons* but differing numbers of *neutrons.*

jovian planets The *giant planets.* They are Jupiter, Saturn, Uranus, and Neptune.

land tide A tide induced in the body of the earth.

latitude The north-south coordinate on the earth.

law of gravity A statement of the dependence of *gravity* on the masses and separation of two objects.

light Same as *electromagnetic radiation.* Usually light means the visible portion of the electromagnetic spectrum.

light curve A graphical record of the variation of the *luminosity* or *magnitude* of an object with time.

light-week The distance that light travels in one week. Approximately 1.5×10^{11} miles.

light-year The distance that light travels in one year. Approximately 6×10^{12} miles.

line of nodes This term usually refers to the line drawn through the two points where the moon's orbit crosses the plane of the *ecliptic.*

lithosphere The outer, rigid shell of the earth, 60 miles thick, containing the crust, continents, and upper mantle. The lithosphere is broken up into several slabs or *plates.*

Local Group The *cluster of galaxies* that contains the *Milky Way Galaxy.*

logarithmic scale Used for the axes of graphs. Equal distances on graphs using logarithmic scales do not represent equal increments in the plotted variable. Instead, equal intervals on these graphs represent constant ratios of the values. If logarithmic scales are used on graphs, it is important to note carefully the actual numerical values recorded along the edge of the graphs.

longitude The longitude of a point on the earth's surface is the angle between the *meridian* passing through that point and the *prime meridian* passing through Greenwich, England.

luminosity The total amount of energy emitted by an astronomical object in one second.

lunar eclipse An eclipse of the moon by the earth. The moon is in the shadow of the earth.

lunar tide The periodic cresting and subsiding of the oceans due to the moon's gravity.

Lyman line Hydrogen *spectral line* for which the *ground state* is the lower level of the *transition* of the *electron.* Lyman lines have their *wavelengths* in the *ultraviolet* region of the *spectrum.*

magnetic broadening Broadening of the *spectral lines* due to the *Zeeman effect.* The stronger the *magnetic field* at the location of the emitting or absorbing atoms, the broader the lines.

magnetic field A magnetic force pervading a region of space.

magnetosphere The region around the earth (or another planet) in which the *magnetic field* of the earth (or other planet) is strong enough to have a measurable effect on the interplanetary gases.

magnetic polarity A specification of the orientation of the north and south poles of the magnetic field.

magnitude A numerical value that indicates the intensity of an object. The larger the numerical value of the magnitude, the less the intensity. See *absolute* and *apparent magnitude.*

Main Sequence A line on the *H-R diagram* that represents the location of the *Main Sequence stars.*

Main Sequence star A star that is burning *hydrogen* in its *core* in a stable fashion.

mantle That portion of the earth's interior that lies between the *crust* and the *core.*

mare (pl. maria) The dark, relatively smooth lava fields first seen on the moon.

mascon Concentration of mass (high-density material) on the moon.

massive galactic halo A cloud of dark luminous matter surrounding many and possibly all galaxies, extending beyond the edge of the visible galaxy, and containing considerably more mass than the visible galaxy at its center. The nature of the matter composing the massive halo has not yet been determined. Massive halos are an important component of the *hidden mass.*

meridian A line circling a spherical body that passes through its two poles.

meteor The luminous streak of light caused by the rapid entry and atmospheric frictional heating of a piece of interplanetary material. Also termed "shooting star."

meteor shower *Meteors* that, in the course of an evening, appear to "shower" out of a certain point in the sky. Caused by the collision of the earth with a swarm of interplanetary particles.

meteorite A *meteor* particle that has fallen to the earth's surface.

microdensitomer A *densitometer* that can record very detailed information from photographic plates.

micrometeorite A *meteorite* so small that the earth's atmosphere can rapidly slow it down, enabling it to float down to the surface without being totally evaporated.

microwaves Electromagnetic waves with wavelengths near 1 centimeter.

Milky Way This is the name of our own *galaxy.*

molecule Two or more atoms that are bound together.

natural width The width of a *spectral line* that is determined only by the *atom's* internal structure. In particular, the natural width excludes effects such as the velocity of the atoms, an external *magnetic field*, and collisions between the atoms.

neap tide The smallest tide. Occurs when the moon is at *first* or *third quarter* and its effect is partly canceled by the *solar tide*.

near-ultraviolet radiation Electromagnetic radiation of wavelength is just shorter than visible radiation.

nebula A term given to any diffuse astronomical object. Examples are *planetary nebula*, extragalactic nebula *(galaxy)*, and *emission nebula* (a type of interstellar gas cloud).

neutrino A particle that has no mass or charge. It is believed to travel at the speed of light. The neutrino and the *photon* are two different types of massless particles. In comparison with a photon, a neutrino can pass through much larger accumulations of matter before being absorbed.

neutron A neutral particle that is approximately 2000 times more massive than an *electron*. The neutron has almost the same mass as a *proton*. The neutron is found in the *nuclei* of all *atoms* except *hydrogen*.

neutron star A star of extremely high density that is composed primarily of *neutrons*. The mass may be a few times the mass of our sun, while its radius is approximately 10 miles.

new moon During this *phase* of the moon the hemisphere facing the earth is unilluminated or nearly so. A small sliver of the illuminated portion may be visible.

Newtonian reflector A reflecting telescope that contains a small mirror angled at 45° to direct the image out of the side of the telescope.

night glow The faint glow of the earth's atmosphere.

nodes The points of intersection of the orbit of a body with a plane.

nova The sudden brightening of a star by a factor of hundreds to thousands. The burst of energy is believed to result from the infall of material onto a star from its binary companion.

nuclear burning The process of *fusion* of *nuclei* with the release of energy.

nuclear energy Energy derived from *fusion* or *fission* of the *nucleus*.

nuclear force A very short-ranged force between neutrons and protons responsible for holding the *nucleus* of an *atom* together. One of the basic physical forces.

nuclear reaction A process of either *fission* or *fusion* in the *nucleus*.

nucleus (of a galaxy) The central region of a *galaxy* that contains an enhanced density of stars.

nucleus (atomic) The central region of an *atom*, usually containing *protons* and *neutrons* in nearly equal numbers.

nucleic acid A long chain of *nucleotides*, an example of which is *DNA*.

nucleotides Molecules that form the building blocks of *nucleic acids*, (e.g., *DNA*).

neutron An electrically neutral particle with the same mass as the *proton*.

nutation The wobble of the axis of rotation of a spinning object. This wobble is superimposed on the *precessional* motion of the axis.

objective The lens or mirror in an optical instrument that forms the *image*. In a refracting telescope the objective is the largest lens.

ocular lens The lens (or grouping of lenses) that the observer looks through to see the image formed by the *objective lens*. The ocular lens magnifies the image.

open cluster Same as a *galactic cluster.*

optical double Two stars that appear to be in close proximity on the *celestial sphere.* However, because one star is actually some distance behind the other, they are not gravitationally associated with each other; they are not a *binary star.*

optical telescope A telescope that is designed to produce an *image* with visible light.

oscillating cosmology A theory that combines the *Big-Bang cosmology* with the notion of a later collapse of the *Universe* and a subsequent Big Bang in a repeated cycle of expansion, collapse, and rebound.

parallax The apparent displacement of an object due to the actual displacement of the observer. For example, over the course of a year, nearby stars appear to shift with respect to the more distant background stars, due to the earth's motion around the sun. The term is also used to signify the number of degrees by which the direction of the object appears to shift.

partial eclipse An *eclipse* of the moon or the sun in which one body is not completely in the shadow of the other. For a *lunar eclipse,* the moon is not completely in the shadow of the earth; for a *solar eclipse,* the moon does not completely eclipse the entire surface of the sun but leaves one edge of the sun visible around the edge of the moon.

parsec Approximately equal to 3.2 *light-years.* A star at a distance of one parsec has a trigonometric *parallax* of one second of arc.

penumbra The grey outer region of a *sunspot.*

periodic table An organized listing of all of the *elements.* The organization of the listing is based on similarities in the chemical properties of the *elements.*

phase (of moon) The different appearance of the moon during the course of a month, as seen from the earth.

photocell An electronic device that converts light *(photons)* into an electric current (electricity). Photocells are used in astronomy to measure the intensity of light.

photodissociation A process using the energy of an absorbed *photon* to break apart a *molecule.*

photographic magnitude The *magnitude* of an object as measured on photographic *plates.*

photometer An astronomical instrument that attaches to a telescope for measuring the apparent *luminosity* of astronomical objects, usually through the use of a photocell.

photon Electromagnetic energy in its particle form. A *quantum* of *electromagnetic radiation.*

photosphere The thin surface layer of the sun from which the majority of the sun's photons escape into space. The visible surface of the sun.

photovisual magnitude A *magnitude* obtained with photographic *plates* through particular filters so that the final sensitivity of the *plate* is similar to the sensitivity of the eye.

plage A bright region on the surface of the sun.

planetarium An instrument that projects correctly positioned images of the stars onto the inside of a darkened room. The building that houses the planetarium instrument is also termed a planetarium.

planetary nebula A shell of glowing gas surrounding a hot central star. That shell of gas was previously ejected by the star itself.

planetesimal A small object circling the sun like a miniature planet.
plate In the field of astronomy, a plate refers to a piece of photosensitized glass that may serve as a permanent photographic record. In the field of geology, a plate is a single segment or slab of the earth's *crust* and upper *mantle.*
plate tectonics Very slow motion of *plates* or slabs in the earth's *lithosphere.* Also, accompanying effects on the earth's surface.
Population I Stars with a wide range of ages and a relatively high abundance of *heavy elements.* In our *galaxy* they are found in the *galactic disc* and particularly in the *spiral arms.*
Population II Relatively old stars with a relatively low abundance of *heavy elements.* In our *galaxy* they are found in the *halo* and in the *nucleus.*
pores Small, dark spots visible on the surface of the sun.
positron Similar to an *electron* (same mass) but positively charged. It is an *antimatter* particle.
precession A slow motion of the axis of rotation of a body causing the axis of rotation to trace out a cone in space.
prime meridian The *meridian* that is defined to be at zero degrees *longitude.*
primordial Original; usually refers to elements appearing in the Universe at the time of the Big Bang.
primordial fireball radiation Radiation filling the Universe at the time of the Big Bang.
prism A wedged-shaped piece of glass that disperses light to produce a *spectrum.*
prominence A region of glowing gas extending far above the surface of the sun.
protein A chain of *amino acids* that controls the *chemical reactions* of life and provides the structure of the cell.
protosun The sun in its collapse phase prior to initiation of hydrogen-burning reactions.
protogalaxy A *galaxy* in its formative stage, still undergoing collapse.
proton A charged particle found in the *nucleus* of an *atom.*
proton-proton cycle A series of *nuclear reactions* that converts *hydrogen* to *helium.* This series uses only *hydrogen* as its initial ingredient.
protostar A star prior to initiation of hydrogen-burning reactions.
pulsar A rapidly rotating *neutron star.* Active areas on the surface of the neutron star emit streams of radiation into space. Because of the rotation, a stream sweeps across the observer's line of sight at regularly timed intervals, creating a succession of pulses or flashes of radiation as seen from the earth.
pulsating star A star that expands and contracts, and simultaneously alters its light intensity in a periodic or semiperiodic fashion.
quantum (of light) *Electromagnetic* energy in its particle form. Same as *photon.*
quasars Objects of stellar appearance with energy output generally exceeding that of galaxies, and with strongly red-shifted spectra; the most luminous and distant object known in the Universe.
Quasi-Stellar Object (QSO) A *quasar* that is not a strong radio source.
Quasi-Stellar Source (QSS) A *quasar* that is a strong radio source.

radial velocity That component of the velocity that lies along the line of sight.
radio astronomy The field of astronomy concerned with the observation and interpretation of *radio waves* of astronomical origin.
radio galaxy A *galaxy* that is a strong source of *radio waves*.
radio telescope A telescope used for focusing and recording *radio waves*.
radio waves *Electromagnetic waves* of long *wavelength*, typically several meters or more.
radioactive An *element* is said to be radioactive if its *nucleus* undergoes spontaneous *fission*.
radioactive elements Those *elements* that are *radioactive*.
rays (lunar) A series of lighter streaks of material that extend radially away from a lunar crater.
red giant A large, luminous star with a cool surface. Red giants are plotted in the upper right-hand corner of the *H-R diagram*.
redshift The displacement of the *wavelength of electromagnetic radiation* to longer wavelengths. The redshift may result from the expansion of the Universe, large relative velocities of separation, or a high gravitational field.
reflecting telescope A telescope whose *objective* is a reflecting mirror.
refraction The bending of light occurring when the light passes from one material to another; for example, from air to glass.
refracting telescope A telescope whose *objective* is a lens.
relativity A theory that describes the nature of matter and the properties of space and time under the conditions of very large constant velocities (the special theory of relativity) or very high accelerations and gravities (the general theory of relativity).
resolution The degree to which fine details can be seen. A high-resolution photograph is one that is sharp.
rift A large crack in the *crust*, often caused by separation of *plates* in the earth's *lithosphere*.
right ascension The *celestial sphere* analogue of *longitude*. The *celestial sphere* is divided into 24 segments or hours. The location for zero hours of right ascension on the *celestial sphere* is determined by astronomical convention.
rille A valley or channel, especially on the moon. Some of the lunar rilles are believed to be carved by flowing lava.
Roche limit A moon that comes too close to its parent planet will be torn apart by the planet's tidal force. The Roche limit is the minimum distance at which the moon can circle its planet and remain intact. Saturn's rings are inside the Roche limit.
rotational broadening The broadening of the *spectral lines* due to the rotation of the astronomical object. The rotation produces a radial velocity.
RR Lyrae variables *Pulsating stars* of low mass whose average absolute *luminosities* are closely the same from *star* to *star*. Knowledge of this and their apparent *luminosities* determines their distances. RR Lyrae variables serve as a useful cosmic yardstick.
satellite An object that is in orbit around another object, usually a planet.
sedimentary rock Rock built up by an accumulation of sediments.
seeing A term that refers to the degree of blurring of a telescope image due to movements of the image induced by the earth's atmosphere.

seismogram A graphical record of the vibrations of the earth measured at the surface of the *crust.*

seismograph A device that produces a *seismogram.*

seismometer The detecting portion of a *seismograph.*

Seyfert galaxy A *galaxy* with an unusually bright, compact *nucleus* with *emission lines* and faint *spiral arms;* a possible connecting link between quasars and normal galaxies.

shock wave A very strong sound-wave impulse. A sound wave of unusually high compression.

sidereal day The time it takes the earth to make one complete rotation on its axis with respect to the fixed stars. Our usual day *(solar day)* is the time it takes the earth to make one complete rotation on its axis with respect to the sun. The solar day is not equal to the sidereal day because our position with respect to the sun is constantly changing; we are in orbit around it.

sidereal time Time based on one complete rotation of the earth on its axis with respect to the fixed stars.

solar day See sidereal day.

solar eclipse An *eclipse* of the sun by the moon. Part of the earth is in the shadow of the moon.

solar mass A unit of weight used by astronomers. For example, a two solar-mass star is twice as massive as the sun. One solar mass is 2×10^{33} grams.

solar system The sun's family. This includes all of the objects orbiting the sun: *planets, satellites, asteroids,* and *comets.*

solar tide Cresting and subsiding of the oceans due to the sun's gravity.

solar time Our "usual" time, based on the average rising and setting times of the sun.

solar wind The outward flow of particles (*electrons, protons,* etc.) from the sun.

spectral type A classification of a stellar *spectrum* based on its overall appearance and the strengths of certain *spectral lines.*

spectrogram A photographic record of a *spectrum.*

spectrograph An instrument that produces *spectrograms.*

spectroheliogram The *image* of the sun obtained with a *spectroheliograph.*

spectroheliograph An astronomical device for photographing the sun using *photons* from the sun of only one particular selected *wavelength.* The resultant photograph reveals the conditions on the sun at the depth at which those particular photons originated.

spectral line An *absorption* or *emission line* in a spectrum.

spectroscope An instrument for directly viewing the *spectrum* of an object with the eye.

spectroscopic binary A pair of stars not separately visible, but known to be a binary because the lines in the composite spectrum shift periodically due to the *Doppler effect.*

spectroscopic parallax The determination of a star's distance using its observed *spectrum* to locate the position of the star on the *H-R diagram,* and hence its *absolute luminosity.*

spectrum The array of colors or intensities of radiation at different *wavelengths* presented in order of their *wavelengths.* The appearance of light after it has been *dispersed* by a *prism* or a *grating.*

spicule Little jets or spikes rising above the surface of the sun. Spicules are best seen as the edge (limb) of the sun.

spiral arm A curved luminous arm extending from the *nucleus* of a *spiral galaxy*. The spiral arms contain an enhanced density of *interstellar material* and young blue stars.

spring tide The highest *tide* during a month, occurring at the *phase* of new or *full moon*, when the moon's tidal force is reinforced by the *solar tide*.

standard candle A light source that is used as a standard of intensity. Astronomers sometimes assume that objects of a certain type have the same luminosity. Using these "standard candles" and the knowledge of the *apparent luminosity*, astronomers may determine their distances.

star A gaseous sphere held together by its own *gravity*; especially, one massive enough to provide internal temperatures needed for nuclear burning. This definition has been expanded to include objects such as *neutron stars*.

Steady-State cosmology A theory of the nature of the Universe that states that the overall properties of the Universe do not change with the passage of time. Combined with the observation of the expansion of the Universe, the Steady-State cosmology implies a constant creation of new matter and energy.

stellar evolution A field of astronomy concerned with the description of the changes in stars as they age, following from their properties from birth to death.

stellar parallax The shift in apparent position of a star, as observed from the earth on opposite sides of its orbit.

stellar wind The *solar wind* of a star.

stratosphere A layer in the earth's atmosphere starting about 10 kilometers (six miles) above the surface.

sunspots A dark, relatively cool spot visible on the surface of the sun, typically 1000 kilometers in diameter. Sunspots are associated with strong *magnetic fields*.

sunspot cycle The 11-year cyclic increase and decrease in the total number of *sunspots* visible on the surface of the sun.

superclusters A collection of galaxies, often numbering thousands of individual galaxies and sometimes including numerous separate clusters of galaxies. Superclusters may extend over distances of several hundred million light-years and are the largest aggregate of matter in the Universe. Neighboring superclusters are separated by regions of comparable size called *voids* which are empty or nearly empty of galaxies.

supergranules Large (30,000 km) *granules* evident near the surface of the sun arising from large subsurface *convection currents*.

supernova A *star* that explodes with a hundred-billionfold increase in luminosity. The explosion is believed to be caused by detonation of the star's *core* or catastrophic collapse and rebound, near the end of the nuclear-burning phase.

supernova remnant The gaseous material expelled by a *supernova* explosion. Often visible as a glowing nebula.

temperature A measure of the random velocities of particles *(atoms, molecules)* comprising an object.

terrestrial planets Mercury, Venus, the earth, and Mars; planets composed mainly of rock iron, resembling the earth.

thermal broadening Broadening of a *spectral line* by the *Doppler shift* resulting from random velocities of the particles comprising the source of the *spectrum*.
third quarter moon A *phase* of the moon in which one half of the visible hemisphere, or one quarter of the complete surface, is illuminated. This *phase* occurs approximately one week after the *phase* of the *full moon*.
tide A bulge produced in a deformable body by the gravity of a nearby object. In particular, the apparent raising and lowering of the level of the ocean as witnessed at a shore.
Titan A moon of Saturn, nearly as large as Mercury and the only satellite known to have a substantial atmosphere.
total eclipse An *eclipse* of the moon for which the moon is entirely within the shadow of the earth, or an eclipse of the sun for which the sun's surface is totally covered by the moon.
triple star Three stars that remain in orbit about each other due to their mutual gravitational attraction.
trigonometric parallax A method of determining the distance to nearby stars by their apparent displacement due to the earth's motion in its orbit around the sun. See *parallax*.
ultraviolet light Light whose *wavelength* is just shorter than *visible light*.
umbra The darkest, central portion of a *sunspot*.
undifferentiated A planet whose composition is the same at all depths is an undifferentiated planet.
Universe The totality of matter, energy, and space.
variable star A star whose *luminosity* or *spectrum* appears to change with time.
vernal equinox A location on the *celestial sphere* where the sun crosses the *celestial equator* in its motion from south to north.
visible light *Electromagnetic radiation* to which the eye is sensitive.
visual magnitude The *magnitude* of an object as measured by the eye (which is more sensitive to red light than the standard astronomical photographic *plates*).
visual binary A *binary* whose component stars can be seen separately.
void A region in which galaxies are rare or missing. Voids separate superclusters of galaxies from neighboring superclusters. Together, voids and superclusters make up the largest scale of inhomogenities in the Cosmos.
volatile (element) An *element* that readily evaporates at normal temperatures.
wavelength The distance between two successive crests or valleys in a wave.
waning moon The *phases* of the moon from *full* through *third quarter* to *new moon*.
white dwarf A small hot star that is near or at the end of its *nuclear burning* phase. White dwarfs are plotted in the lower left-hand corner of the *H-R diagram*.
white light White light is a combination of all *wavelengths* of visible *light*.
window In astronomical usage, a window is a *wavelength* region for which our atmosphere is transparent or partially transparent.
X-rays Radiation with wavelengths shorter than those of *ultraviolet* light.
ylem The initial material of the Universe at the time of the *Big Bang*.

PHOTO CREDITS

END PAPERS: Star charts adapted from *Star Maps for Beginners.* Courtesy C. M. Levitt and Roy K. Marshall
PART ONE – Opener © Erich Lessing/Magnum
CHAPTER 1 – Opener Lick Observatory Photograph **Fig. 1.2** Courtesy Erich Lessing, Magnum **Fig. 1.4** Marrin Robinson **Fig. 1.8** Courtesy Hale Observatories **Fig. 1.9** Marrin Robinson **Fig. 1.13** Marrin Robinson **Fig. 1.16** Marrin Robinson
CHAPTER 2 – Opener Courtesy NASA **Fig. 2.10(a)** and **(b)** Courtesy of the British Museum **Fig. 2.11** Helmut K. Wimmer, The American Museum of Natural History, Hayden Planetarium **Fig. 2.12(a)** and **(b)** Courtesy NASA **Fig. 2.13** Helmut K. Wimmer, The American Museum of Natural History, Hayden Planetarium **Fig. 2.22** By Courtesy of the Trustees of the British Museum **Fig. 2.38** Courtesy Yerkes Observatory
CHAPTER 3 – Opener Courtesy Hale Observatories **Fig. 3.3** Courtesy Hale Observatories **Fig. 3.4** Courtesy Hale Observatories **Fig. 3.5** Arthur D. Code and Theodore E. Houck of the University of Wisconsin – Courtesy Hale Observatories **Fig. 3.6** Courtesy Hale Observatories **Fig. 3.7** Courtesy Lund Observatory, Lund Sweden **Fig. 3.8** Courtesy Lick Observatory **Fig. 3.9** Courtesy Hale Observatories **Fig. 3.10** Courtesy Hale Observatories **Fig. 3.11** Courtesy Lick Observatory **Fig. 3.12** Courtesy Lick Observatory **Fig. 3.13** Courtesy Lick Observatory
PART TWO – Opener The Bettman Archives
CHAPTER 4 – Opener Courtesy NASA **Fig. 4.3** Yerkes Observatory photograph **Fig. 4.4(a)** Courtesy Hale Observatories **(b)** Courtesy NASA **Fig. 4.5(a)** Courtesy Lick Observatory **(b)** Courtesy Hale Observatories **Fig. 4.14(a)** Courtesy Hale Observatories **Fig. 4.15(a)** Gary Ladd **(b)** Gary Ladd **Fig. 4.18** © Association of Universities for Research in Astronomy, Inc. The Kitt Peak National Observatory. Patrick Waddell, Aura Inc. **Fig. 4.19** Astronomy Magazine. Kitt Peak National Observatory **Fig. 4.23** Courtesy Dominion Astrophysical Observatory
CHAPTER 5 – Opener David Moore, Courtesy of Blackstar **Fig. 5.2** Courtesy of Cornell University **Fig. 5.5** The National Radio Astronomy Observatory, operated by Associated Universities, Inc. under contract with the National Science Foundation **Fig. 5.7** Courtesy NASA
CHAPTER 6 – Opener Courtesy Hale Observatories **Fig. 6.15** Courtesy Yerkes Observatory **Fig. 6.16** Courtesy Yerkes Observatory
CHAPTER 7 – Opener (left) Courtesy Danish Information Office, Consulate General of Denmark, New York; (right) Courtesy Princeton University Observatory
CHAPTER 8 – Opener Courtesy Hale Observatories **Fig. 8.2** Courtesy Hale Observatories **Fig. 8.3** Courtesy Hale Observatories
CHAPTER 9 – Opener Courtesy Hale Observatories **Fig. 9.1** Courtesy Hale Observatories **Fig. 9.5** Courtesy Lick Observatory
PART 3 – Opener The Bettman Archives
CHAPTER 10 – Opener Courtesy Hale Observatories **Fig. 10.1** Courtesy Hale Observatories **Fig. 10.4** Courtesy Hale Observatories **Fig. 10.5** Courtesy Hale Observatories **Fig. 10.7** Courtesy Hale Observatories **Fig. 10.9** Courtesy Hale Observatories **Fig. 10.10** Courtesy Hale Observatories **Fig. 10.11** Courtesy Juri and Alar Toomre **Fig. 10.12** Courtesy Hale Observatories **Fig. 10.13** Courtesy Hale Observatories **Fig. 10.14** California Institute of Technology **Fig. 10.15** Courtesy Hale Observatories

CHAPTER 11—Opener © AURA, Inc., Kitt Peak National Observatory **Fig. 11.1** Courtesy Hale Observatories **Fig. 11.5** Courtesy Hale Observatories **Fig. 11.6** Courtesy Hale Observatories **Fig. 11.8** Courtesy Hale Observatories

CHAPTER 12—Opener Courtesy Hale Observatories **Fig. 12.1** Courtesy Hale Observatories **Fig. 12.2** Courtesy Hale Observatories **Fig. 12.3** Courtesy Hale Observatories **Fig. 12.7** Courtesy Hale Observatories **Fig. 12.8** Courtesy Hale Observatories, Maarten Schmidt **Fig. 12.9** Courtesy Hale Observatories **Fig. 12.10** Courtesy Hale Observatories **Fig. 12.12(b)** Data courtesy of P. Henry, Harvard-Smithsonian Center for Astrophysics **Fig. 12.13** Courtesy Hale Observatories **Fig. 12.14** Courtesy J. Tyson, Bell Laboratories **Fig. 12.15** The National Radio Astronomy Observatory, operated by Associated Universities, Inc. under contract with the National Science Foundation

CHAPTER 13—Opener J. R. Eyerman, Life Magazine, © Time, Inc. **Fig. 13.1** Courtesy Hale Observatories

PART 4—Opener The Bettman Archives

CHAPTER 14—Opener Courtesy H. Zirin, Big Bear Solar Observatory **Fig. 14.6(a)** Courtesy Martin Schwarzschild, Princeton University Observatory **Fig. 14.6(b)** Courtesy NASA; **(c)** Courtesy Martin Schwarzschild, Princeton University Observatory; **(d)** Courtesy W. M. Chiplonkar, University of Poona, Poona, India; **(e)** Courtesy Robert B. Leighton, California Institute of Technology **Fig. 14.9** and **14.11** Courtesy Richard B. Dunn, Sacramento Peak Observatory, Air Force Cambridge Research Laboratories **Fig. 14.12** Courtesy Gordon Newkirk, Jr., High Altitude Observatory, Boulder, Colorado **Fig. 14.13** Courtesy Martin Schwarzschild, Princeton University Observatory **Fig. 14.14** Courtesy Hale Observatories **Fig. 14.16** Courtesy Hale Observatories **Fig. 14.20** Courtesy Sara F. Martin and Harry E. Ramsey, Lockheed Solar Observatory **Fig. 14.21** Courtesy NOAA **Fig. 14.22** Courtesy NOAA **Fig. 14.23** Courtesy NOAA **Fig. 14.24** Courtesy Richard B. Dunn, Sacramento Peak Observatory, Air Force Cambridge Research Laboratories **Fig. 14.25** Courtesy Gordon Newkirk, Jr., High Altitude Observatory, Boulder, Colorado **Fig. 14.26** Courtesy Observatoire de Paris, Meudon **Fig. 14.27** Courtesy H. Zirin, Big Bear Solar Observatory, Air Force Cambridge Research Laboratories

CHAPTER 15—Opener Courtesy Hale Observatories **Fig. 15.3** Courtesy NASA **Fig. 15.4** Courtesy Lowell Observatory **Fig. 15.5** Courtesy Lowell Observatory **Fig. 15.8** Courtesy NASA **Fig. 15.9** Courtesy NASA **Fig. 15.10** Courtesy NASA **Fig. 15.11** Courtesy Lowell Observatory **Fig. 15.13** Courtesy Hale Observatories **Fig. 15.17** Courtesy Hale Observatories

CHAPTER 16—Opener Courtesy NASA **Fig. 16.8** Barrett Gallagher

CHAPTER 17—Opener Courtesy NASA **Fig. 17.1** Courtesy NASA **Fig. 17.2** Courtesy NASA **Fig. 17.3** Courtesy U.S. Air Force **Fig. 17.4** Courtesy NASA **Fig. 17.5** Courtesy NASA **Fig. 17.6** Courtesy NASA **Fig. 17.7** Courtesy NASA **Fig. 17.8** Courtesy NASA

CHAPTER 18—Opener Courtesy NASA **Fig. 18.1** Courtesy NASA **Fig. 18.2** NASA

CHAPTER 19—Opener Courtesy NASA **Fig. 19.1** Courtesy NASA **Fig. 19.2** Courtesy NASA **Fig. 19.3** Courtesy NASA **Fig. 19.4** Courtesy NASA **Fig. 19.5** Courtesy NASA **Fig. 19.6** Courtesy NASA **Fig. 19.7** Courtesy NASA **Fig. 19.8** Courtesy NASA **Fig. 19.9** Courtesy NASA **Fig. 19.10** Courtesy NASA

CHAPTER 20—Opener Courtesy J. William Schopf, Elso S. Barghorn, Morton D. Maser and Robert O. Gordon **Fig. 20.4** Courtesy British Information Service

COLOR PLATES

Plates 4–10 Copyright California Institute of Technology **Plate 11** Courtesy U.S. Naval Observatory **Plate 12** © 1979 R. J. Dufour, Rice University, Courtesy Hansen Planetarium **Plate 13** Courtesy NASA **Plate 14** Courtesy Anthony C. S. Readhead and California Institute of Technology **Plates 15–22** Courtesy NASA

INDEX